A Modern Perspective on Type Theory

APPLIED LOGIC SERIES

VOLUME 29

SCOPE OF THE SERIES

Logic is applied in an increasingly wide variety of disciplines, from the traditional subject of philosophy and mathematics to the more recent disciplines of cognitive science, compu ter science, artificial intelligence, and linguistics, leading to new vigor in this ancient subject Kluwer, through its Applied Logic Series, seeks to provide a home for outstanding books an research monographs in applied logic, and in doing so demonstrates the underlying unity an applicability of logic.

The titles published in this series are listed at the end of this volume.

A Modern Perspective on Type Theory

From its Origins until Today

by

FAIROUZ KAMAREDDINE

Heriot Watt University,
Edinburgh, Scotland

TWAN LAAN

Eindhoven, The Netherlands

and

ROB NEDERPELT

Eindhoven University of Technology,
The Netherlands

KLUWER ACADEMIC PUBLISHERS

DORDRECHT / BOSTON / LONDON

A C.I.P. Catalogue record for this book is available from the Library of Congress.

ISBN 1-4020-2334-0 (HB)
ISBN 1-4020-2335-9 (e-book)

Published by Kluwer Academic Publishers,
P.O. Box 17, 3300 AA Dordrecht, The Netherlands.

Sold and distributed in North, Central and South America
by Kluwer Academic Publishers,
101 Philip Drive, Norwell, MA 02061, U.S.A.

In all other countries, sold and distributed
by Kluwer Academic Publishers,
P.O. Box 322, 3300 AH Dordrecht, The Netherlands.

Printed on acid-free paper

Printed in the Netherlands.

Contents

Preface

Preliminaries

We assume that the reader is more or less familiar with the basics of typed λ-calculus and type theory. We give a survey of the most important topics concerning typed λ-calculus in Chapter 4. So, it may be a good idea to read Sections 4b and 4c of that chapter first. We used the historical timeline and presented the λ-calculus (Section 4b) and Pure Type Systems (Section 4c) after our discussion of the prehistory and of Frege, Russell and Ramsey's accounts.

Throughout the book, we use $\mathbb{N}$ to represent the set of the natural numbers (including 0), and $\mathbb{N}^+$ to represent the set $\mathbb{N}$ minus the 0.

Overview of this book

The book is divided into three parts:

Part I

The first part consists of **three chapters** and deals with the evolution of type theory until the 1940s. We study the prehistory of type theory up to 1910 and its development between Russell and Whitehead's *Principia Mathematica* ([145], 1910–1912) and Church's simply typed λ-calculus of 1940. We first argue that the concept of types has always been present in mathematics, though nobody was incorporating them explicitly as such, before the end of the 19th century. Then we proceed by describing how the logical paradoxes entered the formal systems of Frege, Cantor and Peano concentrating on Frege's *Grundgesetze der Arithmetik* for which Russell applied his famous paradox[1] and this led him to introduce the first formulation of a theory of types, the Ramified Type Theory (RTT) (based on

[1] Russell discovered his paradox when he read Cantor's work.

Russell's work [131] of 1908 which succeeded [130]; this work is highly influenced by Frege's concept of types). We present RTT formally using the modern notation for type theory. Ramified types have a double hierarchy: one of *orders* and the other of *types*. Ramified types had limitations and for this reason, Russell and Whitehead added the so-called axiom of reducibility. Ramified types and the reducibility axiom were not fully accepted. This led to various calls for deramification by many including Hilbert and Ackermann, Ramsey and Leon Chwistek. In this book, we concentrate on the simple theory of types (STT) as envisaged by Ramsey, 1926 (and also independently by Hilbert and Ackermann) which is a simplification of the ramified theory of types by removing the orders. In 1940, Church developed, using his λ-calculus and the simple theory of types, the first most influential type theory: Church's simply typed λ-calculus.[2] We present STT and Church's own simply typed λ-calculus ($\lambda_{\to\text{Church}}$[3]) and we finish by comparing RTT, STT and $\lambda_{\to\text{Church}}$.

Chapter 1. In the first chapter we discuss the prehistory of type theory. That is, we study the way in which types implicitly occurred in logic and mathematics before there was an explicit theory of types. We pay special attention to the formalisation of logic that is made in Frege's *Begriffsschrift* [49] and *Grundgesetze der Arithmetik* [52, 56], as in this formalisation many basic ideas are presented that are later used in type theory. Moreover, the system of *Grundgesetze der Arithmetik* is the one for which Russell derives his famous paradox, and this paradox has been the reason for Russell to introduce the first theory of types.

Chapter 2. This first type theory is the subject of the second chapter. Whitehead and Russell present their theory, the Ramified Type Theory (RTT), in an informal way. Several rough descriptions of this theory have been given in the literature (see for instance [124, 70, 29, 31]) but we present a formalisation of RTT that is directly based on the presentation of RTT in Whitehead and Russell's *Principia Mathematica* ([145], 1910-12). The construction of this formalisation is not a simple task. Whitehead and Russell do not present a clear syntax for their so-called *propositional functions* in [145], neither do they make a clear difference between syntax and semantics. We present a formal definition of propositional functions that is faithful to the original ideas exposed in *Principia Mathematica*. A second technical problem is the notion of substitution, which is totally undefined in *Principia Mathematica*. The formalisation of the notion of propositional function

[2]Note that the simple theory of types (1926) existed before the λ-calculus was invented by Church in 1932 and hence before Church's simply typed λ-calculus of 1940. Nevertheless, nowadays, when one refers to simple type theory, one usually means Church's simply typed λ-calculus of 1940. It should be noted furthermore, that Russell's type structure was different from that of Church. The former was set-based with linear sequences of types. The latter was function-based.

[3]We write $\lambda_{\to\text{Church}}$ for the original calculus of Church as presented in [29]. Note that this is different from the calculus $\lambda_{\to}$ used in frameworks like the Barendregt cube and the Pure Type Systems found in [3].

makes it possible to express the notion of substitution of *Principia Mathematica* in terms of λ-calculus. We use techniques from typed and untyped λ-calculus to give a precise description of substitution, and to show that substitution is well-defined as long as we restrict ourselves to well-typed propositional functions of RTT.

Chapter 3. In 1926, Ramsey [124] proposed an important simplification of RTT, the *simple* theory of types (STT). This simple type theory has become the basis for many modern type systems, and for the simply typed λ-calculus of Church [29]. The simplification consisted of the removal of one of the two hierarchies from RTT. The hierarchy of types is maintained, while the hierarchy of orders is removed. In Chapter 3 we discuss this process, known as *deramification.* An important observation of this chapter is that though the orders do not occur in the mainstream of type theories, they still provide an important intuition for logicians. We show that there is a close link between the hierarchy of orders in RTT and the hierarchy of truths that was introduced by Kripke [96]. We also show that Kripke's use of orders is more flexible than Russell's, and that this is due to the fact that orders occur at the semantical level in Kripke's theory, while they occur at the level of syntax in RTT.

Part II

The second part of the book consists of **four chapters** and deals with major developments of type theory since the 1940s. Although type theory was not active in the 1950s, in the 1960s there was a revival of interest in types and their applications to computer science and mathematics. But then, both RTT and $\lambda_{\to\mathrm{Church}}$ turned out to be *too restrictive* for mathematics and computer science where fixed points (to mention one example) play an important role. Since the 1960s, we have seen many new type systems and the influence of types in logic, mathematics and computation continues to grow. Both logicians and computer scientists have developed several branches of typed λ-calculus. Also mathematics has benefitted from typed λ-calculus, especially since 1967 when de Bruijn used his typed-λ-calculus-based AUTOMATH for the analysis and checking of mathematical texts.

Chapter 4. Though type theory clearly served as a method to prevent certain logical paradoxes, the logical system stood apart from the type system until the 1950s. In Chapter 4 we study the ways in which logic can be included in a type system. The various methods are all based on the idea that the proof of a logical implication can be seen as a function. More precisely: A proof of the proposition $A \to B$ is implemented as a function that takes a proof of A as argument, and returns a proof of B. In this way, the *proposition* $A \to B$ can be seen as the *type* of all functions from (proofs of) A to (proofs of) B. Similarly, a *proof* of $A \to B$ becomes a *term* of type $A \to B$. One calls this principle: *propositions as types*, or: *proofs as terms*. Both expressions are abbreviated by PAT. PAT was

discovered independently by different people. In Section 4a we give a historical sketch. Moreover, in this book we try to present, in a uniform framework, various important type systems that were proposed during this century. An important part of this framework is formed by the so-called Pure Type Systems (PTSs). Therefore, a short introduction to typed lambda calculus and PTSs is essential for the understanding of this book. Hence, in Section 4b we present the basic definitions and properties of the λ-calculus. Then in section 4c we introduce a general framework of type systems which includes the most influencial type theories developed since the 1960s including the polymorphic type theory, the dependent type theory and the Pure Type Systems (PTSs). We also introduce the so-called Barendregt cube [3] which is a collection of eight influential type systems. PTSs in general and the cube in particular, will be used to represent in the modern setting, some of the historical systems discussed so far. Moreover, PTSs in general and the cube in particular will be subject to various extensions in Part III of this book.
Chapter 5. In Chapter 5 we describe how the two most important type systems of the pre-PAT-era, RTT and STT, can be described in a PAT style. This gives insight in the various ways in which PAT-implementations can be made.
Chapter 6. In Chapter 6, we return to Russell's orders which we use to place RTT in a context with a modern system of computer mathematics (Constable's Nuprl). Nuprl is based on a modern type theory which differs from Pure Type Systems: Martin-Löf's type theory. As another illustration of the generality of PTSs, we present a complex type system (again Nuprl) as a simple and compact PTS.
Chapter 7. One of the important applications of PAT is the mechanical verification of mathematical proofs. The first tool for such a verification was AUTOMATH. It was developed in the late 1960s. The languages of the various AUTOMATH systems have been studied intensively. In [3], a description of two of the most important systems within the framework of Pure Type Systems is given, but without explanation. In Chapter 7 we study the original language of AUTOMATH and translate it to a PTS format. In doing so, we obtain descriptions similar to those of [3].

Part III

The third part which consists of **three chapters**, deals with some extensions of type theory, especially those related to functions and parameters (arguments).
Chapter 8. In many type theories and lambda calculi, there is no formal possibility to use abbreviations or definitions, i.e., to introduce names for large expressions which can be used several times in a program or a proof. This possibility is essential for practical use, and indeed implementations of Pure Type Systems such as Coq [44] and Nuprl [35] do provide this possibility. Moreover, most implementations of programming languages (Haskell, ML, CAML, etc.) use names for large

expressions via a well-known programming language concept: the so-called *let expressions* or *definitions*. This chapter introduces definitions to Pure Type Systems in two different ways: definitions can be added only to the context or can be added to both the contexts and the syntax of terms. We compare the two notions of definitions studied in this chapter with the definitions of AUTOMATH.

Chapter 9. Our study of the function concept leads to a natural refinement of the Barendregt cube with what we call parameters. This leads to the division of the cube into eight smaller cubes which will enable more accurate presentations of various systems. The refinement of this chapter will be used in the next chapter together with the concept of definitions discussed in Chapter 8, to give a powerful refinement of Pure Type Systems in general and of the Barendregt cube in particular.

Chapter 10. The description of AUTOMATH given in Chapter 7 is precise, but does not take into account two of its most important mechanisms: the definition mechanism and the parameter mechanism. Many other type systems use these mechanisms as well. This motivates us to extend the framework of PTSs with definitions and parameters in Chapter 10. Our extension results in a refinement of the framework of PTSs. In this refined framework, various modern type systems (like LF and ML) can be described in a more precise way than in the PTS framework without definitions and parameters.

Acknowledgements

This book is based on various projects by the authors in the past few years. Of these projects we acknowledge:

- The project "De ontwikkeling van het Typebegrip" ("the evolution of the type concept"), which was financially supported by the Co-operation Centre Tilburg and Eindhoven Universities (SOBU) and which resulted in the PhD thesis by Twan Laan [98]. NWO, Eindhoven University of Technology, the Royal Society and the British Council provided funding for the trips of Twan Laan to Scotland in spring 1995 and winter 1996 and of Fairouz Kamareddine to the Netherlands between 1993 and 2001.

- The project "Advantages of a new lambda-notation", which was financially supported by the Engineering and Physical Sciences Research Council (EPSRC). That project studied amongst other things, the correspondence between Martin-Löf type theory, Russell's ramified type theory and Pure Type Systems [83], the correspondence between Kripke's theories of truth and Russell's orders [82], and also led to the formalisation of definitions in typed lambda calculi by the authors [14, 81] (in collaboration with Roel Bloo).

- The international exchange project(s) supported by Eindhoven University of Technology which allowed Fairouz Kamareddine to make various short and long visits to Eindhoven University of Technology during which she collaborated on various topics elaborated in this book [84, 83, 14, 82, 13, 86, 85, 87].

Many people have influenced the work in this book and we cannot mention them all. In particular, we would like to express our gratitude to Harrie de Swart, and Jos Baeten for making this project possible, for the useful discussions and invaluable support, and for facilitating our exchange visits. Henk Barendregt, Dirk van Dalen, Roger Hindley, Randall Holmes, Jonathan Seldin, and Joe Wells provided much valuable feedback. Joe Wells also provided much valued help with LaTeX for which we are extremely grateful.

Last but not least, we are very grateful for Jane Spurr who handled this book in her very prompt, efficient, friendly and professional manner. Producing this manuscript would have been much harder without you Jane. Thank you.

Fairouz Kamareddine, Twan Laan and Rob Nederpelt
Edinburgh and Eindhoven,
March 2004

Chapter 0

Introduction

Nowadays, type theory has many applications and is used in a lot of different disciplines. Even within logic and mathematics, there are a lot of different type systems. They serve several purposes, and are formulated in various ways. But, before 1903 when Russell first introduced type theory (see the appendix of [130]), there were no earlier formulations of any other type theory. It is only since the second half of the twentieth century that we see explosions of type theories. In this book we follow the evolution of type theory throughout the past century.

Avoiding the paradox in set theory

The explicit and formal use of types (and thus an early form of what is presently called "type theory") was originally intended to prevent the paradoxes that occurred in logic and mathematics at the end of the 19th and the beginning of the 20th century. But it was not the only method developed for this purpose.

Another tool for avoiding the paradox was the fine-tuning of Cantor's Set Theory [23, 24] by Zermelo [147], and the iterative conception of set (see [15]) that resulted from the foundation axiom of Zermelo-Fraenkel's set theory ZF.

In set theory, the axiom of (unrestricted) *comprehension* assumes that each open well-formed expression determines a concept whose extension exists and is the set of all those elements which satisfy the concept. This comprehension axiom is formulated as follows [47, 48]:

(Comprehension) For each open well-formed formula Φ,

$$\exists y \forall x[(x \in y) \Leftrightarrow \Phi(x)] \qquad \text{where } y \text{ is not free in } \Phi(x)$$

Such an unrestricted comprehension leads to a paradox as follows:

Take $\Phi(x)$ to be $\neg(x \in x)$ then
$\exists y \forall x[(x \in y) \Leftrightarrow \neg(x \in x)] \Longrightarrow \exists y[(y \in y) \Leftrightarrow \neg(y \in y)]$

The paradox can be avoided by altering the axioms. The most straightforward such theory is ZF (Zermelo-Fraenkel) where the axioms are made to fit the limitation of size doctrine. As an example, the above comprehension principle is altered to the following so-called *separation axiom*:

(Separation) For each open well formed formula Φ,

$$\exists y \forall x[(x \in y) \Leftrightarrow (x \in z) \wedge \Phi] \qquad \text{where } y \text{ does not occur in } \Phi$$

It is this new axiom which is responsible for the elimination of the paradox: to prove the existence of $\{x : \neg(x \in x)\}$ we need a z big enough which contains $\{x : \neg(x \in x)\}$. But we cannot show the existence of such a z. More precisely the paradox is restricted in ZF as follows:
Take $\Phi(z)$ to be $\neg(z \in z)$, and take $y = \{x : (x \in z) \wedge \neg(x \in x)\}$

- If $(y \in y)$ then $(y \in z)$ and $\neg(y \in y)$ contradiction,
- If $\neg(y \in y)$ then
  - if $(y \in z)$ then $(y \in y)$ contradiction,
  - if $\neg(y \in z)$ then we are fine.

Note however that we still have the syntactical ability to consider whether a set belongs to itself or not, but we are not committed to any set actually belonging to itself. For example,

if we take $\Phi(x)$ to be $x \in x$ then $\exists z \forall x[(x \in z) \Leftrightarrow (x \in \alpha \wedge x \in x)]$;

in this case, although $x \in x$ is well-formed, it is likely to be false in the intended interpretation for any value of x.

In the middle period of the development of ZF, it was felt that the following foundation axiom (which is independent of and consistent with all other axioms of ZF) has to be added:[1]

$$\text{(FA)} \qquad (\exists x)(x \in a) \Rightarrow (\exists x \in a)(\forall y \in x)\neg(y \in a)$$

Despite this, it was clear that in ZF, the foundation axiom does not help in avoiding the paradoxes, it was added as a technical refinement.

[1]This changed in the 1980s when Peter Aczel introduced his non-well-founded set theory which relied on the Anti-Foundation axiom.

As a corollary of (FA), there is no set a which has itself as its only element, for if there was, then taking $x = a$ in the antecedant of (FA) above leads to a contradiction.

Remark 0.1 It is worth pointing out that although very different conceptually, the theories of types (ramified or simple) discussed in Chapter 2 as well as ZF (which includes (FA)), give rise to an *iterative* concept of set. That is, they all require the elements of a set be present before a new set can be constructed [15].

We cannot stop our discussion of set theory without mentioning two particular non-iterative set theories due to Quine. Quine's stratification in his New Foundations NF [120], and his Mathematical Logic ML [121] are sufficiently type-like to merit some discussion. Quine restricted the axiom of comprehension, to obtain the following stratified comprehension principle:

(SCP) $\exists x \forall y[(y \in x) \Leftrightarrow \Phi(y)]$
where x is not free in $\Phi(y)$ and $\Phi(y)$ is stratified.[2]

Quine's NF has attracted a lot of research. Specker [138] refuted the axiom of Choice in NF, Jensen [78] established that NF with Urelements is consistent (even when augmented with Choice, Infinity and unrestricted mathematical induction). However, consistency of NF remains an open problem. Moreover, NF is weak for mathematical induction (for propositions not expressible in type theory). Also, NF is said to lack motivation because its axiom of comprehension is justified only on technical grounds and one's mental image of set theory does not lead to such an axiom. To overcome some of the difficulties, Quine replaced (SCP) by two axioms, one for class existence and one for elementhood. The rule of class existence provides for the existence of the classes of all elements satisfying any condition Φ, stratified or not. The rule of elementhood is such as to provide the elementhood of just those classes which exist for NF. These axioms are as follows:

(Comprehension by a set) $\exists y \forall x(x \in y \Leftrightarrow \Phi(x))$, where x and y range over sets, $\Phi(x)$ is stratified with set variables only in which y does not occur free

(Impredicative comprehension by a class) $\exists y \forall x(x \in y \Leftrightarrow \Phi(x))$, where x ranges over sets, $\Phi(x)$ is any formula in which y does not occur free

[2]Assume a first-order theory where for each primitive predicate $F(x_1, \ldots, x_n)$, we have integer constants $F_1, \ldots, F_n$. A formula Φ in the language of that theory is said to be *stratified* if there is an integer-valued function σ with domain the set of variables appearing in ϕ with the property that in each atomic formula $F(x_{i_1}, \ldots, x_{i_n})$ which appears in ϕ, and each integer $1 \leq j \leq n$, we have $\sigma(x_{i_j}) - \sigma(x_{i_1}) = F_j - F_1$.

ML was liked both for the manipulative convenience we regain in it and the symmetrical universe it furnishes. The earlier version of the first edition of [121] was subject to the Burali-Forti paradox — The well ordered set Ω of all ordinals has an ordinal which is greater than any member of Ω and hence is greater than Ω (see Section 1b4). In the second edition of [121], Quine corrected the axiomatization of ML (following a suggestion of Hao Wang) so that it is demonstrably consistent if NF is consistent. This latter version does not face the Burali-Forti paradox.

Avoiding the paradox in type theory

The approach of avoiding the paradox in type theory is completely different from the set-theoretical approach. First, in the type theoretical approach, it is the language that is altered in order to avoid the paradox, and not the axioms[3]. Moreover, since Church's λ-calculus was extended with simple types in 1940, type theory has continued to focus on the notion of *function* in logic and mathematics. Since 1940, functions have remained one of the main objects of study for type theorists. This book concentrates on the evolution and extensions of types and functions.

We start by giving a historical account as to why formulations of type theory came into being. But our work is not a mere historical description. On the contrary: our goal is not to describe the various type systems that have been developed in their historical setting, but to present them in a modern framework. In this way it becomes clear how the various type systems are related to each other, even if originally those systems are described in very different ways. In addition, this leads to extending modern systems with useful features that existed already in old systems but were omitted in modern ones. Moreove, we can make clear what is

[3]The first two accounts of avoiding the paradox by restricting the language were due to Russell and Poincaré. They both disallowed impredicative specification: only predicative specification (as will be defined below) was to be permitted. Russell's own solution (in [131]) was to adopt the vicious circle principle which can be roughly stated as follows: "*No entity determined by a condition that refers to a certain totality should belong to this totality*". Poincaré (in [119]) took refuge in banning "*les définitions non prédicatives*" which were taken by him to be: *Definitions by a relation between the object to be defined and all individuals of a kind of which either the object itself to be defined is supposed to be a part or other things that cannot be themselves defined except by the object to be defined.* So both Russell and Poincaré required only predicative sets to be considered, where a set $A = \{x : \Phi(x)\}$ is predicative iff Φ contains no variable which can take A as a value. This helps because it is otherwise very easy to get a vicious circle fallacy if we let the arguments of a certain propositional function (or the elements of a set) presuppose the function (or the set) itself. Russell's and Poincaré's solution was to use predicative comprehension, instances of which start with individuals, then generate sets, then new sets and so on as in the following example: Take 0 at level 0, $\{0\}$ at level 1, $\{0, \{0\}\}$ at level 2, and so on. Russell's ramified theory of types in Principia Mathematica applied the vicious circle principle, assuming all the elements of the set before constructing it. This theory obviously overcomes the paradox for the sentence Φ denoting $\neg(y \in y)$ is not predicative.

the essence, or the common basis, of the various modern type theories.

The historical line in this book is, therefore, only part of our *method* of research, and definitely not a *goal* of our research. It is important to stress that we do not give an extensive history of the subject.[4] Other developments deserve attention and we refer the reader especially to Ivor Grattan-Guinness's book [63] for an excellent historical account of many of the concepts discussed in this book. We also refer the reader to the work of Cardone and Hindley [25] on the history of the λ-calculus. Furthermore, Cocchiarella's work in [32, 33] and Landini's book [101] add useful dimensions to our discussion of Frege and Russell's work.

The approach

Following the historical line from Frege (1879) to today, we are confronted with various type systems (e.g., Russell's, Ramsey's, Church's, de Bruijn's, etc.). Often, such a system has already been described in a modern framework, but the relation between the modern description and the original system has not always been made clear. This is particularly the case when the original system is quite far from the modern framework with respect to notation, level of formality and/or purpose. We will focus on such type systems. We describe them within the modern framework in such a way that:

- We respect the ideas and the philosophy underlying the original system;
- We meet contemporary requirements on formality and accuracy.

As basis for our framework we choose the typed λ-calculus, and more specifically, the framework of Pure Type Systems [3] (PTSs for short). There are several reasons for this choice:

- Many type systems have already been placed in this framework (see Example 5.2.4 of [3]);
- PTSs meet contemporary requirements on formality and accuracy;
- PTSs focus on the heart of type theory: function abstraction and application. This makes it possible to compare type systems in a very fundamental way, without being hindered by things that do not touch the heart of the matter;

[4]Curry, in his work on combinatory logic, introduced before 1940 an influential notion of typing that is still used nowadays when one refers to typing à la Curry as opposed to typing à la Church. Similarly, in 1937, Quine in [120] introduced his New Foundations which retained typing axioms, but abandoned the idea of representing types formally as Russell did. Quine's NF presupposes the very simple linear type theory with types 0 for individuals, 1 for sets of individuals, 2 for sets of sets of individuals, etc.

- Though PTSs focus on the heart of type theory, they are easily extendible in several ways. There are already many extensions described in the literature. Below, we list a small number of such extensions:
  - PTSs with definitions, introduced in [137, 14, 81];
  - PTSs with modalities, introduced in [16];
  - PTSs with sum types, see [4];
  - PTSs with quotient types, see [4];
  - PTSs with subset types, see [4];
  - PTSs with parameters, see [84, 13, 87];
  - PTSs with explicit substitutions, see [12];
  - PTSs with Π-reduction, see [90, 81, 86].
- The meta-theory that has been developed for PTSs makes it easier to access, develop and compare meta-theoretic properties of the various original type systems.

By placing several systems in the PTS framework, we also find some omissions in this framework. In particular, there is no extension of PTSs with parametric definitions, while parametric definitions play an important role in the type systems underlying the proof checker AUTOMATH [112]. Extending PTSs with parametric definitions not only opens the possibility of placing AUTOMATH more accurately in the framework of PTSs, it makes it also possible to give a better classification of more modern type systems and their applications. The third part of this book deals with various extensions of PTSs.

Part I

The Evolution of Type Theory until the 1940s

Chapter 1

Prehistory

In this chapter, we discuss the development of type theory before it was actually baptised. This may sound like a contradiction. But types have played an important (though not very apparent) role in mathematics even before the theory of types was explicitly introduced by Russell in 1908 [131]. Moreover, knowledge of the development of logic and mathematics before 1908, and especially of the occurrence of the logical paradoxes at the turn of the century, provides insight in the way in which Russell and others formulated their theories of types.

When the first formalisations of parts of mathematics and logic appeared, the types were left implicit. Cantor's Set Theory [23, 24], Peano's formalisation of the theory of natural numbers in [116], and Frege's *Begriffsschrift* [49] and *Grundgesetze der Arithmetik* [52, 56] did not have a formal type system. The type of an object is indicated by means of natural language ("Let a be a proposition") or is taken for granted. Types were informally present in the background of these theories, but a formal representation of the types was not incorporated: one could say that they were separated from logic and mathematics.

However, even without a formalisation of the notion of types, the introduction of formal language had considerable advantages in the description of mathematical notions. The formalisation made it easier to give a precise definition of important abstract concepts, like the concept of function. The precise formulation allowed for a generalisation of the notion of function to include not only functions that take numbers as an argument, and return a number, but also functions that can take and return other sorts of arguments (like propositions, and even functions). Unfortunately, this also allowed logical paradoxes to enter the formal theory, without the (informal) type mechanism being able to prevent that.

In this chapter we first argue that types have always been present in mathematics, though probably nobody was aware of it before the end of the 19th century (Section 1a). We proceed by describing how the logical paradoxes entered the formal systems of Frege, Cantor and Peano in Section 1b.

The historical remarks in this chapter have been taken from various resources. The most important ones are [11, 39, 67, 94, 118, 127, 146].

1a Paradox threats

The most fundamental idea behind type theory is being able to distinguish between different classes of objects (*types*). Until the end of the 19th century it had hardly ever been necessary to make this ability explicit. The mathematical language itself was predominantly informal, and so was the use of classes of objects.

It is, however, difficult to argue that there were no types before Russell "invented" them in 1908. Already around 325 B.C., Euclid began his *Elements* [45] with the following primitive definitions:

1. A *point* is that which has no part;

2. A *line* is breadthless length.

From these two basic notions of "point" and "line", Euclid defined more complex notions, like the notion of "circle":

15. A *circle* is a plane figure contained by one line such that all the straight lines falling upon it from one point among those lying within the figure are equal to one another.

At first sight, these three observations are mere definitions. But these three pieces of text do not only *define* the notions of point, line and circle, they also show that Euclid *distinguished* between points, lines and circles. Throughout the *Elements*, Euclid always mentioned to which class an object belonged (the class of points, the class of lines, etc.). In doing so, he prevented undesired situations, like the intersection of two points (instead of two lines).

Undesired results? Euclid himself would probably have said: *impossible* results. When talking of an intersection, intuition implicitly forced him to think about the *type* of the objects of which he wanted to construct the intersection. As the intersection of two points is not supported by intuition, he did not even *try* to undertake such a construction.

Euclid's attitude to, and implicit use of type theory was maintained by the mathematicians and logicians of the next twenty-one centuries. From the 19th century on, mathematical systems became less intuitive, for several reasons:

1. The system itself is complex, or abstract. An example is the theory of convergence in real analysis;

2. The system is a formal system, for example, the formalisation of logic in Frege's *Begriffsschrift*;

3. (In the second half of the 20th century:) It is not a human being working with the system, but something with less intuition, in particular: a computer.

We will call these three situations *paradox threats*. In all these cases, there is not enough intuition to activate the (implicitly present) type theory to warn against an impossible situation. One proceeds to reason within the impossible situation and then obtains a result that may be wrong or paradoxical: an *undesired* situation. We mention examples related to the three situations above:

ad 1. The controversial results on convergence of series in analysis obtained in the 17th and 18th century, due to lack of knowledge on what real numbers actually are;

ad 2. The logical paradoxes that arose from self-application of functions. Self-application is intuitively impossible, but this is easily forgotten when working in a formal system in which such self-application can be expressed. The result is undesirable: a logical paradox;

ad 3. An untyped computer program may receive instructions from a not too watchful user to add the number 3 to the string `four` (instead of the number 4). The computer, unaware of the fact that `four` is not a number, starts his calculation. It is not programmed to handle the calculation of $3 + \texttt{four}$. The result of this calculation is unpredictable. The computer may

- give an answer that is clearly wrong (for example, `**`);
- give no answer at all;
- give an answer that is not so clearly wrong (for example, 6).

Especially the last situation is highly undesirable.

The example ad 2 is the main subject of the next section.

1b Paradox threats in formal systems

In the 19th century, the need for a more precise style in mathematics arose. Controversial results had appeared in analysis. Many of these controversies were solved by the work of Cauchy. For instance, he introduced a precise definition of convergence in his *Cours d'Analyse* [26]. Due to the more exact definition of real numbers given by Dedekind [43], the rules for reasoning with real numbers became even more precise.

In 1879, Frege published his *Begriffsschrift* [49], in which he presented the first formalisation of logic. Frege's reasoning was uncommonly precise for those days. Until then, it had been possible to make mathematical and logical concepts more clear by textual refinement in the natural language in which they were described. Frege was not satisfied with this:

> "...I found the inadequacy of language to be an obstacle; no matter how unwieldy the expressions I was ready to accept, I was less and less able, as the relations became more and more complex, to attain the precision that my purpose required."
>
> (*Begriffsschrift*, Preface)

Frege therefore presented a completely formal system, whose

> "first purpose is to provide us with the most reliable test of the validity of a chain of inferences and to point out every presupposition that tries to sneak in unnoticed, so that its origin can be investigated."
>
> (*Begriffsschrift*, Preface)

1b1 Functions and their course of values

The introduction of a very general definition of function was the key to the formalisation of logic. In the Begriffsschrift, Frege defined what we will call the *Abstraction Principle*:

Abstraction Principle 1.1

> *"If in an expression, [...] a simple or a compound sign has one or more occurrences and if we regard that sign as replaceable in all or some of these occurrences by something else (but everywhere by the same thing), then we call the part that remains invariant in the expression a function, and the replaceable part the argument of the function."*
>
> (*Begriffsschrift,* Section 9)

Up to this section in the Begriffsschrift, Frege put no restrictions on what could play the role of an argument. An argument could be a number (as was the situation in analysis), but also a proposition, or a function. Similarly, the result of applying a function to an argument did not necessarily have to be a number. Functions of more than one argument were constructed by a method that is very close to the method presented by Schönfinkel [132] in 1924:

Abstraction Principle 1.2

> *"If, given a function, we think of a sign*[1] *that was hitherto regarded as not replaceable as being replaceable at some or all of its occurrences, then by adopting this conception we obtain a function that has a new argument in addition to those it had before."*
>
> (*Begriffsschrift,* Section 9)

With this definition of function, two of the three possible paradox threats mentioned on p. 10 occurred:

1. The generalisation of the concept of function made the system more abstract and less intuitive. The fact that functions could have different types of arguments is at the basis of the Russell paradox;

2. Frege introduced a *formal* system instead of the informal systems that were used up till then. Type theory, that would be helpful in distinguishing between the different types of arguments that a function might take, was left informal.

So, Frege had to proceed with caution. And so he did, at this stage. He remarked that

> "if the [...] letter [sign] occurs as a function sign, this circumstance [should] be taken into account."
>
> (*Begriffsschrift*, Section 11)

This could be interpreted as if Frege was aware of some typing rule that does not allow to substitute functions for object variables or objects for function variables. In his paper *Function and Concept* [51], Frege more explicitly stated:

> " Now just as functions are fundamentally different from objects, so also functions whose arguments are and must be functions are fundamentally different from functions whose arguments are objects and cannot be anything else. I call the latter first-level, the former second-level."
>
> (*Function and Concept*, pp. 26–27)

[1] We can now regard a sign that previously was considered replaceable as replaceable also in those places in which up to this point it was considered fixed. [footnote by Frege]

A few pages later he proceeded:

> "In regard to second-level functions with one argument, we must make a distinction, according as the role of this argument can be played by a function of one or of two arguments."
>
> (*Function and Concept*, p. 29)

Therefore, we may safely conclude that Frege avoided the two paradox threats in the *Begriffsschrift*. In *Function and Concept* we even see that he was aware of the fact that making a difference between first-level and second-level objects is essential in preventing certain paradoxes:

> "The ontological proof of God's existence suffers from the fallacy of treating existence as a first-level concept."
>
> (*Function and Concept*, p. 27, footnote)

The *Begriffsschrift*, however, was only a prelude to Frege's writings. In *Grundlagen der Arithmetik* [50] he argued that mathematics can be seen as a branch of logic. In *Grundgesetze der Arithmetik* [52, 56] he actually described the elementary parts of arithmetics within an extension of the logical framework that was presented in the *Begriffsschrift*.

Frege approached the paradox threats for a second time at the end of Section 2 of his *Grundgesetze*. There he defined the expression "the function $\Phi(x)$ has the same course-of-values as the function $\Psi(x)$" by

> "the functions $\Phi(x)$ and $\Psi(x)$ always have the same value for the same argument."
>
> (*Grundgesetze*, p. 7)

Note that functions $\Phi(x)$ and $\Psi(x)$ may have equal courses-of-values even if they have different definitions. For instance, let $\Phi(x)$ be $x \wedge \neg x$, and $\Psi(x)$ be $x \leftrightarrow \neg x$, for all propositions x. Then $\Phi(x) = \Psi(x)$ for all x. So $\Phi(x)$ and $\Psi(x)$ are different functions, but have the same course-of-values.

Frege denoted the course-of-values of a function $\Phi(x)$ by $\grave{\varepsilon}\Phi(\varepsilon)$.[2] The definition of equal courses-of-values could therefore be expressed as

$$\grave{\varepsilon}f(\varepsilon) = \grave{\varepsilon}g(\varepsilon) \longleftrightarrow \forall a[f(a) = g(a)]. \tag{1.1}$$

[2]This may well have been the origin of Russell's notation $\hat{x}\Phi(x)$ for the class of objects that have the property Φ. According to a paper by J. B. Rosser [128], the notation $\hat{x}\Phi(x)$ has been at the basis of the current notation $\lambda x.\Phi$ in the λ-calculus. Church is supposed to have written $\wedge x\Phi(x)$ for the function $x \mapsto \Phi(x)$, writing the hat in front of the x in order to distinguish this function from the class $\hat{x}\Phi(x)$. For typographical reasons, the $\wedge$ is supposed to have changed into a λ. On the other hand, J. P. Seldin informed us [135] that he had asked Church about it in 1982, and that Church had answered that there was no particular reason for choosing λ, that some letter was needed and λ happened to have been chosen. Moreover, Curry had told him that Church had a manuscript in which there were many occurrences of λ already in 1929, so three years before the paper [27] appeared.

In modern terminology, we could say that the functions $\Phi(x)$ and $\Psi(x)$ have the same course-of-values if they have the same *graph*.

Frege did not provide a satisfying intuition for the formal notion of course-of-values of a function. He treated courses-of-values as ordinary objects. As a consequence, a function that takes objects as arguments could have its own course-of-values as an argument. In modern terminology: a function that takes objects as arguments can have its own graph as an argument. All essential information of a function is contained in its graph. So intuitively, a system in which a function can be applied to its own graph should have similar possibilities as a system in which a function can be applied to itself. Frege excluded the paradox threats from his system by forbidding self-application, but due to his treatment of courses-of-values these threats were able to enter his system through a back door.

1b2 The Russell Paradox in *Grundgesetze*

In 1902, Russell wrote a letter to Frege [129], in which he informed Frege that he had discovered a paradox in Frege's *Begriffsschrift*. Russell gave his well-known argument, defining the propositional function $f(x)$ by $\neg x(x)$ (in Russell's words: "to be a predicate that cannot be predicated of itself"). He assumed $f(f)$. Then by definition of f, $\neg f(f)$, a contradiction. Therefore: $\neg f(f)$ holds. But then (again by definition of f), $f(f)$ holds. Russell concluded that both $f(f)$ and $\neg f(f)$ hold, a contradiction.

Only six days later, Frege answered Russell that Russell's derivation of the paradox was incorrect [55]. He explained that the self-application $f(f)$ is not possible in the *Begriffsschrift*. $f(x)$ is a function, which requires an *object* as an argument, and a function cannot be an object in the *Begriffsschrift* (see Section 1b1).

In the same letter, however, Frege explained that Russell's argument could be amended to a paradox in the system of his *Grundgesetze*, using the course-of-values of functions. Frege's amendment was shortly explained in that letter, but he added an appendix of eleven pages to the second volume of his *Grundgesetze* in which he provided a very detailed and correct description of the paradox.

The derivation goes as follows (using the same argument as Frege, though replacing Frege's two-dimensional notation by the nowadays more usual one-dimensional notation). First, define the function $f(x)$ by:

$$\neg\forall\varphi[(\grave{\alpha}\varphi(\alpha) = x) \longrightarrow \varphi(x)].$$

Write $K = \grave{\varepsilon}f(\varepsilon)$. By (1.1) we have, for any function $g(x)$,

$$\grave{\varepsilon}g(\varepsilon) = \grave{\varepsilon}f(\varepsilon) \longrightarrow g(K) = f(K)$$

and this implies

$$f(K) \longrightarrow ((\grave{\varepsilon}g(\varepsilon) = K) \longrightarrow g(K)). \tag{1.2}$$

As this holds for any function $g(x)$, we have

$$f(K) \longrightarrow \forall\varphi[\grave{\varepsilon}\varphi(\varepsilon) = K \to \varphi(K)]. \tag{1.3}$$

On the other hand, for any function g,

$$\forall\varphi[\grave{\varepsilon}\varphi(\varepsilon) = K \to \varphi(K)] \longrightarrow ((\grave{\varepsilon}g(\varepsilon) = K) \to g(K)).$$

Substituting $f(x)$ for $g(x)$ results in:

$$\forall\varphi[\grave{\varepsilon}\varphi(\varepsilon) = K \to \varphi(K)] \longrightarrow ((\grave{\varepsilon}f(\varepsilon) = K) \to f(K))$$

and as $\grave{\varepsilon}f(\varepsilon) = K$ by definition of K,

$$\forall\varphi[\grave{\varepsilon}\varphi(\varepsilon) = K \to \varphi(K)] \longrightarrow f(K).$$

Using the definition of f, we obtain

$$\forall\varphi[\grave{\varepsilon}\varphi(\varepsilon) = K \to \varphi(K)] \longrightarrow \neg\forall\varphi[\grave{\varepsilon}\varphi(\varepsilon) = K \to \varphi(K)],$$

hence by reductio ad absurdum,

$$\neg\forall\varphi[\grave{\alpha}\varphi(\alpha) = K \to \varphi(K)],$$

or shorthand:

$$f(K). \tag{1.4}$$

Applying (1.3) results in

$$\forall\varphi[\grave{\alpha}\varphi(\alpha) = K \to \varphi(K)],$$

which implies

$$\neg\neg\forall\varphi[\grave{\alpha}\varphi(\alpha) = K \to \varphi(K)],$$

or shorthand:

$$\neg f(K). \tag{1.5}$$

(1.4) and (1.5) contradict each other.

1b3 How wrong was Frege?

In the history of the Russell paradox, Frege is often depicted as the pitiful person whose system was inconsistent. This suggests that Frege's system was the only one that was inconsistent, and that Frege was very inaccurate in his writings. On these points, history does Frege an injustice.

In fact, Frege's system was much more accurate than other systems of those days. Peano's work, for instance, was less precise on several points:

- Peano hardly paid any attention to logic, especially not to quantification theory;

- Peano did not make a strict distinction between his symbolism and the objects underlying this symbolism. Frege was much more accurate on this point (see also his paper *Über Sinn und Bedeutung* [53]);

- Frege made a strict distinction between a proposition (as an object of interest or discussion) and the *assertion* of a proposition. Frege denoted a proposition, in general, by $-A$, and the assertion of the proposition by $\vdash A$. The symbol $\vdash$ is still widely used in logic and type theory. Peano did not make this distinction and simply wrote A.

Nevertheless, Peano's work was very popular, for several reasons:

- Peano had able collaborators, and in general had a better eye for presentation and publicity. For instance, he bought his own press, so that he could supervise the printing of his journal *Rivista di Matematica* and *Formulaire* [117];

- Peano used a symbolism much more familiar to the notations that were used in those days by mathematicians (and many of his notations, like $\in$ for "is an element of", and $\supset$ for logical implication, are also used in Russell's *Principia Mathematica*, and are actually still in use).

Frege's work did not have these advantages and was hardly read before 1902[3]. In the last paragraph of [54], Frege concluded:

> "...I observe merely that the Peano notation is unquestionably more convenient for the typesetter, and in many cases takes up less room than mine, but that these advantages seem to me, due to the inferior perspicuity and logical defectiveness, to have been paid for too dearly — at any rate for the purposes I want to pursue."
>
> (*Ueber die Begriffschrift des Herrn Peano und meine eigene*, p. 378)

[3]When Peano published his formalisation of mathematics in 1889 [116] he clearly did not know Frege's ***Begriffsschrift***, as he did not mention the work, and was not aware of Frege's formalisation of quantification theory. Peano considered quantification theory to be "abstruse" in [117], on which Frege proudly reacted:

> "In this respect my conceptual notion of 1879 is superior to the Peano one. Already, at that time, I specified all the laws necessary for my designation of generality, so that nothing fundamental remains to be examined. These laws are few in number, and I do not know why they should be said to be abstruse. If it is otherwise with the Peano conceptual notation, then this is due to the unsuitable notation."
>
> ([54], p. 376)

Frege's system was not the only paradoxical one. The Russell paradox can be derived in Peano's system as well, by defining the class

$$K \stackrel{\text{def}}{=} \{x \mid x \notin x\}$$

and deriving $K \in K \longleftrightarrow K \notin K$. In Cantor's Set Theory one can derive the paradox via the same class (or *set*, in Cantor's terminology).

1b4 The importance of Russell's Paradox

Russell's paradox was certainly not the first or only paradox in history. Paradoxes were already widely known in antiquity. The first known paradox is the Achilles paradox of Zeno of Elea. It is a purely mathematical paradox. Due to a precise formulation of mathematics and especially the concept of real numbers, the paradox can now be satisfactorily solved.

The oldest logical paradox is probably the Liar's paradox, also known as the paradox of Epimenides. It can be very shortly formulated by the sentence "This sentence is not true". The paradox was widely known in antiquity. For instance, it is referred to in the Bible (Titus 1:12). It is based on the confusion between language and meta-language.

The Burali-Forti paradox ([22], 1897) is the first of the modern paradoxes. It is a paradox within Cantor's theory on ordinal numbers. Cantor's paradox on the largest cardinal number occurs in the same field. It must have been discovered by Cantor around 1895, but was not published before 1932.

The logicians considered these paradoxes to be out of the scope of logic: the paradoxes based on the Liar's paradox could be regarded as a problem of linguistics, and the paradoxes of Cantor and Burali-Forti occurred in a questionable (in those days highly) part of mathematics: Cantor's Set Theory.

The Russell paradox, however, was a paradox that could be formulated in all the systems that were presented at the end of the 19th century (except for Frege's *Begriffsschrift*). It was at the very basics of logic. It could not be disregarded, and a solution to it had to be found.

Chapter 2

Type theory in Principia Mathematica

When Russell proved Frege's *Grundgesetze* to be inconsistent, Frege was not the only person in trouble. In Russell's letter to Frege (1902), we read:

> "I am on the point of finishing a book on the principles of mathematics"
>
> (*Letter to Frege*, [129])

Therefore, Russell had to find a solution to the paradoxes, before he could finish his book.

His paper *Mathematical logic as based on the theory of types* [131] (1908), in which a first step is made towards the Ramified Theory of Types, started with a description of the most important contradictions that were known up till then, including Russell's own paradox. He then concluded:

> "In all the above contradictions there is a common characteristic, which we may describe as self-reference or reflexiveness. [...] In each contradiction something is said about *all* cases of some kind, and from what is said a new case seems to be generated, which both is and is not of the same kind as the cases of which *all* were concerned in what was said."
>
> (*Ibid.*)

Russell's plan was, therefore, to avoid the paradoxes by avoiding all possible self-references. He postulated the "vicious circle principle":

Vicious Circle Principle 2.1

> "*Whatever involves* all *of a collection must not be one of the collection.*"
>
> ([98], p. 20)

Russell applies this principle very strictly. He implemented it using types, in particular the so-called *ramified* types. The theory presented in *Mathematical logic as based on the theory of types* was elaborated in Chapter II of the Introduction to the famous *Principia Mathematica* [145] (1910-1912). In the *Principia*, Whitehead and Russell founded mathematics on logic, as far as possible. The result was a very formal and accurate build-up of mathematics, avoiding the logical paradoxes.

The logical part of the *Principia* was based on the works of Frege. This was acknowledged by Whitehead and Russell in the preface, and can also be seen throughout the description of Type Theory. The notion of function is based on Frege's Abstraction Principles 1.1 and 1.2, and the *Principia* notation $\hat{x}f(x)$ for a class looks very similar to Frege's $\grave{\varepsilon}f(\varepsilon)$ for course-of-values.

An important difference is that Whitehead and Russell treated functions as first-class citizens. Frege used courses-of-values as a way of speaking about functions (and was confronted with a paradox); in the *Principia* a direct approach was possible. Equality, for instance, was defined for objects as well as for functions by means of Leibniz equality ($x = y$ if and only if $f(x) \leftrightarrow f(y)$ for all propositional functions f — see [145], $*13{\cdot}11$).

The description of the Ramified Theory of Types (RTT) in the *Principia* was, though extensive, still informal. It is clear that Type Theory had not yet become an independent subject. The theory

> "only recommended itself to us in the first instance by its ability to solve certain contradictions"
>
> (*Principia Mathematica*, p. 37)

And though

> "it has also a certain consonance with common sense which makes it inherently credible"
>
> (*Principia Mathematica*, p. 37)

(probably, Whitehead and Russell refer to the implicit, intuitive use of types by mathematicians. See Section 1a), Type Theory was not introduced because it was interesting on its own, but because it had to serve as a tool for logic and mathematics. A formalisation of Type Theory, therefore, was not considered in those days.

Though the description of RTT in the *Principia* was still informal, it was clearly present throughout the work. It was not mentioned very often, but when necessary, Russell made a remark on RTT. This is an important difference with the earlier writings of Frege, Peano and Cantor.

If we want to compare RTT with contemporary type systems, we have to make a formalisation of RTT. Though there are many descriptions of RTT available in the literature (as in Church's work [29, 31], Hilbert and Ackermann's book [70], Ramsey's work [124] and Section 27 of Schütte's book [133]), none of these descriptions presents a formalisation that is both accurate and as close as possible to the ideas of the *Principia*. We will fill up this gap in the literature in the first part of this chapter.

Making such a formalisation is by no means easy:

- Important formal notions, especially the notion of substitution, remained completely unexplained in the *Principia*;
- The accuracy of Frege's work was not present in Russell's. This was already observed by Gödel, who said that the precision of Frege was lost in the writings of Russell, and who, due to the informality of some basic notions of the *Principia*, had to give his paper [61] the title *Über formal unentscheidbare Sätze der Principia Mathematica und* verwandter *Systeme*.

In Section 1b1 we saw that Frege generalised the notion of function from analysis. For Russell's formalisation of mathematics within logic, a special kind of these functions was needed: the so-called *propositional* functions. A propositional function (pf) always returns a *proposition* when it is applied to suitable arguments. In Section 2a, we introduce a formalised version of these pfs. This makes it possible to compare pfs with other formal systems, like the λ-calculus, and to give a precise definition of substitution.

In Section 2b we give a formalisation of Russell's notion of ramified type (Section 2b1), followed by a formal definition of the notion *the pf f is of type t* (Section 2b2). We motivate this definition (Section 2b3) by referring to passages in the *Principia*. As the formalisation of pf is precise enough to be translated to the λ-calculus, we can make a comparison between RTT and current type systems.

Thanks to our formal notation and its relation to the λ-calculus, we are able to prove properties of RTT in an easy way, using properties of modern type systems. This will be done in Section 2c. Due to the new notation it is relatively easy to see that we have proved variants of well-known theorems from Type Theory, like Strong Normalisation, Free Variable Lemma, Strengthening Lemma, Unicity of Types and Subterm Lemma.

In Section 2d we answer in full detail the question which pfs are typable. We also make a comparison between our notion of typable pf, and the corresponding notion in the *Principia*, and conclude that these two notions of typable pf coincide.

2a Principia's propositional functions

In this section we present a formalisation of the propositional functions (pfs) of the *Principia*. In Section 2a1 we give a syntax that is as close as possible to the ideas of the *Principia*. Intuition about this syntax is provided in Section 2a2 by translating pfs into λ-terms. In Section 2a4 we define some related notions that are needed in the rest of the chapter. We devote a special section 2a5 to the notion of substitution. This notion is clearly present in the *Principia*, but not formally defined. Due to the translation to λ-calculus of Section 2a2, we are able to give a precise definition.

2a1 Definition

The definition of propositional function in the *Principia* is as follows:

> "By a "propositional function" we mean something which contains a variable x, and expresses a *proposition* as soon as a value is assigned to x."
>
> (*Principia Mathematica*, p. 38)

Pfs are, however, *constructed* from propositions[1] with the use of the Abstraction Principles: they arise when in a proposition one or more occurrences of a sign are replaced by a variable. Therefore we have to begin our formalisation with certain basic propositions, certain basic signs, and signs that indicate a replaceable object. For this purpose we use

- A set $\mathcal{A}$ of *individual symbols* (the basic signs);
- A set $\mathcal{V}$ of *variables* (the signs that indicate replaceable objects);
- A set $\mathcal{R}$ of *relation symbols* together with a map $\mathfrak{a} : \mathcal{R} \to \mathbb{N}^+$ indicating the *arity* of each relation-symbol (these are used to form the basic propositions).

We want to have a sufficient supply of individual symbols, variables and relation symbols and therefore assume that $\mathcal{A}$ and $\mathcal{V}$ are infinite (but countable), and that $\{R \in \mathcal{R} \mid \mathfrak{a}(R) = n\}$ is infinite (but countable) for each $n \in \mathbb{N}^+$. We assume that $\{\mathtt{a}_1, \mathtt{a}_2, \ldots\} \subseteq \mathcal{A}$, $\{\mathtt{x}, \mathtt{y}, \mathtt{z}, \mathtt{x}_1, \ldots\} \subseteq \mathcal{V}$ and $\{\mathtt{R}, \mathtt{S}, \ldots\} \subseteq \mathcal{R}$. We use $a_1, a_2, \ldots$ as metavariables over $\mathcal{A}$; $x, y, z, x_1, \ldots$ as metavariables over $\mathcal{V}$ and $R, S, \ldots$ as metavariables over $\mathcal{R}$. For technical reasons we assume that there is an *order* (e.g.

[1] Of course, the *stratification* notion plays a crucial role in the formulations of propositions and formulae and this notion has been used in other modern works (e.g. by Leivant in [102]).

alphabetical) on $\mathcal{V}$. We write $x < y$ if x is ordered before y, and not equal to y (so: $<$ is *strict*). In particular, we assume that

$$\mathtt{x} < \mathtt{x}_1 < \ldots \mathtt{y} < \mathtt{y}_1 < \ldots \mathtt{z} < \mathtt{z}_1 \ldots$$

and: for each x there is a y with $x < y$.

Definition 2.2 (Atomic propositions) A list of symbols of the form $R(a_1, \ldots, a_{\mathfrak{a}(R)})$ is called an *atomic proposition.*

Other names used for these atomic propositions in the *Principia* are *elementary judgements* and *elementary propositions* (cf. [145], pp. xv, 43–45, and 91).

Propositional functions in *Principia Mathematica* are generated from atomic propositions by two means:

- The use of logical connectives and quantifiers;
- Abstraction from (earlier generated) propositional functions, using the abstraction principles.

This leads to the following formal definition of propositional function. Examples are given in Example 2.5 and intuition is provided in Section 2a2.

Definition 2.3 (Propositional functions) We define a collection $\mathcal{P}$ of *propositional functions* (pfs), and for each element f of $\mathcal{P}$ we simultaneously define the collection $\text{FV}(f)$ of *free variables* of f:

1. If $i_1, \ldots, i_{\mathfrak{a}(R)} \in \mathcal{A} \cup \mathcal{V}$ then $R(i_1, \ldots, i_{\mathfrak{a}(R)}) \in \mathcal{P}$.
   $\text{FV}\big(R(i_1, \ldots, i_{\mathfrak{a}(R)})\big) \stackrel{\text{def}}{=} \{i_1, \ldots, i_{\mathfrak{a}(R)}\} \cap \mathcal{V}$;
2. If $f, g \in \mathcal{P}$ then $f \vee g \in \mathcal{P}$ and $\neg f \in \mathcal{P}$.
   $\text{FV}(f \vee g) \stackrel{\text{def}}{=} \text{FV}(f) \cup \text{FV}(g)$; $\text{FV}(\neg f) \stackrel{\text{def}}{=} \text{FV}(f)$;
3. If $f \in \mathcal{P}$ and $x \in \text{FV}(f)$ then $\forall x[f] \in \mathcal{P}$.
   $\text{FV}(\forall x[f]) \stackrel{\text{def}}{=} \text{FV}(f) \setminus \{x\}$;
4. If $n \in \mathbb{N}$ and $k_1, \ldots, k_n \in \mathcal{A} \cup \mathcal{V} \cup \mathcal{P}$, then $z(k_1, \ldots, k_n) \in \mathcal{P}$.
   $\text{FV}(z(k_1, \ldots, k_n)) \stackrel{\text{def}}{=} \{z, k_1, \ldots, k_n\} \cap \mathcal{V}$.
   If $n = 0$ then we write $z()$ in order to distinguish the pf $z()$ from the variable z[2];
5. All pfs can be constructed by using the construction-rules 1, 2, 3 and 4 above.

[2]It is important to note that a variable is not a pf. See for instance [130], Chapter VIII: "The variable", p. 94 of the 7th impression.

We use the letters f, g, h as meta-variables over $\mathcal{P}$.

Note that in clause 4. of the above definition, the variable binding in pf arguments of terms $z(k_1, \ldots, k_n)$ may be quite unexpected. We explain this feature in detail in Section 2a3 and especially in Remark 2.10.

Definition 2.4 (Propositions) A propositional function f is a *proposition* if $\mathrm{FV}(f) = \varnothing$.

Example 2.5 We give some examples of (higher-order) pfs of the form $z(k_1, \ldots, k_n)$ in ordinary mathematics. To keep the link with mathematics clear, we use some extra logical connectives like $\leftrightarrow$ and $\rightarrow$.

1. The pfs $\mathtt{z}(\mathtt{x})$ and $\mathtt{z}(\mathtt{y})$ in the definition of equality according to Leibniz: By definition $\mathtt{x} = \mathtt{y}$ if and only if

$$\forall \mathtt{z}[\mathtt{z}(\mathtt{x}) \leftrightarrow \mathtt{z}(\mathtt{y})];$$

2. The pfs $\mathtt{z}(0)$, $\mathtt{z}(\mathtt{x})$ and $\mathtt{z}(\mathtt{y})$ in the formulation of the principle of mathematical induction:

$$\begin{aligned} \forall \mathtt{z}[\mathtt{z}(0) \quad &\rightarrow \quad (\forall \mathtt{x} \forall \mathtt{y}[\mathtt{z}(\mathtt{x}) \rightarrow (\mathtt{S}(\mathtt{x}, \mathtt{y}) \rightarrow \mathtt{z}(\mathtt{y}))]) \\ &\rightarrow \quad \forall \mathtt{x}[\mathtt{z}(\mathtt{x})]] \end{aligned}$$

(we suppose that the relation symbol $\mathtt{S}$ represents the successor function: $\mathtt{S}(x, y)$ holds if and only if y is the successor of x);

3. $\mathtt{z}()$ in the formulation of the law of the excluded middle:

$$\forall \mathtt{z}[\mathtt{z}() \vee \neg \mathtt{z}()].$$

2a2 Principia's propositional functions as λ-terms

The binding structure and the notion of free variable of pfs become more clear if we translate pfs to λ-terms. Moreover, such a translation will be useful at several places in this chapter, for instance when we give a definition of substitution.

We first translate one of the examples of Example 2.5. Then we give a formal definition of the translation that we have in mind. After that we provide additional remarks and intuition on pfs.

Example 2.6 Consider the pf $f \equiv \forall \mathtt{z}[\mathtt{z}(\mathtt{x}) \leftrightarrow \mathtt{z}(\mathtt{y})]$ of Example 2.5.1. Two objects x and y are Leibniz-equal if and only if they share the same properties. These objects are represented by the variables $\mathtt{x}$ and $\mathtt{y}$. The variable $\mathtt{z}$ is a variable for properties of objects, in other words: predicates over objects. Such a predicate is a function that takes the object as argument, and returns a truth value. The expression $\mathtt{z}(\mathtt{x})$ indicates that the predicate that is taken for $\mathtt{z}$ must be applied to the object that is taken for $\mathtt{x}$. Therefore, we translate $\mathtt{z}(\mathtt{x})$ by an application of $\mathtt{z}$ to $\mathtt{x}$ in the λ-calculus: $\mathtt{zx}$. Similarly we translate the expression $\mathtt{z}(\mathtt{y})$ by $\mathtt{zy}$.

Just as in [29], we can interpret logical connectives as functions. Therefore we can translate $\mathtt{z}(\mathtt{x}) \leftrightarrow \mathtt{z}(\mathtt{y})$ by the λ-term $\leftrightarrow(\mathtt{zx})(\mathtt{zy})$. We also handle the translation of universal quantification as in [29], hence $\forall\mathtt{z}[\ldots]$ translates to $\forall(\lambda\mathtt{z}.\ldots)$. As an effect we get a λ-term $\forall(\lambda\mathtt{z}.\leftrightarrow(\mathtt{zx})(\mathtt{zy}))$ with two free variables, $\mathtt{x}$ and $\mathtt{y}$. But we want to have a *function* taking two *arguments*. This can be solved by a double λ-abstraction. The final result is $\lambda\mathtt{x}.\lambda\mathtt{y}.\forall(\lambda\mathtt{z}.(\leftrightarrow(\mathtt{zx})(\mathtt{zy})))$.

We remark that the pf f has two free variables, $\mathtt{x}$ and $\mathtt{y}$. These two free variables correspond to the two arguments that the propositional function takes, and therefore to the two λ-abstractions that are at the front of the translation of f.

In the following definition, we translate the propositional functions to λ-terms in a similar way as we did in Example 2.6. Let $f \in \mathcal{P}$ and let $x_1 < \cdots < x_m$ be the free variables of f. We define a λ-term $\overline{f}$. We do this in such a way that $\overline{f} \equiv \lambda x_1.\cdots\lambda x_m.F$, where F is a λ-term that is not of the form $\lambda x.F'$. To keep notations uniform, we also give translations $\overline{a}$ for $a \in \mathcal{A}$ and $\overline{x}$ for $x \in \mathcal{V}$. To keep notations short, we use $\lambda_{i=1}^{m} x_i.F$ as shorthand for $\lambda x_1.\cdots\lambda x_m.F$.

Definition 2.7

- $\overline{a} \stackrel{\text{def}}{=} a$ for $a \in \mathcal{A}$;
- $\overline{x} \stackrel{\text{def}}{=} x$ for $x \in \mathcal{V}$;
- Now assume $f \in \mathcal{P}$ has free variables $x_1 < \cdots < x_m$. Use induction on the structure of f:
  - $f \equiv R(i_1,\ldots,i_{\mathfrak{a}(R)})$. Then $\overline{f} \stackrel{\text{def}}{=} \lambda_{i=1}^{m} x_i.R\overline{i_1}\cdots\overline{i_{\mathfrak{a}(R)}}$;
  - $f \equiv f_1 \vee f_2$. We can assume that for $j = 1,2$, $\overline{f_j} \equiv \lambda_{i=1}^{m_j} y_i^j.F_j$, where $y_1^j < \cdots < y_{m_j}^j$ are the free variables of f_j.
    Then $\overline{f} \stackrel{\text{def}}{=} \lambda_{i=1}^{m} x_i.\vee F_1 F_2$.
    If $f \equiv \neg f'$ then we can assume that $\overline{f'} \equiv \lambda_{i=1}^{m} x_i.F$, because $x_1 < \cdots < x_m$ are the free variables of f'.
    Let $\overline{f} \stackrel{\text{def}}{=} \lambda_{i=1}^{m} x_i.\neg F$;
  - $f \equiv z(k_1,\ldots,k_n)$. Let $\overline{f} \equiv \lambda_{i=1}^{m} x_i.z\overline{k_1}\cdots\overline{k_n}$;
  - $f \equiv \forall x[f']$. We can assume that $\overline{f} \equiv \lambda_{i=1}^{j-1} x_i.\lambda x.\,\lambda_{i=j}^{m} x_i.F$, because $x_1,\ldots,x_m,x$ are the free variables of f'.
    Define $\overline{f} \equiv \lambda_{i=1}^{m} x_i.\forall(\lambda x.F)$.

Example 2.8

f	$\overline{f}$
$\mathtt{R(x)}$	$\lambda\mathtt{x.Rx}$
$\mathtt{z(R(x),S(a))}$	$\lambda\mathtt{z.z(\lambda x.Rx)(Sa)}$
$\mathtt{z_1(a)} \vee \mathtt{z_2()}$	$\lambda\mathtt{z_1}.\lambda\mathtt{z_2}.\vee(\mathtt{z_1 a})\mathtt{z_2}$
$\mathtt{z(y(R(x)))}$	$\lambda\mathtt{z.z(\lambda y.y(\lambda x.Rx))}$
$\forall\mathtt{x[R(x)]}$	$\forall(\lambda\mathtt{x.Rx})$

By induction on the structure of f one can prove the following properties of $\overline{f}$:

Lemma 2.9 (Properties of $^{-}$) *Let $f \in \mathcal{P}$.*

1. $\text{FV}(\overline{f}) = \varnothing$;
2. $\overline{f}$ *is in β-normal form;*
3. $\overline{f}$ *is a λI-term;*
4. *If $x_1 < \cdots < x_m$ are the free variables of f, then $\overline{f} \equiv \lambda_{i=1}^{m} x_i.F$, where F is not of the form $\lambda x.F'$.* ⊠

Observe that we use FV for indicating both the free variables of a pf and the free variables of a λ-term. We take care that it will always be clear in which meaning we use FV. In the above definition we also assume familiarity with the notion of λI-term (see Definition 4.10 and [2]): in λI, terms of the form $\lambda x.F$ are only allowed if x appears as a free variable in F.

2a3 Principia's pfs and their translation in λ-calculus

We make some remarks on the definition of propositional function (Definition 2.3).

Remark 2.10 We show that the propositional functions of Definition 2.3 are indeed objects that exist in the theory of Russell.

1. In Rule 1 we describe the atomic propositions, and the atomic propositions in which one or more individuals have been replaced by variables due to one or more applications of the abstraction principles. The abstraction principles are not only present in the works of Frege, but also in the *Principia* (cf. for instance ∗9·14 and ∗9·15);
2. Rule 2 describes the use of the logical connectives $\vee$ and $\neg$. These logical connectives are also used in the *Principia*. Implication[3], conjunction[4] and

[3] cf. *Principia*, ∗1·01, p. 94

[4] cf. *Principia*, ∗3·01, p. 107

logical equivalence[5] are defined in terms of negation and disjunction. In examples, we sometimes use symbols for implication, conjunction and logical equivalence as abbreviations;

3. Rule 3 describes the use of the universal quantifier. It is explicitly stated in the *Principia* (cf. pp. 14–16) that the pf $\forall x[f]$ can only be constructed if f is a pf that contains x as a variable. Existential quantification[6] is defined in terms of negation and universal quantification;

4. Rule 4 is also an instantiation of the abstraction principle. The pfs that can be constructed by using the construction-rules 1–3 only are exactly the pfs of what in these days would be called first-order predicate logic. With rule 4, higher-order pfs can be constructed. This is based on the following idea. Let f be a (fixed) pf in which $k_1, \ldots, k_n$ occur. We can interpret f as an *instantiation* of a function that has taken arguments $k_1, \ldots, k_n$. We now generalise this to $z(k_1, \ldots, k_n)$, representing *any* function z taking these arguments. Such a construction is also explicitly present in the *Principia*:

> "the first matrices[7] that occur are those whose values are of the forms $\varphi x, \psi(x, y), \chi(x, y, z, \ldots)$, *i.e.* where the arguments, however many there may be, are all individuals. Such [propositional] functions we will call '*first-order* functions.' We may now introduce a notation to express 'any first-order function.' "
>
> (*Principia Mathematica*, p. 51)

Remark 2.11 The definition of free variable needs some special attention. We must notice that, for instance,

$$\mathrm{FV}(\mathtt{z}(\mathtt{R}(\mathtt{x}), \mathtt{S}(\mathtt{a}))) = \{\mathtt{z}\} \text{ and not } \{\mathtt{x}, \mathtt{z}\}.$$

The reason for this is that the notion of free variable should harmonise with the intuitive notion of "argument place" of Frege and Russell. As was indicated in Remark 2.10.4, $\mathtt{z}$ represents an arbitrary function that takes $\mathtt{R}(\mathtt{x})$ and $\mathtt{S}(\mathtt{a})$ as arguments and returns a proposition. This means that we do not have to supply an argument for $\mathtt{x}$ "by hand". As soon as we feed a suitable[8] argument f to $\mathtt{z}$ in $\mathtt{z}(\mathtt{R}(\mathtt{x}), \mathtt{S}(\mathtt{a}))$, f will take the arguments $\mathtt{R}(\mathtt{x})$ and $\mathtt{S}(\mathtt{a})$, and return a proposition.

This idea is also clearly reflected in the translation of $\mathtt{z}(\mathtt{R}(\mathtt{x}), \mathtt{S}(\mathtt{a}))$ to the λ-term $\lambda\mathtt{z}.\mathtt{z}(\lambda\mathtt{x}.\mathtt{Rx})(\mathtt{Sa})$. The variable $\mathtt{x}$ is bound in a subterm $\lambda\mathtt{x}.\mathtt{Rx}$ that is an argument to the variable $\mathtt{z}$. The full λ-term is a function of $\mathtt{z}$ only. See Example 2.24.

---

[5] cf. *Principia*, ∗4·01, p. 117

[6] cf. *Principia*, ∗10·01, p. 140

[7] see Remark 2.12 [footnote of the authors].

[8] At this stage, we cannot provide a formalisation of "suitable". This can only be done after we have introduced types, and formalised the notion "the pf f is of type t"

Remark 2.12 It appears that there is also an alternative way of constructing pfs in the *Principia.* Whitehead and Russell distinguish between quantifier-free pfs (so-called *matrices*, i.e. the pfs that can be constructed using construction-rules 1, 2 and 4). Then they form pfs by defining that

- Any matrix is a pf;
- If f is a pf and $x \in \text{FV}(f)$ then $\forall x[f]$ is a pf with free variables $\text{FV}(f) \setminus \{x\}$.

This definition is a little different from our Definition 2.3, as a pf of the form $z(\forall x[f])$ is not a matrix and therefore not a pf according to this alternative definition. Nevertheless we feel that Whitehead and Russell intended to give our Definition 2.3. In the *Principia* ([145], ∗54) they define the natural number 0 as the propositional function $\forall \mathbf{x}[\neg \mathbf{z}(\mathbf{x})]$[9]. In defining the principle of induction on natural numbers, one needs to express the property "0 has the property y", or: $y(0)$. But $y(0)$ is not a pf according to this alternative definition, as 0 contains quantifiers.

Therefore we feel that our Definition 2.3, which is also based on the definition of function by Frege and on the definition of propositional function on p. 38 of the *Principia*, is the definition that was meant by Whitehead and Russell.

Remark 2.13 Note that pfs as such do not yet obey to the vicious circle principle 2.1! For example, $\neg\mathbf{z}(\mathbf{z})$ (the pf that is at the basis of the Russell paradox) is a pf. In Section 2b we will assign types to some pfs, and it will be shown (Remark 2.66) that no type can be assigned to the pf $\neg\mathbf{z}(\mathbf{z})$.

Remark 2.14 Before we make further developments of the theory based on pfs, we must decide which of the two syntaxes introduced above shall be used in the sequel. It looks attractive to use the syntax of the λ-calculus:

- This syntax is well-known;
- It is used for many other type systems, so it makes the comparison of ramified type theory with modern type systems easier;
- There is a lot of meta-theory on typed and untyped λ-calculus. This can be useful when proving certain properties of the formalisation of the ramified theory of types that is to be introduced in the next sections;

[9] This definition is based on Frege's definition in *Grundlagen der Arithmetik* [50] (1884). See [145], vol. II, p. 4. In [50], the natural number n is defined as the class of predicates f for which there are exactly n objects a for which $f(a)$ holds. Hence 0 is the class of predicates f for which $f(a)$ does not hold for any object a. So 0 can be described by the pf $\forall \mathbf{x}[\neg \mathbf{z}(\mathbf{x})]$

- The syntax of the λ-calculus gives a better look on the notion of free variable than the syntax of pfs.

Nevertheless, we shall only indirectly use λ-calculus for our further study of the ramified type theory in this Chapter. We have several reasons for that:

- There are much more λ-terms than there are pfs. More precisely, the mapping $\bar{\ }$ is not surjective. As we want to study the theory of *Principia Mathematica* as precisely as possible, we only want to study the propositional functions, which are directly related to the syntax used by Russell and Whitehead. Not using pf-syntax may result in a system in which it is not clear which term belongs to the original ramified type theory and which term does not;

- The syntax of the λ-calculus is strongly curried. This would give problems in the definition of substitution. In a pf $\mathtt{R}(\mathtt{x}, \mathtt{y})$ we may want to substitute some object a for $\mathtt{y}$ without substituting anything for $\mathtt{x}$. In the λ-calculus, substitution should be translated to application followed by β-reduction to β-normal form. If we want to substitute something for $\mathtt{y}$ in the translation $\lambda\mathtt{x}.\lambda\mathtt{y}.\mathtt{Rxy}$ of $\mathtt{R}(\mathtt{x}, \mathtt{y})$, we have to substitute something for $\mathtt{x}$ first. Choosing a different representation of propositional functions does not help: the representation $\lambda\mathtt{y}.\lambda\mathtt{x}.\mathtt{Ryx}$ would have given problems if we wanted to substitute something for $\mathtt{x}$ without substituting something for $\mathtt{y}$;

- The translation of pfs to the λ-calculus makes it possible to use the metatheory and the intuition of the λ-calculus when we need it without losing control over the original system.

2a4 Principia's related notions

We proceed our discussion of pfs by defining a number of related notions. If a pf $z(k_1, \ldots, k_n)$ takes an argument f for the variable z, the list $k_1, \ldots, k_n$ indicates what should be substituted for the free variables of f. We therefore call this list the list of *parameters* of $z(k_1, \ldots, k_n)$. A formal definition:

Definition 2.15 (Parameters) Assume f is a pf, and $k \in \mathcal{A} \cup \mathcal{V} \cup \mathcal{P}$. We define, inductively, the notion k *is a parameter of* f, and write $\mathrm{PAR}(f)$ for the set of parameters of f.

- $\mathrm{PAR}\big(R(i_1, \ldots, i_{\mathfrak{a}(R)})\big) \stackrel{\mathrm{def}}{=} \{i_1, \ldots, i_{\mathfrak{a}(R)}\}$;

- $\mathrm{PAR}(f_1 \vee f_2) \stackrel{\mathrm{def}}{=} \mathrm{PAR}(f_1) \cup \mathrm{PAR}(f_2)$ and $\mathrm{PAR}(\neg f) \stackrel{\mathrm{def}}{=} \mathrm{PAR}(f)$;

- $\mathrm{PAR}(\forall x[f]) \stackrel{\mathrm{def}}{=} \mathrm{PAR}(f)$;

- $\text{PAR}(z(k_1,\ldots,k_n)) \stackrel{\text{def}}{=} \{k_1,\ldots,k_n\}$.

Note that $\mathtt{x}$ is not a parameter of $\mathtt{z}(\mathtt{R}(\mathtt{x}),\mathtt{S}(\mathtt{a}))$, but it is a *recursive* parameter according to the following definition:

Definition 2.16 (Recursive parameters) Assume f is a pf. We define, inductively, the set of *recursive parameters* of f, $\text{RP}(f)$:

- $\text{RP}\big(R(i_1,\ldots,i_{\mathfrak{a}(R)})\big) \stackrel{\text{def}}{=} \{i_1,\ldots,i_{\mathfrak{a}(R)}\}$;
- $\text{RP}(\neg f) \stackrel{\text{def}}{=} \text{RP}(f)$; $\text{RP}(f_1 \vee f_2) \stackrel{\text{def}}{=} \text{RP}(f_1) \cup \text{RP}(f_2)$;
- $\text{RP}(\forall x[f]) \stackrel{\text{def}}{=} \text{RP}(f)$;
- $\text{RP}(z(k_1,\ldots,k_n)) \stackrel{\text{def}}{=} \{k_1,\ldots,k_n\} \cup \bigcup_{k_i \in \mathcal{P}} \text{RP}(k_i)$.

Another important notion is the notion of α-equality[10]. We want the pfs $\mathtt{R}(\mathtt{x})$ and $\mathtt{R}(\mathtt{y})$ to be the same. However, we want the pfs $\mathtt{S}(\mathtt{x},\mathtt{y})$ and $\mathtt{S}(\mathtt{y},\mathtt{x})$ to be different. The reason for this is the alphabetical order of the variables $\mathtt{x}$, $\mathtt{y}$. As $\mathtt{x} < \mathtt{y}$, we will consider $\mathtt{x}$ to be the "first" variable of the pfs $\mathtt{S}(\mathtt{x},\mathtt{y})$ and $\mathtt{S}(\mathtt{y},\mathtt{x})$, and $\mathtt{y}$ the "second" variable. The place of the "first" variable in $\mathtt{S}(\mathtt{x},\mathtt{y})$, however, is different from the place of the "first" variable in $\mathtt{S}(\mathtt{y},\mathtt{x})$.[11] We therefore present the following definition of α-equality:

Definition 2.17 (α-equality) Let f and g be pfs. We say that f and g are *α-equal*, notation $f =_\alpha g$, if there is a bijection $\varphi : \mathcal{V} \to \mathcal{V}$ such that

- g can be obtained from f by replacing each variable that occurs in f by its φ-image;
- $x < y$ iff $\varphi(x) < \varphi(y)$.

This definition corresponds to the definition of α-equality in λ-calculus in the following way:

Lemma 2.18 *Let $f, g \in \mathcal{P}$. $f =_\alpha g$ if and only if $\overline{f} =_\alpha \overline{g}$.* ⊠

[10] Historically, it is not correct to use this terminology when discussing the Type Theory of the *Principia*, which dates from the first decade of this century. The term α-equality originates from Curry and Feys' book *Combinatory Logic* [40], which appeared only in 1958. In this book, conversion rules for the λ-calculus are numbered with Greek letters α, β. The rule numbered with α is now known as α-conversion; the rule numbered with β is now known as β-conversion. In earlier papers of Church, Rosser and Kleene, these rules were numbered with Roman capitals I, II, and the terminology α-conversion, β-conversion, was not used.

[11] Compare this with their equivalents in the λ-calculus $\lambda\mathtt{x}.\lambda\mathtt{y}.\mathtt{Sxy}$ and $\lambda\mathtt{x}.\lambda\mathtt{y}.\mathtt{Syx}$, which are not α-equal, either. We do not want to use the λ-notation for determining which variable is "first" and which is "second", for reasons to be explained in Remarks 2.14 and 2.31.

Sometimes, we are not that precise, and want the pfs $\mathtt{S(x,y)}$ and $\mathtt{S(y,x)}$ to be α-equal. This can be a consideration especially if we are not interested in which free variable is "first" and which is "second". We call this weakened notion of α-equality: α_P-equality (α-equality modulo permutation):

Definition 2.19 (α-equality modulo permutation) Let f and g be pfs. We say that f and g are *α_P-equal*, notation $f =_{\alpha_P} g$, if there is a bijection $\varphi : \mathcal{V} \to \mathcal{V}$ such that g can be obtained from f by replacing each variable that occurs in f by its φ-image.

2a5 Principia's substitution

As function construction in *Principia Mathematica* can be compared to β-expansion plus removing an argument in the λ-calculus, this suggests that instantiation in the *Principia* must be comparable to application plus β-reduction in the λ-calculus. In [97] we showed that this is indeed the case. There, we gave a laborious definition of instantiation using the syntax of and the intuition behind pfs. We showed that this definition is faithful to the original ideas of the *Principia* and that it can be imitated in λ-calculus using a translation similar to the one in Definition 2.7. This allows us to give a definition of substitution for pfs that is based on that imitation in λ-calculus, as we do below.

As was argued in Remark 2.14, the mapping $f \mapsto \overline{f}$ is not perfectly suited for a definition of substitution. This was due to the currying of the λ-abstractions that are at the front of the term $\overline{f}$. We therefore take a slightly different notation and remove these front abstractions from $\overline{f}$:

Definition 2.20 Let $f \in \mathcal{P}$ with free variables $x_1 < \cdots < x_m$. Then $\overline{f} \equiv \lambda_{i=1}^{m} x_i.F$ for some λ-term F. Let $\widetilde{f} \stackrel{\text{def}}{=} F$.

Example 2.21

f	$\widetilde{f}$
$\mathtt{R(x)}$	$\mathtt{Rx}$
$\mathtt{z(R(x),S(a))}$	$\mathtt{z(\lambda x.Rx)(Sa)}$
$\mathtt{z_1(a)} \vee \mathtt{z_2()}$	$\vee(\mathtt{z_1 a})\mathtt{z_2}$
$\mathtt{z(y(R(x)))}$	$\mathtt{z(\lambda y.y(\lambda x.Rx))}$
$\forall \mathtt{x[R(x)]}$	$\forall(\lambda \mathtt{x.Rx})$

The mapping $f \mapsto \widetilde{f}$ has similar properties as $f \mapsto \overline{f}$ (cf. Lemma 2.9):

$$\begin{array}{ccc} f[x_1,\ldots,x_n{:=}g_1,\ldots,g_n] & = & h \\ \big\downarrow & & \big\uparrow \\ (\lambda x_1\cdots x_n.\widetilde{f})\overline{g_1}\cdots\overline{g_n} & \twoheadrightarrow_\beta^{\mathrm{nf}} & \widetilde{h} \end{array}$$

Figure 2.1: Substitution via β-reduction

Lemma 2.22 (Properties of $\widetilde{\ }$)

1. $\mathrm{FV}(f) = \mathrm{FV}\left(\widetilde{f}\right)$;
2. $\widetilde{f}$ *is in* β*-normal form for all* f;
3. $\widetilde{f}$ *is a* λI*-term for all* f;
4. $\overline{f}$ *is a closure (see Definition 4.9) of* $\widetilde{f}$;
5. *If* $\widetilde{f} =_\alpha \widetilde{g}$, *then* $f =_\alpha g$. ⊠

With the λ-notation we can rely on the notions of β-reduction and β-normal form to give the following definition of substitution:

Definition 2.23 (Substitution) Let $f \in \mathcal{P}$, assume $x_1,\ldots,x_n$ are distinct variables, and $g_1,\ldots,g_n \in \mathcal{A}\cup\mathcal{V}\cup\mathcal{P}$. Assume that the λ-term

$$(\lambda x_1\cdots x_n.\widetilde{f})\overline{g_1}\cdots\overline{g_n}$$

has a β-normal form H. Assume $h \in \mathcal{P}$ such that $\widetilde{h} \equiv H$ (If such an h exists, it is unique due to Lemma 2.22.5). Then $f[x_1,\ldots,x_n{:=}g_1,\ldots,g_n] \stackrel{\mathrm{def}}{=} h$.

We sometimes abbreviate $f[x_1,\ldots,x_n{:=}g_1,\ldots,g_n]$ to $f[x_i{:=}g_i]_{i=1}^n$ or $f[\vec{x}{:=}\vec{g}]$.

So substitution in RTT can be seen as application plus β-reduction to β-normal form in the λ-calculus. Definition 2.23 is schematically reflected in Figure 2.1. Notice that $f[x_1,\ldots,x_n{:=}g_1,\ldots,g_n]$ should be seen as a *simultaneous* substitution of $g_1,\ldots,g_n$ for $x_1,\ldots,x_n$. As the $\overline{g_i}$s are either closed λ-terms, or individuals, or variables, it is no problem to define this simultaneous substitution via a list of applications that results in a list of *consecutive* substitutions.

Example 2.24

1. $\mathtt{S}(\mathtt{x}_1)[\mathtt{x}_1{:=}\mathtt{a}_1] \equiv \mathtt{S}(\mathtt{a}_1)$, as $(\lambda\mathtt{x}_1.\mathtt{S}\mathtt{x}_1)\mathtt{a}_1 \rightarrow_\beta^{\mathrm{nf}} \mathtt{S}\mathtt{a}_1$;

2. $\mathtt{S}(\mathtt{x}_1)[\mathtt{x}_2{:=}\mathtt{a}_2] \equiv \mathtt{S}(\mathtt{x}_1)$, as $(\lambda \mathtt{x}_2.\mathtt{S}\mathtt{x}_1)\mathtt{a}_2 \rightarrow_\beta^{\mathrm{nf}} \mathtt{S}\mathtt{x}_1$;

3. $\mathtt{z}(\mathtt{S}(\mathtt{x}_1), \mathtt{x}_2, \mathtt{a}_2)[\mathtt{x}_1{:=}\mathtt{a}_1] \equiv \mathtt{z}(\mathtt{S}(\mathtt{x}_1), \mathtt{x}_2, \mathtt{a}_2)$ as $(\lambda \mathtt{x}_1.\mathtt{z}(\lambda \mathtt{x}_1.\mathtt{S}\mathtt{x}_1)\mathtt{x}_2\mathtt{a}_2)\mathtt{a}_1 \rightarrow_\beta^{\mathrm{nf}} \mathtt{z}(\lambda \mathtt{x}_1.\mathtt{S}\mathtt{x}_1)\mathtt{x}_2\mathtt{a}_2$.
This shows that the λ-notation is more precise and convenient with respect to free variables. In $\mathtt{z}(\mathtt{S}(\mathtt{x}_1), \mathtt{x}_2, \mathtt{a}_2)$, it is not immediately clear whether $\mathtt{x}_1$ is a free variable or not and one might tend to write

$$\mathtt{z}(\mathtt{S}(\mathtt{x}_1), \mathtt{x}_2, \mathtt{a}_2)[\mathtt{x}_1{:=}\mathtt{a}_1] \equiv \mathtt{z}(\mathtt{S}(\mathtt{a}_1), \mathtt{x}_2, \mathtt{a}_2).$$

The λ-notation is more explicit in showing that $\mathtt{x}_1 \notin \mathrm{FV}(\mathtt{z}(\mathtt{S}(\mathtt{x}_1), \mathtt{x}_2, \mathtt{a}_2))$;

4. $\mathtt{z}(\mathtt{R}(\mathtt{a}), \mathtt{S}(\mathtt{a}))[\mathtt{z}{:=}\mathtt{z}_1() \vee \mathtt{z}_2()] \equiv \mathtt{R}(\mathtt{a}) \vee \mathtt{S}(\mathtt{a})$, as $(\lambda \mathtt{z}.\mathtt{z}(\mathtt{Ra})(\mathtt{Sa}))(\lambda \mathtt{z}_1\mathtt{z}_2.\vee \mathtt{z}_1\mathtt{z}_2) \rightarrow_\beta$ $(\lambda \mathtt{z}_1\mathtt{z}_2.\vee \mathtt{z}_1\mathtt{z}_2)(\mathtt{Ra})(\mathtt{Sa}) \twoheadrightarrow_\beta^{\mathrm{nf}} \vee(\mathtt{Ra})(\mathtt{Sa})$;

5. $\mathtt{x}_2(\mathtt{x}_1, \mathtt{R}(\mathtt{x}_1))[\mathtt{x}_2{:=}\mathtt{x}_4(\mathtt{x}_3)] \equiv \mathtt{R}(\mathtt{x}_1)$ as $(\lambda \mathtt{x}_2.\mathtt{x}_2\mathtt{x}_1(\lambda \mathtt{x}_1.\mathtt{R}\mathtt{x}_1))(\lambda \mathtt{x}_3\mathtt{x}_4.\mathtt{x}_4\mathtt{x}_3) \twoheadrightarrow_\beta$ $(\lambda \mathtt{x}_3\mathtt{x}_4.\mathtt{x}_4\mathtt{x}_3)\mathtt{x}_1(\lambda \mathtt{x}_1.\mathtt{R}\mathtt{x}_1) \twoheadrightarrow_\beta$ $(\lambda \mathtt{x}_1.\mathtt{R}\mathtt{x}_1)\mathtt{x}_1 \rightarrow_\beta^{\mathrm{nf}} \mathtt{R}\mathtt{x}_1$.

Remark 2.25 $f[x_1, \ldots, x_n{:=}g_1, \ldots, g_n]$ is not always defined. For its existence we need:

- The existence of the normal form H in Definition 2.23. For instance, this normal form does not exist if we choose $n = 1$, $f \equiv \mathtt{x}_1(\mathtt{x}_1)$ and $g_1 \equiv \mathtt{x}_1(\mathtt{x}_1)$: then we obtain for the calculation of $f[\mathtt{x}_1{:=}g_1]$ the famous λ-term $(\lambda \mathtt{x}_1.\mathtt{x}_1\mathtt{x}_1)(\lambda \mathtt{x}_1.\mathtt{x}_1\mathtt{x}_1)$;
- The existence of a (unique) h such that $\widetilde{h} \equiv H$. For instance, if we take $n = 1$, $f \equiv \mathtt{z}(\mathtt{a})$ (with $\mathtt{z} \in \mathcal{V}$ and $\mathtt{a} \in \mathcal{A}$) and $g_1 \equiv \mathtt{a}$, then $H \equiv \mathtt{aa}$ and there is no $h \in \mathcal{P}$ such that $\widetilde{h} \equiv \mathtt{aa}$.

In Section 2c2 we will prove that, as long as we are within the type system RTT (to be introduced in Section 2b), both H and h always exist uniquely (Corollary 2.73). Until then, the notation $f[x_1, \ldots, x_n{:=}g_1, \ldots, g_n] = h$ implicitly assumes that the substitution *exists*.

Remark 2.26 If we compute a substitution $f[x_1, \ldots, x_n{:=}g_1, \ldots, g_n]$, we have to reduce the λ-term $(\lambda x_1 \cdots x_n.\widetilde{f})\bar{g_1} \cdots \bar{g_n}$ to its β-normal form (if there is any). One might wonder whether this is too restrictive: In a reduction path to this normal form, there may be an intermediate result H that could be interpreted as the final result of the substitution $f[\vec{x}{:=}\vec{g}]$. However, this never happens, as any term that can be interpreted as such a result is always of the form $\widetilde{h}$, and is therefore always in β-normal form (Lemma 2.22.2).

Remark 2.27 The alphabetical order of the variables plays a crucial role in the substitution process, as it determines in which order the free variables of a pf f are curried in the translation $\overline{f}$. For example, take the substitutions $\mathtt{z}(\mathtt{a},\mathtt{b})[\mathtt{z}{:=}\mathtt{R}(\mathtt{x},\mathtt{y})]$ and $\mathtt{z}(\mathtt{a},\mathtt{b})[\mathtt{z}{:=}\mathtt{R}(\mathtt{y},\mathtt{x})]$. The result of the first one is obtained via the normal form of $(\lambda\mathtt{z}.\mathtt{zab})(\lambda\mathtt{xy}.\mathtt{Rxy})$, which is equal to $\mathtt{Rab}$, translated: $\mathtt{R}(\mathtt{a},\mathtt{b})$. The second one is calculated via $(\lambda\mathtt{z}.\mathtt{zab})(\lambda\mathtt{xy}.\mathtt{Ryx})$, resulting in $\mathtt{Rba}$ and $\mathtt{R}(\mathtt{b},\mathtt{a})$.

Remark 2.28 Now that substitution has been properly defined, we could define that f is an *abstraction* of g if:

"there are $x_1,\ldots,x_n \in \mathrm{FV}(f)$ and $h_1,\ldots,h_n \in \mathcal{A}\cup\mathcal{P}$ such that $f[\vec{x}{:=}\vec{h}] \equiv g$, (or, in λ-calculus notation: $(\lambda x_1\cdots x_n.\widetilde{f})\overline{h_1}\cdots\overline{h_n} \twoheadrightarrow_\beta \widetilde{g}$)".

This means that the set of abstractions of a pf g is comparable to the set of β-expansions of the λ-term $\widetilde{g}$.

Some elementary calculation with substitutions can be done using the following lemma:

Lemma 2.29

1. *Assume* $(f_1 \vee f_2)[\vec{x}{:=}\vec{h}]$ *exists. Then* $f_j[\vec{x}{:=}\vec{h}]$ *exists for* $j = 1, 2$, *and*
$$(f_1 \vee f_2)[\vec{x}{:=}\vec{h}] \equiv (f_1[\vec{x}{:=}\vec{h}]) \vee (f_2[\vec{x}{:=}\vec{h}]);$$
2. *Assume* $(\neg f)[\vec{x}{:=}\vec{h}]$ *exists. Then* $f[\vec{x}{:=}\vec{h}]$ *exists, and*
$$(\neg f)[\vec{x}{:=}\vec{h}] \equiv \neg(f[\vec{x}{:=}\vec{h}]);$$
3. *Assume* $(\forall x{:}t^a[f])[\vec{x}{:=}\vec{h}]$ *exists, and* $x \notin \vec{x}$. *Then* $f[\vec{x}{:=}\vec{h}]$ *exists, and*
$$(\forall x{:}t^a[f])[\vec{x}{:=}\vec{h}] \equiv \forall x{:}t^a[f[\vec{x}{:=}\vec{h}]];$$
4. *Assume* $z(k_1,\ldots,k_n)[z{:=}f]$ *exists, and* $x_1 < \cdots < x_n$ *are the free variables of* f. *Then* $f[\vec{x}{:=}\vec{k}]$ *exists, and*
$$z(k_1,\ldots,k_n)[z{:=}f] \equiv f[\vec{x}{:=}\vec{k}];$$
5. *Assume* $z(k_1,\ldots,k_n)[\vec{x}{:=}\vec{h}]$ *exists,* $z \equiv x_p$, *and* $y_1 < \cdots < y_n$ *are the free variables of* $k_p \in \mathcal{P}$. *Define* $k'_i \equiv h_j$ *if* $k_i \equiv x_j$, *and* $k'_i \equiv k_i$ *otherwise. Then* $k_p[\vec{y}{:=}\vec{k'}]$ *exists, and*
$$z(k_1,\ldots,k_n)[\vec{x}{:=}\vec{h'}] \equiv k_p[y_j{:=}\vec{k'}].$$

PROOF: Directly from the definition of substitution. ⊠

2b The Ramified Theory of Types RTT

After we have formalised the notion of propositional function in Section 2a we now give a precise description of the type theory underlying the *Principia*. First we explicitly introduce types (Section 2b1 — there is no such introduction in *Principia*), and then we formalise the notion "the propositional function f has type t" (Section 2b2).

2b1 Types

Types in the *Principia* have a double hierarchy: one of (simple) types and one of orders. We start by introducing the first hierarchy. Then, we extend this hierarchy with orders, resulting in the ramified types of the *Principia*.

Simple types

As we saw in Section 1b, Frege already distinguished between objects, functions that take objects as arguments, and functions that take functions as arguments. He also made a distinction between functions that take one and functions that take two arguments (see the quotations from *Function and Concept* on p. 13). In the *Principia*, Whitehead and Russell use a similar principle. Whilst Frege's argument for this distinction was only that functions are fundamentally different from objects, and that functions taking objects as arguments are fundamentally different from functions taking functions as arguments, Whitehead and Russell are more precise:

> "[The difference between objects and propositional functions] arises from the fact that a [propositional] function is essentially an ambiguity, and that, if it is to occur in a definite proposition, it must occur in such a way that the ambiguity has disappeared, and a wholly unambiguous statement has resulted."
>
> (*Principia Mathematica*, p. 47)

There is no definition of "type" in the *Principia*, only a definition of "being of the same type":[12]

> "*Definition of "being of the same type."* The following is a step-by-step definition, the definition for higher types presupposing that for lower types. We say that u and v "are of the same type" if

[12] See Definition 2.2 for the notion of elementary proposition. In the *Principia*, an elementary pf is a pf that has elementary propositions as values, when it takes suitable arguments.

1. both are individuals,
2. both are elementary [propositional] functions[13] taking arguments of the same type,
3. u is a pf and v is its negation,
4. u is $\varphi\hat{x}$[14] or $\psi\hat{x}$, and v is $\varphi\hat{x} \vee \psi\hat{x}$, where $\varphi\hat{x}$ and $\psi\hat{x}$ are elementary pfs,
5. u is $(y).\varphi(\hat{x}, y)$[15] and v is $(z).\psi(\hat{x}, z)$, where $\varphi(\hat{x}, \hat{y})$, $\psi(\hat{x}, \hat{y})$ are of the same type,
6. both are elementary propositions,
7. u is a proposition and v is $\sim u$[16], or
8. u is $(x).\varphi x$ and v is $(y).\psi y$, where $\varphi\hat{x}$ and $\psi\hat{x}$ are of the same type."

(*Principia Mathematica*, $*9{\cdot}131$, p. 133)

The definition has to be seen as the definition of an equivalence relation. For instance, assume that $\varphi\hat{x}$, $\psi\hat{x}$ and $\chi\hat{x}$ are elementary pfs. Then by rule 4, $\varphi\hat{x}$ and $\varphi\hat{x} \vee \psi\hat{x}$ are of the same type, and so are $\varphi\hat{x}$ and $\varphi\hat{x} \vee \chi\hat{x}$. By (implicit) transitivity, $\varphi\hat{x} \vee \psi\hat{x}$ and $\varphi\hat{x} \vee \chi\hat{x}$ are of the same type.

The definition seems rather precise at first sight. But there are several remarks to be made:

- The notion "being of the same type" seems to be defined for pfs taking one argument only. On the other hand, rules 2 and 5 suggest that such a definition should be extended to pfs taking two arguments. How this should be done is not made explicit;

- According to this definition, $\mathtt{z}_1() \vee \neg\mathtt{z}_1()$ is not of the same type as $\mathtt{z}_1()$. Rules 2 and 4 are the only rules by which one could derive that $\mathtt{z}_1()$ and $\mathtt{z}_1() \vee \neg\mathtt{z}_1()$ are of the same type. But if we want to use these rules, $\mathtt{z}_1()$ must be an elementary pf, which it is not: It can take the argument $\forall\mathtt{x}[\mathtt{R}(\mathtt{x})]$, which has as result the proposition $\forall\mathtt{x}[\mathtt{R}(\mathtt{x})]$. This is not an elementary proposition and therefore $\mathtt{z}_1()$ is not an elementary pf.

---

[13]The term *elementary functions* refers to a pf that has only elementary propositions as value, when it takes suitable (well-typed) arguments. See *Principia*, p. 92.

[14]Whitehead and Russell use $\varphi\hat{x}$ to denote that φ is a pf that has, amongst others, x as a free variable. Similarly, they use $\varphi(\hat{x}, \hat{y})$ to indicate that φ has x, y amongst its free variables.

[15]Whitehead and Russell write $(x).\varphi(x)$ where we would write $\forall x[\varphi]$.

[16]$\sim u$ is *Principia* notation for $\neg u$.

So there are quite some omissions in this definition. However, the intention of the definition is clear: pfs that take a different number of arguments, or that take arguments of different types, cannot be of the same type.

In order to express what is meant by "being of the same type", we explain first what these types "are". The notion "being of the same type" can then be replaced by "having the same type". The notion of simple type as defined below is due to Ramsey [124] (1926). Historically, it is incorrect to give Ramsey's definition of simple types before Russell's definition of ramified types, as Russell's definition is of an earlier date, and Ramsey's definition is in fact based on Russell's ideas and not the other way around. On the other hand, the ideas behind simple types were already explained by Frege (see the quotes from *Function and Concept* on page 13). Moreover, knowledge of the intuition behind simple types will make it easier to understand the ramified ones.[17] Therefore we present Ramsey's definition first.

Definition 2.30 (Ramsey's simple types)

1. 0 is a simple type;
2. If $t_1, \ldots, t_n$ are simple types, then also $(t_1, \ldots, t_n)$ is a simple type. $n = 0$ is allowed: then we obtain the simple type $()$;
3. All simple types can be constructed using rules 1 and 2.

We use $t, u, t_1, \ldots$ as metavariables over simple types.

Here, $(t_1, \ldots, t_n)$ is the type of pfs that should take n arguments (have n free variables), the ith argument having type t_i. The type $()$ stands for the type of the propositions, and the type 0 stands for the type of the individuals.

Remark 2.31 To formalise the notion of ith argument that a pf takes, we use the alphabetical order on variables that was introduced in Section 2a. The ith argument taken by a pf will be substituted for the ith free variable of that pf, according to the alphabetical order.

Now it becomes clear why we considered the alphabetical order of variables in the definition of α-equality (Definition 2.17): we want α-equal pfs to have the same type. However, if f has type (t_1, t_2) and two free variables $x < y$, and g is the same as f except that the roles of x and y have been switched, then g will have type (t_2, t_1). Therefore we demand that the renaming of variables must maintain the alphabetical order. See also Remark 2.47.7.

---

[17] See [72, 73] for a further discussion of the difference between simple and ramified type theory, especially in connection with Quine's new foundations for which there is a consistency result for its predicative version (and hence one can get models of predicative type theory in which very strong versions of "systematic ambiguity" hold). In particular, [73] contains a discussion of the relationship between a predicative *linear* type scheme (with types indexed by the natural numbers) and the full ramified type scheme of *Principia*.

Example 2.32 The propositional function $\mathtt{R(x)}$ should have type (0), as it takes one individual as argument.

The propositional function $\mathtt{z(R(x), S(a))}$ takes one argument. This argument must be a pf that can take $\mathtt{R(x)}$ as its first argument (so this first argument must be of type (0)), and a proposition (of type $()$) as its second argument. We conclude that in $\mathtt{z(R(x), S(a))}$, we must substitute pfs of type $((0), ())$ for $\mathtt{z}$. Therefore, $\mathtt{z(R(x), S(a))}$ has type $(((0), ()))$.

The intuition presented in Remark 2.31 and Example 2.32 will be formalised in Definition 2.44. Theorem 2.58 shows that this formalisation follows the intuition.

Just as propositional functions can be translated to λ-terms, simple types can be translated to types of the simply typed λ-calculus of Church (see [29], and Section 3b2).

Definition 2.33 (Translating simple types to $\lambda_{\rightarrow\mathbf{Church}}$-types) We define a type $T(t)$ for each simple type t by induction:

1. $T(0) \stackrel{\text{def}}{=} \iota$;
2. $T((t_1, \ldots, t_n)) \stackrel{\text{def}}{=} T(t_1) \rightarrow \cdots \rightarrow T(t_n) \rightarrow o$.

A simple type t of Definition 2.30 has the same interpretation as its translation $T(t)$. Moreover, T is injective:

Lemma 2.34 *If t and u are simple types, then $T(t) = T(u)$ if and only if $t = u$.*

PROOF: Induction on the definition of simple type. ⊠

Notation 2.35 From now on we will use a slightly different notation for quantification in pfs. Instead of $\forall x[f]$ we now explicitly mention the type (say: t) over which is quantified: $\forall x{:}t[f]$. We do the same with the translations of pfs to λ-calculus: instead of $\lambda x.F$ we write $\lambda x{:}T(t).F$.

Ramified types

Up to now, the type of a pf only depends on the types of the arguments that it can take. In the *Principia*, a second hierarchy is introduced by regarding also the types of the variables that are bound by a quantifier (see *Principia*, pp. 51–55). Whitehead and Russell consider, for instance, the propositions $\mathtt{R(a)}$ and $\forall \mathtt{z}{:}()[\mathtt{z}() \vee \neg \mathtt{z}()]$ to be of a different level. The first is an atomic proposition, while the latter is based on the pf $\mathtt{z}() \vee \neg \mathtt{z}()$. The pf $\mathtt{z}() \vee \neg \mathtt{z}()$ involves an arbitrary proposition $\mathtt{z}$, therefore $\forall \mathtt{z}{:}()[\mathtt{z}() \vee \neg \mathtt{z}()]$ quantifies over all propositions $\mathtt{z}$. According to the vicious circle principle 2.1, $\forall \mathtt{z}{:}()[\mathtt{z}() \vee \neg \mathtt{z}()]$ cannot belong to this collection of propositions.

This problem is solved by dividing types into *orders* (not to be confused with the *alphabetical order* on the variables). An order is simply a natural number. Basic propositions are of order 0, and in $\forall \mathsf{z}{:}()[\mathsf{z}() \vee \neg\mathsf{z}()]$ we must mention the order of the propositions over which is quantified. The pf $\forall \mathsf{z}{:}()^n[\mathsf{z}() \vee \neg\mathsf{z}()]$ quantifies over all propositions of order n, and has order $n+1$.

The division of types into orders gives *ramified* types.

Definition 2.36 (Ramified types)

1. 0^0 is a ramified type;

2. If $t_1^{a_1}, \ldots, t_n^{a_n}$ are ramified types, and $a \in \mathbb{N}$, $a > \max(a_1, \ldots, a_n)$, then $(t_1^{a_1}, \ldots, t_n^{a_n})^a$ is a ramified type (if $n = 0$ then take $a \geq 0$);

3. All ramified types can be constructed using rules 1 and 2.

If t^a is a ramified type, then a is called the *order* of t^a.

Remark 2.37 In $(t_1^{a_1}, \ldots, t_n^{a_n})^a$, we demand that $a > a_i$ for all i. This is because a pf of this type presupposes all the elements of type $t_i{}^{a_i}$, and therefore must be of an order that is higher than a_i.

Example 2.38 We give some examples of ramified types:

- 0^0;
- $(0^0)^1$;
- $\left((0^0)^1, (0^0)^4\right)^5$;
- $\left(0^0, ()^2, \left(0^0, (0^0)^1\right)^2\right)^7$.

$\left(0^0, \left(0^0, (0^0)^2\right)^2\right)^7$ is not a ramified type.

Ramified types can also be translated to types of the simply typed λ-calculus. However, we lose the orders if we do so.

Definition 2.39 (Translating ramified types to $\lambda_{\to}$Church-types) We define a type $T(t)$ for each ramified type t by induction:

1. $T(0^0) \stackrel{\text{def}}{=} \iota$;

2. $T\left((t_1^{a_1}, \ldots, t_n^{a_n})^a\right) \stackrel{\text{def}}{=} T(t_1) \to \cdots \to T(t_n) \to o$.

In the rest of this chapter we simply speak of *types* when we mean *ramified* types, as long as no confusion arises.

In the type $(0^0)^1$, all orders are "minimal", i.e., not higher than strictly necessary. This is, for instance, not the case in the type $(0^0)^2$. Types in which all orders are minimal are called *predicative* and play a special role in the Ramified Theory of Types. A formal definition:

Definition 2.40 (Predicative types)

1. 0^0 is a predicative type;
2. If $t_1{}^{a_1}, \ldots, t_n{}^{a_n}$ are predicative types, and $a = 1 + \max(a_1, \ldots, a_n)$ (take $a = 0$ if $n = 0$), then $(t_1^{a_1}, \ldots, t_n^{a_n})^a$ is a predicative type;
3. All predicative types can be constructed using rules 1 and 2 above.

The mapping T is injective when restricted to predicative types:

Lemma 2.41 *If t^a and u^b are predicative types, then $T(t^a) = T(u^b)$ if and only if $t^a = u^b$.*

PROOF: Induction on the definition of predicative type. ⊠

2b2 Formalisation of the Ramified Theory of Types

In this section we formalise the intuition on types presented in Example 2.32 and Definition 2.33 together with the intuition on orders that was given just above Definition 2.36. Before we can do this we must introduce some additional terminology.

In the pf $\mathtt{R}(\mathbf{x})$ we implicitly assume that $\mathbf{x}$ is a variable for which objects of type 0 must be substituted. For our formalisation we want to make the information on the type of a variable explicit. We do this by storing this information in the so-called *contexts*. Contexts, common in modern type systems, are not used in the *Principia*.

Definition 2.42 (Contexts) Let $x_1, \ldots, x_n \in \mathcal{V}$ be distinct variables, and assume $t_1^{a_1}, \ldots, t_n^{a_n}$ are ramified types. Then $\{x_1{:}t_1^{a_1}, \ldots, x_n{:}t_n^{a_n}\}$ is a *context*. The set $\{x_1, \ldots, x_n\}$ is called the *domain* of the context and is denoted by $\mathrm{dom}(\{x_1{:}t_1^{a_1}, \ldots, x_n{:}t_n^{a_n}\})$. We will use Greek capitals Γ, Δ as meta-variables over contexts.

The pfs $\mathbf{z}_1(\mathbf{y}_1)$ and $\mathbf{z}_2(\mathbf{y}_2)$ are α-equal, according to Definition 2.17. But in a context $\Gamma \equiv \left\{\mathbf{y}_1{:}0^0, \mathbf{z}_1{:}(0^0)^1, \mathbf{y}_2{:}(0^0)^1, \mathbf{z}_2{:}\left((0^0)^1\right)^2\right\}$ one does not want to see $\mathbf{z}_1(\mathbf{y}_1)$ and $\mathbf{z}_2(\mathbf{y}_2)$ as equal, as the types of $\mathbf{y}_1$ and $\mathbf{y}_2$ differ, and the types of $\mathbf{z}_1$ and $\mathbf{z}_2$ differ as well. Therefore, we introduce a more restricted version of α-equality:

Definition 2.43 Let Γ be a context and f and g pfs. We say that f and g are α_Γ-*equal*, notation $f =_{\alpha,\Gamma} g$, if there is a bijection $\varphi : \mathcal{V} \to \mathcal{V}$ such that

- g can be obtained from f by replacing each variable that occurs in f by its φ-image;
- $x < y$ iff $\varphi(x) < \varphi(y)$;
- $x{:}t \in \Gamma$ iff $\varphi(x){:}t \in \Gamma$.

We will now define what we mean by $\Gamma \vdash f : t^a$, or, in words: f is of type t^a in the context Γ.[18] In this definition we will try to follow the line of the Principia as much as possible. If $\Gamma \equiv \varnothing$ then we will write $\vdash f : t^a$.

We explain some aspects of the following definition in Section 2b3.

Definition 2.44 (Ramified Theory of Types: RTT**)** The *judgement* $\Gamma \vdash f : t^a$ is inductively defined as follows:

1. **(start)** For all a:
$$\vdash a : 0^0.$$
For all atomic pfs f:
$$\vdash f : ()^0;$$

2. **(connectives)** Assume $\Gamma \vdash f{:}(t_1^{a_1}, \ldots, t_n^{a_n})^a$, $\Delta \vdash g{:}(u_1^{b_1}, \ldots, u_m^{b_m})^b$, and $x < y$ for all $x \in \mathrm{dom}(\Gamma)$ and $y \in \mathrm{dom}(\Delta)$. Then
$$\Gamma \cup \Delta \vdash f \vee g : \left(t_1^{a_1}, \ldots, t_n^{a_n}, u_1^{b_1}, \ldots, u_m^{b_m}\right)^{\max(a,b)};$$
$$\Gamma \vdash \neg f : (t_1^{a_1}, \ldots, t_n^{a_n})^a;$$

3. **(abstraction from parameters)** If $\Gamma \vdash f : (t_1^{a_1}, \ldots, t_m^{a_m})^a$, $t_{m+1}^{a_{m+1}}$ is a *predicative* type[19], $g \in \mathcal{A} \cup \mathcal{P}$ is a parameter of f, $\Gamma \vdash g : t_{m+1}^{a_{m+1}}$, and $x < y$ for all $x \in \mathrm{dom}(\Gamma)$, then
$$\Gamma' \vdash h : (t_1^{a_1}, \ldots, t_{m+1}^{a_{m+1}})^{\max(a, a_{m+1}+1)}.$$
Here, h is a pf obtained by replacing all parameters g' of f which are α_Γ-equal to g by y. Moreover, Γ' is the subset of the context $\Gamma \cup \{y : t_{m+1}^{a_{m+1}}\}$ such that $\mathrm{dom}(\Gamma')$ contains exactly all the variables that occur in h[20];

[18] The symbol $\vdash$ in $\Gamma \vdash f : t^a$ is the same symbol that Frege used to assert a proposition. It enters Type Theory in 1934 [38], via Curry's combinatory logic. Curry defines a functionality combinator F in such a way that $\mathsf{F}XYf$ holds, exactly if f is a function from X to Y. To denote the assertion of $\mathsf{F}XYf$, Curry uses Frege's symbol $\vdash$.

[19] The restriction to predicative types only is based on *Principia*, pp. 53–54.

[20] In Lemma 2.56 we prove that this context always exists.

4. **(abstraction from pfs)** If $\Gamma \vdash f : (t_1^{a_1},\ldots,t_m^{a_m})^a$, $(t_1^{a_1},\ldots,t_m^{a_m})^a$ is a *predicative* type[19], $x < z$ for all $x \in \mathrm{dom}(\Gamma)$, and $y_1 < \cdots < y_n$ are the free variables of f, then

$$\Gamma' \vdash z(y_1,\ldots,y_n) : (t_1^{a_1},\ldots,t_m^{a_m},(t_1^{a_1},\ldots,t_m^{a_m})^a)^{a+1},$$

where Γ' is the subset of $\Gamma \cup \{z{:}(t_1^{a_1},\ldots,t_m^{a_m})^a\}$ such that $\mathrm{dom}(\Gamma') = \{y_1,\ldots,y_n,z\}$[20];

5. **(weakening)** If Γ, Δ are contexts, $\Gamma \subseteq \Delta$, and $\Gamma \vdash f : t^a$, then also $\Delta \vdash f : t^a$;

6. **(substitution)** If y is the ith free variable in f (according to the order on variables), and $\Gamma \cup \{y : t_i^{a_i}\} \vdash f : (t_1^{a_1},\ldots,t_n^{a_n})^a$, and $\Gamma \vdash k : t_i^{a_i}$ then

$$\Gamma' \vdash f[y{:=}k] : (t_1^{a_1},\ldots,t_{i-1}^{a_{i-1}},t_{i+1}^{a_{i+1}},\ldots,t_n^{a_n})^b.$$

Here, $b = 1 + \max(a_1,\ldots,a_{i-1},a_{i+1},\ldots,a_n,c)$, and

$$c = \max\{j \mid \forall x{:}t^j \text{ occurs in } f[y{:=}k]\}$$

(if $n = 1$ and $\{j \mid \forall x{:}t^j \text{ occurs in } f[y{:=}k]\} = \varnothing$ then take $b = 0$) and once more, Γ' is the subset of $\Gamma \cup \{y : t_i^{a_i}\}$ such that $\mathrm{dom}(\Gamma')$ contains exactly all the variables that occur in $f[y{:=}k]$[20];

7. **(permutation)** If y is the ith free variable in f (according to the order on variables), and $\Gamma \cup \{y{:}t_i^{a_i}\} \vdash f : (t_1^{a_1},\ldots,t_n^{a_n})^a$, and $x < y'$ for all $x \in \mathrm{dom}(\Gamma)$, then

$$\Gamma' \vdash f[y{:=}y'] : (t_1^{a_1},\ldots,t_{i-1}^{a_{i-1}},t_{i+1}^{a_{i+1}},\ldots,t_n^{a_n},t_i^{a_i})^a.$$

Γ' is the subset of $\Gamma \cup \{y{:}t_i^{a_i}, y'{:}t_i^{a_i}\}$ such that $\mathrm{dom}\Gamma'$ contains exactly all the variables that occur in $f[y{:=}y']$[20];

8. **(quantification)** If y is the ith free variable in f (according to the order on variables), and $\Gamma \cup \{y{:}t_i^{a_i}\} \vdash f : (t_1^{a_1},\ldots,t_n^{a_n})^a$, then

$$\Gamma \vdash \forall y{:}t_i^{a_i}[f] : (t_1^{a_1},\ldots,t_{i-1}^{a_{i-1}},t_{i+1}^{a_{i+1}},\ldots,t_n^{a_n})^a.$$

Definition 2.45 (Legal propositional functions) A pf f is called *legal*, if there is a context Γ and a ramified type t^a such that $\Gamma \vdash f : t^a$.

Remark 2.46 In our attempt to faithfully implement Russell's ramified theory of types in the above definition, we face a limitation in the terms typable by our system. For example, it is not possible to type the pf $x(\forall y.(y() \vee \neg y()))$ or $x_1(x_2(x_1))$.[21] In fact, Russell intended (cf. page 165 of *Principia*) that non-predicative orders in his hierarchy are always obtained from predicative ones by quantification. Rule 8 of the above definition is the only one which creates non-predicative types but the increase of order is only at the top level of the type. This means that we cannot type terms $z(k_1, \ldots, k_n)$ where one of the k_i's happens to be of non-predicative type. In fact, Theorem 2.84 will prove that terms $z(k_1 \ldots k_n)$ are typable only if the k_i's can be assigned predicative types. This may be considered as a serious restriction but our aim is to faithfully represent Russell's ramified theory of types.

A drawback to our system is that, without the ability to assign non-predicative types to variables, one cannot even *state* the axiom of reducibility. Russell himself may have noted the need for variables with non-predicative types when he introduced on page 165 of *Principia* a convention for variable functions without assigned order which he used in the formal statement of the axiom of reducibility. However, Russell did not allow quantification over such variables. In this book we ignore the representation of the axiom of reducibility. In Section 3a2, we discuss the controversial nature of this axiom leading to the deramification and we leave the extension of our formalisation of Russell's ramified type theory to include the reducibility axiom as future work.

Finally, based on our above discussion, note that the third and fourth types given in Example 2.38 cannot be assigned as types to a legal pf in the sense of Definition 2.44. Future extensions must also address these examples.

2b3 Discussion and examples

We will make some remarks on Definition 2.44. First of all, we motivate the eight rules of Definition 2.44 by referring to passages in the *Principia*. Then we make some technical remarks, and give some examples of how the rules work. It will be made clear that the substitution rule is problematic, because substitution is not clearly defined in the *Principia*.

Remark 2.47 We will motivate RTT (Definition 2.44) by referring to the Principia:

1. Individuals and elementary judgements (atomic propositions) are, also in the *Principia*, the basic ingredients for creating legal pfs;[22]

2. We can see rule 2 "at work" in $*12$, p. 163 of the *Principia*[23]:

[21] We are grateful for Randall Holmes for drawing our attention to this point.

[22] As for individuals: see *Principia*, $*9$, p. 132, where "Individual" is presented as a primitive idea. As for elementary judgements: See *Principia*, Introduction, pp. 43-45.

[23] In the *Principia*, Whitehead and Russell write $\varphi!x$ instead of φx to indicate that φx is not only (what we would call) a pf, but even a *legal* pf.

> "We can build up a number of new formulas, such as [...] $\varphi!x \vee \varphi!y$, $\varphi!x \vee \psi!x$, $\varphi!x \vee \psi!y$, [...] and so on."
>
> (*Principia Mathematica*, $*$12, p. 163))

The restriction about contexts that we make in rule 2 has technical reasons and is not made in the *Principia*. It will be discussed in Remark 2.49;

3. Rule 3 is justified by $*9{\cdot}14$ and $*9{\cdot}15$ in the *Principia*. It is an instantiation of the abstraction principles 1.1 and 1.2 for functions that was already proposed by Frege. In Frege's definition one does not have to replace all parameters g' that are α_Γ-equal to g, but one can also take some of these parameters. In Section 2d we show that this is not a serious restriction.

   The restriction to *predicative* types is in line with the *Principia* (cf. *Principia*, pp. 53–54);

4. Rule 4 is based on the Introduction of *Principia* where pfs are built, and

   > "the first matrices that occur are those whose values are of the forms $\varphi x, \psi(x,y), \chi(x,y,z,\dots)$, *i.e.* where the arguments, however many there may be, are all individuals. Such [propositional] functions we will call '*first-order* functions.' We may now introduce a notation to express 'any first-order function.' "
   >
   > (*Principia Mathematica*, p. 51)

   This quote from the *Principia* is an instance of Frege's abstraction principles, and so is rule 4 of our formalisation. It results in second order pfs, and the process can be iterated to obtain pfs of higher orders.

   Rule 4 makes it possible to introduce variables of higher order. In fact, leaving out rule 4 would lead to first-order predicate logic, as without rule 4 it is impossible to introduce variables of types that differ from 0^0.

   The use of predicative types only is inspired by the *Principia*, again;

5. The weakening rule cannot be found in the *Principia*, because no formal contexts are used there. It is implicitly present, however: the addition of an extra variable to the set of variables does not affect the well-typedness of pfs that were already constructed;

6. The rule of substitution is based on $*9{\cdot}14$ and $*9{\cdot}15$ of the *Principia*, and can be seen as an inverse of the abstraction operators in rule 3 and 4. Since we do not yet know whether the substitution $f[y{:=}k]$ exists or not, we limit the use of rule 6 to the cases in which the substitution exists. In Section 2c2 we show that it always exists if the premises of rule 6 are fulfilled;

7. In the system above, the (sequential) order of the t_is is related to the alphabetic order of the free variables of the pf f that has type $(t_1, \ldots, t_n)$ (see the remark before Definition 2.17, Remark 2.31, and Theorem 2.58). This alphabetic order plays a role in the clear presentation of results like Theorem 2.58, and in the definition of substitution (see Remark 2.27).

   With rule 7 we want to express that the order of the t_is in $(t_1, \ldots, t_n)$ and the alphabetic order of the variables are not characteristics of the *Principia*, but are only introduced for the technical reasons explained in this remark. This is worked out in Corollary 2.59;

8. Notice that in the quantification rule, both f and $\forall x{:}t_i^{a_i}.f$ have order a. The intuition is that the order of a propositional function f equals one plus the maximum of the orders of *all* the variables (either free or bound by a quantifier) in f. This is in line with the *Principia*: see [145], page 53. See also the introduction to Definition 2.36, and the proof of Lemma 2.60 below.

Remark 2.48 Rules 3 and 4 are a restricted version of the abstraction principles of Frege, with less power. It is, for instance, not possible to imitate all the abstractions of Remark 2.10 by using rules 3 and 4 only. But in combination with the other rules, rule 3 and 4 are sufficient (see Example 2.54 for the cases of Remark 2.10, and Section 2d, especially Theorem 2.84).

Remark 2.49 In rule 2 of RTT, we make the assumption that the variables of Γ must all come before the variables of Δ. The reason for this is that we want to prevent undesired results like

$$\mathtt{x}_1{:}0^0 \vdash \mathtt{R}_1(\mathtt{x}_1) \vee \mathtt{R}_2(\mathtt{x}_1) : (0^0, 0^0)^1.$$

In fact, $\mathtt{R}_1(\mathtt{x}_1) \vee \mathtt{R}_2(\mathtt{x}_1)$ has *only one* free variable, so its type should be $(0^0)^1$ and not $(0^0, 0^0)^1$ (see Example 2.53, second part). For technical reasons (the order of the $t_i^{a_i}$s; see also Theorem 2.58) we strengthen the assumption such that for $x \in \mathrm{dom}(\Gamma)$ and $y \in \mathrm{dom}(\Delta)$, $x < y$ must hold.

As Whitehead and Russell do not have a formal notation for types, they do not forbid this kind of constructions in the *Principia*. In Lemma 2.82 we show that our limitation to contexts with disjoint domains as made in rule 2 is not a real limitation: all the desired judgements can still be derived for contexts with non-disjoint domains.

Remark 2.50 In both rules 3 and 4 we see that it is necessary to introduce at least one new variable. It is, for instance, not possible to interpret the proposition $\mathtt{R}(\mathtt{a})$ as a (constant) pf of type $(0^0)^1$. This is in line with the abstraction principles of Frege and Russell. In Frege's definition 1.1, for example, it is explicitly mentioned that the object that is to be replaced occurs at least once in the expression.

Translated to λ-calculus this means that the *Principia* have λI-terms, only. See also Lemma 2.9.3 and Lemma 2.22.3.

Remark 2.51 Contexts as used in RTT contain, in a sense, too much information: not only information on all free variables, but also information on non-free variables (Cf. rules 3, 6 and 7. The set of non-free variables contains more than only the variables that are bound by a quantifier. For example, in the pf $\mathtt{z}(\mathtt{R}(\mathtt{x}))$, $\mathtt{x}$ is neither free, nor bound by a quantifier).

Remark 2.52 The system is based on the abstraction principles of Frege. In a context Γ, one cannot introduce a variable of a certain type t unless one has a pf (or an individual) f that has type t in Γ. This is different from modern, λ-calculus based systems, where one can introduce a variable of a type u without knowing whether or not there are terms of this type u.

We give some examples, in order to illustrate how our system works. Example 2.53 shows applications of the rules. Example 2.54 makes a link between the intuitive notion of abstraction that was explained in Remark 2.10 and the abstraction rules 3 and 4 of our system.

We will use a notation of the form $\dfrac{X_1 \cdots X_n}{Y} N$, indicating that from the judgements $X_1, \ldots, X_n$, we can infer the judgement Y by using the RTT-rule of Definition 2.44 with number N. As usual, this is called a *derivation step*. Subsequent derivation steps give a *derivation*. A derivation *of a judgement* Y is a derivation tree with Y as root (the final conclusion). The types in the examples below are all predicative (as a pf of impredicative type must have a quantifier, and the examples below are quantifier-free). To avoid too much notation, we omit the orders.

Example 2.53

- $\vdash \mathtt{S}(\mathtt{a}_1, \mathtt{a}_2) : ()$;
- $\dfrac{\vdash \mathtt{R}_1(\mathtt{a}_1) : () \qquad \vdash \mathtt{R}_2(\mathtt{a}_1) : ()}{\vdash \mathtt{R}_1(\mathtt{a}_1) \vee \mathtt{R}_2(\mathtt{a}_1) : ()} 2$

  but *not*:

$$\frac{\mathtt{x}_1 : 0 \vdash \mathtt{R}_1(\mathtt{x}_1) : (0) \qquad \mathtt{x}_1 : 0 \vdash \mathtt{R}_2(\mathtt{x}_1) : (0)}{\mathtt{x}_1 : 0 \vdash \mathtt{R}_1(\mathtt{x}_1) \vee \mathtt{R}_2(\mathtt{x}_1) : (0, 0)} 2$$

($\mathtt{x}_1 \not< \mathtt{x}_1$ because $<$ is strict). To obtain $\mathtt{R}_1(\mathtt{x}_1) \vee \mathtt{R}_2(\mathtt{x}_1)$ we must make a different start:

$$\frac{\dfrac{\vdash \mathtt{R}_1(\mathtt{a}_1) : () \qquad \vdash \mathtt{R}_2(\mathtt{a}_1) : ()}{\vdash \mathtt{R}_1(\mathtt{a}_1) \vee \mathtt{R}_2(\mathtt{a}_1) : ()} 2 \qquad \vdash \mathtt{a}_1 : 0}{\mathtt{x}_1 : 0 \vdash \mathtt{R}_1(\mathtt{x}_1) \vee \mathtt{R}_2(\mathtt{x}_1) : (0)} 3;$$

- $$\frac{\begin{array}{l} x_1 : 0, x_2 : 0, z_1 : ((0),(0)) \vdash z_1(R(x_1), R(x_2)) : (((0),(0))) \\ x_1 : 0, x_2 : 0, z_1 : ((0),(0)) \vdash R(x_1) : (0) \end{array}}{z_1 : ((0),(0)), z_2 : (0) \vdash z_1(z_2, z_2) : (((0),(0)),(0))}\,3$$

  As $R(x_1)$ is α-equal to $R(x_2)$ in the context, both $R(x_1)$ and $R(x_2)$ are replaced by the newly introduced variable z_2;

- $$\frac{x_1 : 0, x_2 : 0 \vdash S(x_1, x_2) : (0,0)}{x_1 : 0, x_2 : 0, z : (0,0) \vdash z(x_1, x_2) : (0,0,(0,0))}\,4;$$

- $$\frac{\dfrac{x_1 : 0 \vdash R_1(x_1) \vee R_2(x_1) : (0) \qquad \vdash a_1 : 0}{\vdash R_1(a_1) \vee R_2(a_1) : ()}\,6;}{x_1 : 0 \vdash R_1(a_1) \vee R_2(a_1) : ()}\,5$$

- $$\frac{\begin{array}{l} x_1 : 0, x_2 : 0, x_3 : (0,0) \vdash R(x_1) \vee \neg x_3(x_1, x_2) : (0,0,(0,0)) \\ x_1 : 0, x_2 : 0 \vdash T(x_1, x_1, x_2) : (0,0) \end{array}}{x_1 : 0, x_2 : 0 \vdash R(x_1) \vee \neg T(x_1, x_1, x_2) : (0,0)}\,6$$

  $T(x_1, x_1, x_2)$ is substituted for x_3.

Example 2.54 We give a formal derivation of the examples of the abstraction rules that were given in Remark 2.10. Again, we omit the orders.

- Constructing $z(a) \vee S(a)$ from $R(a) \vee S(a)$ cannot be done with the use of rule 4 only. The following derivation is correct:

  $$\frac{\dfrac{\dfrac{\vdash a{:}0}{z{:}(0) \vdash a{:}0}\,5 \qquad \dfrac{\dfrac{\vdash a{:}0 \qquad \vdash R(a){:}()}{x{:}0 \vdash R(x){:}(0)}\,3}{x{:}0, z{:}(0) \vdash z(x){:}(0,(0))}\,4}{z{:}(0) \vdash z(a){:}((0))}\,6 \qquad \vdash S(a){:}()}{z{:}(0) \vdash z(a) \vee S(a) : ((0))}\,2.$$

  To obtain $z(a)$ instead of $z()$, we must transform $R(a)$ into a pf $R(x)$ by abstracting from a. Then we can construct $z(x)$ by abstraction from pfs (rule 4). In this way, the "frame" for $z(a)$ is of the right form. Substituting a for x gives $z(a)$ (and "neutralises" the application of rule 3 at the top of the derivation). Simply applying rule 4 on the judgement $\vdash R(a) : ()$ does not work: it results in $z() \vdash z() : (())$;

- Constructing $z_1() \vee z_2()$ is easier: $z_1()$ can be obtained by abstracting from $R(a)$, and $z_2()$ similarly from $S(a)$. Result:

  $$\frac{\dfrac{\vdash R(a){:}()}{z_1{:}() \vdash z_1(){:}(())}\,4 \qquad \dfrac{\vdash S(a){:}()}{z_2{:}() \vdash z_2(){:}(())}\,4}{z_1{:}(), z_2{:}() \vdash z_1() \vee z_2() : ((),())}\,2.$$

We see that in fact two abstractions are needed to construct this pf: we must abstract from $\mathtt{R(a)}$ as an instance of the pf $\mathtt{z_1}()$, and from $\mathtt{S(a)}$ as an instance of the pf $\mathtt{z_2}()$. As rule 4 does not work on parts of pfs, these abstractions have to be made before we use rule 2. Applying rule 4 on $\vdash \mathtt{R(a)} \vee \mathtt{S(a)} : ()^0$ would result in $\mathtt{z} : () \vdash \mathtt{z}() : (())$;

- We can extend the derivation of $\mathtt{z_1}{:}(), \mathtt{z_2}{:}() \vdash \mathtt{z_1}() \vee \mathtt{z_2}() : ((),())$ to obtain a type for $\mathtt{z(R(a), S(a))}$:

$$\dfrac{\dfrac{\dfrac{\mathtt{x_1}{:}(), \mathtt{x_2}{:}() \vdash \mathtt{x_1}() \vee \mathtt{x_2}() : ((),())}{\mathtt{x_1}{:}(), \mathtt{x_2}{:}(), \mathtt{z}{:}((),()) \vdash \mathtt{z(x_1, x_2)}{:}((),(),((),()))}\,4}{\mathtt{x_2}{:}(), \mathtt{z}{:}((),()) \vdash \mathtt{z(R(a), x_2)} : ((),((),()))}\,6}{\mathtt{z}{:}((),()) \vdash \mathtt{z(R(a), S(a))} : (((),()))}\,6$$

(for reasons of space, we omitted the premises

$\mathtt{z}{:}((),()), \mathtt{x_2}{:}() \vdash \mathtt{R(a)}{:}()$ and $\mathtt{z}{:}((),()) \vdash \mathtt{S(a)}{:}()$

of the first and second application of the substitution rule);

- For the derivation of the type of $\mathtt{z(R(x), S(a))}$ we first make a derivation of the "frame" $\mathtt{z(y_1, y_2)}$ of this pf:

$$\dfrac{\dfrac{\dfrac{\dfrac{\vdash \mathtt{a}{:}0}{\mathtt{y_1}{:}(0) \vdash \mathtt{a}{:}0}\,5 \qquad \dfrac{\dfrac{\vdash \mathtt{a}{:}0 \qquad \vdash \mathtt{R(a)}{:}()}{\mathtt{x}{:}0 \vdash \mathtt{R(x)}{:}(0)}\,3}{\mathtt{x}{:}0, \mathtt{y_1}{:}(0) \vdash \mathtt{y_1(x)}{:}(0,(0))}\,4}{\mathtt{y_1}{:}(0) \vdash \mathtt{y_1(a)}{:}((0))}\,6 \qquad \dfrac{\vdash \mathtt{R(a)}{:}()}{\mathtt{y_2}{:}() \vdash \mathtt{y_2}(){:}(())}\,4}{\mathtt{y_1}{:}(0), \mathtt{y_2}{:}() \vdash \mathtt{y_1(a)} \vee \mathtt{y_2}() : ((0),())}\,2}{\mathtt{y_1}{:}(0), \mathtt{y_2}{:}(), \mathtt{z}{:}((0),()) \vdash \mathtt{z(y_1, y_2)} : ((0),(),((0),()))}\,4.$$

Then we derive $\mathtt{x}{:}0 \vdash \mathtt{R(x)}{:}(0)$ and $\vdash \mathtt{S(a)}{:}()$, and after applying the weakening rule, we can substitute $\mathtt{R(x)}$ for $\mathtt{y_1}$ and $\mathtt{S(a)}$ for $\mathtt{y_2}$. As a result, we get

$$\mathtt{z}{:}((0),()), \mathtt{x}{:}0 \vdash \mathtt{z(R(x), S(a))} : (((0),())).$$

Example 2.55 In the example below, the orders are important:

$$\dfrac{\dfrac{\dfrac{\vdash \mathtt{R(a)} : ()^0 \qquad \dfrac{\vdash \mathtt{R(a)} : ()^0}{\vdash \neg\mathtt{R(a)} : ()^0}\,2}{\vdash \mathtt{R(a)} \vee \neg\mathtt{R(a)} : ()^0}\,2}{\mathtt{z}{:}()^0 \vdash \mathtt{z}() \vee \neg\mathtt{z} : (()^0)^1}\,4}{\vdash \forall \mathtt{z}{:}()^0[\mathtt{z}() \vee \neg\mathtt{z}()] : ()^1}\,8.$$

We see that $\forall \mathtt{z}{:}()^0[\mathtt{z}()^0 \vee \neg\mathtt{z}()]$ does not have a predicative type. This is the case because this pf has a bound variable $\mathtt{z}$ that is of a higher order than the order of any free variable (as there are no free variables here). Therefore, the order of this pf is determined by the order bound variable $\mathtt{z}$.

We still need to prove that the contexts in the conclusions of rules 3, 4 and 6 exist. This follows from the following Lemma:

Lemma 2.56 *Assume* $\Gamma \vdash f : t^a$. *Then*

1. **(Free variable lemma)** *All variables of* f *that are not bound by a quantifier are in* $\mathrm{dom}(\Gamma)$*;*

2. **(Strengthening lemma)** *If* Δ *is the (unique) subset of* Γ *such that* $\mathrm{dom}(\Delta)$ *contains exactly all the variables of* f *that are not bound by a quantifier, then* $\Delta \vdash f : t^a$.

PROOF: An easy induction on the definition of $\Gamma \vdash f : t^a$. ⊠

2c Properties of RTT

2c1 Types and free variables

In this section we treat some meta-properties of RTT. Using the λ-notation for pfs, we can often refer to known results in typed λ-calculus[24].

Theorem 2.57 (First Free Variable Theorem)
Let $f \in \mathcal{P}$*;* $k_1, \ldots, k_n \in \mathcal{A} \cup \mathcal{V} \cup \mathcal{P}$.

$$\mathrm{FV}(f[x_1, \ldots, x_n{:=}k_1, \ldots, k_n]) =$$

$$(\mathrm{FV}(f) \setminus \{x_1, \ldots, x_n\}) \cup \{k_i \in \mathcal{V} \mid x_i \in \mathrm{FV}(f)\}.$$

PROOF: Write $h \equiv f[x_1, \ldots, x_n{:=}k_1, \ldots, k_n]$. We know that $\widetilde{h}$ is the β-normal form of the λ-term $(\lambda_{i=1}^{n} x_i.\widetilde{f})\overline{k_1} \cdots \overline{k_n}$. By Lemma 2.22.3 and Lemma 2.9.4 we know that $\widetilde{f}, \overline{k_1}, \ldots, \overline{k_n}$ are all λI-terms. We conclude that $\widetilde{f}[x_1{:=}\overline{k_1}] \cdots [x_n{:=}\overline{k_n}]$ is also a λI-term. As

$$\left(\mathop{\lambda}_{i=1}^{n} x_i.\widetilde{f}\right) \overline{k_1} \cdots \overline{k_n} \twoheadrightarrow_\beta \widetilde{f}[x_1{:=}\overline{k_1}] \cdots [x_n{:=}\overline{k_n}]$$

[24] The meta-properties can also be proved directly, without λ-calculus: see [97].

we have by the Church-Rosser Theorem that $\widetilde{f}[x_1{:=}\overline{k_1}]\cdots[x_n{:=}\overline{k_n}] \twoheadrightarrow_\beta \widetilde{h}$, and therefore:

$$\begin{array}{rcl}
\mathrm{FV}(h) & \stackrel{(lem\ 2.22.1)}{=} & \mathrm{FV}(\widetilde{h}) \\
& \stackrel{(1)}{=} & \mathrm{FV}\left(\widetilde{f}[x_1{:=}\overline{k_1}]\cdots[x_n{:=}\overline{k_n}]\right) \\
& = & \left(\mathrm{FV}\left(\widetilde{f}\right)\setminus\{x_1,\ldots,x_n\}\right)\cup\bigcup\{\mathrm{FV}\left(\overline{k_i}\right) \mid x_i\in\mathrm{FV}\left(\widetilde{f}\right)\} \\
& \stackrel{(2)}{=} & \left(\mathrm{FV}\left(\widetilde{f}\right)\setminus\{x_1,\ldots,x_n\}\right)\cup\{\overline{k_i}\in V \mid x_i\in\mathrm{FV}\left(\widetilde{f}\right)\} \\
& \stackrel{(lem\ 2.22.1)}{=} & (\mathrm{FV}(f)\setminus\{x_1,\ldots,x_n\})\cup\{k_i\in\mathcal{V}\mid x_i\in\mathrm{FV}(f)\}.
\end{array}$$

At (1) we use that $\widetilde{f}[x_1{:=}\overline{k_1}]\cdots[x_n{:=}\overline{k_n}]$ is a λI-term that β-reduces to $\widetilde{h}$; at (2) we use the fact that $\mathrm{FV}\left(\overline{k_i}\right)=\varnothing$ whenever $k_i\in\mathcal{A}\cup\mathcal{P}$ (by definition of $\overline{k_i}$). ⊠

Theorem 2.58 (Second Free Variable Theorem) *Assume that we can derive* $\Gamma\vdash f : (t_1^{a_1},\ldots,t_n^{a_n})^a$, *and* $x_1<\cdots<x_m$ *are the free variables of* f. *Then* $m=n$ *and* $x_i : t_i^{a_i}\in\Gamma$ *for all* $i\le n$.

PROOF: An easy induction on $\Gamma\vdash f : (t_1^{a_1},\ldots,t_n^{a_n})^a$. For rules 6 and 7, use Theorem 2.57. ⊠

We can now prove a corollary that we promised in Remark 2.47.7:

Corollary 2.59 *Assume that we can derive* $\Gamma\vdash f : (t_1^{a_1},\ldots,t_n^{a_n})^a$ *and let* φ *be a bijection* $\{1,\ldots,n\}\to\{1,\ldots,n\}$. *Then there is a context* Γ' *and a pf* f' *which is* α_P*-equal to* f *such that*

$$\Gamma'\vdash f' : \left(t_{\varphi(1)}^{a_{\varphi(1)}},\ldots,t_{\varphi(n)}^{a_{\varphi(n)}}\right)^a.$$

PROOF: By the second Free Variable Theorem, we can assume that f has n free variables $x_1<\cdots<x_n$, and that $x_i{:}t_i^{a_i}\in\Gamma$ for all $i\in\{1,\ldots,n\}$.
Take n new free variables $z_1<\cdots<z_n$ such that $z_i>y$ for all $y\in\mathrm{dom}(\Gamma)$.
Now apply rule 7 of RTT n times. ⊠

We can also prove unicity of types and unicity of orders. Orders are unique in the following sense:

Lemma 2.60 *Assume* $\Gamma\vdash f : t^a$. *If* x *occurs in* f *and* $x : u^b\in\Gamma$, *then* u^b *is predicative. Moreover, if also* $\Gamma\vdash f : t'^{a'}$, *then* $a=a'$.

PROOF: By induction on the derivation of $\Gamma\vdash f : t^a$ one shows that a variable x that occurs in f always has a predicative type in Γ, and that both a and a' equal one plus the maximum of the orders of all the (free and non-free) variables that occur in f. ⊠

Corollary 2.61 (Unicity of Types for pfs) *Assume Γ is a context, f is a pf, $\Gamma \vdash f : t^a$ and $\Gamma \vdash f : u^b$. Then $t^a \equiv u^b$.*

PROOF: $t \equiv u$ follows from Theorem 2.58; $a = b$ from Lemma 2.60. ⊠

Remark 2.62 We cannot omit the context Γ in Corollary 2.61. For example, the pf $\mathtt{z}(\mathtt{x})$ can have different types in different contexts, as is illustrated by the following derivations (we have omitted the orders as they can be calculated via Lemma 2.60):

$$\cfrac{\cfrac{\vdash \mathtt{R}(\mathtt{a}_1) : () \qquad \vdash \mathtt{a}_1 : 0}{\mathtt{x} : 0 \vdash \mathtt{R}(\mathtt{x}) : (0)}3}{\mathtt{x} : 0, \mathtt{z} : (0) \vdash \mathtt{z}(\mathtt{x}) : (0, (0))}4$$

versus

$$\cfrac{\cfrac{\vdash \mathtt{R}(\mathtt{a}_1) : ()}{\mathtt{x} : () \vdash \mathtt{x}() : (())}4}{\mathtt{x} : (), \mathtt{z} : (()) \vdash \mathtt{z}(\mathtt{x}) : ((), (()))}4.$$

Theorem 2.58 and Corollary 2.61 show that our system RTT makes sense, in a certain way: The type of a pf only depends on the context and does not depend on the way in which we derived the type of that pf.

As a corollary of Corollary 2.61 we find:

Corollary 2.63 *If $\Gamma \vdash f : t^a$, $\Gamma \vdash k : u^b$, $x{:}u^b \in \Gamma$ and $\Gamma \vdash f[x{:=}k] : t'^{a'}$ then $a \geq a'$.*

PROOF: If $x \notin \text{FV}(f)$ then $f \equiv f[x{:=}k]$ and the corollary follows from Unicity of Types (Corollary 2.61). If $x \in \text{FV}(f)$ then the variables that occur in $f[x{:=}k]$, occur either in f or in k, and as the order of k is smaller than the order of f ($x \in \text{FV}(f)$, so $b < a$), the corollary follows from the proof of Lemma 2.60. ⊠

2c2 Strong normalisation

We investigate the problem whether there exists (in the situation of Definition 2.44.6) a pf h such that $h \equiv f[y{:=}k]$. We show that this is the case, in Corollary 2.73.

The nonexistence of $f[\vec{x}{:=}\vec{k}]$ can have two reasons:

- The λ-term $(\lambda_{i=1}^{n} x_i.\widetilde{f})\overline{k_1} \cdots \overline{k_n}$ has no β-normal form;
- The λ-term $(\lambda_{i=1}^{n} x_i.\widetilde{f})\overline{k_1} \cdots \overline{k_n}$ has β-normal form H, but there is no $h \in \mathcal{P}$ such that $\widetilde{h} \equiv H$.

However, these two things do not occur if we use substitution under the restrictions of Definition 2.44.6. For a proof of this we use the simply typed λ-calculus of Church [29]. This is not only of help for the existence-proof of h, but also shows that RTT can be seen as a subsystem of $\lambda_{\to}$Church. However, we remark that the orders of RTT are lost in the embedding to $\lambda_{\to}$Church. A definition of Church's calculus is given in Section 3b2. We translated the propositional functions and the ramified types of RTT to terms and types of $\lambda_{\to}$Church in Definitions 2.7, 2.20, and 2.39. We now extend the mapping T of Definition 2.39 to contexts:

Definition 2.64 We define a standard context Γ_0 in $\lambda_{\to}$Church, in which type information on $\vee$, $\neg$, $\forall$ and elements of $\mathcal{A}$ and $\mathcal{R}$ is stored:

$$\begin{aligned}\Gamma_0 \stackrel{\text{def}}{=} & \{\neg : o \to o, \vee : o \to o \to o\} \cup \\ & \{a : \iota \mid a \in \mathcal{A}\} \cup \\ & \{\forall_{t^a} : T(t^a) \to o \mid t^a \text{ is a ramified type}\} \cup \\ & \{R : \underbrace{\iota \to \ldots \to \iota}_{m \text{ times } \iota} \to o \mid R \in \mathcal{R}, \mathfrak{a}(R) = m\}.\end{aligned}$$

If Γ is a context in RTT then $T(\Gamma) \stackrel{\text{def}}{=} \Gamma_0 \cup \{x : T(t^a) \mid x{:}t^a \in \Gamma\}$.

In particular, $T(\varnothing) = \Gamma_0$.

Theorem 2.65 (Typability in RTT implies typability in $\lambda_{\to}$Church)
If $\Gamma \vdash f : t^a$ then

1. $T(\Gamma) \vdash_{\lambda_{\to}\mathit{Church}} \widetilde{f} : o$;
2. $T(\varnothing) \vdash_{\lambda_{\to}\mathit{Church}} \overline{f} : T(t^a)$.

PROOF: A straightforward induction on $\Gamma \vdash f : t^a$ with the use of Theorem 2.58 and the Subject Reduction property for λ-Church (Lemma 4.34). ⊠

Remark 2.66 Observe that the above theorem immediately excludes the pf that leads to the Russell paradox from the well-typed pfs: If $\neg \mathsf{z}(\mathsf{z})$ were legal then the λ-term $\neg(\mathsf{z}\mathsf{z})$ would be typable in λ-Church, which is not the case (see [3]).

Using the strong normalisation of λ-Church (Theorem 4.40), it is easy to solve the first problem:

Theorem 2.67 *Take $i \leq n$. Assume that $\Gamma \cup \{y{:}t_i^{a_i}\} \vdash f : (t_1^{a_1}, \ldots, t_n^{a_n})^a$ and that $\Gamma \vdash k : t_i^{a_i}$ (so: the preconditions of rule 6 of* RTT *are fulfilled). Then $(\lambda y{:}T(t_i^{a_i}).\widetilde{f})\overline{k}$ is strongly normalising.*

PROOF: The theorem is easy for $k \in \mathcal{A}$, so assume $k \in \mathcal{P}$. By Theorem 2.65.2, $T(\varnothing) \vdash \overline{k} : T(t_i^{a_i})$, and by weakening $T(\Gamma) \vdash \overline{k} : T(t_i^{a_i})$. As, by Theorem 2.65.1, $T(\Gamma) \cup \{y{:}T(t_i^{a_i})\} \vdash \widetilde{f} : o$ and therefore $T(\Gamma) \vdash \lambda y{:}T(t_i^{a_i}).\widetilde{f} : T(t_i^{a_i}) \to o$, we have $T(\Gamma) \vdash (\lambda y{:}T(t_i^{a_i}).\widetilde{f})\overline{k} : o$; so the term $(\lambda y{:}T(t_i^{a_i}).\widetilde{f})\overline{k}$, being a typable term in $\lambda_{\to\text{Church}}$, is strongly normalising. ⊠

The second problem is harder to tackle: substitution (Definition 2.23) is defined in terms of λ-calculus, and not every λ-term H has an equivalent h in $\mathcal{P}$ with $\widetilde{h} \equiv H$. This makes it hard to see what happens, especially in case of a substitution

$$z(h_1, \ldots, h_m)[x_1, \ldots, z, \ldots, x_n{:=}k_1, \ldots, k, \ldots, k_n].$$

For this substitution we must calculate the β-normal form of

$$(\lambda x_1{:}\tau_1 \cdots z{:}\tau \cdots x_n{:}\tau_n.z\overline{h_1} \cdots \overline{h_m})\overline{k_1} \cdots \overline{k} \cdots \overline{k_n}.$$

This term reduces to $\overline{k}H_1 \cdots H_m$ for some β-normal H_js. If $k \in \mathcal{P}$ then this new term may not be in β-normal form, and it is not clear what will be the final result (cf. Examples 2.24.4 and 2.24.5.).

The problem clearly has to do with the special structure of λ-terms H for which there is a (legal) $h \in \mathcal{P}$ with $\widetilde{h} \equiv H$. Such terms $\widetilde{h}$ have one important property: All variables are either arguments of functions, or they are applied to the maximal number of arguments that is possible according to the type of that variable. For instance: if a term $\widetilde{h}$ is of the form $zH_1 \cdots H_m$, then the type of z will be of the form $\tau_1 \to \cdots \to \tau_m \to o$.

We call such terms *fully applied* and give the following formal definition:

Definition 2.68 (Fully applied λ-terms) Let Γ be a $\lambda_{\to\text{Church}}$-context, and let M be a Γ-legal term of type t. Write $M \equiv M_0M_1 \cdots M_m$, where M_0 is either a variable or a term of the form $\lambda x{:}\tau.M_0'$. We define the notion *M is Γ-fully applied* by induction on the length of M:

- If M_0 is a variable then M is Γ-fully applied if M has type o in Γ, and for $1 \leq i \leq m$, either M_i is Γ-fully applied, or M_i is a variable;
- If $M_0 \equiv \lambda x{:}\tau.M_0'$ then M is Γ-fully applied if M_0' is $(\Gamma, x{:}\tau)$-fully applied, and for $1 \leq i \leq m$, either M_i is Γ-fully applied, or M_i is a variable.

If it is clear which context Γ is used, we just write *fully applied* instead of *Γ-fully applied.*

It will be shown that for each legal propositional function f, $\widetilde{f}$ and $\overline{f}$ are fully applied. This can be done by induction on the derivation of $\Gamma \vdash f : t^a$. For the substitution case, we need some additional properties of fully applied terms.

Lemma 2.69 *If M is $(\Gamma, y{:}\tau, \Delta)$-fully applied, and N is (Γ, Δ)-fully applied, then $M[y{:=}N]$ is (Γ, Δ)-fully applied.*

PROOF: Induction on the structure of M.

- $M \equiv xM_1 \cdots M_m$.
  - $x \equiv y$. Then $M[y{:=}N] \equiv NM_1[y{:=}N] \cdots M_m[y{:=}N]$. Distinguish:
    - $N \equiv zN_1 \cdots N_n$. Notice that $\Gamma, \Delta \vdash N : o$. This means that $\Gamma, y{:}\tau, \Delta \vdash y : o$, and therefore $m = 0$ and $M[y{:=}N] \equiv N$, thus: $M[y{:=}N]$ is fully applied;
    - $N \equiv (\lambda x{:}v.N')N_1 \cdots N_n$. As N is fully applied, N' is fully applied, and the N_js are either variables or they are fully applied. By induction, $M_i[y{:=}N]$ is either a variable or fully applied for $1 \leq i \leq m$. This means that

$$(\lambda x{:}v.N')N_1 \cdots N_n M_1[y{:=}N] \cdots M_m[y{:=}N]$$

      is fully applied;
  - $x \not\equiv y$. By induction, $M_i[y{:=}N]$ is either a variable or fully applied for $1 \leq i \leq m$. By the Substitution Lemma (4.30):
    $xM_1[y{:=}N] \cdots M_m[y{:=}N]$ has type o in (Γ, Δ).
    This means that $xM_1[y{:=}N] \cdots M_m[y{:=}N]$ is fully applied;
- $M \equiv (\lambda x{:}v.M')M_1 \cdots M_m$. By induction, $M'[y{:=}N]$ is $(\Gamma, \Delta, x{:}v)$-fully applied, and $M_i[y{:=}N]$ is either a variable, or fully applied. Therefore $M[y{:=}N]$ is fully applied.

⊠

Lemma 2.70 *If M is Γ-fully applied and $M \rightarrow_\beta M'$, then M' is Γ-fully applied.*

PROOF: Induction on the structure of M.

- $M \equiv xM_1 \cdots M_m$. The reduction must occur in a term M_i, say $M_i \rightarrow_\beta M_i'$. As M_i cannot be a variable, M_i is fully applied. By the induction hypothesis, M_i' is fully applied. M has type o, so by Subject Reduction Lemma 4.34, M' has type o. Hence M' is fully applied;
- $M \equiv (\lambda x{:}v.M_0)M_1 \cdots M_m$. If the reduction occurs within $M_1, \ldots, M_m$ or M_0 then we can give a similar argument as in (1). Now assume $M' \equiv M_0[x{:=}M_1]M_2 \cdots M_m$. Observe: M_0 is $(\Gamma, x{:}v)$-fully applied. If M_1 is a variable, then clearly $M_0[x{:=}M_1]$ is Γ-fully applied. If M_1 is not a variable then M_1 is Γ-fully applied, so by Lemma 2.69, $M_0[x{:=}M_1]$ is Γ-fully applied. Distinguish:

- $M_0[x{:=}M_1] \equiv yN_1 \cdots N_n$. As $M_0[x{:=}M_1]$ is fully applied, it has type o. This means that $m = 1$, and that $M' \equiv M_0[x{:=}M_1]$ is fully applied;
- $M_0[x{:=}M_1] \equiv (\lambda y{:}\tau.N)N_1 \cdots N_n$. As $M_0[x{:=}M_1]$ is fully applied, N is fully applied, and the N_js are either variables or fully applied terms. Therefore $M' \equiv (\lambda y{:}\tau.N)N_1 \cdots N_n M_2 \cdots M_m$ is fully applied.

⊠

Now we can prove:

Lemma 2.71 *Let $f \in \mathcal{P}$. If $\Gamma \vdash f : t^a$ then $\widetilde{f}$ is $T(\Gamma)$-fully applied, and $\overline{f}$ is $T(\varnothing)$-fully applied.*

PROOF: We prove that $\widetilde{f}$ is $T(\Gamma)$-fully applied. Then it easily follows that $\overline{f}$ is $T(\varnothing)$-fully applied. We use induction on the derivation of $\Gamma \vdash t^a$. All cases are easy to check, except for the abstraction-from-parameters and the substitution cases:

3. (abstraction from parameters) We use notations as in Definition 2.44.3. By induction on the structure of f it is shown that if $\widetilde{f}$ and $\widetilde{g}$ are fully applied, then $\widetilde{h}$ is fully applied;
6. (substitution) We use notations as in Definition 2.44.6. Let $h \equiv f[y{:=}k]$. If $k \in \mathcal{A}$ then h is just f in which all free occurrences of y have been replaced by k. It is then easy to see that $\widetilde{h}$ is fully applied. Now suppose $k \in \mathcal{P}$. By the induction hypothesis, $\widetilde{f}$ and $\widetilde{g}$ are fully applied. This means that $(\lambda y{:}\widetilde{f})\widetilde{g}$ is fully applied. This term β-reduces to $\widetilde{h}$. By Lemma 2.70, $\widetilde{h}$ is fully applied.

⊠

We now know that each legal pf f gives rise to fully applied λ-terms $\widetilde{f}$ and $\overline{f}$. This is of great help in showing that substitutions always exist in case of the application of rule 6 of Definition 2.44. We first show that each substitution in RTT that gives rise to a β-reduction path starting with a fully applied λ-term, really exists. Then it is easy to show that substitutions always exist in the situation of Definition 2.44.6.

Lemma 2.72 *Let $f \in \mathcal{P}$, $k_1, \ldots, k_n \in \mathcal{A} \cup \mathcal{V} \cup \mathcal{P}$, and $(\lambda_{i=1}^n x_i{:}t_i.\widetilde{f})\overline{k_1} \cdots \overline{k_n}$ be $T(\Gamma)$-fully applied, where Γ is a RTT-context. Then there is $h \in \mathcal{P}$ such that $h \equiv f[x_1, \ldots, x_n{:=}k_1, \ldots, k_n]$.*

PROOF: Notice that $(\lambda_{i=1}^n x_i{:}t_i.\widetilde{f})\overline{k_1} \cdots \overline{k_n}$ is a legal term of $\lambda_{\to\text{Church}}$, and therefore strongly normalising. Let q be the length of the longest reduction path of this term. Use induction on q, so assume that the lemma has been proved for all $q' < q$ (write IH1 for the induction hypothesis). We use induction on the structure of f (and write IH2 for this induction hypothesis). Some cases can be handled directly, for other cases we need the help of Lemma 2.70.

1. $f \equiv R(i_1, \ldots, i_{\mathfrak{a}(R)})$. Define

$$i'_j \equiv \begin{cases} k_\ell & \text{if } i_j \equiv x_\ell; \\ i_j & \text{if } i_j \notin \{x_1, \ldots, x_n\}. \end{cases}$$

Observe that $i'_j \in \mathcal{A} \cup \mathcal{V}$: otherwise there would have been $k_\ell \in \mathcal{P}$ such that $i_j \equiv x_\ell$. But as R has type $\iota \to \ldots \to \iota \to o$, x_ℓ must have type ι, and therefore $\overline{k_\ell}$ has type ι, which means that k_ℓ cannot be a pf (Theorem 2.58, Theorem 2.65).

Let $f' \equiv R(i'_1, \ldots, i'_{\mathfrak{a}(R)})$. As none of the i'_js is a pf, we have $f' \in \mathcal{P}$. Observe:

$$\left(\mathop{\lambda}_{i=1}^{n} x_i{:}t_i.\widetilde{f}\right) \overline{k_1} \cdots \overline{k_n} \twoheadrightarrow_\beta \widetilde{f'},$$

so $f[x_1, \ldots, x_n{:=}k_1, \ldots, k_n]$ exists and is equal to f';

2. $f \equiv f_1 \vee f_2$. Notice: $\widetilde{f}$ is fully applied and $\widetilde{f} \equiv \vee \widetilde{f_1}\,\widetilde{f_2}$, so $\widetilde{f_1}$ and $\widetilde{f_2}$ are fully applied. Therefore, $(\lambda_{i=1}^{n} x_i{:}t_i.\widetilde{f_j})\overline{k_1} \cdots \overline{k_n}$ is fully applied for $j = 1, 2$. By the induction hypothesis[25], there are $h_1, h_2 \in \mathcal{P}$ such that

$$f_j[x_1, \ldots, x_n{:=}k_1, \ldots, k_n] \equiv h_j$$

for $j = 1, 2$. This means that

$$\left(\mathop{\lambda}_{i=1}^{n} x_i{:}t_i.\widetilde{f}\right) \overline{k_1} \cdots \overline{k_n} \twoheadrightarrow_\beta \vee \widetilde{h_1}\,\widetilde{h_2}$$

and therefore $f[x_1, \ldots, x_n{:=}k_1, \ldots, k_n] \equiv h_1 \vee h_2$.

A similar proof can be given for $f \equiv \neg f'$;

3. $f \equiv \forall x{:}t^{\mathfrak{a}}[f']$. Notice that $\widetilde{f} \equiv \forall_{t^{\mathfrak{a}}}(\lambda x{:}T(t^{\mathfrak{a}}).\widetilde{f'})$. As $\widetilde{f}$ is fully applied, $\widetilde{f'}$ is fully applied as well. This means that $(\lambda_{i=1}^{n} x_i{:}t_i.\widetilde{f'})\overline{k_1} \cdots \overline{k_n}$ is fully applied. By the induction hypothesis[25], there is $h \in \mathcal{P}$ such that $f'[x_1, \ldots, x_n{:=}k_1, \ldots, k_n] \equiv h$. This means that

$$\left(\mathop{\lambda}_{i=1}^{n} x_i{:}t_i.\widetilde{f}\right) \overline{k_1} \cdots \overline{k_n} \twoheadrightarrow_\beta \forall_{t^{\mathfrak{a}}} \lambda x{:}t^{\mathfrak{a}}.\widetilde{h}$$

and therefore $f[x_1, \ldots, x_n{:=}k_1, \ldots, k_n] \equiv \forall x{:}t^{\mathfrak{a}}[h]$. As $\widetilde{f}, \widetilde{k_1}, \ldots, \widetilde{k_n}$ are all λI-terms, $x \in \mathrm{FV}(\widetilde{h})$, so $x \in \mathrm{FV}(h)$, which means that $\forall x{:}t^{\mathfrak{a}}[h]$ is indeed a pf;

[25] Observe that the longest reduction path of $(\lambda x_1{:}t_1 \cdots x_n{:}t_n.\widetilde{f_j})\overline{k_1} \cdots \overline{k_n}$ has a length $\leq q$. If the length is equal to q then use IH2; otherwise use IH1.

4. $f \equiv z(h_1, \ldots, h_m)$. If $z \notin \{x_1, \ldots, x_n\}$ then we can give a proof similar to the case $f \equiv R(i_1, \ldots, i_{\alpha(R)})$. Now assume $z \equiv x_p$. Define

$$h'_j \equiv \begin{cases} k_\ell & \text{if } h_j \equiv x_\ell; \\ h_j & \text{if } h_j \notin \{x_1, \ldots, x_n\}. \end{cases}$$

Notice that $\widetilde{f}$ is fully applied. As $\widetilde{f}$ starts with a variable (this is due to the definition of $\widetilde{f}$), it has type o. Therefore,

$$\left(\mathop{\lambda}_{i=1}^{n} x_i{:}t_i.\widetilde{f}\right)\overline{k_1}\cdots\overline{k_n}$$

has type o as well. Observe:

$$\left(\mathop{\lambda}_{i=1}^{n} x_i{:}t_i.\widetilde{f}\right)\overline{k_1}\cdots\overline{k_n} \twoheadrightarrow_\beta \overline{k_p}\overline{h'_1}\cdots\overline{h'_m}.$$

Write $K \equiv \overline{k_p}\overline{h'_1}\cdots\overline{h'_m}$. K is fully applied (Lemma 2.70), and has type o (Subject Reduction Lemma 4.34). Observe that $\overline{k_p} \equiv \lambda_{y=1}^{q} y_j{:}v_j.\widetilde{k_p}$. Hence $\widetilde{k_p}$ is fully applied. By definition of $\widetilde{k_p}$, $\widetilde{k_p}$ starts with a variable. Therefore, $\widetilde{k_p}$ has type o. As K has type o as well, we have that $q = m$, and that K represents the substitution

$$k_p[y_1, \ldots, y_m{:=}h'_1, \ldots, h'_m].$$

The longest reduction path of K is shorter than the longest reduction path of

$$\left(\mathop{\lambda}_{i=1}^{n} x_i{:}t_i.\widetilde{f}\right)\overline{k_1}\cdots\overline{k_n}$$

so we can apply the induction hypothesis IH1 and conclude that there is $h \in \mathcal{P}$ such that $k_p[y_1, \ldots, y_m{:=}h'_1, \ldots, h'_m] \equiv h$. But then also $f[x_1, \ldots, x_n{:=}k_1, \ldots, k_n] \equiv h$. ⊠

With this lemma it is easy to show that substitution always exists in the case of RTT-rule 6 of Definition 2.44.

Theorem 2.73 (Existence of substitution) *If $f \in \mathcal{P}$, y is the ith free variable in f, $\Gamma \cup \{y : t_i^{a_i}\} \vdash f : (t_1^{a_1}, \ldots, t_n^{a_n})^a$, and $\Gamma \vdash k : t_i^{a_i}$, then $f[y{:=}k]$ exists.*

PROOF: Notice that $\widetilde{f}$ and $\overline{k}$ are fully applied. Therefore $(\lambda y{:}T(t_i^{a_i}).\widetilde{f})\overline{k}$ is fully applied. By Lemma 2.72, $f[y{:=}k]$ exists. ⊠

2c3 Subterm property

The technique of fully applied λ-terms that was used in Section 2c2 to prove the existence of substitution can also be used to prove another important property of type systems for RTT: the Subterm Property. This property states that if a propositional function is typable, then its recursive parameters (see Definition 2.16) are typable as well. If all recursive parameters of a legal pf f are typable, we say that f has the subterm property:

Definition 2.74 Assume $\Gamma \vdash f : t^a$. If for all $h \in \mathrm{RP}(f) \cap \mathcal{P}$ there is $\Delta \supseteq \Gamma$ and a predicative type u^b such that $\Delta \vdash h : u^b$, then f has the *subterm property.* Notation: $\mathrm{SP}(f)$.

Just as in Section 2c2, we prove by induction on the derivation $\Gamma \vdash f : t^a$ that all legal pfs have the subterm property. Again, all cases are easy, except for the substitution rule 6 of Definition 2.44. This case can be solved using similar techniques as in Section 2c2.

Lemma 2.75 *Let $f \in \mathcal{P}$, $k_1, \ldots, k_n \in \mathcal{A} \cup \mathcal{V} \cup \mathcal{P}$, and $(\lambda_{i=1}^{n} x_i{:}t_i.\tilde{f})\overline{k_1} \cdots \overline{k_n}$ be fully applied with respect to $T(\Gamma)$, where Γ is a RTT-context. If $\mathrm{SP}(f)$ and $\mathrm{SP}(k_i)$ for all $k_i \in \mathcal{P}$, then*

1. $\mathrm{RP}\Big(f[\vec{x}{:=}\vec{k}]\Big) \subseteq \mathrm{RP}(f) \cup \{k_i \mid 1 \leq i \leq n\} \cup \bigcup_{k_i \in \mathcal{P}} \mathrm{RP}(k_i)$;
2. $\mathrm{SP}\Big(f[\vec{x}{:=}\vec{k}]\Big)$.

PROOF: Clearly, (2) follows from (1). We prove (1) by induction on the length of the reduction path of $(\lambda_{i=1}^{n} x_i{:}t_i.\tilde{f})\overline{k_1} \cdots \overline{k_n}$. We use induction on the structure of f, and only treat the interesting case: $f \equiv z(h_1, \ldots, h_m)$, and $z \equiv x_p$. As in the proof of Lemma 2.72, we define

$$h'_j \equiv \begin{cases} k_\ell & \text{if } h_j \equiv x_\ell; \\ h_j & \text{if } h_j \not\in \{x_1, \ldots, x_n\}, \end{cases}$$

and prove that $f[x_1, \ldots, x_n{:=}k_1, \ldots, k_n] \equiv k_p[y_1, \ldots, y_m{:=}h'_1, \ldots, h'_m]$. As the reduction path of $(\lambda_{j=1}^{m} y_j{:}u_j.\tilde{k}_p)\overline{h'_1} \cdots \overline{h'_m}$ is shorter than the reduction path of $(\lambda_{i=1}^{n} x_i{:}t_i.\tilde{f})\overline{k_1} \cdots \overline{k_n}$, we can use the induction hypothesis:[26]

$$\begin{array}{rcl} \mathrm{RP}\Big(f[\vec{x}{:=}\vec{k}]\Big) & \equiv & \mathrm{RP}\Big(k_p[\vec{y}{:=}\vec{h}]\Big) \\ & \subseteq & \mathrm{RP}(k_p) \cup \{h'_j \mid 1 \leq j \leq m\} \cup \displaystyle\bigcup_{h'_j \in \mathcal{P}} \mathrm{RP}(h'_j) \\ & \subseteq & \mathrm{RP}(f) \cup \{k_i \mid 1 \leq i \leq n\} \cup \displaystyle\bigcup_{k_i \in \mathcal{P}} \mathrm{RP}(k_i). \end{array}$$

⊠

[26]Note that $\mathrm{RP}(k_p) \subseteq \bigcup_{k_i \in \mathcal{P}} \mathrm{RP}(k_i)$, $\{h'_j \mid 1 \leq j \leq m\} \subseteq \mathrm{RP}(f) \cup \{k_i \mid 1 \leq i \leq n\}$ and $\bigcup_{h'_j \in \mathcal{P}} \mathrm{RP}(h'_j) \subseteq \mathrm{RP}(f) \cup \bigcup_{k_i \in \mathcal{P}} \mathrm{RP}(k_i)$.

As a corollary we get:

Corollary 2.76 (Subterm Lemma) *If* $\Gamma \vdash f : t^a$ *then* $\textsc{sp}(f)$.

PROOF: Induction on $\Gamma \vdash f : t^a$. All cases are easily checked except for the substitution rule, which is proved with Lemma 2.75. ⊠

2d Legal propositional functions

We recall Definition 2.45: a pf f is called *legal* if $\Gamma \vdash f : t^a$ for some Γ and t^a. We will check whether this definition of legal pf coincides with the definition of formula that was given in the *Principia*. For this purpose we prove a number of lemmas concerning the relation between legal pfs and predicative types.

We do not distinguish between pfs that are α_P-equal, nor between types $(t_1, \ldots, t_n)$ and $(t_{\varphi(1)}, \ldots, t_{\varphi(n)})$ for a bijection φ. This is justified by Corollary 2.59 and by the fact, that pfs that are α_P-equal are supposed to be the same in the *Principia* too.

We define the notion "up to α_P-equality" formally:

Definition 2.77 Let $f \in \mathcal{P}$, Γ a context, t^a a type. *f is of type t^a in the context Γ up to α_P-equality*, notation $\Gamma \vdash f : t^a(\mathrm{mod}\,\alpha_P)$, if there is $f' \in \mathcal{P}$, a context Γ' and a bijection $\varphi : \mathcal{V} \to \mathcal{V}$ such that

- $\Gamma' \vdash f' : t^a$;
- f' and f are α_P-equal via the bijection φ;
- $\Gamma' = \{\varphi(x){:}u^b \mid x{:}u^b \in \Gamma\}$.

We say that f is *legal* in the context Γ *up to α_P-equality* if there is a type u^b such that $\Gamma \vdash f : u^b(\mathrm{mod}\,\alpha_P)$. We say that f is *legal up to α_P-equality* if there is a context Γ such that f is legal in Γ up to α_P-equality.

The following lemma states that all predicative types are "inhabited":

Lemma 2.78 *If t^a is predicative then there are f, Γ such that $\Gamma \vdash f : t^a$.*

PROOF: We use induction on predicative types.

The case $t = 0^0$ is trivial.

Now assume $t = (t_1^{a_1}, \ldots, t_m^{a_m})^a$. By induction there are f_i and Γ_i such that $\Gamma_i \vdash f_i : t_i^{a_i}$ for all $i \leq m$. Take a fixed i. We shall find a context Δ_i and a legal pf g_i such that $\Delta_i \vdash g_i{:}(t_i^{a_i})^{a_i+1}$. Distinguish two cases:

- $t_i^{a_i} = 0^0$. Then make the following derivation:

$$\frac{\Gamma_i \vdash \mathtt{R}(f_i) : ()^0 \qquad \Gamma_i \vdash f_i : 0^0}{\Gamma_i, z_i{:}0^0 \vdash \mathtt{R}(z_i) : (0^0)^1} 3.$$

Write $\Delta_i \equiv \Gamma_i \cup \{z_i{:}0^0\}$, and $g_i \equiv \mathtt{R}(z_i)$, then $\Delta_i \vdash g_i{:}(t_i^{a_i})^{a_i+1}$;

- $t_i \neq 0^0$, say $t_i^{a_i} = (u_1^{b_1}, \ldots, u_n^{b_n})^{a_i}$. Because of Theorem 2.58, f_i has n free variables, say $x_1 < \cdots < x_n$, such that $x_j{:}u_j^{b_j} \in \Gamma_i$. Now use rule 4 of RTT:

$$\frac{\Gamma_i \vdash f_i : t_i^{a_i}}{\Gamma_i' \vdash z_i(x_1, \ldots, x_n) : (u_1^{b_1}, \ldots, u_n^{b_n}, t_i^{a_i})^{a_i+1}} 4 \tag{2.1}$$

where $\Gamma_i' \equiv \{x_j{:}u_j^{b_j} \mid 1 \leq j \leq n\} \cup \{z_i{:}t_i^{a_i}\}$. Use rule 8 n times:

$$\{z_i{:}t_i^{a_i}\} \vdash \forall x_1{:}u_1^{b_1}[\cdots \forall x_n{:}u_n^{b_n}[z_i(x_1, \ldots, x_n)] \cdots] : (t_i^{a_i})^{a_i+1}.$$

Write $\Delta_i \equiv \{z_i{:}t_i^{a_i}\}$ and

$$g_i = \forall x_1{:}u_1^{b_1}[\cdots \forall x_n{:}u_n^{b_n}[z_i(x_1, \ldots, x_n)] \cdots].$$

For arbitrary $i = 1, \ldots, m$ we now have: $\Delta_i \vdash g_i : (t_i^{a_i})^{a_i+1}$.

We can assume that $x < y$ for $x \in \mathrm{dom}(\Delta_i)$, $y \in \mathrm{dom}(\Delta_j)$ with $i < j$. Write $\Delta \equiv \Delta_1 \cup \ldots \cup \Delta_m$. Now apply rule 2 consecutively $m-1$ times, to obtain:

$$\Delta \vdash g_1 \vee (g_2 \vee \ldots \vee (g_{m-1} \vee g_m) \ldots) : (t_1^{a_1}, \ldots, t_m^{a_m})^{\max(a_1, \ldots, a_m)+1}.$$

Notice that $(t_1^{a_1}, \ldots, t_m^{a_m})^a$ is predicative, so $a = \max(a_1, \ldots, a_m) + 1$. ⊠

Remark 2.79 From a modern point of view, this is a remarkable lemma. Many modern type systems are based on the principle of propositions-as-types (see Section 4a). In such systems types represent propositions, and terms inhabiting such a type represent proofs of that proposition. In a propositions-as-types based system in which all types are inhabited, all propositions are provable. Such a system would be (logically) inconsistent. RTT is not based on propositions-as-types, and there is nothing paradoxical or inconsistent in the fact that all RTT-types are inhabited.

This lemma can be generalised to some non-predicative types:

Corollary 2.80 *If $(t_1^{a_1}, \ldots, t_m^{a_m})^a$ is a type such that the $t_i^{a_i}$ are all predicative, then there are f and Γ such that $\Gamma \vdash f : (t_1^{a_1}, \ldots, t_m^{a_m})^a$.*

PROOF: With Lemma 2.78 we can construct g_1 and Δ_1 such that

$$\Delta_1 \vdash g_1 : (t_1^{a_1}, \dots, t_m^{a_m})^{\max(a_1,\dots,a_m)+1}.$$

Let u^a be a predicative type of order a. Determine, again with Lemma 2.78, g_2 and Δ_2 such that $\Delta_2 \vdash g_2 : u^a$. Assume $x_1 < \cdots < x_n$ are the free variables of g_2 and $x_j{:}u_j^{b_j} \in \Delta_2$. Notice that $u^a = (u_1^{b_1}, \dots, u_n^{b_n})^a$ (Theorem 2.58). Apply rule 8 and weakening n times to obtain:

$$\Delta_2 \vdash \forall x_1{:}u_1^{b_1}[\cdots \forall x_n{:}u_n^{b_n}[g_2]\cdots] : ()^a.$$

We can assume that $x < y$ for all $x \in \text{DOM}(\Delta_1)$ and all $y \in \text{DOM}(\Delta_2)$, so we can use rule 2 to conclude:

$$\Delta_1 \cup \Delta_2 \vdash g_1 \vee \forall x_1{:}u_1^{b_1}[\cdots \forall x_n{:}u_n^{b_n}[g_2]\cdots] : (t_1^{a_1}, \dots, t_m^{a_m})^a.$$

⊠

We can use the techniques of the preceding proof to show $z(k_1, \dots, k_m)$ is legal if $k_1, \dots, k_m$ are either legal pfs or variables, and z is "fresh".

Lemma 2.81 *If $k_1, \dots, k_n \in \mathcal{A} \cup \mathcal{V} \cup \mathcal{P}$, $t^a = (t_1^{a_1}, \dots, t_n^{a_n})^a$ is a predicative type, $\Gamma \vdash k_i : t_i^{a_i}$ for all $k_i \in \mathcal{A} \cup \mathcal{P}$ and $k_i : t_i^{a_i} \in \Gamma$ for all $k_i \in \mathcal{V}$, and $z \in \mathcal{V} \setminus \mathrm{dom}(\Gamma)$, then $z(k_1, \dots, k_n)$ is legal in the context $\Gamma \cup \{z : t^a\}$ (up to α_P-equality).*

PROOF: First, we make a derivation of

$$\{z{:}t^a\} \cup \{x_i{:}t_i^{a_i} \mid 1 \le i \le n\} \vdash z(x_1, \dots, x_n) : (t_1^{a_1}, \dots, t_n^{a_n}, t^a)^{a+1} (\mathrm{mod}\, \alpha_P),$$

similarly to the derivation of (2.1) in the proof of Lemma 2.78. Next, find (with Lemma 2.78) $k'_1, \dots, k'_n$ such that

- $k'_i \equiv k_i$ if $k_i \in \mathcal{A} \cup \mathcal{P}$;
- $k'_i \in \mathcal{A} \cup \mathcal{P}$ has type $t_i^{a_i}$ in a context Δ_i if $k_i \in \mathcal{V}$;
- Δ, the union of the contexts Γ and the Δ_is, is a context;
- For $k_i \in \mathcal{V}$: k'_i and k'_j are α_Γ-equal if and only if $k_i \equiv k_j$.[27]

Apply rule 6 n times (as in the proof of Lemma 2.78), and where necessary the weakening rule, to obtain:

$$\{z{:}t^a\} \cup \Delta \vdash z(k'_1, \dots, k'_n) : (t^a)^{a+1} (\mathrm{mod}\, \alpha_P).$$

Now introduce, with rule 3, new variables for the k'_i which are not equal to k_i, to obtain a legal pf that is α_P-equal to $z(k_1, \dots, k_m)$. ⊠

[27] One might wonder whether there are enough pfs of one type that are not α_Γ-equal. Lemma 2.78 provides only one pf for each type. But if we have that pf, say k, then we use rule 2 of RTT to create $\neg k$, $\neg\neg k$, $\neg\neg\neg k$, etc. k, $\neg k$, $\neg\neg k$,... are all α_Γ-different and of the same type as k.

It is also not hard to show that $f \vee g$ is legal if f and g are (see also Remark 2.49):

Lemma 2.82 *If f and g are legal in contexts Γ_1 and Γ_2, respectively, and $\Gamma_1 \cup \Gamma_2$ is a context, then $f \vee g$ is legal in the context $\Gamma_1 \cup \Gamma_2$ (up to α_P-equality).*

PROOF: For reasons of clarity, we again leave out the orders of the ramified types. Note that we can not simply apply rule 2 of RTT, as the contexts Γ_1 and Γ_2 may not obey to the condition on them in rule 2.

Assume that $\Gamma_1 \vdash f : (t_1, \ldots, t_m)$, $\Gamma_2 \vdash g : (u_1, \ldots, u_n)$. Assume also that $x_1 < \cdots < x_m$ and $y_1 < \cdots < y_n$ are the free variables of f and g, respectively.

Write $t = (t_1, \ldots, t_m)$ and $u = (u_1, \ldots, u_n)$.

Take variables $x'_1 < \ldots x'_m < z_1 < y'_1 < \cdots < y'_n < z_2$ not occurring in the domain of $\Gamma_1 \cup \Gamma_2$; let

$$\Delta_1 = \{z_1{:}t, x'_1{:}t_1, \ldots x'_m{:}t_m\};$$

$$\Delta_2 = \{z_2{:}u, y'_1{:}u_1, \ldots y'_n{:}u_n\}.$$

Similarly to the derivation (2.1) in the proof of Lemma 2.78, we can derive

$$\Delta_1 \vdash z_1(x'_1, \ldots, x'_m) : (t_1, \ldots, t_m, t);$$

$$\Delta_2 \vdash z_2(y'_1, \ldots, y'_n) : (u_1, \ldots, u_n, u).$$

As Δ_1 and Δ_2 obey to the conditions of rule 2 of RTT, we can derive

$$\Delta_1 \cup \Delta_2 \vdash z_1(x'_1, \ldots, x'_m) \vee z_2(y'_1, \ldots, y'_n) : (t_1, \ldots, t_m, t, u_1, \ldots, u_n, u).$$

With similar techniques as in the proof of Lemma 2.81 we can now derive

$$\Gamma_1 \cup \Gamma_2 \cup \{z_1{:}t, z_2{:}u\} \vdash z_1(x_1, \ldots, x_m) \vee z_2(y_1, \ldots, y_n) : v (\mathrm{mod}\, \alpha_P)$$

for a certain type v (notice that the sets $\{x'_1, \ldots, x'_m\}$ and $\{y'_1, \ldots, y'_n\}$ do not overlap, whilst the sets $\{x_1, \ldots, x_m\}$ and $\{y_1, \ldots, y_n\}$ may overlap). Use rule 6 twice: Substitute f for z_1 and substitute g for z_2. This gives a derivation of $f \vee g$ in the context $\Gamma_1 \cup \Gamma_2$ (mod α_P). ⊠

The following lemma is easy to prove and will be used in the proof of the main result of this section.

Lemma 2.83 *If $R(i_1, \ldots, i_{\mathfrak{a}(R)})$ is a pf with free variables $x_1 < \cdots < x_m$, then it is legal in the context $\{x_j{:}0 \mid 1 \leq j \leq m\}$.*

PROOF: Write $f = R(i_1, \ldots, i_{\mathfrak{a}(R)})$. Let $a_1, \ldots, a_m \in \mathcal{A}$ be m different individuals that do not occur in f, and replace each variable x_j in f by a_j, calling the result f'. By the first rule of RTT, f' is legal in the empty context. Reintroducing the variables $x_1, \ldots, x_m$ (by applying rule 3 of RTT m times) for the individuals $a_1, \ldots, a_m$, respectively, we obtain that f is legal in the context $\{x_j{:}0 \mid 1 \leq j \leq m\}$. ⊠

Finally, we can give a characterisation of the legal pfs:

Theorem 2.84 (Legal pfs in RTT) *Let $f \in \mathcal{P}$. f is legal (mod α_P) if and only if:*

- *$f \equiv R(i_1, \ldots, i_{\mathfrak{a}(R)})$, or*
- *$f \equiv z(k_1, \ldots, k_n)$, $z \not\equiv k_j$ for all $k_j \in \mathcal{V}$ and z does not occur in any $k_j \in \mathcal{P}$, and there is Γ with $\mathrm{FV}(f) \subseteq \mathrm{DOM}(\Gamma)$ and for all $k_j \in \mathcal{P}$, $\Gamma \vdash k_j{:}t_j^{a_j}$ for some predicative type $t_j^{a_j}$, or*
- *$f \equiv \neg f'$ and f' is legal (mod α_P) or*
- *$f \equiv f_1 \vee f_2$ and there are Γ_i and $t_i^{a_i}$ such that $\Gamma_i \vdash f_i{:}t_i^{a_i}$ (mod α_P) for $i = 1, 2$ and $\Gamma_1 \cup \Gamma_2$ is a context, or*
- *$f \equiv \forall x{:}t^a.f'$ and f' is legal.*

PROOF: Use induction on the structure of f:

- $f \equiv R(i_1, \ldots, i_{\mathfrak{a}(R)})$. This is Lemma 2.83;
- $f \equiv z(k_1, \ldots, k_n)$.
  "$\Leftarrow$" is Lemma 2.81.
  "$\Rightarrow$": f is legal, so there is Γ with $\Gamma \vdash f : t^a$.
  As $T(\Gamma) \vdash_{\lambda_{\to\text{Church}}} z\overline{k_1} \cdots \overline{k_n} : o$ (Theorem 2.65), $z : (u_1^{b_1}, \ldots, u_n^{b_n})^b$ for a predicative type $(u_1^{b_1}, \ldots, u_n^{b_n})^b$, and $T(\Gamma) \vdash_{\lambda_{\to\text{Church}}} \overline{k_j} : T(u_j^{b_j})$. If $z \equiv k_j$ then $z \equiv \overline{k_j}$ and hence $z{:}u_j^{b_j} \in \Gamma$, which is impossible. By Corollary 2.76, each $k_j \in \mathcal{P}$ is typable in Γ, and as $T(\Gamma) \vdash_{\lambda_{\to\text{Church}}} \overline{k_j} : T(u_j^{b_j})$ and the type of k_j is predicative, $\Gamma \vdash k_j{:}u_j^{b_j}$. Notice that $b_j < b$, so it is impossible that z occurs in a $k_j \in \mathcal{P}$;
- "$\Leftarrow$" is Rule 2 of RTT (for $\neg$) and Lemma 2.82 (for $\vee$). "$\Rightarrow$" (for $\vee$; the proof for $\neg$ is similar): Let Δ be the context containing all the variables of f (also those that are bound by a quantifier; we can assume that different quantifiers bind different variables) and their types. f is built from several pfs of the form $R(i_1, \ldots, i_{\mathfrak{a}(R)})$ and $z(k_1, \ldots, k_m)$ (we will call these pfs the *constituents* of f), and the logical connectives $\neg$, $\vee$ and $\forall$.

Reasoning as in the "$\Rightarrow$" part of the first two cases of the proof of this lemma, we can show the preconditions for Lemma 2.83 (for constituents of the form $R(i_1, \ldots, i_{\mathfrak{a}(R)})$) and Lemma 2.81 (for constituents of the form $z(k_1, \ldots, k_m)$). Applying these Lemmas, we find that any constituent h of f is typable in Δ. Using Rule 2 of RTT (for $\neg$), Rule 8 of RTT (for $\forall$) and Lemma 2.82 (for $\vee$), we find that $f_1 \vee f_2$ itself is typable;

- "$\Leftarrow$" is Rule 8 of RTT. "$\Rightarrow$" is similar to "$\Rightarrow$" in the previous case. ⊠

We can now answer the question whether our legal pfs (as defined in Definition 2.45) are the same as the formulas of the *Principia*.

First of all, we must notice that all the legal pfs from Definition 2.45 are also formulas of the *Principia*: This was motivated in Remark 2.47.

Moreover, we proved (in 2.84) that if f is a pf, then the only reasons why f cannot be legal (according to Definition 2.45) are:

- There is a constituent $z(k_1, \ldots, k_m)$ of f in which z occurs in one of the k_i's;
- There is a constituent $z(k_1, \ldots, k_m)$ of f and a $j \in \{1, \ldots, m\}$ such that k_j is a pf, but not a legal pf;
- f contains two non-overlapping constituents f_1, f_2 that cannot be typed in one and the same context;
- There is a legal constituent $z(k_1, \ldots, k_m)$ of f which is not of predicative type.

Pfs of the first type cannot be legal in the *Principia*, because of the vicious circle principle. The same holds for pfs of the second type, because also in the *Principia*, parameters cannot be untyped. The third problem is a non-issue in the *Principia*. Formal contexts are not present in the *Principia*, but have been introduced in this chapter to make a precise analysis of RTT possible. Propositional functions of the *Principia* are always constructed in one, implicitly defined, context.[28] A formula,

[28] It is worth remarking that it is possible to formalize *Principia* without resorting to explicit contexts at all. For example, Randall Holmes has an implementation which constructs contexts from the structures of the terms analysed [74]. Following Randall Holmes, the price of this is that the types deduced for terms by his checker are polymorphic: for STT this isn't a problem at all (it's an advantage); Holmes expresses that in RTT, the handling of polymorphic types was quite difficult – he had to allow orders defined in terms of the unknown orders of polymorphic types. Further, in RTT, the type checker had to be much smarter than the STT checker, because it had to be able to deduce identity between polymorphic types in order to successfully infer types for quite simple terms (such as the "definition of equality" $(\forall x.(x(y) \leftrightarrow x(z)))$, where the two variables y and z are both polymorphic, and one has to be careful to determine that they have the same type (because they are in the same argument of the same unknown pf x) before attempting the final type-checking of the term: if one is not careful about the order in which things are done, two incompatible types for x will be deduced depending on unknown and possibly different orders for y and z).

therefore, cannot contain two non-overlapping constituents that cannot be typed in the same context. This excludes pfs of the third type. As to the fourth type, it represents Russell's assumption that non-predicative orders in his hierarchy are always obtained from predicative ones by generalization (i.e., by quantification). Of course Russell's assumption is not true of terms $z(k_1, \ldots, k_n)$ where one of the k_i's happens to be of non-predicative type. This means that both our system and Russell's intended system are not able to type such terms.

We conclude that we have described the legal pfs of the *Principia Mathematica* with the formal system RTT.

We now give some refinements of Theorem 2.84 that will be useful in later chapters:

Theorem 2.85 *Assume* $\Gamma \vdash f : t^a$.

- *If* $f \equiv R(i_1, \ldots, i_{\mathfrak{a}(R)})$ *and* $x \in \mathrm{FV}(f)$ *then* $x{:}0^0 \in \Gamma$;
- *If* $f \equiv z(k_1, \ldots, k_m)$ *then there are* $u_1^{b_1}, \ldots, u_m^{b_m}, b$ *such that*
  - $z{:}\left(u_1^{b_1}, \ldots, u_m^{b_m}\right)^b \in \Gamma$;
  - $\Gamma \vdash k_i{:}u_i^{b_i}$ *for* $k_i \in \mathcal{A} \cup \mathcal{P}$;
  - $k_i{:}u_i^{b_i} \in \Gamma$ *for* $k_i \in \mathcal{V}$.

PROOF:

- By Theorem 2.65, $T(\Gamma) \vdash_{\lambda_{\rightarrow}\text{Church}} Ri_1 \cdots i_{\mathfrak{a}(R)} : o$. This means that $x{:}\iota \in T(\Gamma)$. Therefore, $x{:}0^0 \in \Gamma$;
- Let $u_1^{b_1}, \ldots, u_m^{b_m}, b$ be as in the proof of Theorem 2.84. We only need to check that $k_i{:}u_i^{b_i} \in \Gamma$ for $k_i \in \mathcal{V}$. We already know that $k_i{:}T(u_i^{b_i}) \in T(\Gamma)$, and as $u_i^{b_i}$ is predicative, and the type of the variable k_i in Γ must be predicative as well, we have $k_i{:}u_i^{b_i} \in \Gamma$. ⊠

Conclusions

In this chapter we gave a formalisation of the Ramified Theory of Types. Some of the main ideas underlying this theory were already present in Frege's Abstraction Principles 1.1 and 1.2.

RTT not only prevents the paradoxes of Frege's *Grundgesetze der Arithmetik*, but also guarantees the well-definedness of substitution, as we have shown in Corollary 2.73. This second problem was not realized in the *Principia*, where substitution did not even have a proper definition.

	in RTT	in λ-calculus
Strengthening Lemma	2.56.2	4.35
Free Variable Lemma	2.56.1	4.26
Unicity of Types	2.61	4.36
Subterm Lemma	2.76	4.33
Strong Normalisation	2.73	4.40

Figure 2.2: Comparison of the properties of RTT and modern typed λ-calculus

There is a close relation between substitution in *Principia* and β-reduction in the λ-calculus (Definition 2.23). RTT has characteristics that are also the basic properties of modern type systems for the λ-calculus. See Figure 2.2. As there is no real reduction in RTT, we don't have an equivalent of the Subject Reduction property. However, the fact that the Free Variable property (Theorem 2.58) is maintained under substitution can be seen as a (very weak) form of Subject Reduction Lemma 4.34.

Expressing Russell's propositional functions in the λ-calculus has made it possible to compare these pfs with λ-terms. We found that pfs can be seen as λ-terms, but in a rather simple way:

- A pf is always a λI-term, i.e. if $\lambda x{:}A.B$ is a subterm of the translation $\widetilde{f}$ of a pf f, then $x \in \mathrm{FV}(B)$;
- The translation of a pf always results in a fully applied λ-term in β-normal form;
- Substitution in the *Principia* can be seen as application plus β-reduction to normal form.

Although the description of the Ramified Theory of Types in the *Principia* is very informal, it is remarkable that an accurate formalisation of this system can be made (see Theorem 2.84 and the discussion that follows it). The formalisation shows that Russell and Whitehead's ideas on the notion of types, though very informal to modern standards, must have been very thorough and to the point.[29]

Apart from the orders, RTT is a subsystem of $\lambda_{\to\text{Church}}$, [29] via the embeddings $\bar{\ }$ of Section 2a2 and T of Section 2c2. There are, however, important differences between the way in which the type of a pf is determined in RTT, and the way in which the type of a λ-term is determined in $\lambda_{\to\text{Church}}$. The rules of RTT, and the method of deriving the types of pfs that was presented in Section 2d, have

[29] Another related approach for formalising RTT is due to Holmes [74]. Holmes starts from [85] and provides an interesting implementation and a system which are both closer to Russell's own notation than to the modern notation as is our approach in this book.

a bottom-up character: one can only introduce a variable of a certain type in a context Γ, if there is a pf that has that type in Γ. In $\lambda_{\rightarrow\text{Church}}$, one can introduce variables of any type without wondering whether such a type is inhabited or not.

Church's $\lambda_{\rightarrow\text{Church}}$ is more general than RTT in the sense that Church does not only describe (typable) *propositional* functions. In $\lambda_{\rightarrow\text{Church}}$, also functions of type $\tau \rightarrow \iota$ (where ι is the type of individuals) can be described, and functions that take such functions as arguments, etc..

A characteristic of RTT that is maintained in many modern type systems is the syntactic nature of the system: type and order of a pf are determined on purely syntactical grounds. No attention is paid to the interpretation of such a pf. This is remarkable, as the propositions $\forall \mathtt{x}{:}0^0[\mathtt{R}(\mathtt{x})]$ and $\forall \mathtt{x}{:}0^0[\mathtt{R}(\mathtt{x})] \vee \forall \mathtt{z}{:}()^9[\mathtt{z}() \wedge \neg\mathtt{z}()]$ are logically equivalent in most logics[30], though they are of different types (the former pf has type $()^1$ and the latter has type $()^{10}$). In Section 3c we show that other viewpoints are possible besides this concentration on syntax.

[30] At least in all the logical systems that Russell had in mind when he wrote the *Principia*

Chapter 3

Deramification

In this chapter we discuss the development of type theory in the period between the appearance of *Principia Mathematica* (1910-1912) and Church's formulation of the Simple Theory of Types [29] in 1940.

In Section 3a we show that RTT was not a very easy system to work with. Ramsey [124], and Hilbert and Ackermann [70], simplified the system by removing the orders. The result is known as the *Simple Theory of Types* (STT).

Nowadays, STT is known via Church's formalisation in λ-calculus. However, STT already existed (1926) before λ-calculus did (1932), and is therefore not inextricably bound up with λ-calculus. In Section 3b we show how we can obtain a formalisation of STT directly from the formalisation of RTT that was presented in Chapter 2 by simply removing the orders. Most of the properties that were proved for RTT hold for STT as well, including Unicity of Types and Strong Normalisation. The proofs are all similar to the proofs that were given for RTT. We also make a comparison between Church's formalisation in λ-calculus and the formalisation of STT that is obtained from RTT. It appears that Church's system is much more than only a formalisation. Because of the λ-calculus it is more expressive.

The removal of orders from type theory may suggest that orders are to blame for the restrictiveness of RTT, and that the concept of order is problematic. In Section 3c we show that this is not necessarily the case. We introduce a system KTT, based on Kripke's Hierarchy of Truths [96], that has an approach completely opposite to STT. Whilst STT is *order*-free, and *types* play the main role, Kripke's Hierarchy of Truths is *type*-free, and *orders* play an important, though not a restrictive, role. The main difference between Kripke's and Russell's notion of order is that Russell's classification is purely syntactical, whilst Kripke's is essentially semantical. We show that RTT can be embedded in KTT (Section 3c2), and that there is a straightforward relation between the orders in RTT and the hierarchy of truths of KTT.

3a History of the deramification

3a1 The problematic character of RTT

The main part of the *Principia* is devoted to the development of logic and mathematics using the legal pfs of the ramified type theory. It appears that RTT is not easy to use. The main reason for this is the implementation of the so-called *ramification*: the division of simple types into orders. We illustrate this with two examples:

Example 3.1 (Equality) One tends to define the notion of *equality* in the style of Leibniz ([58]):

$$\mathtt{x} =_{\mathrm{L}} \mathtt{y} \stackrel{\mathrm{def}}{\leftrightarrow} \forall \mathtt{z}[\mathtt{z}(\mathtt{x}) \leftrightarrow \mathtt{z}(\mathtt{y})],$$

or in words: Two individuals are equal if and only if they have exactly the same properties.

Unfortunately, in order to express this general notion in our formal system, we have to incorporate *all* pfs $\forall \mathtt{z} : (0^0)^n[\mathtt{z}(\mathtt{x}) \leftrightarrow \mathtt{z}(\mathtt{y})]$ for $n > 1$, and this cannot be expressed in one pf.

The ramification does not only influence definitions in logic. Some important mathematical concepts cannot be defined any more. Here is an example from analysis:

Example 3.2 (Real numbers and least upper bounds) Dedekind constructed the real numbers from the rationals using the so-called Dedekind cuts. In this construction, a real number is a set r of rationals such that

- $r \neq \varnothing$;
- $r \neq \mathbb{Q}$;
- If $x \in r$ and $y < x$ then $y \in r$;
- If $x \in r$ then there is $y \in r$ with $x < y$.

For instance, the real number $\frac{1}{2}$ is represented by the set $\{x \in \mathbb{Q} \mid 2x < 1\}$, and the real number $\sqrt{2}$ is represented by the set $\{x \in \mathbb{Q} \mid x < 0 \text{ or } x^2 < 2\}$.

If we take $\mathbb{Q}$ as the set of individuals $\mathcal{A}$, and assume that the binary relation $<$ on $\mathbb{Q}$ is an element of $\mathcal{R}$, the set of relations, we can see real numbers as unary predicates f over $\mathbb{Q}$ such that

$$\begin{gathered}\exists \mathtt{x}{:}0^0[\mathtt{z}(\mathtt{x})] \wedge \exists \mathtt{x}{:}0^0[\neg \mathtt{z}(\mathtt{x})] \wedge \\ \forall \mathtt{x}{:}0^0[\forall \mathtt{y}{:}0^0[\mathtt{z}(\mathtt{x}) \rightarrow \mathtt{y} < \mathtt{x} \rightarrow \mathtt{z}(\mathtt{y})]] \wedge \\ \forall \mathtt{x}{:}0^0[\mathtt{z}(\mathtt{x}) \rightarrow \exists \mathtt{y}{:}0^0[\mathtt{z}(\mathtt{y}) \wedge \mathtt{x} < \mathtt{y}]]\end{gathered} \tag{3.1}$$

holds if we substitute f for z. We will abbreviate the predicate (3.1) (with the free variable z) as $\mathbb{R}$. It has type $\left((0^0)^1\right)^2$, and real numbers can be seen as pfs of type $(0^0)^1$. We will, for shortness of notation, write $\mathbb{R}(f)$ for $\mathbb{R}[\mathrm{z}{:=}f]$, so $\mathbb{R} \equiv \mathbb{R}(\mathrm{z})$. A real number r is smaller than or equal to another real number r' if for all x with $r(x)$, also $r'(x)$ holds. We write, shorthand, $r \leq r'$ if r is smaller than or equal to r'.

In traditional mathematics, the above would define a system that obeys the traditional axioms for real numbers. In particular, the theorem of the least upper bound holds for this system. This theorem states that each non-empty subset of $\mathbb{R}$ with an upper bound has a least upper bound. In our formalism:

$$\forall \mathrm{v}{\subseteq}\mathbb{R}\left[\begin{array}{c}\left(\begin{array}{c}\exists \mathrm{z}_1{\in}\mathbb{R}[\mathrm{v}(\mathrm{z}_1)] \wedge \\ \exists \mathrm{z}_2{\in}\mathbb{R}\forall \mathrm{z}_3{\in}\mathbb{R}[\mathrm{v}(\mathrm{z}_3) \to \mathrm{z}_3 \leq \mathrm{z}_2]\end{array}\right) \\ \to \\ \exists z_1 \in \mathbb{R}\left[\begin{array}{c}\forall \mathrm{z}_2{\in}\mathbb{R}[\mathrm{v}(\mathrm{z}_2) \to \mathrm{z}_2 \leq \mathrm{z}_1] \wedge \\ \forall \mathrm{z}_3{\in}\mathbb{R}\left[\begin{array}{c}\forall \mathrm{z}_4{\in}\mathbb{R}[\mathrm{v}(\mathrm{z}_4) \to \mathrm{z}_4 \leq \mathrm{z}_3] \\ \to \mathrm{z}_1 \leq \mathrm{z}_3\end{array}\right]\end{array}\right]\end{array}\right]$$

(We write, shorthand, $\forall v{\subseteq}\mathbb{R}[g]$ to denote

$$\forall v{:}\left((0^0)^1\right)^2[\forall \mathrm{u}{:}(0^0)^1[v(\mathrm{u}) \to \mathbb{R}(\mathrm{u})] \to g],$$

and $\forall z{\in}\mathbb{R}[g]$ to denote $\forall z{:}(0^0)^1[\mathbb{R}(z) \to g]$). If we try to prove this theorem within the system of Dedekind as formulated in the *Principia*-language RTT, we have to specify a type t^a for the variable z_1. As z_1 must be a real number, its type must be $(0^0)^1$. If we give a proof of the theorem, and construct some object f that should be the least upper bound of a set of real numbers V, f will depend on V. Therefore, a general description of f will have a variable v for V in it. As v is of order 2, f must be of order 3 or more. Therefore, f cannot be a real number, since real numbers have order 1. This makes it impossible to give a constructive proof of the theorem of the least upper bound within a ramified type theory.

This is a consequence of the fact that it is not possible in RTT to give a definition of an object that refers to the class to which this object belongs (because of the Vicious Circle Principle). Such a definition is called an *impredicative* definition. The relation with the notion of impredicative *type* is immediate:[1] an object defined

[1] This terminology is again the one assumed by *Principia* and not everyone agrees with it. There is actually no problem with the formulation of "impredicative" types from a predicative standpoint: objects of these types are predicatively respectable. An object of truly impredicative type would be defined using quantifiers over *its own* type or even higher types (as is allowed in simple type theory), and would not be typable in the ramified theory at all.

by an impredicative definition is of a higher order than the order of the elements of the class to which this object should belong. This means that the defined object has an *impredicative* type.

Nowadays we would consider the use of the Vicious Circle Principle too strict. The impredicative definition of f is a matter of *syntax*, whilst the existence of the object f has to do with *semantics*.[2] The fact that we are not able to give a predicate definition of f does not imply that such an object does not exist. Here we must remark that Russell and Whitehead did not make a distinction between syntax and semantics in the *Principia*.[3] Therefore they had to interpret the Vicious Circle Principle in the strict way above.

3a2 The Axiom of Reducibility

Russell and Whitehead tried to solve these problems with the so-called *axiom of reducibility*.

Axiom 3.3 (Axiom of Reducibility) *For each formula f, there is a formula g with a* predicative *type such that f and g are (logically) equivalent.*

Accepting this axiom, one may define equality on formulas of order 1 only:

$$\mathtt{x} =_1 \mathtt{y} \stackrel{\text{def}}{=} \forall \mathtt{z} : (0^0)^1[\mathtt{z}(\mathtt{x}) \leftrightarrow \mathtt{z}(\mathtt{y})].$$

If f is a function of type $(0^0)^n$ for some $n > 1$, and $\mathtt{a}$ and $\mathtt{b}$ are individuals for which the Leibniz equality $\mathtt{a} =_L \mathtt{b}$ holds then $f(\mathtt{a}) \leftrightarrow f(\mathtt{b})$ holds: With the Axiom of Reducibility we can determine a predicative function g (so of type $(0^0)^1$), equivalent to f. As g has order 1, $g(\mathtt{a}) \leftrightarrow g(\mathtt{b})$ holds. And because f and g are equivalent, also $f(\mathtt{a}) \leftrightarrow f(\mathtt{b})$ holds. This solves the problem of Example 3.1. A similar solution gives, in Example 3.2, the proof of the theorem of the least upper bound.

The validity of the Axiom of Reducibility has been questioned from the moment it was introduced. In the introduction to the 2nd edition of the *Principia*, Whitehead and Russell admit:

[2]This is obviously a point on which one might disagree. For example, Randall Holmes is unconvinced by the remarks about "syntax" and "semantics". He believes that the syntactical criteria of ramified type theory are a correct implementation of the Vicious Circle Principle and that the Vicious Circle Principle is best understood as a criterion appropriate for *definitions*. According to Randall Holmes, if instances of abstraction or comprehension principles are to be thought of as definitions, then impredicative abstraction or comprehension is indeed questionable and hence the conclusion to be drawn is that abstraction or comprehension axioms are not definitions, but assertions of matters of fact (so "semantic" rather than "syntactic"), and so are not subject to the Vicious Circle Principle (it is not that it should be applied in a more lenient way, but that it does not apply at all). But as long as the Vicious Circle Principle is to be applied, syntactical criteria are appropriate: what a correct definition is should be a matter of syntax.

[3]Though the basic ideas for this were already present in the works of Frege. See for instance *Über Sinn und Bedeutung* [53].

"This axiom has a purely pragmatic justification: it leads to the desired results, and to no others. But clearly it is not the sort of axiom with which we can rest content."

(*Principia Mathematica*, p. xiv)

Moreover, Weyl states that

"if the properties are constructed there is no room for an axiom here; it is a question which ought to be decided on ground of the construction"

(Mathematics and Logic: A brief survey serving as preface to a review of "The Philosophy of Bertrand Russell", p. 5)

and that

"with his axiom of reducibility Russell therefore abandoned the road of logical analysis and turned from the constructive to the existential-axiomatic standpoint."

(*Ibid.*, p. 6)

With the more modern developments of logic in our mind, we could add the following objection, associated to Weyl's argument above, against the Axiom. The Axiom of Reducibility states that for each f, there is a predicative g that is logically equivalent to f. The function g is something at *object* level, but the statement "f is logically equivalent to g" is a statement at a higher level than the object level. Pfs exist (at least, in the syntactic construction) independently from the existence of the notion of logical equivalence.

Moreover, there is more than one notion of logical equivalence, corresponding to the various kinds of logic that have been developed, or could have been developed. It would be remarkable if one Axiom of Reducibility would provide predicative pfs f for any kind of logic that is available, or can be thought of, and indeed, this is not true as shown by the following trivial example:

Example 3.4 We consider the so-called "bureaucratic logic". This logic has a set of axioms A, and no derivation rules at all. In short, a proposition g is true if and only if $g \in A$. Take, for the sake of the argument,

$$A = \{g \in \mathcal{P} \mid \vdash g{:}()^0\},$$

so A is the set of all predicative propositions. In this system, a proposition g is true if and only if it is predicative. If f is an impredicative proposition, then so is $f \leftrightarrow g$, for any proposition g. Therefore, $f \leftrightarrow g$ is false for any proposition g, in particular for any predicative proposition g. So the Axiom of Reducibility does not hold in bureaucratic logic.

Though Weyl [144] made an effort to develop analysis within the Ramified Theory of Types (but without the Axiom of Reducibility), and various parts of mathematics can be developed within RTT and without the Axiom[4], the general attitude towards RTT (without the axiom) was that the system was too restrictive, and that a better solution had to be found.

3a3 Deramification

The first impulse to such a solution was given by Ramsey in 1926 [124]. He recalls that the Vicious Circle Principle 2.1 was postulated in order to prevent the paradoxes. Though all the paradoxes were prevented by this Principle, Ramsey considers it essential to divide them into two parts:

1. One group of paradoxes is removed

   > "by pointing out that a propositional function cannot significantly take itself as argument, and by dividing functions and classes into a hierarchy of types according to their possible arguments."
   >
   > (The Foundations of Mathematics, p. 356)

   This means that a class can never be a member of itself. The paradoxes solved by introducing the hierarchy of types (but not orders), like the Russell paradox, and the Burali-Forti paradox, are called *logical* or *syntactical* paradoxes;

2. The second group of paradoxes is excluded by the hierarchy of orders. These paradoxes (like the Liar's paradox, and the Richard paradox) are based on the confusion of language and meta-language. These paradoxes are, therefore, not of a purely mathematical or logical nature. When a proper distinction between object language (the pfs of the system RTT, for example) and meta-language is made, these so-called *semantical* paradoxes disappear immediately.

Ramsey agrees with the part of the theory that eliminates the syntactic paradoxes. This part is in fact RTT without the orders of the types. The second part, the hierarchy of orders, does not gain Ramsey's support: if a proper distinction between object-language and meta-language is made, the semantic paradoxes disappear. Moreover, by accepting the hierarchy in its full extent one either has to accept the Axiom of Reducibility or reject ordinary real analysis. Ramsey is supported in his view by Hilbert and Ackermann [70]. They all suggest a *deramification* of the

[4]See [76], where many algebraic notions are developed within the Nuprl Proof Development System, a proof checker based on the hierarchy of types and orders of RTT without the Axiom of Reducibility.

theory, i.e. leaving out the orders of the types. When making a proper distinction between language and meta-language, the deramification will not lead to a re-introduction of the (semantic) paradoxes.

The solution proposed by Ramsey, and Hilbert and Ackermann, looks better than the Axiom of Reducibility. Nevertheless, both deramification and the Axiom of Reducibility are violations of the Vicious Circle Principle, and reasons (of a more fundamental character than "they do not lead to a re-introduction of the semantic paradoxes" and "it leads to the desired results, and to no others") why these violations can be harmlessly made must be given. Gödel [62] fills in this gap. He points out that whether one accepts this second principle or not, depends on the philosophical point of view that one has with respect to logical and mathematical objects:

> "it seems that the vicious circle principle [...] applies only if the entities involved are constructed by ourselves. In this case there must clearly exist a definition (namely the description of the construction) which does not refer to a totality to which the object defined belongs, because the construction of a thing can certainly not be based on a totality of things to which the thing to be constructed itself belongs. If, however, it is a question of objects that exist independently of our constructions, there is nothing in the least absurd in the existence of totalities containing members, which can be described only by reference to this totality."

(Russell's mathematical logic)

The remark puts the Vicious Circle Principle back from a proposition (a statement that is either true or false, without any doubt) to a philosophical principle that will be easily accepted by, for instance, intuitionists (for whom mathematics is a pure mental construction) or constructivists, but that will be rejected, at least in its full strength, by mathematicians with a more platonic point of view.

Gödel is supported in his ideas by Quine [123], sections 34 and 35. Quine's criticism on impredicative definitions (for instance, the definition of the least upper bound of a nonempty subset of the real numbers with an upper bound) is not on the definition of a special symbol, but rather on the very assumption of the *existence* of such an object at all. Quine continues by stating that even for Poincaré, who was an opponent of impredicative definitions and deramification, one of the doctrines of classes is that they are there "from the beginning". So, even for Poincaré there should be no evident fallacy in impredicative definitions.

The deramification has played an important role in the development of type theory. In 1932 and 1933, Church presented his (untyped) λ-calculus [27, 28].

In 1940 he combined this theory with a deramified version of Russell's theory of types to the system that is known as the *simply typed λ-calculus*[5].

3b The Simple Theory of Types STT

3b1 Constructing the Simple Theory of Types from RTT

So far, we have seen the development of type theory since the appearance of *Principia Mathematica* (1910-1912) went through a process of deramification where Ramsey [124], and Hilbert and Ackermann [70], simplified the Ramified Theory of Types by removing the orders. The result is known as the *Simple Theory of Types* (STT).

Nowadays, STT is known via Church's formalisation in λ-calculus. However, STT already existed (1926) before λ-calculus did (1932), and is therefore not inextricably bound up with λ-calculus. In this section we show how we can obtain a formalisation of STT directly from the formalisation of RTT that was presented in Section 2 by simply removing the orders. Most of the properties that were proved for RTT hold for STT as well, including Unicity of Types and Strong Normalisation. The proofs are all similar to the proofs that were given for RTT. We also make a comparison between Church's formalisation in λ-calculus and the formalisation of STT that is obtained from RTT. It appears that Church's system is much more than only a formalisation. Because of the λ-calculus, Church's system is more expressive.[6]

It is straightforward to carry out the deramification as it was originally proposed by Ramsey, Hilbert and Ackermann: We take the formalisation of RTT that was presented in Chapter 2, and leave out all the orders and the references to orders (including the notions of predicative and impredicative types). The system we obtain in this way will be denoted STT. The types used in the system are the simple types of Definition 2.30.

---

[5]Thus, the adjective *simple* is used to distinguish the theory from the more complicated — both in its construction with a double hierarchy and in its use — *ramified* theory. The classification "simple", therefore, has nothing to do with the fact that STT, formulated with λ-calculus as described in [29], is the simplest system of the Barendregt cube (see Section 4c).

[6]The removal of orders from type theory may suggest that orders are to blame for the restrictiveness of RTT, and that the concept of order is problematic. In Section 3c we will show that this is not necessarily the case by introducing a system KTT, based on Kripke's Hierarchy of Truths [96], that has an approach completely opposite to STT. Whilst STT is *order*-free, and *types* play the main role, Kripke's Hierarchy of Truths is *type*-free, and *orders* play an important, though not a restrictive, role. The main difference between Kripke's and Russell's notion of order is that Russell's classification is purely syntactical, whilst Kripke's is essentially semantical. In Section 3c2 we will show that RTT can be embedded in KTT and that there is a straightforward relation between the orders in RTT and the hierarchy of truths of KTT.

The following definitions, lemmas, theorems and corollaries, including their proofs, can be adapted to STT without any problems: Definition 2.42, Definition 2.43, Definition 2.44, Definition 2.45, Lemma 2.56, Theorem 2.57 (first free variable theorem), Theorem 2.58 (second free variable theorem), Corollary 2.59, Corollary 2.61 (unicity of types), Definition 2.64, Theorem 2.65, Theorem 2.67, Lemma 2.71, Lemma 2.72, Theorem 2.73 (existence of substitution), Definition 2.74, Lemma 2.75 and Corollary 2.76 (subterm lemma).

The description of legal pfs for STT follows the same line as in Section 2d, with straightforward adaptions of Definition 2.77, Lemma 2.78 (now, all simple types are inhabited), Lemma 2.81, Lemma 2.82, Lemma 2.83, and finally Theorem 2.84 (characterisation of legal pfs):

Theorem 3.5 *Let $f \in \mathcal{P}$. f is legal (mod α) if and only if:*

- $f \equiv R(i_1, \ldots, i_{\mathfrak{a}(R)})$, *or*
- $f \equiv z(k_1, \ldots, k_n)$, $z \not\equiv k_j$ *for all* $k_j \in \mathcal{V}$ *and* z *does not occur in any* $k_j \in \mathcal{P}$, *and there is* Γ *with* $\text{FV}(f) \subseteq \text{DOM}(\Gamma)$ *and for all* $k_j \in \mathcal{P}$, $\Gamma \vdash k_j{:}t_j$, *or*
- $f \equiv \neg f'$ *and* f' *is legal (mod* α*) or* $f \equiv f_1 \vee f_2$, *there are* Γ_i *and* t_i *such that* $\Gamma_i \vdash f_i{:}t_i$ *(mod* α*) and* $\Gamma_1 \cup \Gamma_2$ *is a context, or*
- $f \equiv \forall x{:}t.f'$ *and* f' *is legal.* ⊠

A comparison between the formalisations of STT and RTT can easily be made using Theorems 3.5 and 2.84. We find that

- All RTT-legal pfs are (when the ramified types behind the quantifiers are replaced by their corresponding simple types) STT-legal;
- A STT-legal pf f is RTT-legal, except when f contains a subformula of the form $z(k_1, \ldots, k_n)$, where one or more of the k_js are not RTT-legal or can only be typed in RTT by an impredicative type.

3b2 Church's simply typed λ-calculus $\lambda_{\to\text{Church}}$

We give a definition of the simply typed λ-calculus as introduced by Church [29] in 1940.

The types and terms in the original presentation of $\lambda_{\to\text{Church}}$ are a bit different from the presentation in [3]. We give some explanation after repeating the original definition:

Definition 3.6 (Types of $\lambda_{\to\text{Church}}$) The types of $\lambda_{\to\text{Church}}$ are defined as follows:

- ι and o are types;
- If α and β are types, then so is $\alpha \rightarrow \beta$.

We denote the set of simple types by $\mathcal{T}$.

ι represents the type of individuals; o is the type of propositions. $\alpha \rightarrow \beta$ is the type of functions with domain α and range β. We use $\alpha, \beta, \ldots$ as meta-variables over types. $\rightarrow$ associates to the right: $\alpha \rightarrow \beta \rightarrow \gamma$ denotes $\alpha \rightarrow (\beta \rightarrow \gamma)$.

Definition 3.7 (Terms of $\lambda_{\rightarrow\mathbf{Church}}$) The terms of $\lambda_{\rightarrow\mathrm{Church}}$ are the following:

- $\neg$, $\wedge$, $\forall_\alpha$ for each type α, and $\imath_\alpha$ for each type α, are terms;
- A variable is a term;
- If A, B are terms, then so is AB;
- If A is a term, and x a variable, then $\lambda x{:}\alpha.A$ is a term.

Definition 3.8 (Contexts of $\lambda_{\rightarrow\mathbf{Church}}$) A *context* in $\lambda_{\rightarrow\mathrm{Church}}$ is a set $\{x_1{:}\alpha_1, \ldots x_n{:}\alpha_n\}$ where the x_i are distinct variables and the α_i are types.

Some terms are typable (legal) in $\lambda_{\rightarrow\mathrm{Church}}$, according to the following derivation rules:

Definition 3.9 (Typing rules of $\lambda_{\rightarrow\mathbf{Church}}$) The judgement $\Gamma \vdash A : \alpha$ holds if it can be derived using the following rules:

- $\Gamma \vdash \neg : o \rightarrow o$;
  $\Gamma \vdash \wedge : o \rightarrow o \rightarrow o$;
  $\Gamma \vdash \forall_\alpha : (\alpha \rightarrow o) \rightarrow o$;
  $\Gamma \vdash \imath_\alpha : (\alpha \rightarrow o) \rightarrow \alpha$;
- $\Gamma \vdash x : \alpha$ if $x{:}\alpha \in \Gamma$;
- If $\Gamma, x{:}\alpha \vdash A : \beta$ then $\Gamma \vdash (\lambda x{:}\alpha.A) : \alpha \rightarrow \beta$;
- If $\Gamma \vdash A : \alpha \rightarrow \beta$ and $\Gamma \vdash B : \alpha$ then $\Gamma \vdash (AB) : \beta$.

We use $\vdash_{\lambda_{\rightarrow\mathrm{Church}}}$ if we need to distinguish derivability in $\lambda_{\rightarrow\mathrm{Church}}$ from derivability in other type systems.

The simply typed λ-calculus can be seen as a Pure Type System, and therefore has the properties of Pure Type Systems [3]. To adapt the simply typed λ-calculus to a Pure Type System, some amendments are made:

- The two basic types 0, o are replaced by an infinite set of *type variables*;
- The constants $\neg$, $\wedge$, $\forall_\alpha$ and ι_α are not introduced in the PTS-presentation.

These adaptions do not seriously affect the system and are only used to make $\lambda_{\to\text{Church}}$ fit in the PTS-framework.

3b3 Comparison of RTT with $\lambda_{\to\text{Church}}$

Apart from the orders, RTT is a subsystem of $\lambda_{\to\text{Church}}$ via the embeddings $\bar{\ }$ of Section 2a2 and the mapping T of Definition 2.39. There are, however, important differences between the way in which the type of a pf is determined in RTT, and the way in which the type of a λ-term is determined in λ-Church. The rules of RTT, and the method of deriving the types of pfs that was presented in Section 2d, have a bottom-up character: one can only introduce a variable of a certain type in a context Γ, if there is a pf that has that type in Γ. In $\lambda_{\to\text{Church}}$, one can introduce variables of any type without wondering whether such a type is inhabited or not.

Church's $\lambda_{\to\text{Church}}$ is more general than RTT in the sense that Church does not only describe (typable) *propositional* functions. In $\lambda_{\to\text{Church}}$, also functions of type $\tau \to \iota$ (where ι is the type of individuals) can be described, and functions that take such functions as arguments, etc..

We summarise the results concerning RTT and $\lambda_{\to\text{Church}}$:

- Simple types can be translated to $\lambda_{\to\text{Church}}$-types (see Definition 2.33). Moreover, the translation mapping T is injective both on simple types (Lemma 2.34) and on predicative types (Lemma 2.41).
- Ramified types can be translated to $\lambda_{\to\text{Church}}$-types (see Definition 2.39).
- Moreover, we have related typing in RTT to that of Church's $\lambda_{\to\text{Church}}$ (see Theorem 2.65).

3b4 Comparison of STT with $\lambda_{\to\text{Church}}$

Nowadays, the Simple Theory of Types is often identified with Church's formalisation of it in [29]. The definition of $\lambda_{\to\text{Church}}$ that was given there is repeated in Section 3b2.

We make the following remarks with respect to $\lambda_{\to\text{Church}}$ and the Simple Theory of Types.

Remark 3.10 We see that the constants $\neg$, $\wedge$, $\forall_\alpha$ and ι_α are terms. This may need some explanation for the modern reader.

- Church considers $\neg$ and $\wedge$ to be functions. The function $\neg$ takes a proposition as argument, and returns a proposition; similarly $\wedge$ takes two propositions as arguments, and returns a proposition. In Definition 3.9, we see that $\neg$ and $\wedge$ are assigned the corresponding types $o \to o$ and $o \to o \to o$;

- More remarkable: $\forall_\alpha$ and $\imath_\alpha$ are just terms, and do not act as binding operators. The usual variable binding of $\forall_\alpha$ and $\imath_\alpha$ is obtained via λ-abstraction: instead of $\forall x{:}\alpha[f]$, Church writes $\forall_\alpha(\lambda x{:}\alpha.f)$. In this way, $\forall_\alpha$ is a function that takes a propositional function of type $\alpha \to o$ as argument, and returns a proposition (a term of type o). In Definition 3.9, $\forall_\alpha$ obtains the corresponding type $(\alpha \to o) \to o$. Similarly, the choice operator $\imath_\alpha$ takes a propositional function of type $\alpha \to o$ as argument, and returns a term of type α. The term $\imath x{:}\alpha.f$, or in Church's notation: $\imath_\alpha(\lambda x{:}\alpha.f)$, has as interpretation: the (unique) object t of type α for which $f[x{:=}t]$ holds. Correspondingly, the type of $\imath_\alpha$ is $(\alpha \to o) \to \alpha$.

The mappings T for types and $\bar{\ }$ for terms (see Definition 2.39 and Definition 2.7), adapted for STT, make it possible to compare STT with $\lambda_{\to\text{Church}}$.

Regarding the types, we find that T gives an injective correspondence between types of STT and $\lambda_{\to\text{Church}}$. T is clearly not surjective, as $T(t)$ is never of the form $\alpha \to \iota$ (this follows directly from Definition 2.33). This indicates an important difference between STT and $\lambda_{\to\text{Church}}$. In RTT and STT, functions (other than propositional functions) have to be defined via relations (and this is the way it is done in *Principia Mathematica*). The value of such a function f, described via the relation R, for a certain value a is described using the $\imath$-operator: $\imath y.R(a,y)$ (to be interpreted as: the unique y for which $R(a,y)$ holds). Things get even more complicated if one realizes that the $\imath$-operator is not a part of the syntax used in *Principia Mathematica*, but an abbreviation with a not so straightforward translation (see [145], pp. 66–71). In $\lambda_{\to\text{Church}}$, as everywhere in λ-calculus, functions (both propositional functions and other ones) are first-class citizens, which means that the construction with the $\imath$-operator is not the first tool to be used when constructing a function. If one has an algorithm (a λ-term) that describes the function f, the value of f for the argument a can be easily described via the term fa. And even if such an algorithm is not at hand, one can use the $\imath$-operator, which is part of the syntax of $\lambda_{\to\text{Church}}$. This makes $\lambda_{\to\text{Church}}$ much easier to use for the formalisation of logic and mathematics than RTT and STT.

Regarding the terms, $\bar{\ }$ provides an injective correspondence between terms of STT and $\lambda_{\to\text{Church}}$. Again, this mapping is not surjective, for several reasons:

- T is not surjective. As there is no t with $T(t) = \iota \to \iota$, there cannot be a legal pf f such that $\overline{f} \equiv \lambda \mathbf{x}{:}\iota.\mathbf{x}$ (cf. Theorem 2.65.2 adapted for STT);

- We already observed that $\overline{f}$ is a λI-term for all $f \in \mathcal{P}$. $\lambda_{\to\text{Church}}$ also allows terms like $\lambda \mathbf{x}{:}\alpha.\mathbf{y}$;
- If $\overline{f} \equiv zH_1 \cdots H_n$ for some $z \in \mathcal{V}$ and some terms $H_1, \ldots, H_n$, the H_is must be either *closed* λ-terms, or variables, or individuals. This means that there is no $f \in \mathcal{P}$ such that $\overline{f} \equiv \lambda \mathbf{z}{:}o{\to}o.\lambda \mathbf{x}{:}\iota.\mathbf{z}(\mathbf{Rx})$, since $\mathbf{Rx}$ contains the free variable $\mathbf{x}$ and is neither a variable nor an individual;
- We remark that $\overline{f}$ is always a closed λ-term, so there is no $f \in \mathcal{P}$ such that $\overline{f} \equiv \mathbf{x}$;
- It has already been remarked that the ι-operator is part of the syntax of $\lambda_{\to\text{Church}}$, and this is not the case in STT and RTT.

The discussion above makes clear that $\lambda_{\to\text{Church}}$ is a far more expressive system than RTT and STT. Type-theoretically, it generalises the idea of function types of Frege and Russell from propositional functions to more general functions.

Philosophically, there is another important difference between STT and $\lambda_{\to\text{Church}}$. The systems STT and RTT have a strong bottom-up approach: To type a higher-order pf one has to start with propositions of order 0. Only by applying the abstraction principles, it is possible to obtain higher-order pfs. In $\lambda_{\to\text{Church}}$, one can introduce a variable of a higher-order type at once, without having to refer to terms of lower order.

3c Are orders to be blamed?

The historical success of the deramification makes it attractive to conclude that the ramification of Russell's theory is to be "blamed" for the restrictiveness of RTT: the orders were just an emergency measure, and by removing them from the theory, everything works fine. This reasoning, however, is a bit too fast. Orders still play a role in logic, and they provide a useful intuition to describe how complicated a certain proposition is (for example "first-order", or "second-order").

Moreover, we feel that there are reasons to criticise not the *concept* of order, but Russell's *definition* of order. Russell's classification of pfs in types and orders is purely syntactic. This is quite harmless as far as (simple) types are concerned: the number of arguments that a propositional function takes is a notion that can be reasonably described by syntactic methods.[7]

For orders, the syntactic classification is more questionable. Look for instance at the pfs $f \equiv \forall \mathbf{x}{:}0^0[\mathbf{R}(\mathbf{x})]$ and $g \equiv \forall \mathbf{x}{:}0^0[\mathbf{R}(\mathbf{x})] \vee \forall \mathbf{z}{:}()^9[\mathbf{z}() \vee \neg\mathbf{z}()]$. According

[7]The only criticism that one could have is that Russell's method excludes the so-called "constant" functions, i.e. functions of which the outcome is independent of one or more of the arguments it takes. We saw this in our translation of pfs to λ-terms: all the translations were λI-terms (see Lemma 2.22.3).

to the *Principia*, g is a proposition of order 10, because g contains a variable of order 9. On the other hand, f, a proposition of order 1, is logically equivalent to g. So, the *interpretation* of g does not essentially involve the variable $\mathbf{z}$ of order 9. We could therefore argue that the order of g, for semantic reasons, should not be higher than 1, the order of f.

In the forthcoming sections we show that there are workable systems that do have an order-like hierarchy and nevertheless are not restrictive. The system that we present however, does not make a syntactic classification of propositional functions into orders, but a semantic one.

The system is based on Kripke's paper [96]. In 1936, Tarski [141] proved that introducing a truth-predicate T in a first-order language leads to contradictions. Such a predicate T is true for objects that are encodings of true propositions, and false for objects that are encodings of false propositions. For this reason, Tarski distinguishes between object-language and meta-language, and the truth predicate for propositions of the object language occurs only at the meta-level. For a truth predicate for propositions in meta-language one needs a meta-meta-language, etcetera.

Kripke however, allows a restricted truth-predicate in a first-order language. The restrictions on this predicate are such that no contradictions occur. The construction of Kripke's truth predicate has remarkable similarities with the use of orders in the Ramified Theory of Types, and we show that RTT can be embedded within Kripke's Theory of Truths KTT. It is even possible to define a notion of order for pfs of RTT, based on the construction of Kripke's truth predicate. An important difference is that this new notion of order is (partially) based on the interpretation of a pf, whilst Russell's definition of order is purely syntactic.

In Section 3c1 we describe KTT. In Section 3c2 we embed RTT in KTT and show that Russell's syntactic approach is much more restrictive than Kripke's semantic approach.

3c1 Kripke's Theory of Truth KTT

In this section, we shortly describe Kripke's Theory of Truth: KTT (see [96]). Kripke expresses higher-order formulas within a first-order language, using the fact that many interesting languages are rich enough to express their own syntax via a Gödel Numbering.

In the rest of this Section 3c, L is a first-order language. $\mathcal{A}$ is the set of constant symbols in L, $\mathcal{F}$ is the set of function symbols in L, and $\mathcal{R}$ represents the set of predicate symbols of L. We assume that $\mathfrak{M} = \langle \boldsymbol{A}, [\![\ ]\!]\rangle$ is a model for L, where $[\![\ ]\!]$ is an interpretation function for the symbols of L.

Let us also assume two subsets S_1 and S_2 of $\boldsymbol{A}$ such that $S_1 \cap S_2 = \varnothing$. Kripke extends L by adding a monadic predicate T. The main idea is to interpret T as a unary "truth predicate" $\boldsymbol{T}$ such that S_1 contains the elements $\boldsymbol{a}$ of $\boldsymbol{A}$ that are (codes of) formulas which we consider to be "true", and S_2 contains those $\boldsymbol{a} \in \boldsymbol{A}$ that are (codes of) formulas which we consider to be "false". This extension of the model is denoted as $\mathfrak{M}(S_1, S_2)$. We do not demand that $S_1 \cup S_2 = \boldsymbol{A}$, hence $\boldsymbol{T}$ may be a *partially* defined predicate.

Definition 3.11 (Logical Truth for KTT**)** Assume s to be an assignment function $\mathcal{V} \to \boldsymbol{A}$. We define $\mathfrak{M} \models f[s]$ as follows[8]:

f	$\mathfrak{M} \models f[s]$	$\mathfrak{M} \models \neg f[s]$
$R(a_1, \ldots, a_m)$	$(a_1(s), \ldots, a_m(s)) \in [\![R]\!]$	$(a_1(s), \ldots, a_m(s)) \notin [\![R]\!]$
$g_1 \wedge g_2$	$\mathfrak{M} \models g_1[s]$ and $\mathfrak{M} \models g_2[s]$	$\mathfrak{M} \models ((\neg g_1) \vee (\neg g_2))[s]$
$g_1 \vee g_2$	$\mathfrak{M} \models g_1[s]$ or $\mathfrak{M} \models g_2[s]$	$\mathfrak{M} \models ((\neg g_1) \wedge (\neg g_2))[s]$
$\forall x[g]$	$\mathfrak{M} \models g[s[x{:=}\boldsymbol{a}]]$ for all $\boldsymbol{a} \in \boldsymbol{A}$	$\mathfrak{M} \models (\exists x[\neg g])[s]$
$\exists x[g]$	$\mathfrak{M} \models g[s[x{:=}\boldsymbol{a}]]$ for an $\boldsymbol{a} \in \boldsymbol{A}$	$\mathfrak{M} \models (\forall x[\neg g])[s]$
$\neg\neg g$	$\mathfrak{M} \models g[s]$	$\mathfrak{M} \models (\neg g)[s]$

Here, $R \in \mathcal{R}$ has arity m; $a, a_1, \ldots, a_m$ are terms of L, and g, g_1, g_2 are formulas of L. Moreover, $s[x{:=}\boldsymbol{a}]$ is the assignment function that assigns $\boldsymbol{a}$ to x, and $s(y)$ to any variable $y \in \mathcal{V} \setminus \{x\}$. Now let $S_1, S_2 \subseteq \boldsymbol{A}$ such that $S_1 \cap S_2 = \varnothing$. $L(\mathrm{T})$ is the extension of L with the monadic predicate T. We extend the definition of $\mathfrak{M} \models f$ to $\mathfrak{M}(S_1, S_2) \models f$ by putting

$$\mathfrak{M}(S_1, S_2) \models (\mathrm{T}(a))[s] \text{ iff } a(s) \in S_1$$

and

$$\mathfrak{M}(S_1, S_2) \models (\neg\mathrm{T}(a))[s] \text{ iff } a(s) \in S_2.$$

It is important (and easy) to notice that the extension of L and $\mathfrak{M}$ to $L(\mathrm{T})$ and $\mathfrak{M}(S_1, S_2)$ is conservative:

Lemma 3.12 *Assume f is a sentence in L, and $S_1, S_2 \subseteq \boldsymbol{A}$ such that $S_1 \cap S_2 = \varnothing$. Then $\mathfrak{M} \models f$ if and only if $\mathfrak{M}(S_1, S_2) \models f$.* ⊠

The predicate T cannot express truth in a direct way. This is because T is a predicate in a first order language, and therefore can only take terms (objects) as arguments, and not propositions. However, there is an indirect way in which T can express truth: enumerate all formulas of $L(\mathrm{T})$.

From now on, we assume that we have an injective, primitive recursive function $\langle\ \rangle$ from the formulas (including non-closed formulas) of the first order language

[8]Notice that even though this definition is different from Tarski's definition, especially with respect to the definition of $\mathfrak{M} \models \neg f$, it is easy to prove the equivalence of both definitions. This is because all primitive predicates of L are totally defined. We took this definition however, as we need to extend it for the partial predicate T.

$L(\mathtt{T})$ to the terms of L. If f is a sentence then we can form the proposition $\mathtt{T}(\langle f \rangle)$, which expresses the truth of f. But we can also form the proposition $\mathtt{T}(\langle \mathtt{T}(\langle f \rangle)\rangle)$, so we can discuss the truth of the truth of f, etcetera. This makes it possible to express higher-order propositions.

As announced, we use the sets S_1 and S_2 to establish the truth of such higher-order propositions. Actually, we build a hierarchy of sets $S_{\alpha,1}$ and $S_{\alpha,2}$ for ordinals α. We will see that this hierarchy has much in common with Russell's hierarchy of orders.

Definition 3.13 For any ordinal α we define a pair of sets $(S_{\alpha,1}, S_{\alpha,2})$ and a model $\mathfrak{M}_\alpha$.

- $S_{0,1} \stackrel{\text{def}}{=} \varnothing$; $S_{0,2} \stackrel{\text{def}}{=} \varnothing$; $\mathfrak{M}_0 \stackrel{\text{def}}{=} \mathfrak{M}(S_{0,1}, S_{0,2})$;
- If $S_{\alpha,1}$, $S_{\alpha,2}$ and $\mathfrak{M}_\alpha$ have been defined, then we define:

$$\begin{aligned} S_{\alpha+1,1} &\stackrel{\text{def}}{=} \{[\![\langle f \rangle]\!] \mid f \text{ is a sentence and } \mathfrak{M}_\alpha \models f\} \\ S_{\alpha+1,2} &\stackrel{\text{def}}{=} \{[\![\langle f \rangle]\!] \mid f \text{ is a sentence and } \mathfrak{M}_\alpha \models \neg f\} \\ \mathfrak{M}_{\alpha+1} &\stackrel{\text{def}}{=} \mathfrak{M}(S_{\alpha+1,1}, S_{\alpha+1,2}); \end{aligned}$$

- If α is a limit ordinal and $S_{\beta,1}$, $S_{\beta,2}$ and $\mathfrak{M}_\beta$ have been defined for all $\beta < \alpha$, then

$$\begin{aligned} S_{\alpha,i} &\stackrel{\text{def}}{=} \bigcup_{\beta<\alpha} S_{\beta,i}; \\ \mathfrak{M}_\alpha &\stackrel{\text{def}}{=} \mathfrak{M}(S_{\alpha,1}, S_{\alpha,2}). \end{aligned}$$

The proof of the following lemma is not difficult. Yet, the lemma plays a crucial role in the rest of this Section.

Lemma 3.14 (Conservation of Knowledge (1))
If $\alpha < \beta$ then $S_{\alpha,1} \subseteq S_{\beta,1}$ and $S_{\alpha,2} \subseteq S_{\beta,2}$.

PROOF: Induction on β. The only non-trivial case is $\beta = \beta' + 1$. By the induction hypothesis, $S_{\alpha,i} \subseteq S_{\beta',i}$ for all $\alpha < \beta'$, so it suffices to prove that $S_{\beta',i} \subseteq S_{\beta,i}$.

We only give the proof for the case $S_{\beta,1}$; the proof for $S_{\beta,2}$ is similar.

So assume $\boldsymbol{a} \in S_{\beta',1}$. Determine $\gamma < \beta'$ and a sentence f such that $[\![\langle f \rangle]\!] = \boldsymbol{a}$ and $\mathfrak{M}_\gamma \models f$. Use induction on the definition of $\mathfrak{M}_\gamma \models f$. We treat only one case, the others are trivial: Assume $f \equiv \mathtt{T}(a')$ for some term a' of $L(\mathtt{T})$. Then $[\![a']\!] \in S_{\gamma,1}$ by definition of $\mathfrak{M}_\gamma \models f$. By the induction hypothesis on α we know: $S_{\gamma,1} \subseteq S_{\beta',1}$, so: $[\![a']\!] \in S_{\beta',1}$. By definition, this means $\mathfrak{M}_{\beta'} \models f$. Hence $[\![\langle f \rangle]\!] \in S_{\beta'+1,1}$, and this means $\boldsymbol{a} \in S_{\beta,1}$. ⊠

Corollary 3.15 (Conservation of Knowledge (2))
If $\alpha < \beta$ *and* $\mathfrak{M}_\alpha \models f$ *then* $\mathfrak{M}_\beta \models f$. ⊠

Remark 3.16 It is not the case that $\mathfrak{M}_\alpha \not\models f$ implies $\mathfrak{M}_\beta \not\models f$ for $\alpha < \beta$. For instance, let $f = \forall \mathtt{x}[\mathtt{R}(\mathtt{x}) \vee \neg\mathtt{R}(\mathtt{x})]$. Notice that $\mathfrak{M}_0 \models f$. Therefore, $\langle f\rangle \in S_{1,1}$, so $\mathfrak{M}_1 \models \mathtt{T}(\langle f\rangle)$. But $S_{0,1} = \varnothing$, so $\mathfrak{M}_0 \not\models \mathtt{T}(\langle f\rangle)$.

We prove that the theories with the new predicate T are all consistent:

Lemma 3.17 *Let* α *be an ordinal.*

1. *For all formulas* f *of* $L(\mathtt{T})$ *and for all assignments* s,

$$\mathfrak{M}_\alpha \not\models (f \wedge \neg f)[s];$$

2. $S_{\alpha,1} \cap S_{\alpha,2} = \varnothing$.

PROOF: We use induction on α. So assume the lemma has been proven for all $\beta < \alpha$ (IH1).

1. Use induction on the structure of f. So assume the lemma has been proven for all subformulas g of f (IH2). We treat three cases only (the other cases are similar).
   - $f = R(a_1, \dots, a_m)$. If $\mathfrak{M}_\alpha \models (f \wedge \neg f)[s]$ then $(a_1(s), \dots, a_m(s)) \in [\![R]\!]$ and $(a_1(s), \dots, a_m(s)) \notin [\![R]\!]$, which is impossible;
   - $f = g_1 \wedge g_2$. If $\mathfrak{M}_\alpha \models (f \wedge \neg f)[s]$ then $\mathfrak{M}_\alpha \models (g_1 \wedge g_2)[s]$, and so $\mathfrak{M}_\alpha \models g_j[s]$ for $j = 1, 2$. Moreover, $\mathfrak{M}_\alpha \models (\neg(g_1 \wedge g_s))[s]$, so $\mathfrak{M}_\alpha \models ((\neg g_1) \vee (\neg g_2))[s]$, and hence $\mathfrak{M}_\alpha \models (\neg g_j)[s]$ for $j = 1$ or $j = 2$. Therefore $\mathfrak{M}_\alpha \models (g_j \wedge \neg g_j)[s]$ for $j = 1$ or $j = 2$, which contradicts (IH2);
   - $f = \mathtt{T}(a)$ for a term a. If $\mathfrak{M}_\alpha \models (f \wedge \neg f)[s]$ then $a(s) \in S_{\alpha,1}$ and $a(s) \in S_{\alpha,2}$, which contradicts (IH1);
2. If $\boldsymbol{a} \in S_{\alpha_1} \cap S_{\alpha_2}$, then determine a formula g such that $[\![\langle g\rangle]\!] = \boldsymbol{a}$, and $\beta_1, \beta_2 < \alpha$ such that $\mathfrak{M}_{\beta_1} \models g$ and $\mathfrak{M}_{\beta_2} \models \neg g$. Let $\beta = \max(\beta_1, \beta_2)$. Then $\beta < \alpha$ and because of Conservation of Knowledge 3.15, $\mathfrak{M}_\beta \models g \wedge \neg g$. This contradicts (IH1). ⊠

We can see the construction of the models $\mathfrak{M}_\alpha$ as a process of obtaining knowledge. At the initial stage 0, $\mathtt{T}(\langle f\rangle)$ is not defined for any formula f. There is no knowledge at all.

By applying the definition of truth given for $\mathfrak{M}_0$, we obtain knowledge. Some sentences f can be judged true: $\mathfrak{M}_0 \models f$. We store the code of f in $S_{1,1}$. Some other sentences g can be judged false: $\mathfrak{M}_0 \models \neg g$. The code of g is stored in $S_{1,2}$. Note that it is not possible to judge all the sentences. For instance, neither $\mathfrak{M}_0 \models \forall \mathtt{x}[\mathtt{T}(\mathtt{x}) \vee \neg \mathtt{T}(\mathtt{x})]$ nor $\mathfrak{M}_0 \models \neg\forall \mathtt{x}[\mathtt{T}(\mathtt{x}) \vee \neg \mathtt{T}(\mathtt{x})]$ holds, so $[\![\langle \forall \mathtt{x}[\mathtt{T}(\mathtt{x}) \vee \neg \mathtt{T}(\mathtt{x})]\rangle]\!]$ neither belongs to $S_{1,1}$, nor to $S_{1,2}$.

The knowledge we have obtained is expressed by the behaviour of the predicate T in $\mathfrak{M}_1$. At stage 1 we know more about T than at stage 0 $S_{0,1} = S_{0,2} = \varnothing$, but $S_{1,1} \neq \varnothing$ and $S_{1,2} \neq \varnothing$. Hence more sentences can be judged true or false. We store the codes of sentences that were judged "true" at level 1 in $S_{2,1}$ and codes of the sentences that were judged "false" at level 1 in $S_{2,2}$. The Lemma on Conservation of Knowledge 3.14 guarantees that this process only extends our knowledge, i.e.:

- Sentences that were judged to be true at level 0 remain true at level 1;
- Sentences that were judged to be false at level 0 remain false at level 1.

By iterating this process we arrive at the levels $2, 3, \ldots, \omega, \omega + 1, \ldots$. One might expect that for each sentence f there is an ordinal α for which $f \in S_{\alpha,1} \cup S_{\alpha_2}$. This is not the case. There are sentences of which the truth will never be established. See the forthcoming example 3.34.

3c2 RTT in KTT

Both in RTT and in KTT we are confronted with a hierarchy. Russell constructs a hierarchy by dividing propositions and propositional functions into different orders, taking care that a propositional function f can only depend on objects of a lower order than the order of f.

Kripke does not make this distinction beforehand. He has only one truth-predicate (T), but decisions about the truth of propositions are split into different levels: At the first level only decisions about propositions that do not involve any knowledge about T are made (for example: the proposition $\mathtt{R}(\mathtt{a}_1)$, but also the proposition $(\mathtt{R}(\mathtt{a}_1) \vee \neg \mathtt{R}(\mathtt{a}_1)) \vee \mathtt{T}(\mathtt{a}_2))$. At the second level decisions about propositions involving T for codes of first-level propositions are made, and so on.

Before we can compare RTT with KTT, we must give a formal definition of logical truth for pfs of RTT. After that, we investigate the similarity between both hierarchies by describing RTT within KTT. Then, we investigate in which way RTT is more restrictive with respect to self-reference than KTT.

Logical truth for RTT in Tarski's style

As KTT uses Tarski's notion of logical truth, we use a similar notion for RTT. This definition of logical truth is quite informal. For example, the first clause "If $(a_1, \ldots, a_m) \in R$ then RTT $\models R(a_1, \ldots, a_m)$" requires the symbol R to be already fully interpreted and to denote a relation independently of any Tarskian assignment function. This is in line with Russell, who did not make distinction between the syntactic symbol R in "$\models R(a_1,\ldots,a_m)$" and the semantic use of R in "$(a_1,\ldots,a_m)\in R$". We take care that it will always be clear whether we use a symbol in its syntactic or in its semantic way.

We must remark that the definition of logical truth for RTT is not due to Russell and cannot be found in *Principia Mathematica*. In *Principia*, Russell and Whitehead use a notion of truth based on derivations in natural deduction style. But in order to make clear the similarities between RTT and KTT, we must use the notion of truth used in KTT in RTT as well.

Definition 3.18 (Logical Truth for RTT) Let Γ be an RTT-context with domain $\mathcal{V}$. Let $f\in\mathcal{P}$ be a legal pf in Γ with free variables $x_1,\ldots,x_n$ of types $t_1^{a_1},\ldots,t_n^{a_n}$. Let $s:\mathcal{V}\to\mathcal{P}$ be such that $s(x_i):t_i^{a_i}\in\Gamma$. We define RTT $\models_\Gamma f[s]$ by induction on the order of f. We give this definition by induction on the structure of f.

- RTT $\models_\Gamma R(i_1,\ldots,i_{\mathfrak{a}(R)})[s]$ if $(i_1(s),\ldots,i_{\mathfrak{a}(R)}(s))\in R$;
- RTT $\models_\Gamma (g_1\vee g_2)[s]$ if RTT $\models_\Gamma g_1[s]$ or RTT $\models_\Gamma g_2[s]$;
- RTT $\models_\Gamma (\neg g)[s]$ if RTT $\not\models_\Gamma g[s]$;
- $f=z(k_1,\ldots,k_n)$. The order of $f[z{:=}s(z)]$ is lower than the order of f. Therefore we can define: RTT $\models_\Gamma f[s]$ if RTT $\models_\Gamma (f[z{:=}s(z)])[s]$;
- $f=\forall x{:}t^a[g]$. The order of g is equal to the order of $\forall x{:}t^a[g]$, so we can assume that RTT $\models_{\Gamma'} g[s']$ has already been defined for contexts Γ' and valuations s'. Therefore we can define RTT $\models_\Gamma (\forall x{:}t^a[g])\,[s]$ if for all $h\in\mathcal{P}\cup\mathcal{A}$ for which $\Gamma\setminus\{x{:}t^a\}\vdash h:t^a$, RTT $\models_{\Gamma[x:t^a]} g[s[x{:=}h]]$.

Here $\Gamma[x{:}t^a]$ is the same context as Γ, except that we now assign the type t^a to the variable x. We write RTT $\models f$ instead of RTT $\models_\Gamma f$ if it is clear which context Γ is used.

RTT embedded in KTT

To embed RTT in a first order language L, we have to cope with two technical problems:

1. We need to encode the notion of and the manipulation with (higher-order) propositional functions into a first-order language. The manipulation is particularly important with respect to substitution, which in the higher-order situation is much more complicated than in the first order case (cf. the definition of substitution 2.23);

2. In Russell's theory, it is only possible to quantify over a part of all propositions. This makes it impossible to translate, for instance, the proposition $\forall \mathtt{p}:()^1[\mathtt{p}() \vee \neg \mathtt{p}()]$ directly to $\forall \mathtt{x}[\mathtt{T}(\mathtt{x}) \vee \neg \mathtt{T}(\mathtt{x})]$, as the quantifier in the latter also quantifies over (codes of) higher-order propositions.

As we do not want RTT-contexts to be involved in this coding, we assume that each variable in RTT (implicitly) has a superscript t, indicating its ramified type. We only consider the legal propositional functions of RTT, and given a context Γ it is always possible to assign a unique type to each free variable in such a pf (cf. Lemma 2.56.1). Therefore we can do without contexts, as the types of the variables are now clear from the function in which they occur. For reasons of clarity, we will not explicitly write this superscript, as long as no confusion arises.

We propose the following solutions to the problems sketched above (we first give the definition and afterwards explain our thoughts behind it):

Definition 3.19 Extend the language $L(\mathtt{T})$ with for each ramified type t a monadic predicate $\mathtt{Typ}_t$, and for each $n \in \mathbb{N}$ a $(n+1)$-ary function $\mathtt{App}_n$. We code the typable propositional functions f of $\mathcal{P}$ as formulas $\overline{f}$ in this extension.[9] We do this by induction on the structure of f.

- If $f \equiv R(i_1, \ldots, i_{\mathfrak{a}(R)})$, then f is present in the original language L and we take $\overline{f} \stackrel{\text{def}}{=} f$;
- If $f \equiv f_1 \vee f_2$, then $\overline{f} \stackrel{\text{def}}{=} \overline{f_1} \vee \overline{f_2}$;
- If $f \equiv \neg f'$, then $\overline{f} \stackrel{\text{def}}{=} \neg \overline{f'}$;
- If $f \equiv \forall x{:}u[f']$, then $\overline{f} \stackrel{\text{def}}{=} \forall x[\neg \mathtt{Typ}_u(x) \vee \overline{f'}]$;
- If $f \equiv z(k_1, \ldots, k_m)$, write $K_i \equiv \langle \overline{k_i} \rangle$ for $k_i \in \mathcal{P}$[10], and $K_i \equiv k_i$ for $k_i \in \mathcal{A} \cup \mathcal{V}$. Define $\overline{f} \stackrel{\text{def}}{=} \mathtt{T}(\mathtt{App}_m(z, K_1, \ldots, K_m))$.

Notation 3.20 To keep notations uniform, we sometimes want to speak about $\langle \overline{x} \rangle$ when we only intend to mention x, for $x \in \mathcal{V}$, and about $\langle \overline{a} \rangle$ when only meaning a, for $a \in \mathcal{A}$. Hence, we formally define: $\langle \overline{x} \rangle \stackrel{\text{def}}{=} x$ and $\langle \overline{a} \rangle \stackrel{\text{def}}{=} a$ for all $x \in \mathcal{V}$ and all $a \in \mathcal{A}$.

We now give a formal interpretation of the newly introduced predicate symbol $\mathtt{Typ}_t$ and function symbol $\mathtt{App}_n$. We take $\mathcal{A}$ as domain of our model, so: $\boldsymbol{A} = \mathcal{A}$. This corresponds with the fact that Russell did not make a distinction between syntax and semantics. The following definition is also based on this fact:

[9] This mapping $\overline{}$ is different from the mapping $\overline{}$ that was used in Chapter 2.

[10] Observe: $\langle$ and $\rangle$ do not belong to the syntax of L extended with $\mathtt{T}$, $\mathtt{Typ}_t$ and $\mathtt{App}_n$. We define K_i to be the encoding of the translation of k_i, and not the list of symbols started with $\langle$, followed by the list of symbols that represent $\overline{k_i}$, and closed by $\rangle$.

Definition 3.21 We define the function $[\![\ ]\!]$:

- $[\![a]\!] = a$ for all $a \in \mathcal{A}$;
- $[\![R]\!] = R$ for all $R \in \mathcal{R}$;
- $[\![\mathtt{Typ}_{0^0}]\!] \stackrel{\text{def}}{=} \mathcal{A}$, and for $t \neq 0^0$, $[\![\mathtt{Typ}_t]\!] = \{[\![\langle\overline{f}\rangle]\!] \mid f \in \mathcal{P}, f : t\}$;
- We do not give a full semantics of the function $\mathtt{App}_n$. This is because we need $\mathtt{App}_n$ and its semantics only in some special cases. Assume $n \in \mathbb{N}$, $f \in \mathcal{P}$ is of type $(t_1, \ldots, t_n)$ and has free variables $x_1 < \cdots < x_n$. Also assume $k_1, \ldots, k_n \in \mathcal{A} \cup \mathcal{P}$, $k_i : t_i$ We define:

$$[\![\mathtt{App}_n]\!]([\![\langle\overline{f}\rangle]\!], [\![\langle\overline{k_1}\rangle]\!], \ldots, [\![\langle\overline{k_n}\rangle]\!]) \stackrel{\text{def}}{=} [\![\Big\langle\overline{f[x_1, \ldots, x_n{:=}k_1, \ldots, k_n]}\Big\rangle]\!].$$

Together, $\mathcal{A}$ and $[\![\ ]\!]$ form a model $\mathfrak{R}$ for the translation of RTT in KTT.

We make some remarks with respect to these definitions.

Remark 3.22 It is clear that the newly introduced functions $\mathtt{App}_n$ can be used for carrying out substitutions, thus solving the first of the technical problems stated at the beginning of this subsection. The predicates $\mathtt{Typ}_t$ (typability with type t) solve the second problem, as can be seen in the definition of $\overline{\forall x{:}u[f]}$.

Remark 3.23 Our work is related to (but independent of) Paul Gilmore's work on NaDSet 1. NaDSet 1 is a theory of generalised abstraction which makes n-ary predication a primitive of the system, with the unary truth predicate being trivially definable upon this basis. For a useful connection between KTT and NaDSet 1, the reader is referred to [47].

Remark 3.24 The extensions of $L(\mathtt{T})$ with the relation symbol $\mathtt{Typ}_t$ and the function symbol $\mathtt{App}_n$ are of a mere technical character. Therefore, we think that we can still speak of an *embedding* of RTT within KTT.

Below, we work in two systems: RTT and KTT. These systems have a different notion of substitution, though they use the same notation for expressing substitution. From the context however, it will always be clear which kind of substitution is meant.

The language $L(\mathtt{T})$ extended with $\mathtt{Typ}_t$ and $\mathtt{App}_n$ is an example of the languages described in Section 3c1, and we can construct $\mathfrak{R}_\alpha$ for each ordinal α as described in that section.

Substitution in the language KTT is ordinary first-order substitution. Higher-order substitutions in KTT can be obtained via the application operators $\mathtt{App}_n$. For future results, it is essential to know that the combination of first-order substitution and the operators App_n in KTT is compatible with the higher-order substitution for RTT that was defined in Definition 2.23. This is shown in the following Lemma (we write $f[x_i{:=}g_i]_{i=1}^n$ as shorthand for $f[x_1,\ldots,x_n{:=}g_1,\ldots,g_n]$):

Lemma 3.25 (Substitution Lemma) *Let g be a legal propositional function such that $\mathrm{FV}(g) = \{x_1,\ldots,x_m\}$, and $k_1,\ldots,k_m \in \mathcal{A} \cup \mathcal{P}$ such that x_i and k_i have the same type for all i. Let p be at least the order of $g[x_i{:=}k_i]_{i=1}^{n-1}$, and q at least the order of $g[x_i{:=}k_i]_{i=1}^n$. Then*

$$\mathfrak{N}_p \models \overline{g[x_i{:=}k_i]_{i=1}^{n-1}}\left[x_i{:=}\langle\overline{k_i}\rangle\right]_{i=n}^m$$

if and only if

$$\mathfrak{N}_q \models \overline{g[x_i{:=}k_i]_{i=1}^{n}}\left[x_i{:=}\langle\overline{k_i}\rangle\right]_{i=n+1}^m.$$

PROOF: We prove the following two statements

1. $$\mathfrak{N}_p \models \overline{g[x_i{:=}k_i]_{i=1}^{n-1}}\left[x_i{:=}\langle\overline{k_i}\rangle\right]_{i=n}^m$$

   if and only if

   $$\mathfrak{N}_q \models \overline{g[x_i{:=}k_i]_{i=1}^{n}}\left[x_i{:=}\langle\overline{k_i}\rangle\right]_{i=n+1}^m;$$

2. $$\mathfrak{N}_p \models \overline{\neg g[x_i{:=}k_i]_{i=1}^{n-1}}\left[x_i{:=}\langle\overline{k_i}\rangle\right]_{i=n}^m$$

   if and only if

   $$\mathfrak{N}_q \models \overline{\neg g[x_i{:=}k_i]_{i=1}^{n}}\left[x_i{:=}\langle\overline{k_i}\rangle\right]_{i=n+1}^m.$$

It is necessary to prove these two statements instead of the single statement of the Lemma because of the special way in which we defined $\mathfrak{N}_\alpha \models \neg f$. We use induction on the order of g. So assume (induction hypothesis A) that the substitution lemma is proved for all g' with an order smaller than the order of g. Use induction on the structure of $g[x_i{:=}k_i]_{i=1}^{n-1}$ (induction hypothesis B).

- $g[x_i{:=}k_i]_{i=1}^{n-1} \equiv R(i_1,\ldots,i_{\mathfrak{a}(R)})$.

  1. Note: $\overline{g[x_i{:=}k_i]_{i=1}^{n-1}}\left[x_i{:=}\langle\overline{k_i}\rangle\right]_{i=n}^m \equiv \overline{g[x_i{:=}k_i]_{i=1}^{n}}\left[x_i{:=}\langle\overline{k_i}\rangle\right]_{i=n+1}^m$. As the truth of $R(a_1,\ldots,a_{\mathfrak{a}(R)})$ can always be established at level 0 (so in $\mathfrak{N}_0$), there is nothing to be proved;
  2. Similar;

- $g[x_i{:=}k_i]_{i=1}^{n-1} \equiv g_1 \vee g_2$.

1. The following statements are equivalent:

$$
\begin{aligned}
&\mathfrak{R}_p \models \overline{g_1 \vee g_2}\left[x_i{:=}\langle\overline{k_i}\rangle\right]_{i=n}^{m};\\
&\mathfrak{R}_p \models (\overline{g_1} \vee \overline{g_2})\ \left[x_i{:=}\langle\overline{k_i}\rangle\right]_{i=n}^{m};\\
&\mathfrak{R}_p \models \overline{g_1}\left[x_i{:=}\langle\overline{k_i}\rangle\right]_{i=n}^{m} \vee \overline{g_2}\left[x_i{:=}\langle\overline{k_i}\rangle\right]_{i=n}^{m};\\
&\mathfrak{R}_p \models \overline{g_j}\left[x_i{:=}\langle\overline{k_i}\rangle\right]_{i=n}^{m} \text{ for some } j \in \{1,2\}.
\end{aligned}
\tag{3.2}
$$

As the order of g_j is at most the order of $g[x_i{:=}k_i]_{i=1}^{n-1}$, and the order of $g_j[x_n{:=}k_n]$ is at most the order of $g[x_i{:=}k_i]_{i=1}^{n}$, we can apply induction hypothesis B. Therefore, (3.2) is equivalent to the following statements:

$$
\begin{aligned}
\mathfrak{R}_q &\models \overline{g_j[x_n{:=}k_n]}\left[x_i{:=}\langle\overline{k_i}\rangle\right]_{i=n+1}^{m} \text{ for some } j \in \{1,2\};\\
\mathfrak{R}_q &\models \overline{g_1[x_n{:=}k_n]}\left[x_i{:=}\langle\overline{k_i}\rangle\right]_{i=n+1}^{m} \vee\\
&\qquad \overline{g_2[x_n{:=}k_n]}\left[x_i{:=}\langle\overline{k_i}\rangle\right]_{i=n+1}^{m};\\
\mathfrak{R}_q &\models \left(\overline{g_1[x_n{:=}k_n]} \vee \overline{g_2[x_n{:=}k_n]}\right)\ \left[x_i{:=}\langle\overline{k_i}\rangle\right]_{i=n+1}^{m};\\
\mathfrak{R}_q &\models \overline{(g_1 \vee g_2)[x_n{:=}k_n]}\left[x_i{:=}\langle\overline{k_i}\rangle\right]_{i=n+1}^{m};
\end{aligned}
$$

2. Similar;

- $g[x_i{:=}k_i]_{i=1}^{n-1} \equiv \neg g'$.

1. This is induction hypothesis B(2) on g';
2. The following statements are equivalent:

$$
\begin{aligned}
&\mathfrak{R}_p \models \overline{\neg\neg g'}\left[x_i{:=}\langle\overline{k_i}\rangle\right]_{i=n}^{m};\\
&\mathfrak{R}_p \models \neg\neg\overline{g'}\left[x_i{:=}\langle\overline{k_i}\rangle\right]_{i=n}^{m};\\
&\mathfrak{R}_p \models \overline{g'}\left[x_i{:=}\langle\overline{k_i}\rangle\right]_{i=n}^{m}.
\end{aligned}
\tag{3.3}
$$

By induction hypothesis B(1) on g', (3.3) is equivalent to the following statements:

$$
\begin{aligned}
&\mathfrak{R}_q \models \overline{g'[x_n{:=}k_n]}\left[x_i{:=}\langle\overline{k_i}\rangle\right]_{i=n+1}^{m};\\
&\mathfrak{R}_q \models \neg\neg\overline{g'[x_n{:=}k_n]}\left[x_i{:=}\langle\overline{k_i}\rangle\right]_{i=n+1}^{m};\\
&\mathfrak{R}_q \models \overline{\neg\neg g'[x_n{:=}k_n]}\left[x_i{:=}\langle\overline{k_i}\rangle\right]_{i=n+1}^{m};
\end{aligned}
$$

- $g[x_i{:=}k_i]_{i=1}^{n-1} \equiv \forall x{:}t[g']$.

1. The following statements are equivalent:

$$\mathfrak{N}_p \models \overline{\forall x{:}t[g']}\left[x_i{:=}\langle \overline{k_i}\rangle\right]_{i=n}^{m};$$
$$\mathfrak{N}_p \models \forall x[\neg\mathtt{Typ}_t(x) \vee \overline{g'}]\left[x_i{:=}\langle \overline{k_i}\rangle\right]_{i=n}^{m};$$
$$\mathfrak{N}_p \models \forall x\left[\neg\mathtt{Typ}_t(x) \vee \overline{g'}\left[x_i{:=}\langle \overline{k_i}\rangle\right]_{i=n}^{m}\right];$$
$$\mathfrak{N}_p \models \overline{g'}\left[x_i{:=}\langle \overline{k_i}\rangle\right]_{i=n}^{m}\,[x{:=}\langle \overline{h}\rangle] \text{ for all } h : t. \tag{3.4}$$

By induction hypothesis B(1) on g', (3.4) is equivalent to the following statements:

$$\mathfrak{N}_q \models \overline{g'[x_n{:=}k_n]}\left[x_i{:=}\langle \overline{k_i}\rangle\right]_{i=n+1}^{m}\,[x{:=}\langle \overline{h}\rangle] \text{ for all } h : t;$$
$$\mathfrak{N}_q \models \forall x\left[\neg\mathtt{Typ}_t(x) \vee \overline{g'[x_n{:=}k_n]}\left[x_i{:=}\langle \overline{k_i}\rangle\right]_{i=n+1}^{m}\right];$$
$$\mathfrak{N}_q \models \forall x\left[\neg\mathtt{Typ}_t(x) \vee \overline{g'[x_n{:=}k_n]}\right]\,\left[x_i{:=}\langle \overline{k_i}\rangle\right]_{i=n+1}^{m};$$
$$\mathfrak{N}_q \models \overline{\forall x{:}t[g'[x_n{:=}k_n]]}\left[x_i{:=}\langle \overline{k_i}\rangle\right]_{i=n+1}^{m};$$
$$\mathfrak{N}_q \models \overline{\forall x{:}t[g'][x_n{:=}k_n]}\left[x_i{:=}\langle \overline{k_i}\rangle\right]_{i=n+1}^{m};$$

2. Similar;

- $g[x_i{:=}k_i]_{i=1}^{n-1} \equiv z(h_1, \ldots, h_r)$.

1. If $x_n \not\equiv z$, then

$$\overline{g}\left[x_i{:=}\langle \overline{k_i}\rangle\right]_{i=n}^{m} \equiv \overline{g[x_n{:=}k_n]}\left[x_i{:=}\langle \overline{k_i}\rangle\right]_{i=n+1}^{m},$$

and nothing is to be proved. So assume $z \equiv x_n$. Let $y_1 < \cdots < y_r$ be the free variables of k_n. The following statements are equivalent:

$$\mathfrak{N}_p \models \overline{x_n(h_1, \ldots, h_r)}\left[x_i{:=}\langle \overline{k_i}\rangle\right]_{i=n}^{m};$$
$$\mathfrak{N}_p \models \mathtt{T}\left(\mathtt{App}_r\left(x_n, \langle \overline{h_1}\rangle, \ldots, \langle \overline{h_r}\rangle\right)\right)\,\left[x_i{:=}\langle \overline{k_i}\rangle\right]_{i=n}^{m};$$
$$\mathfrak{N}_p \models \mathtt{T}\left(\mathtt{App}_r\left(\langle \overline{k_n}\rangle, \langle \overline{h_1}\rangle, \ldots, \langle \overline{h_r}\rangle\right)\right)\,\left[x_i{:=}\langle \overline{k_i}\rangle\right]_{i=n+1}^{m}. \tag{3.5}$$

Let $h'_j \equiv k_s$ if $h_j \equiv x_s$ and $h'_j \equiv h_j$ if $h_j \notin \{x_{n+1}, \ldots, x_m\}$. Then (3.5) is equivalent to the following statements:

$$\mathfrak{N}_p \models \mathtt{T}\left(\mathtt{App}_r\left(\langle \overline{k_n}\rangle, \left\langle \overline{h'_1}\right\rangle, \ldots, \langle \overline{h'_r}\rangle\right)\right);$$
$$\mathfrak{N}_{p-1} \models \overline{k_n[y_j{:=}h'_j]_{j=1}^{r}};$$
$$\mathfrak{N}_{p-1} \models \overline{k_n[y_j{:=}h_j]_{j=1}^{r}[x_i{:=}k_i]_{i=n+1}^{m}}. \tag{3.6}$$

We can use induction hypothesis A: the order of k_n is smaller than the order of $z(h_1,\ldots,h_r)$ and therefore smaller than the order of g (see Corollary 2.63).). Thus (3.6) is equivalent to the following statements:

$$\mathfrak{N}_q \models \overline{k_n[y_j{:=}h_j]_{j=1}^{r}}\left[x_i{:=}\langle\overline{k_i}\rangle\right]_{i=n+1}^{m};$$
$$\mathfrak{N}_q \models \overline{z(h_1,\ldots,h_r)[x_n{:=}k_n]}\left[x_i{:=}\langle\overline{k_i}\rangle\right]_{i=n+1}^{m};$$

2. Similar. ⊠

Corollary 3.26 *Assume g is a pf of order p and $g[x{:=}k]$ is a proposition of order q. If $\mathfrak{N}_q \models \overline{g[x{:=}k]}$ then $\mathfrak{N}_p \models \overline{g}\left[x{:=}\langle\overline{k}\rangle\right]$.* ⊠

Remark 3.27 We have actually proved a stronger fact:

Assume g is a propositional function of order m and $g[x{:=}k]$ is a proposition of order n. If $\mathfrak{N}_n \models \overline{g[x{:=}k]}$ then $\mathfrak{N}_p \models \overline{g}[x{:=}\langle\overline{k}\rangle]$, where $p = min(m, n{+}1)$.

This tells us more about the role of the predicate T: Although a substitution may lower the order of a propositional function by more than one, only one application of the T-predicate is involved (hence only one level in the hierarchy of truths). However, in the theorem below we only need the (weaker) form in which we presented the Substitution Lemma originally.

We now prove the main theorem of this section.

Theorem 3.28 (Embedding of RTT in KTT) *Let Γ be an RTT-context with domain $\mathcal{V}$. Let $f \in \mathcal{P}$ be a legal pf in Γ with free variables $x_1,\ldots,x_n$ of types $t_1^{\mathfrak{a}_1},\ldots,t_n^{\mathfrak{a}_n}$. Let $s : \mathcal{V} \to \mathcal{P}$ be an assignment function such that $s(x_i) : t_i^{\mathfrak{a}_i} \in \Gamma$. Let n be the order of $f[s] \equiv f[x_i{:=}s(x_i)]_{i=1}^{m}$.*
Then RTT $\models f[s]$ if and only if $\mathfrak{N}_n \models \overline{f[s]}$.

PROOF:

⇒ As in the proof of Lemma 3.25, we have to deal with the special way in which we defined $\text{KTT}_n \models \neg f$. Therefore, we simultaneously prove:

1. If RTT $\models f[s]$ then $\mathfrak{N}_n \models \overline{f[s]}$;
2. If RTT $\models \neg f[s]$ then $\mathfrak{N}_n \models \overline{\neg f[s]}$.

The proof follows the same induction structure as the definition of RTT $\models f[s]$ (Definition 3.18).

- $f \equiv R(a_1,\ldots,a_{\mathfrak{a}(R)})$ for some $R \in \mathcal{R}$ and some $a_1,\ldots,a_{\mathfrak{a}(R)} \in \mathcal{A} \cup \mathcal{V}$. Notice that $\overline{f} \equiv f$ and that $\overline{f[s]} \equiv f[s]$. Write

$$f[s] \equiv R(a_1',\ldots,a_{\mathfrak{a}(R)}')$$

for certain $a'_1, \ldots, a'_{\mathfrak{a}(R)} \in \mathcal{A}$. As RTT $\models R(a'_1, \ldots, a'_{\mathfrak{a}(R)})$, we know that $(a'_1, \ldots, a'_{\mathfrak{a}(R)}) \in R$, hence $\mathfrak{N}_n \models R(a'_1, \ldots, a'_{\mathfrak{a}(R)})$. The proof is similar for $\neg f$;

- $f \equiv g_1 \vee g_2$.
  Then g_1 and g_2 are legal. Assume n_j is the order of $g_j[s]$. Notice that $n_j \leq n$.
  First assume RTT $\models f[s]$. As

$$f[s] \equiv (g_1 \vee g_2)[s] \overset{(lem\ 2.29.1)}{=} g_1[s] \vee g_2[s],$$

  we have RTT $\models g_j[s]$ for $j = 1$ or $j = 2$. By the induction hypothesis on the structure of f: $\mathfrak{N}_{n_j} \models \overline{g_j[s]}$, and as $n_j \leq n$: $\mathfrak{N}_n \models \overline{g_j[s]}$. Therefore, $\mathfrak{N}_n \models \overline{g_1[s]} \vee \overline{g_2[s]}$. Now observe that

$$\overline{f[s]} \equiv \overline{(g_1 \vee g_2)[s]} \overset{(lem\ 2.29.1)}{=} \overline{g_1[s]} \vee \overline{g_2[s]}.$$

  Assume RTT $\models \neg f[s]$. By a similar argument, we find that RTT $\models \neg g_j[s]$ for $j = 1$ and $j = 2$. By the induction hypothesis on the structure of f, $\mathfrak{N}_{n_j} \models \overline{\neg g_j[s]}$ for $j = 1, 2$, so $\mathfrak{N}_n \models \neg \overline{g_1[s]} \wedge \neg \overline{g_2[s]}$, or in other words:

$$\mathfrak{N}_n \models \neg \left(\overline{g_1[s]} \vee \overline{g_2[s]} \right).$$

  Observe that $\neg \left(\overline{g_1[s]} \vee \overline{g_2[s]} \right) \equiv \overline{\neg f[s]}$;
- $f \equiv \neg g$.
  If RTT $\models f[s]$ then use the induction hypothesis on the structure of g to get $\mathfrak{N}_n \models \overline{\neg g[s]}$, hence $\mathfrak{N}_n \models \overline{f[s]}$.
  If RTT $\models \neg f[s]$, then RTT $\models g[s]$, so by induction on the structure of g, $\mathfrak{N}_n \models \overline{g[s]}$, so $\mathfrak{N}_n \models \neg\neg \overline{g[s]}$, so $\mathfrak{N}_n \models \overline{\neg f[s]}$;
- $f \equiv \forall x{:}t[g]$.
  If RTT $\models f[s]$ then for all $k{:}t$, RTT $\models g[s[x{:=}k]]$, where $s[x{:=}k]$ is the assignment function which assigns k to x, and $s(y)$ to every $y \in \mathcal{V} \setminus \{x\}$. By the induction hypothesis on the structure of f, we know that for all $k : t$, $\mathfrak{N}_{n_k} \models \overline{g[s[x{:=}k]]}$, where n_k is the order of $g[s[x{:=}k]] \equiv g[x_i{:=}s(x_i)][x{:=}k]$.
  By Corollary 3.26 we have: For all $k : t$,

$$\mathfrak{N}_n \models \overline{g[x_i{:=}s(x_i)]_{i=1}^{m}}[x{:=}\langle \overline{k} \rangle].$$

  Hence, for all $a \in \mathcal{A}$,

$$\mathfrak{N}_n \models \left(\neg \mathtt{Typ}_t(x) \vee \overline{g[x_i{:=}s(x_i)]_{i=1}^{m}} \right) [x{:=}a].$$

So $\mathfrak{N}_n \models \forall x \left[\neg \mathrm{Typ}_t(x) \vee \overline{g[x_i{:=}s(x_i)]_{i=1}^m}\right]$. Observe:

$$\begin{aligned}\overline{f[s]} &\equiv \overline{(\forall x{:}t[g])\,[s]}\\ &\equiv \overline{\forall x{:}t\,[g[x_i{:=}s(x_i)]_{i=1}^m]}\\ &\equiv \forall x\left[\neg \mathrm{Typ}_t(x) \vee \overline{g[x_i{:=}s(x_i)]_{i=1}^m}\right].\end{aligned}$$

The argument for RTT $\models \neg f$ is similar;

- $f \equiv z(h_1, \ldots, h_p)$. Determine q such that $z \equiv x_q$. Write $h'_j \equiv h_j$ if $h_j \notin \{x_1, \ldots, x_m\}$; $h'_j \equiv s(x_\ell)$ if $h_j \equiv x_\ell$. Assume that $s(x_q)$ has free variables $y_1 < \cdots < y_r$. Observe:

$$f[x_i{:=}s(x_i)]_{i=1}^m \equiv s(x_q)[y_j{:=}h'_j]_{j=1}^r.$$

So if RTT $\models f[x_i{:=}s(x_i)]_{i=1}^m$ then RTT $\models s(x_q)[y_j{:=}h'_j]_{j=1}^r$.
As the order of $s(x_q)$ is smaller than the order of f, we can use induction on the order of f to obtain (with Conservation of Knowledge) that: $\mathfrak{N}_n \models \overline{s(x_q)[y_j{:=}h'_j]_{j=1}^r}$, which is equivalent to: $\mathfrak{N}_n \models \overline{f[x_i{:=}s(x_i)]_{i=1}^m}$.
The proof for RTT $\models \neg f[x_i{:=}s(x_i)]_{i=1}^m$ is similar;

$\Leftarrow$ This is easily shown now by contraposition. Assume, for the sake of the argument, that RTT $\models f$ does not hold. Then RTT $\models \neg f$, so by the $\Rightarrow$ part of the theorem (that was proved above), $\mathfrak{N}_n \models \neg \overline{f}$. So, if $\mathfrak{N}_n \models \overline{f}$ then RTT $\models f$.

⊠

This theorem clearly shows the relation between the orders in RTT and the levels of truth in KTT. The heart of the proof of Theorem 3.28 is in the proof of case $x_n(h_1, \ldots, h_r)$ of the Substitution Lemma 3.25 (via its corollary 3.26). This is the only place in the proof where the properties of the predicate T are used. It is understandable that these properties must be used at exactly this place when we look at the definition of propositional functions and the typing rules for propositional functions. Exactly the possibility of constructing a propositional function of the form $x_n(h_1, \ldots, h_r)$ makes it possible to arrive at higher-order propositional functions and higher-order propositions. So exactly at this spot, Kripke's predicate T must appear, in order to raise one level in KTT as well.

One might expect to need the properties of T in the proof of the case $f \equiv z(h_1, \ldots, h_p)$ of Theorem 3.28 as well. But we see that this is not the case. This is understandable: we do not consider the truth of $z(h_1, \ldots, h_p)$ itself, but of $z(h_1, \ldots, h_p)[x_i{:=}s(x_i)]_{i=1}^m$.
Furthermore, we do not work with the order of $z(h_1, \ldots, h_p)$, but with n, the order of $z(h_1, \ldots, h_p)[x_i{:=}s(x_i)]_{i=1}^m$.
The shift to a lower order has to do with the orders of $z(h_1, \ldots, h_p)$ and $s(x_q)$, but

not with the orders of $z(h_1,\ldots,h_p)[x_i{:=}s(x_i)]_{i=1}^m$ and $s(x_q)[y_j{:=}h'_j]_{j=1}^r$. These last two propositions are syntactically equivalent and therefore of the same order.

Corollary 3.29 *If $f \in \mathcal{P}$ is a legal proposition of order n, then* RTT $\models f$ *if and only if* $\mathfrak{R}_n \models \overline{f}$. ⊠

Corollary 3.30 (Conservativity of $\mathfrak{R}_\omega$ over RTT**)**
RTT $\models f$ *if and only if* $\mathfrak{R}_\omega \models \overline{f}$. ⊠

We cannot improve the result of Theorem 3.28 in general: For all n, there are propositions f of order n in RTT whose code is provable at level $\mathfrak{R}_n$ in KTT, but not at any lower level.

Theorem 3.31 *Let $n > 0$, and let f_n be the nth-order-proposition*

$$\forall \mathtt{p}{:}()^{n-1}[\mathtt{p}() \vee \neg\mathtt{p}()].$$

Then:

$$\mathfrak{R}_m \models \overline{f_n} \text{ if and only if } m \geq n.$$

PROOF:

$\Leftarrow$ follows from Theorem 3.28 and Lemma 3.14;

$\Rightarrow$ is by induction on n. Observe that

$$\overline{f_n} \equiv \forall \mathtt{p}[\neg\mathtt{Typ}_{()^{n-1}}(\mathtt{p}) \vee (\mathtt{T}(\mathtt{Appo}(\mathtt{p})) \vee \neg\mathtt{T}(\mathtt{Appo}(\mathtt{p})))].$$

- $n = 1$. Let g be any proposition of order 0 in RTT. Then $\mathfrak{R}_0 \models \mathtt{Typ}_{()^0}(\langle\overline{g}\rangle)$ but as T is completely undefined at level 0 ($S_{0,1} = S_{0,2} = \varnothing$),
$$\mathfrak{R}_0 \not\models \mathtt{T}(\mathtt{Appo}(\langle\overline{g}\rangle)) \vee \neg\mathtt{T}(\mathtt{Appo}(\langle\overline{g}\rangle)).$$
Hence, $\mathfrak{R}_0 \not\models \overline{f_1}$;
- Assume the theorem has been proved for $n-1$. Assume $m < n$ and $\mathfrak{R}_m \models \overline{f_n}$. By definition of $\models$, we have:
$$\mathfrak{R}_{n-1} \models \mathtt{T}(\mathtt{Appo}(\langle\overline{f_{n-1}}\rangle)) \vee \neg\mathtt{T}(\mathtt{Appo}(\langle\overline{f_{n-1}}\rangle)),$$
and for reasons of consistency: $\mathfrak{R}_m \models \mathtt{T}(\mathtt{Appo}(\langle\overline{f_{n-1}}\rangle))$. Therefore, $\mathfrak{R}_m \models \mathtt{T}(\overline{\langle f_{n-1}\rangle})$, so, by the definition of T: $\mathfrak{R}_{m-1} \models \overline{f_{n-1}}$, which contradicts the induction hypothesis, as $m-1 < n-1$. ⊠

There are however, for any n, propositions f of order n in RTT for which $\mathfrak{R}_m \models \overline{f}$ or $\mathfrak{R}_m \models \overline{\neg f}$ can already be established for $m < n$.

Example 3.32 Consider a proposition $g \equiv g_1 \vee g_2$ where g_1 is a true proposition of order m and g_2 is any proposition of order $n > m$. As g_1 is true in RTT, we have $\mathfrak{R}_m \models \overline{g_1}$, and therefore $\mathfrak{R}_m \models \overline{g}$.

The restrictiveness of Russell's theory

We illustrate the different approaches of Russell and Kripke by an example given by Kripke himself in [96].

Example 3.33 Two politicians, Wim and Frits are quarrelling about who is telling the truth and who is lying. Of course, Wim states that anything said by Frits is untrue (A), and Frits argues that any statement of Wim is false (B). The utterances (A) and (B) can be complete nonsense, but they can also be meaningful. This does not only depend on the (syntactic) structure of (A) and (B), but also on their semantics, that is: on the utterances of Wim and Frits (which may be more than only (A) and (B)):

1. Assume, (A) is the only statement that Wim makes, and (B) is the only statement that was made by Frits. Then (A) and (B) are nonsense. More precise, there is no reason to believe that (A) is true, and there is no reason either to believe that (B) is true. Namely: if we want to prove that (A) is true, we must show first that all statements of Frits are false, in other words: that (B) is false. But in order to establish the falsehood of (B) we must first find a true statement of Wim, that is: we must prove (A). Summarising: The truth of (A) can only be established if the truth of (A) has already been established before. So the truth of (A) will never be established.

   Similarly, we show that the falsehood of (A) will never be established, and that neither truth nor falsehood of (B) will ever be established;

2. Now assume that (B) is still the only statement made by Frits, but that Wim has not only uttered (A), but also argues that that one equals one (C). Statement (C) is clearly true. This means that Frits has been lying. Therefore, Frits' only statement (B) is false, hence Wim's statement (A) is true.

We formalise this situation as follows. We assume that the first order language L contains at least three relation symbols, $\mathtt{W}$, F and $=$. Moreover, we assume to have an individual symbols $\mathtt{1}$. $\mathtt{W}$ will be interpreted as the set of (codes of the) utterances of Wim, F shall represent the set of (codes of the) utterances of Frits. In this way, we can encode the expression (A) by

$$A \equiv \forall \mathtt{x}[\neg \mathtt{F}(\mathtt{x}) \vee \neg \mathtt{T}(\mathtt{x})].$$

Here, T is the truth predicate as introduced earlier in this Section. Similarly, (B) is encoded by

$$B \equiv \forall \mathtt{x}[\neg \mathtt{W}(\mathtt{x}) \vee \neg \mathtt{T}(\mathtt{x})]$$

and (C) is encoded by

$$C \equiv \mathtt{1} = \mathtt{1}.$$

We model the situations 1 and 2 above as follows:

1. In the first situation we take $\mathbb{N}$ as our domain. The semantics of $\mathtt{W}$ must represent the set of utterances of Wim, so we let $\boldsymbol{W} = [\![\mathtt{W}]\!] = \{[\![\langle A\rangle]\!]\}$. Similarly, $\boldsymbol{F} = [\![\mathtt{F}]\!] = \{[\![\langle B\rangle]\!]\}$. Further we let $[\![\mathtt{1}]\!] = 1$. This gives us a model $\mathfrak{M}^1$ for L. We can build a hierarchy $\mathfrak{M}^1_\alpha$ (for ordinals α) as explained in Section 3c1. We show that there is no ordinal α for which $\mathfrak{M}^1_\alpha \models A$.

   Assume that α is the smallest ordinal for which it holds that $\mathfrak{M}^1_\alpha \models A$. Then for all $n \in \mathbb{N}$ and all assignment functions $s : \mathcal{V} \to \mathbb{N}$ we have $\mathfrak{M}^1_\alpha \models (\neg\mathtt{F}(\mathtt{x}) \vee \neg\mathtt{T}(\mathtt{x}))[s[\mathtt{x}{:=}n]]$. Hence, for all $n \in \mathbb{N}$, either $n \notin \boldsymbol{F}$ or $n \in S_{\alpha,2}$. Note that $[\![\langle B\rangle]\!] \in \boldsymbol{F}$. Therefore: $[\![\langle B\rangle]\!] \in S_{\alpha,2}$. But then there is $\beta < \alpha$ such that $\mathfrak{M}^1_\beta \models \neg B$. Using the definition of $\mathfrak{M}^1_\beta \models \neg B$, this means that there is $n \in \mathbb{N}$ such that $n \in \boldsymbol{W}$ and $n \in S_{\beta,1}$. The only candidate for such n is $[\![\langle A\rangle]\!]$, as this is the only element of $\boldsymbol{W}$. Hence: $[\![\langle A\rangle]\!] \in S_{\beta,1}$. Therefore, there is $\gamma < \beta$ for which $\mathfrak{M}^1_\gamma \models A$. This is a contradiction because $\gamma < \alpha$, and α is the smallest ordinal for which $\mathfrak{M}^1_\alpha \models A$.

   In a similar way we can show that for all α, $\mathfrak{M}^1_\alpha \not\models \neg A$, $\mathfrak{M}^1_\alpha \not\models B$, and $\mathfrak{M}^1_\alpha \not\models \neg B$. This corresponds to our earlier conclusions that the sentences (A) and (B) are nonsense if they are the only utterances of Wim and Frits;

2. For the second situation, we change the model $\mathfrak{M}^1$ to a model $\mathfrak{M}^2$ by replacing $\boldsymbol{W}$ by $\{[\![\langle A\rangle]\!], [\![\langle C\rangle]\!]\}$. Notice that $\mathfrak{M}^2_0 \models C$, because $1 = 1$. Therefore, $[\![\langle C\rangle]\!] \in S_{1,1}$, so $\mathfrak{M}^2_1 \models \mathtt{T}(\langle C\rangle)$. As $[\![\langle C\rangle]\!] \in \boldsymbol{W}$, we also have $\mathfrak{M}^2_1 \models \mathtt{W}(\langle\mathtt{C}\rangle)$. Therefore, $\mathfrak{M}^2_1 \models \neg(\neg\mathtt{W}(\mathtt{x}) \vee \neg\mathtt{T}(\mathtt{x}))[s[\mathtt{x}{:=}[\![\langle C\rangle]\!]]]$, for any assignment function $s : \mathcal{V} \to \mathbb{N}$. Hence $\mathfrak{M}^2_1 \models \neg B$.

   This shows that Frits' statement is indeed false, but it also shows that Wim's statement (A) is true at level 2: As $\mathfrak{M}^2_1 \models \neg B$, $[\![\langle B\rangle]\!] \in S_{2,2}$, and this implies that $\mathfrak{M}^2_2 \models A$.

In Russell's terminology it would not be possible to type expressions like A and B at all. The expression A involves B, and therefore has to be of higher order than B. Similarly, B involves A, so it has to be of a higher order than A.

This indicates an important difference between RTT and KTT: Kripke allows much more expressions to be included in the system. In some situations these expressions will never obtain any truth-value (like A and B in the first example), but in other situations (so with other definitions of the primitive predicates) the same expressions will get a truth-value. Kripke concludes:

> "it would be fruitless to look for an *intrinsic*[11] criterion that will enable us to sieve out – as meaningless, or ill-formed – those sentences which lead to paradox"
>
> ([96], p. 692)

[11] Italics of Kripke

Example 3.34 Another, more formal, example of a proposition f in KTT for which there is no $g \in \mathcal{P}$ with $\overline{g} \equiv f$ is the proposition

$$f \stackrel{\text{def}}{=} \forall \mathtt{x}[\mathtt{T}(\mathtt{x}) \vee \neg\mathtt{T}(\mathtt{x})].$$

Notice that this is an impredicative proposition. It expresses that all propositions are either true or false, including f itself.

Assume, for the sake of the argument, that $\overline{g} \equiv f$. Let m be the order of g. Determine whether RTT $\models g$ or RTT $\models \neg g$. We give the argument for the case RTT $\models g$; the argument for $\neg g$ is easy. If RTT $\models g$ then by Corollary 3.29, $\mathfrak{R}_m \models \forall \mathtt{x}[\mathtt{T}(\mathtt{x}) \vee \neg\mathtt{T}(\mathtt{x})]$. This implies $\mathfrak{R}_m \models \mathtt{T}(f_m) \vee \neg\mathtt{T}(f_m)$, where f_m is as in Theorem 3.31. By definition of T this means $\mathfrak{R}_{m-1} \models f_m$ or $\mathfrak{R}_{m-1} \models \neg f_m$, both contradicting Theorem 3.31.

3c3 Orders and types

RTT is based on a double hierarchy: One of types and one of orders. This double hierarchy is too restrictive. It is possible to develop logic and mathematics within RTT, but we saw that the proof of the theorem of the least upper bound, which is fundamental in real analysis, cannot be given. The origin of the problem is the use of the so-called predicative and impredicative propositional functions.

It is therefore interesting to notice the relation between orders in RTT and levels of truth in KTT, as formulated in Theorem 3.28. It shows that Kripke's system can be regarded as a system based on RTT of which not the *orders*, but the *types* have been removed. In this way, KTT can be seen as a system that is dual to the simple theory of types.

KTT however, has a more subtle approach than many type theories as it does not exclude any, possibly "paradoxical", expression from the syntax, which is the usual type-theoretic approach. If an expression is paradoxical, it will not get a truth value at any level α of the hierarchy of Truths. Whether an expression is paradoxical or not does not only depend on its syntactic structure, but also on the domain $\boldsymbol{A}$ and the relations of $\boldsymbol{R}$ on $\boldsymbol{A}$ (see Example 3.33). So paradoxes are only excluded at the level of semantics.

The discussion above shows that the orders of RTT are not to be blamed for the restrictiveness of RTT. KTT is a system which contains orders but has only few restrictions towards self-application. It is the *combination* of orders and types that makes RTT restrictive.

The special structure of KTT makes it possible to define a notion of *semantic* order in RTT:

Definition 3.35 Let $f \in \mathcal{P}$ have type t^a. The *semantic order* of f is the smallest natural number n for which either $\mathfrak{R}_n \models \overline{f}$ or $\mathfrak{R}_n \models \neg\overline{f}$.

By Theorem 3.28, the semantic order of f is always smaller than or equal to its (syntactic) order.

Conclusions

We saw in Section 3a1 that the Ramified Theory of Types is very restrictive for the description of mathematics within logic, because it is not possible to formulate impredicative definitions in RTT.

This was already realised by Russell and Whitehead, who tried to solve this by postulating the Axiom of Reducibility 3.3. This axiom has been criticised from the moment it was written down, both by Russell and Whitehead themselves and by others. Ramsey, Hilbert and Ackermann deramify RTT: They remove the orders. They observe that this does not lead to known paradoxes as long as a proper distinction between language and metalanguage is made.

Gödel and Quine observe that the deramification does not violate the Vicious Circle Principle, as long as one accepts that objects and pfs exist independently of our constructions.

So historically speaking, one could say that the orders were blamed for the restrictiveness of RTT. In Section 3c we showed that this is not correct. We used the formalisation of RTT that was given in Chapter 2 to compare RTT with Kripke's Theory of Truth KTT. We established the relation between Russell's hierarchy of orders and Kripke's hierarchy of truth-levels. In particular we showed that:

1. A proposition of RTT of order n is true if and only if its interpretation is true at level n in Kripke's Truth Hierarchy (Theorem 3.28);

2. The truth of some propositions of order n of RTT cannot be established in KTT at a level of truth hierarchy smaller than n (Theorem 3.31). Yet for some other propositions, it can be established at an earlier level (Example 3.32).

We also saw that Russell's theory is more restrictive than Kripke's. On the one hand, all propositional functions of RTT can be coded in Kripke's Truth Theory; on the other hand there are formulas of Kripke's theory that cannot be expressed in RTT, respecting both hierarchies.

We feel that the orders are not to be blamed (alone) for the restrictiveness of RTT. KTT clearly has a structure with orders (see Definition 3.35); nevertheless it is possible to give impredicative definitions (see Example 3.34).

Russell excludes all propositional functions that might lead to paradoxical situations beforehand. Kripke does not exclude them, though it is not guaranteed that each proposition gets a truth value. This may depend on the model chosen (see Example 3.33).

Whether the orders should be blamed or not, the main line in the history continues with non-ramified theories. For example, Church's combination of λ-calculus with simple type theory, the basis for most modern type systems, has no orders. In Chapter 6, we discuss again the role and benefits of orders.

Part II

Propositions as Types, Pure Type Systems, AUTOMATH

Chapter 4

Propositions as Types and Pure Type Systems

In this chapter we discuss the notions of Propositions as Types and Proofs as Terms (both abbreviated as PAT), the lambda calculus and Pure Type Systems (PTSs). The PAT principle has played an important role in the development of Type Theory after the Second World War. It opened the possibility to use Type Theory not only as a restrictive method (to prevent paradoxes) but also as a constructive method. Many proof checkers and theorem provers, like AUTOMATH [112], Coq [44], Nuprl [35], LF [64], use the PAT principle. Lambda calculus was introduced by Church [27, 28], as a formalisation of the notion of function. With this formal notation he could formulate his set of postulates for the foundation of logic. Kleene and Rosser [92] showed that Church's set of postulates was inconsistent. The lambda calculus itself, however, appeared to be a very useful tool. In Chapter 2 of this book we showed that it is much more clear and accurate than the notion of (propositional) function as introduced by Russell and Whitehead in Principia Mathematica [145].

Nowadays, [2] is the standard work for (untyped) lambda calculus. We present the basic definitions and properties of the λ-calculus in Section 4b.

Being a suitable framework for the formalisation of functions, it is not surprising that lambda calculus appeared to be an excellent tool for formalising the Simple Theory of Types [29]. In Section 3b2 we gave a short description of Church's formalisation. This formalisation is at the basis of most modern type theories and especially at the basis of PTSs. PTSs were introduced by Terlouw [142] and Berardi [10], providing a general framework in which many type systems can be described. Section 4c presents the definition of PTSs, Section 4c1 presents the Barendregt cube which consists of a general framework of eight influential PTSs, and Section 4c2 discusses the most important meta-properties as described in [3].

4a Propositions as Types and Proofs as Terms (PAT)

In the first three chapters we discussed type systems the way they were initially designed, namely to prevent the logical paradoxes. But although the systems of both Russell and Church have some logical symbols in them (like $\vee$, $\forall$), these theories themselves cannot be seen as a logical system. If one wants to make logical derivations, one has to build a logical system on top of one of these type systems.

However, type theory nowadays also plays an important role in logic in a different way: It can be used as a logical system itself. This use of type theory is generally known as "propositions as types" or "proofs as terms". As we will see in this section, both expressions only partially cover the idea of using type theory as a logical system. As they both abbreviate to PAT, we will use this abbreviation to indicate both "propositions as types" and "proofs as terms".

"Proofs as terms" already suggests an important advantage of using type theory as a logical system: In this method proofs are first-class citizens of the logical system, whilst for many other logical systems, proofs are rather complex objects outside the logic (for example: derivation trees), and therefore cannot be easily manipulated.

Below we mention some origins of the PAT principle.

4a1 Intuitionistic logic

The idea of PAT originates in the formulation of intuitionistic logic. Though it is not correct that "intuitionistic logic" is simply the logic that is used in intuitionistic mathematics[1], there are frequently occurring constructions in intuitionistic math-

[1] "Intuitionistic logic" is standard terminology for "logic without the law of the excluded middle". The terminology suggests that it is "the logic that is used in intuitionism". However, intuitionism, that is: the philosophy of Brouwer and the mathematics based on that philosophy, declares mathematics to be independent of logic. According to that philosophy, a proof of a mathematical theorem is a method to read that theorem as a tautology. The fact that one needs a list of tautologies before the proof of more complicated theorems becomes clear, only indicates that the constructions we make are too complicated to be comprehended immediately. Mathematics itself however, is a construction in one's mind, independent of logic:

> "Een logische opbouw der wiskunde, onafhankelijk van de wiskundige intuïtie, is onmogelijk — daar op die manier slechts een taalgebouw wordt verkregen, dat van de eigenlijke wiskunde onherroepelijk gescheiden blijft — en bovendien een contradictio in terminis — daar een logisch systeem, zoo goed als de wiskunde zelf, de wiskundige oer-intuïtie nodig heeft"
>
> (*Over de Grondslagen der Wiskunde* [17], p. 180)

(A logical construction of mathematics, independent of the mathematical intuition, is impossible — for by this method no more is obtained than a linguistic structure, which irrevocably remains separated from mathematics — and moreover it is a contradictio in terminis — because a logical system needs the basic intuition of mathematics as much as mathematics itself needs it. [Translation from [69]]).

ematics that have a logical counterpart. One of these constructions is the proof of an implication. Heyting [68] describes the proof of an implication $a \Rightarrow b$ as: Deriving a solution for the problem b from the problem a. Kolmogorov [95] is even more explicit, and describes a proof of $a \Rightarrow b$ as the construction of a method that transforms each proof of a into a proof of b. This means that a proof of $a \Rightarrow b$ can be seen as a *(constructive) function* from the proofs of a to the proofs of b. In other words, the proofs of the proposition $a \Rightarrow b$ form exactly the set of functions from the set of proofs of a to the set of proofs of b. This suggests to identify a proposition with the set of its proofs. Now *types* are used to represent these sets of proofs. An element of such a set of proofs is represented as a *term* of the corresponding type. This means that propositions are interpreted as *types*, and proofs of a proposition a as *terms of type a*.

4a2 The discovery of PAT: Curry

PAT was, independently from Heyting and Kolmogorov, discovered by Curry and Feys [40]. In paragraph 8C of [40], Curry describes the so-called F-objects, which correspond more or less to the simple types of Church in [29]. As a basis, a list of primitive objects $\vartheta_1, \vartheta_2, \ldots$ is chosen. All these primitive objects are F-objects. Moreover, if α and β are F-objects, then so is $\mathsf{F}\alpha\beta$. Here, F is a new symbol. $\mathsf{F}\alpha\beta$ must be interpreted as the class of functions from α to β. If α is an F-object, then the statement $\vdash \alpha X$ must be interpreted as "the object X belongs to α". The following *rule*-F is adopted: If $\vdash \mathsf{F}XYZ$ and $\vdash XU$ then $\vdash Y(ZU)$. The intuitive meaning of this rule is: If Z belongs to $\mathsf{F}XY$ and U belongs to X, then ZU belongs to Y. This rule immediately corresponds to the application-rule of λ-Church (see Section 3b2).

Earlier in [40], Curry has introduced the combinator P, which is the implication combinator. $\mathsf{P}XY$ can be interpreted as the proposition "if X then Y". The combinator P comes together with a *rule*-P: If $\vdash \mathsf{P}XY$ and $\vdash X$ then $\vdash Y$. Curry notices that this rule has similar behaviour as rule-F.

Curry is the first one to give a formalisation of PAT. For each F-object α he defines a corresponding proposition α^P as follows: $\vartheta_i^P \equiv \vartheta_i$ and $(\mathsf{F}\alpha\beta)^P \equiv \mathsf{P}\alpha^P\beta^P$. Remark that Curry's function $\alpha \mapsto \alpha^P$ is in fact an embedding of types in propositions (so a types-as-propositions embedding instead of a propositions-as-types embedding).

Curry then derives the following theorem, where $\mathsf{F}_m X_1 \cdots X_m Y$ is an abbreviation of $\mathsf{F}X_1(\mathsf{F}X_2(\ldots(\mathsf{F}X_m Y)\ldots))$:

> "If $\vdash \mathsf{F}_m\xi_1 \cdots \xi_m\eta X$ then $\vdash (\mathsf{F}_m\xi_1 \cdots \xi_m\eta)^P$. Moreover, if $\vdash \mathsf{F}_m\xi_1 \cdots \xi_m\eta X$ is derivable from the premises $\vdash \alpha_i a_i$ $(i = 1, \ldots, p)$ then $\vdash (\mathsf{F}_m\xi_1 \cdots \xi_m\eta)^P$ is derivable from the premises $\vdash \alpha_i^P$ $(i = 1, \ldots, p)$."
>
> ([40], paragraph 9E, Theorem 1)

In other words: If there is (under certain type conditions $a_i{:}\alpha_i$) an object X that is a function taking arguments of types $\xi_1, \ldots, \xi_m$, resulting in an object of type η, then the corresponding theorem is derivable (if we presuppose α_i^P). Or in short: The types-as-propositions embedding $\alpha \mapsto \alpha^P$ is sound.

The converse of the theorem holds as well:

> "If $\vdash (\mathsf{F}_m\xi_1 \cdots \xi_m\eta)^P$ is derivable by rule-P from the premises $\vdash \alpha_i^P$, then for each derivation of this fact and each assignment of $a_1, \ldots, a_p$ to $\alpha_1, \ldots, \alpha_p$ respectively there exists an X such that $\vdash \mathsf{F}_m\xi_1 \cdots \xi_m\eta X$ is derivable from the premises $\vdash \alpha_i a_i$ $(i = 1, \ldots, p)$ by rule-F alone."
>
> ([40], paragraph 9E, Theorem 2)

In other words: The types-as-propositions embedding $\alpha \mapsto \alpha^P$ is complete.

The treatment of PAT in [40] is mainly directed towards Propositions as Types. Proofs as terms are implicitly present in the theory of [40]: The term X in the proof of Theorem 1 of [40] can be seen as a proof of the proposition $(\mathsf{F}_m\xi_1 \cdots \xi_m\eta)^P$. But this is not made explicit in [40].

Example 4.1 We show the deduction of the proposition $A \to A$ from the logical axioms:

$X \to Y \to X$[2] (the K-*axiom*)

$(X \to Y \to Z) \to (X \to Y) \to X \to Z$ (the S-*axiom*),

both in the style of the combinator P and in the PAT-style. Both derivations correspond to the derivation of the proposition $A \to A$ in natural deduction style, with the use of modus ponens, and the K- and S-axioms only:

$$\dfrac{\dfrac{\vdash (A \to (A \to A) \to A) \to (A \to A \to A) \to A \to A \qquad \vdash A \to (A \to A) \to A}{\vdash (A \to A \to A) \to A \to A} \qquad \vdash A \to A \to A}{\vdash A \to A}.$$

- We use $\mathsf{P}_m X_1 \cdots X_m Y$ as an abbreviation for

  $$\mathsf{P}X_1(\mathsf{P}X_2(\ldots(\mathsf{P}X_m Y)\ldots)).$$

  So $\mathsf{P}_m X_1 \cdots X_m Y$ can be interpreted as the proposition

  $$X_1 \to X_2 \to \cdots X_m \to Y.$$

  In this notation, Rule-P looks as follows:

  $$\dfrac{\vdash \mathsf{P}_{m+1}X_0 \cdots X_m Y \qquad \vdash X_0}{\vdash \mathsf{P}_m X_1 \cdots X_m Y}.$$

[2] We assume that $\to$ is associative to the right, i.e. $X \to Y \to Z$ denotes $X \to (Y \to Z)$ and not $(X \to Y) \to Z$.

For terms X, Y, Z, we take the following axioms:

(K): $\vdash \mathsf{P}_2 XYX$;

(S): $\vdash \mathsf{P}_3(\mathsf{P}_2 XYZ)(\mathsf{P}XY)XZ$.

Let A be a term. From the axioms we derive $\vdash \mathsf{P}AA$, using rule-P:

$$\cfrac{\cfrac{\vdash \mathsf{P}_3(\mathsf{P}_2 A(\mathsf{P}AA)A)(\mathsf{P}A(\mathsf{P}AA))AA \qquad \vdash \mathsf{P}_2 A(\mathsf{P}AA)A}{\vdash \mathsf{P}_2(\mathsf{P}A(\mathsf{P}AA))AA} \qquad \vdash \mathsf{P}A(\mathsf{P}AA)}{\vdash \mathsf{P}AA};$$

- In PAT-style, the situation is similar. Now we do not use any axioms, but we use some standard combinators. The combinator K (which can be compared to the λ-term $\lambda \mathbf{xy.x}$) has type $\mathsf{F}_2 XYX$, for arbitrary F-objects X, Y (a term can have more than one type in Curry's theory). K can be seen as a "proof" of the axiom $(\mathsf{F}_2 XYX)^P$. This is indicated by putting K behind the axiom:

  $$(\mathsf{F}_2 XYX)^P \mathsf{K}.$$

  The combinator S, comparable to the λ-term $\lambda \mathbf{xyz.xz(yz)}$, has type

  $$\mathsf{F}_3(\mathsf{F}_2 XYZ)(\mathsf{F}XY)XZ$$

  for arbitrary F-objects X, Y, Z. S can be seen as a "proof" of the axiom $(\mathsf{F}_3(\mathsf{F}_2 XYZ)(\mathsf{F}XY)XZ)^P$. This is denoted as

  $$(\mathsf{F}_3(\mathsf{F}_2 XYZ)(\mathsf{F}XY)XZ)^P \mathsf{S}.$$

  The derivation above now translates to:

  $$\cfrac{\cfrac{\vdash \mathsf{F}_3(\mathsf{F}_2 A(\mathsf{F}AA)A)(\mathsf{F}A(\mathsf{F}AA))AA\mathsf{S} \qquad \vdash \mathsf{F}_2 A(\mathsf{F}AA)A\mathsf{K}}{\vdash \mathsf{F}_2(\mathsf{F}A(\mathsf{F}AA))AA(\mathsf{SK})} \qquad \vdash \mathsf{F}A(\mathsf{F}AA)\mathsf{K}}{\vdash \mathsf{F}AA(\mathsf{SKK})}.$$

  The conclusion of this derivation can be read as: SKK is a function from A to A, or, with PAT in mind: SKK is a proof of the proposition $A \to A$.

Both derivations correspond to the derivation of the proposition $A \to A$ in natural deduction style, with the use of modus ponens, and the K- and S-axioms only:

$$\cfrac{\cfrac{\vdash (A \to (A \to A) \to A) \to (A \to A \to A) \to A \to A \qquad \vdash A \to (A \to A) \to A}{\vdash (A \to A \to A) \to A \to A} \qquad \vdash A \to A \to A}{\vdash A \to A}.$$

4a3 The discovery of PAT: Howard

Howard [75] follows the argument of Curry and Feys [40] and combines it with Tait's discovery of the correspondence between cut elimination and β-reduction of λ-terms [140].

Example 4.2 The idea is as follows. Consider the following derivation in natural deduction style of a proposition B:

$$\dfrac{\dfrac{\begin{array}{c}[A]\\ \boxed{\mathfrak{D}_1}\\ B\end{array}}{A \to B} \qquad \begin{array}{c}\boxed{\mathfrak{D}_2}\\ A\end{array}}{B}$$

Here, $[A]$ denotes that the assumption A has been discharged at the point where we concluded $A \to B$ from B. $\mathfrak{D}_1$ is a derivation with some assumptions of A, and conclusion B, whilst $\mathfrak{D}_2$ is a derivation with conclusion A. The derivation $\mathfrak{D}_2$ can be used to replace the assumptions of A in derivation $\mathfrak{D}_1$. This means that we can transform the derivation to:

$$\begin{array}{c}\boxed{\mathfrak{D}_2}\\ A\\ \boxed{\mathfrak{D}_1}\\ B\end{array}$$

where copies of $\mathfrak{D}_2$ have replaced the assumptions A in $\mathfrak{D}_1$.

We can decorate the two derivations above with λ-terms that represent proofs. This results in the following two deductions:

$$\dfrac{\dfrac{\begin{array}{c}[x{:}A]\\ \boxed{\mathfrak{D}_1}\\ T : B\end{array}}{(\lambda x{:}A.T) : (A \to B)} \qquad \begin{array}{c}\boxed{\mathfrak{D}_2}\\ S : A\end{array}}{((\lambda x{:}A.T)S) : B}$$

and

$$\begin{array}{c}\boxed{\mathfrak{D}_2}\\ S : A\\ \boxed{\mathfrak{D}_1}\\ T[x{:=}S] : B\end{array}$$

The assumption of A is represented by a variable x of type A. This is a natural idea: the variable expresses the idea "assume we have some proof of A". The derivation $\mathfrak{D}_1$ is represented by a λ-term T, in which the variable x may occur (we can use the assumption A in derivation $\mathfrak{D}_1$). Then the term $\lambda x{:}A.T$ exactly represents a proof of $A \to B$: it is a function that transforms any proof x of A into a proof T of B. As $\mathfrak{D}_2$ is a derivation of A (assume, S is a proof term of A), we can apply $\lambda x{:}A.T$ to S, obtaining a proof $(\lambda x{:}A.T)S$ of B.

Now substituting the derivation $\mathfrak{D}_2$ for the assumptions of A in $\mathfrak{D}_1$ is nothing more than replacing the assumption "assume we have some proof of A" by the explicit proof S, or in other words: substituting S for x. This results in a term T, where each occurrence of x has been replaced by S: the λ-term $T[x{:=}S]$.

We see that the proof transformation exactly corresponds to the β-reduction

$$(\lambda x{:}A.T)S \to_\beta T[x{:=}s]$$

This is the first time that proofs are treated as λ-terms. Howard doesn't call these λ-terms "proofs" but "constructions".

Moreover, Howard's treatment of PAT pays attention to both Propositions as Types (following the line of Curry and Feys) and Proofs as Terms (by using λ-terms to represent proofs, thus following the interpretation of logical implication as given by Heyting).

Howard's discovery dates from 1969, but was not published until 1980.

4a4 The discovery of PAT: de Bruijn

Independently of Curry and Feys and Howard, we find a variant of PAT in the first AUTOMATH system of de Bruijn (AUT-68 [112], [19]). Though de Bruijn was probably influenced by Heyting (see [21] in [112], p. 211), his ideas arose independently from Curry, Feys and Howard. This can be clearly seen in Section 2.4 of [18], where propositions as types (or better: Proofs as terms) is implemented in the following way, differing from the method of Curry and Howard.

First, a constant `bool` is introduced. `bool` is a type: The type of propositions. If b is a term of type `bool` (so b is a proposition), then $\mathtt{true}(b)$ is a primitive notion of type `type`. $\mathtt{true}(b)$ represents the type of the proofs of b. So, a proof of proposition b is of type $\mathtt{true}(b)$ and not of type b (since propositions themselves are no types) With this "bool-style" implementation (as it was called by de Bruijn in [21]) in mind, it becomes clear why de Bruijn prefers the terminology "proofs as terms" to "propositions as types": In the bool-style implementation, propositions are not represented as types. Only the class of proofs of such a proposition is represented as a type. Proofs however, are represented as terms, just as in Howard's implementation of PAT. So in the bool-style implementation, the link between proposition and type is not as direct as the link between proof and term. The implementation of Howard (called "prop-style" by de Bruijn) does not make any distinction between a proposition and the type of its proofs.

The bool-style implementation has as advantage that one does not need a higher order lambda calculus to construct predicate logic. In relatively weak AUTOMATH systems such as AUT-68 one usually finds a "bool-style" implementation of PAT. It would be impossible to give a "prop-style" implementation in such a system as its λ-calculus is not strong enough to support it. In AUTOMATH systems with a more powerful λ-calculus we also find "prop-style" implementations. See [111] for a global description of how prop-style implementations are made in AUTOMATH.

Another advantage of the bool-style implementation is that one does not depend on a fixed interpretation of the logical connectives. One is free to define ones own logical system (and it is possible to base that system on the Brouwer-Heyting-Kolmogorov interpretation of the logical connectives, just like the prop-style implementation of PAT). This has been one of the reasons for de Bruijn to implement PAT in a bool-style way (see [21]).

Though the bool-style implementation has disappeared from later AUTOMATH systems, it is still in use in the Edinburgh Logical Framework [64], and the systems proposed in certain formulations of the Calculus of Constructions (e.g., by Streicher [139]).

4b Lambda calculus

We give a description of typed lambda terms. This description follows the line of [3], as it mainly serves as a description of Pure Type Systems (PTSs).

Definition 4.3 Let $\mathbb{V}$ be a set of variables and $\mathbb{C}$ a set of constants. The set $\mathcal{T}(\mathbb{V},\mathbb{C})$ (shorthand: $\mathcal{T}$, if it is clear which sets $\mathbb{V}$ and $\mathbb{C}$ are used) of typed lambda terms with variables from $\mathbb{V}$ and constants from $\mathbb{C}$ is defined by the following abstract syntax:

$$\mathcal{T} ::= \mathbb{V} \mid \mathbb{C} \mid \mathcal{T}\mathcal{T} \mid \lambda\mathbb{V}{:}\mathcal{T}.\mathcal{T} \mid \Pi\mathbb{V}{:}\mathcal{T}.\mathcal{T}.$$

If x does not occur in B then $\Pi x{:}A.B$ is sometimes denoted by $A \to B$.

We assume that $\mathbb{V}$ and $\mathbb{C}$ are countably infinite and are disjoint. We use $\equiv$ to denote syntactical equality between typed lambda terms.

We use x, y, z, α, β as meta-variables over $\mathbb{V}$. In examples, we sometimes want to use some specific elements of $\mathbb{V}$; we use typewriter-style to denote such specific elements. So: $\mathtt{x}$ is a specific element of $\mathbb{V}$; while x is a meta-variable *over* $\mathbb{V}$. The variables $\mathtt{x}$, $\mathtt{y}$, $\mathtt{z}$ are assumed to be *distinct* elements of $\mathbb{V}$ (so $\mathtt{x} \not\equiv \mathtt{y}$ etc.), while meta-variables $x, y, z, \ldots$ may refer to variables in the object language that are syntactically equal. We use $A, B, C, \ldots, a, b, \ldots$ as meta-variables over $\mathcal{T}$.

A term $\lambda x{:}A.b$ has as intuitive interpretation the function that assigns $b[x{:=}a]$ (the term b in which each occurrence of x has been replaced by a) to each a that belongs to (is an element of, has type) A. If b has type B, then $\lambda x{:}A.b$ is a function from A to B. $A \to B$ should be interpreted as the type of functions from A to B. This means: $\lambda x{:}A.b$ has type $A \to B$.

Example 4.4 $\lambda \mathtt{x}{:}A.\mathtt{x}$ is the identity function on A, and has type $A \to A$.

In some situations, we allow that the type B of b *depends on* the variable x. In that case, $b[x{:=}a]$ is of type $B[x{:=}a]$ for a of type A. Then $\lambda x{:}A.b$ is a function with domain A and range $\bigcup_{a:A} B[x{:=}a]$, with the special property that the function value for $a{:}A$, $b[x{:=}a]$, belongs to the subset $B[x{:=}a]$ of $\bigcup_{a:A} B[x{:=}a]$. The type of such functions will be represented by $\Pi x{:}A.B$.

Example 4.5 The polymorphic identity $\lambda \mathtt{y}{:}{*}.\lambda \mathtt{x}{:}\mathtt{y}.\mathtt{x}$ has as type

$$\Pi \mathtt{y}{:}{*}.\mathtt{y} \to \mathtt{y}.$$

We could have written $\Pi \mathtt{y}{:}{*}.\Pi \mathtt{x}{:}\mathtt{y}.\mathtt{y}$ instead of $\Pi \mathtt{y}{:}{*}.\mathtt{y} \to \mathtt{y}$. Here $*$ can be interpreted as the class of types.

Remark 4.6 The term $\lambda \mathtt{y}{:}{*}.\lambda \mathtt{x}{:}\mathtt{y}.\mathtt{x}$ in Example 4.5 is a function of two variables: $\mathtt{y}$ and $\mathtt{x}$. The function is constructed by repeated λ-abstraction. The first λ-abstraction (over $\mathtt{x}$) leads to a function of one variable: $\lambda \mathtt{x}{:}\mathtt{y}.\mathtt{x}$, and another λ-abstraction (over $\mathtt{y}$) leads to the desired function of two variables. The use of repeated λ-abstraction in order to represent functions of more than one variable is called "currying" after H. B. Curry, though currying was already discovered by Schönfinkel in 1924 [132], before Curry discovered it, and the basic ideas for currying can already be found in the works of Frege, which date from 1879 (see Section 1b1 of this book).

The following notational conventions allow us to reduce the number of brackets in terms:

Notation 4.7

- We write $\lambda \vec{x}{:}\vec{A}.B$, or $\lambda_{i=1}^{n}\, x_i{:}A_i.A$, as shorthand for

$$\lambda x_1{:}A_1.(\lambda x_2{:}A_2.(\cdots(\lambda x_n{:}A_n.A)\cdots));$$

- We use $\Pi \vec{x}{:}\vec{A}.B$, or $\Pi_{i=1}^{n} x_i{:}A_i.A$, as shorthand for

$$\Pi x_1{:}A_1.(\Pi x_2{:}A_2.(\cdots(\Pi x_n{:}A_n.A)\cdots));$$

- We write $AB_1 \cdots B_n$ as shorthand for

$$(\cdots((AB_1)B_2)\cdots B_n).$$

Definition 4.8 For $A \in \mathcal{T}$ we define $\text{FV}(A)$, the set of *free variables* of A, as follows:

- $\text{FV}(c) = \varnothing$ for $c \in \mathbb{C}$;
- $\text{FV}(x) = \{x\}$ for $x \in \mathbb{V}$;
- $\text{FV}(A_1 A_2) = \text{FV}(A_1) \cup \text{FV}(A_2)$;
- $\text{FV}(\lambda x{:}A_1.A_2) = (\text{FV}(A_1) \setminus \{x\}) \cup \text{FV}(A_2)$;
- $\text{FV}(\Pi x{:}A_1.A_2) = (\text{FV}(A_1) \setminus \{x\}) \cup \text{FV}(A_2)$.

Definition 4.9 If $\text{FV}(a) = \{x_1, \ldots, x_n\}$ and $A_1, \ldots, A_n$ is a list of terms then $\lambda_{i=1}^{n}\, x_i{:}A_i.a$ is a *closure* of a.

Notice that there may be many different closures of one and the same term a.

A subset of the set of λ-terms that will be used in this book is the set of the so-called λI-terms:

Definition 4.10 Let $\mathbb{V}$ be a set of variables and $\mathbb{C}$ a set of constants. The set of λI-terms $\mathcal{T}_I$ over $\mathbb{V}$ and $\mathbb{C}$ is defined as follows:

- If $v \in \mathbb{V}$ then $v \in \mathcal{T}_I$; if $c \in \mathbb{C}$ then $c \in \mathcal{T}_I$;
- If $A, B \in \mathcal{T}_I$ then $AB \in \mathcal{T}_I$;
- If $A, b \in \mathcal{T}_I$ and $x \in \text{FV}(b)$ then $\lambda x{:}A.b \in \mathcal{T}_I$;
- If $A, B \in \mathcal{T}_I$ then $\Pi x{:}A.B \in \mathcal{T}_I$.

So within the set of λI-terms, a λ-abstraction $\lambda x{:}A.b$ can only be made if the term b really *depends* on the variable x. This means that *constant* functions, and functions of more variables that are constant in one or more of their variables, are excluded from the set of λI-terms.

Terms that are equal up to a change of bound variables are considered to be syntactically equal. This allows us to assume the so-called *Barendregt Convention*:

Convention 4.11 (Barendregt Convention) Bound variables will be chosen to be different from free variables. For instance, we write $(\lambda \mathbf{y}{:}A.\mathbf{y})\mathbf{x}$ instead of $(\lambda \mathbf{x}{:}A.\mathbf{x})\mathbf{x}$. Moreover, we use different bound names for different bound variables.

Once this variable convention has been assumed we can define substitution in a straightforward manner (whereas the definition in [40] is more complicated, and a formal definition of substitution is completely absent in [27, 28] and [30]):

Definition 4.12 (Substitution) We define $A[x{:=}B]$ by induction on the structure of A:

- $y[x{:=}B] \equiv \begin{cases} B & \text{if } y \equiv x; \\ y & \text{if } y \not\equiv x; \end{cases}$
- $(A_1 A_2)[x{:=}B] \equiv A_1[x{:=}B]A_2[x{:=}B]$;
- $(\lambda y{:}A_1.A_2)[x{:=}B] \equiv \lambda y{:}A_1[x{:=}B].A_2[x{:=}B]$;
- $(\Pi y{:}A_1.A_2)[x{:=}B] \equiv \Pi y{:}A_1[x{:=}B].A_2[x{:=}B]$.

We use the abbreviation $A[x_i{:=}B_i]_{i=m}^n$ to denote

$$A[x_m{:=}B_m]\cdots[x_n{:=}B_n].$$

If $m > n$ then $A[x_i{:=}B_i]_{i=m}^n$ denotes A. We also use the notation $A[\vec{x}{:=}\vec{B}]$ for $A[x_i{:=}B_i]_{i=1}^n$.

Definition 4.13 (Compatibility) Let $\to$ be a relation on lambda terms. We say that $\to$ is compatible if whenever $A \to A'$ we get:

$$\begin{array}{cc} AB \to A'B & BA \to BA' \\ \lambda x{:}A.B \to \lambda x{:}A'.B & \lambda x{:}B.A \to \lambda x{:}B.A' \\ \Pi x{:}A.B \to \Pi x{:}A'.B & \Pi x{:}B.A \to \Pi x{:}B.A' \end{array}$$

On lambda terms we have the notion of *β-reduction.*

Definition 4.14 (β-reduction) The relation $\to_\beta$ is described by the contraction rule

$$(\lambda x{:}A_1.A_2)B \to_\beta A_2[x{:=}B]$$

and the compatibility rules of Definition 4.13.

$\twoheadrightarrow_\beta$ is the smallest reflexive and transitive relation that includes $\to_\beta$; $=_\beta$ is the smallest reflexive, symmetric and transitive relation that includes $\to_\beta$. By $A \twoheadrightarrow_\beta^+ B$ we indicate that $A \twoheadrightarrow_\beta B$, but $A \not\equiv B$.

A term that has no subterm of the form $(\lambda x{:}A_1.A_2)B$ is a term in *β-normal form*, or a *normal form* if no confusion arises. We write $A \to_\beta^{\mathrm{nf}} B$ if $A \to_\beta B$ and B is in β-normal form. Similarly, $A \twoheadrightarrow_\beta^{\mathrm{nf}} B$ if $A \twoheadrightarrow_\beta B$ and B is in β-normal form.

The most important property of $\to_\beta$ is the so-called *Church-Rosser property*:

Theorem 4.15 *If $A \twoheadrightarrow_\beta B_1$ and $A \twoheadrightarrow_\beta B_2$ then there is C such that $B_1 \twoheadrightarrow_\beta C$ and $B_2 \twoheadrightarrow_\beta C$.* ⊠

There are numerous proofs of this theorem in the literature. The most well-known is via the Strip Lemma (see [2], Chapter 11), another short and elegant proof is given by Tait and Martin-Löf [104], also described in Chapter 3 of [2].

In this book we see many variants on the basic lambda terms of Definition 4.3, for instance lambda terms with parameters in Chapter 10. It is easy to prove that these variants have the Church-Rosser property for $\to_\beta$ as well.

4c Pure Type Systems

Pure Type Systems (PTSs) were introduced by Berardi [10] and Terlouw [142] as a general framework in which many current type systems can be described. The framework is a generalisation of the well-known Barendregt cube.

The description below is based on [3].

Terms of PTSs are as given in Definition 4.3. That is:

Definition 4.16 (Terms) Let $\mathbb{V}$ be a set of variables and $\mathbb{C}$ a set of constants. Assume that $\mathbb{V}$ and $\mathbb{C}$ are countably infinite and are disjoint. The set $\mathcal{T}(\mathbb{V},\mathbb{C})$ (shorthand: $\mathcal{T}$) is given by:

$$\mathcal{T} ::= \mathbb{V} \mid \mathbb{C} \mid \mathcal{T}\mathcal{T} \mid \lambda\mathbb{V}{:}\mathcal{T}.\mathcal{T} \mid \Pi\mathbb{V}{:}\mathcal{T}.\mathcal{T}.$$

We use the same notations and conventions of Section 4b.

For PTSs, we simply assume β-reduction of Definition 4.14. That is:

Definition 4.17 (Reduction) The relation $\to_\beta$ is described by the contraction rule

$$(\lambda x{:}A_1.A_2)B \to_\beta A_2[x{:=}B]$$

and the compatibility rules of Definition 4.13.

$=_\beta$ is the smallest reflexive, symmetric and transitive relation that includes $\to_\beta$.

Definition 4.18 (PTSs Specification) A *specification* for PTSs is a triple $(\boldsymbol{S},\boldsymbol{A},\boldsymbol{R})$, such that $\boldsymbol{S} \subseteq \mathbb{C}$, $\boldsymbol{A} \subseteq \boldsymbol{S}\times\boldsymbol{S}$ and $\boldsymbol{R} \subseteq \boldsymbol{S}\times\boldsymbol{S}\times\boldsymbol{S}$. The specification is called *singly sorted* if $\boldsymbol{A}$ is a (partial) function $\boldsymbol{S} \to \boldsymbol{S}$, and $\boldsymbol{R}$ is a (partial) function $\boldsymbol{S}\times\boldsymbol{S} \to \boldsymbol{S}$. $\boldsymbol{S}$ is called the set of *sorts*, $\boldsymbol{A}$ is the set of *axioms*, and $\boldsymbol{R}$ is the set of (Π-formation) *rules* of the specification.

Definition 4.19 (Contexts) A *context* is a finite (possibly empty) list of variable declarations $x_1{:}A_1,\ldots,x_n{:}A_n$ (shorthand: $\vec{x}{:}\vec{A}$). We call $\{x_1,\ldots,x_n\}$ the *domain* $\text{DOM}\left(\vec{x}{:}\vec{A}\right)$ of the context. The *empty context* is denoted $\langle\rangle$. If Γ,Δ are contexts then we write $\Gamma \subseteq \Delta$ if all declarations in Γ are also in Δ.

We use Γ, Δ as meta-variables for contexts.
Barendregt convention 4.11 is also extended to contexts.
Substitution can be extended to contexts:

Definition 4.20 We define $\Gamma[x{:=}A]$ by induction on the length of Γ:

- $\langle\rangle[x{:=}A] \equiv \langle\rangle$;
- $(\Gamma', y{:}B)[x{:=}A] \equiv \begin{cases} \Gamma', y{:}B & \text{if } x \equiv y; \\ \Gamma'[x{:=}A], y{:}B[x{:=}A] & \text{if } x \not\equiv y. \end{cases}$

Definition 4.21 (Pure Type Systems) Let $\mathfrak{S} = (\boldsymbol{S}, \boldsymbol{A}, \boldsymbol{R})$ be a specification. The Pure Type System $\lambda\mathfrak{S}$ describes in which ways judgements $\Gamma \vdash_{\mathfrak{S}} A : B$ (or $\Gamma \vdash A : B$, if it is clear which $\mathfrak{S}$ is used) can be derived. $\Gamma \vdash A : B$ states that A has type B in context Γ.

(axiom) $$\langle\rangle \vdash s_1 : s_2 \qquad (s_1, s_2) \in \boldsymbol{A}$$

(start) $$\frac{\Gamma \vdash A : s}{\Gamma, x{:}A \vdash x : A} \qquad x \notin \mathrm{DOM}(\Gamma)$$

(weak) $$\frac{\Gamma \vdash A : B \qquad \Gamma \vdash C : s}{\Gamma, x{:}C \vdash A : B} \qquad x \notin \mathrm{DOM}(\Gamma)$$

(Π) $$\frac{\Gamma \vdash A : s_1 \qquad \Gamma, x{:}A \vdash B : s_2}{\Gamma \vdash (\Pi x{:}A.B) : s_3} \qquad (s_1, s_2, s_3) \in \boldsymbol{R}$$

(λ) $$\frac{\Gamma, x{:}A \vdash b : B \qquad \Gamma \vdash (\Pi x{:}A.B) : s}{\Gamma \vdash (\lambda x{:}A.b) : (\Pi x{:}A.B)}$$

(appl) $$\frac{\Gamma \vdash F : (\Pi x{:}A.B) \qquad \Gamma \vdash a : A}{\Gamma \vdash Fa : B[x{:=}a]}$$

(conv) $$\frac{\Gamma \vdash A : B \qquad \Gamma \vdash B' : s \qquad B =_\beta B'}{\Gamma \vdash A : B'}$$

Definition 4.22 (Statement, judgement, Legal context and term) Let Γ be a context, $A, B, C \in \mathcal{T}$.

1. $A : B$ is called a *statement*. A and B are its *subject* and *predicate* respectively.

2. $\Gamma \vdash A : B$ is called a *judgement*.
   We write $\Gamma \vdash A : B : C$ to denote $\Gamma \vdash A : B \wedge \Gamma \vdash B : C$.

3. Γ is called *legal* if $\exists A_1, B_1 \in \mathcal{T}$ such that $\Gamma \vdash A_1 : B_1$.

4. A is called a Γ-*term* if $\exists B_1 \in \mathcal{T}[\Gamma \vdash A : B_1 \vee \Gamma \vdash B_1 : A]$.
   A is called *legal* if $\exists \Gamma_1[A$ is a Γ_1-term$]$.

An important class of examples of PTSs is formed by the eight PTSs of the so-called Barendregt cube. These systems all have $\{*, \Box\}$ as set of sorts, and $*{:}\Box$ as only axiom, but they differ on the Π-formation rules that are allowed (see the next section). Another PTS that occurs in this book is Luo's Extended Calculus of Constructions ECC (see [3]). This is a PTS with

$$\begin{aligned} \boldsymbol{S} &= \mathbb{N}; \\ \boldsymbol{A} &= \{(n, n+1) \mid n \in \mathbb{N}\}; \\ \boldsymbol{R} &= \{(m, 0, 0) \mid m \in \mathbb{N}\} \cup \{(m, n, r) \mid 0 \leq m, n \leq r\}. \end{aligned}$$

This is indeed an extension of λC (write $*$ for 0 and $\Box$ for 1).

4c1 The Barendregt cube

In [3], Barendregt proposes a framework, now often called the Barendregt cube, in which eight important and well-known type systems are presented in a uniform way. This makes a detailed comparison of these systems possible. The weakest system $\lambda_{\rightarrow}$ of the cube is a simplified version of Church's simply typed λ-calculus $\lambda_{\rightarrow\text{Church}}$ [29], and the strongest system is the Calculus of Constructions λC [37]. Girard's well-known System F [60] figures on the cube between $\lambda_{\rightarrow}$ and λC. In addition, all the systems of the cube are PTSs. Moreover, via the Propositions-as-Types principle (see [75]), many logical systems can be described in the systems of the cube, see [59].

The cube has two sorts $*$ (the set of types) and $\Box$ (the set of kinds) with the unique axiom $* : \Box$. If $A : *$ (resp. $A : \Box$) we say A is a type (resp. a kind). Each system of the cube has its own set $\boldsymbol{R}$ which must contain $(*, *, *)$ and which should satisfy: $\boldsymbol{R} \subseteq \{(*, *, *), (*, \Box, \Box), (\Box, *, *), (\Box, \Box, \Box)\}$. We use (s_1, s_2) to denote (s_1, s_2, s_2).

Note that as there are only two sorts, $*$ and $\Box$, and as each set $\boldsymbol{R}$ must contain $(*, *)$, there are only eight possible different systems of the cube. With the rule (Π), an important aspect of the cube is that it provides a factorisation of the expressive power of the Calculus of Constructions into three features: *polymorphism*, *type constructors*, and *dependent types*:

- $(*, *)$ is the basic rule that forms types. All type systems of the cube have this rule.

- $(\Box, *)$ is the rule that takes care of polymorphism. The sytem F (also known as $\lambda 2$), which is due independently to Girard and Reynolds, is the weakest system on the cube that features this rule.

- $(\Box, \Box)$ takes care of type constructors. The system $\lambda\underline{\omega}$ is the weakest system on the cube that features this rule.

$\lambda\!\to$	$(*,*)$			
$\lambda 2$	$(*,*)$	$(\Box,*)$		
λP	$(*,*)$		$(*,\Box)$	
$\lambda\underline{\omega}$	$(*,*)$			$(\Box,\Box)$
$\lambda P2$	$(*,*)$	$(\Box,*)$	$(*,\Box)$	
$\lambda\omega$	$(*,*)$	$(\Box,*)$		$(\Box,\Box)$
$\lambda P\underline{\omega}$	$(*,*)$		$(*,\Box)$	$(\Box,\Box)$
λC	$(*,*)$	$(\Box,*)$	$(*,\Box)$	$(\Box,\Box)$

Figure 4.1: Different type formation conditions

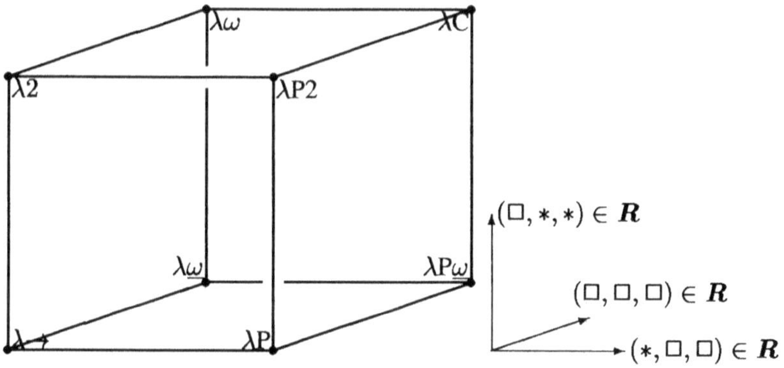

Figure 4.2: The Barendregt cube

- $(*,\Box)$ takes care of term dependent types. The system λP is the weakest system on the cube that features this rule.

In this section we shortly repeat the definition of the systems in the cube. For background information the reader may consult [3].

Terms and reduction in the cube are as in Definitions 4.16 and 4.17. The next definition shows that a cube specification is a special case of a PTS specification.

Definition 4.23 (Specifications in the cube) Let (s_1, s_2) denote (s_1, s_2, s_2). A *cube specification* is a triple $(\boldsymbol{S}, \boldsymbol{A}, \boldsymbol{R})$, such that:

- $\boldsymbol{S} = \{*, \Box\} \subseteq \mathbb{C}$,
- $\boldsymbol{A} = \{* : \Box\} \subseteq \boldsymbol{S} \times \boldsymbol{S}$ and
- $\boldsymbol{R} \subseteq \{(*,*), (*,\Box), (\Box,*), (\Box,\Box)\} \subseteq \boldsymbol{S} \times \boldsymbol{S} \times \boldsymbol{S}$ such that $(*,*) \in \boldsymbol{R}$.

We use the same notations as Definition 4.18. Note that the only difference between one cube specification and another is in the set of rules $\boldsymbol{R}$.

Note that, as all the systems of the cube have the same sets of sorts $\boldsymbol{S}$ and axioms $\boldsymbol{A}$, it is enough to represent each system by its set of rules $\boldsymbol{R}$ instead of using all the specification $(\boldsymbol{S}, \boldsymbol{A}, \boldsymbol{R})$.

Definitions 4.19, 4.20 and 4.22 remain unchanged for the cube. Now we come to the definition of the different systems of the cube (which is a special case of the definition of Pure Type Systems 4.21. As the only difference between one cube specification and another is in the set of Rules $\boldsymbol{R}$, we can represent a cube system simply by its set of rules $\boldsymbol{R}$ instead of using the triple $(\boldsymbol{S}, \boldsymbol{A}, \boldsymbol{R})$.

Definition 4.24 (Systems of the Barendregt cube)
Let $(\boldsymbol{S}, \boldsymbol{A}, \boldsymbol{R})$ be a cube specification. The type system $\lambda\boldsymbol{R}$ describes in which ways judgments $\Gamma \vdash_{\boldsymbol{R}} A : B$ (or $\Gamma \vdash A : B$, if it is clear which $\boldsymbol{R}$ is used) can be derived. $\Gamma \vdash A : B$ states that A has type B in context Γ. The typing rules are inductively defined as follows:

(axiom) $$\langle\rangle \vdash * : \square$$

(start) $$\frac{\Gamma \vdash A : s}{\Gamma, x{:}A \vdash x : A} \qquad x \notin \text{DOM}(\Gamma)$$

(weak) $$\frac{\Gamma \vdash A : B \qquad \Gamma \vdash C : s}{\Gamma, x{:}C \vdash A : B} \qquad x \notin \text{DOM}(\Gamma)$$

(Π) $$\frac{\Gamma \vdash A : s_1 \qquad \Gamma, x{:}A \vdash B : s_2}{\Gamma \vdash (\Pi x{:}A.B) : s_2} \qquad (s_1, s_2) \in \boldsymbol{R}$$

(λ) $$\frac{\Gamma, x{:}A \vdash b : B \qquad \Gamma \vdash (\Pi x{:}A.B) : s}{\Gamma \vdash (\lambda x{:}A.b) : (\Pi x{:}A.B)}$$

(appl) $$\frac{\Gamma \vdash F : (\Pi x{:}A.B) \qquad \Gamma \vdash a : A}{\Gamma \vdash Fa : B[x{:=}a]}$$

(conv) $$\frac{\Gamma \vdash A : B \qquad \Gamma \vdash B' : s \qquad B =_\beta B'}{\Gamma \vdash A : B'}$$

There are eight different possibilities for $\boldsymbol{R}$ leading to the systems in Figure 4.1.

The dependencies between these systems can be depicted in the Barendregt cube (see Figure 4.2). Furthermore, the systems in the cube are related to other type systems as is shown in the overview of Figure 4.3.

System	Related system	Names, references
$\lambda_{\to}$	λ^{τ}	simply typed λ-calculus; [29], [2] (Appendix A), [71] (Chapter 14)
$\lambda 2$	F	second order typed λ-calculus; [60], [126]
λP	AUT-QE[3]	[19]
	LF[4]	[64]
$\lambda P2$		[103]
$\lambda\underline{\omega}$	POLYREC	[125]
$\lambda\omega$	$F\omega$	[60]
λC	CC	Calculus of Constructions; [37]

Figure 4.3: Systems of the cube

4c2 Metaproperties of PTSs

Pure Type Systems (and hence all the systems of the cube) have some important meta-properties, which we describe below. Throughout this section, $\vdash$ denotes derivability in a PTS with a certain specification $\mathfrak{S} = (\boldsymbol{S}, \boldsymbol{A}, \boldsymbol{R})$.

Lemma 4.25 (Restricted Weakening) *If $\Gamma \vdash A : B$, we may assume the derivation of $\Gamma \vdash A : B$ to contain only applications of the rule (weak) that are of the form*

$$\frac{\Gamma \vdash u : B \qquad \Gamma \vdash C : s}{\Gamma, x{:}C \vdash u : B}$$

where $u \in \mathbb{V} \cup \mathbb{C}$. ⊠

Lemma 4.26 (Free Variable Lemma) *Let $\Gamma \equiv x_1{:}A_1, \ldots, x_n{:}A_n$ be legal, say $\Gamma \vdash B : C$. Then*

1. *The x_i are distinct;*
2. $\text{FV}(B), \text{FV}(C) \subseteq \text{DOM}(\Gamma)$;
3. $\text{FV}(A_i) \subseteq \{x_1, \ldots, x_{i-1}\}$ *for* $1 \leq i \leq n$. ⊠

Lemma 4.27 (Start Lemma) *Let Γ be legal. Then*

1. $\Gamma \vdash s_1 : s_2$ *for all* $(s_1, s_2) \in \boldsymbol{A}$;

[3] A more precise study of AUT-QE, respecting the parameter structure of AUT-QE, shows that AUT-QE can be positioned a little bit higher in the cube: exactly inbetween λP and λC. See Chapter 10, especially Section 10e4. — footnote by the authors.

[4] In Chapter 10, Section 10e2, we show that the practical use of LF does not use the full power of λP. In the refinement of the Barendregt cube presented there, we show that the use of LF in practice corresponds to a system that is inbetween $\lambda_{\to}$ and λP. — footnote by the authors.

2. *$\Gamma \vdash x : A$ for all $(x{:}A) \in \Gamma$.* ⊠

Lemma 4.28 (Transitivity Lemma) *Let Γ, Δ be contexts. Assume Γ is legal, $\Gamma \vdash x{:}A$ for all $(x{:}A) \in \Delta$, and $\Delta \vdash B : C$. Then $\Gamma \vdash B : C$.* ⊠

Lemma 4.29 (Thinning Lemma) *If Δ is legal, $\Gamma \subseteq \Delta$ and $\Gamma \vdash A : B$, then $\Delta \vdash A : B$.* ⊠

Lemma 4.30 (Substitution Lemma) *If $\Gamma, x{:}A, \Delta \vdash B : C$ and $\Gamma \vdash D : A$ then $\Gamma, \Delta[x{:=}D] \vdash B[x{:=}D] : C[x{:=}D]$.* ⊠

Lemma 4.31 (Generation Lemma)

1. *If $\Gamma \vdash c : C$ for a $c \in \mathbb{C}$ then there is $s \in \boldsymbol{S}$ such that $C =_\beta s$ and $(c{:}s) \in \boldsymbol{A}$;*
2. *If $\Gamma \vdash x : C$ then there is $s \in \boldsymbol{S}$ and $B =_\beta C$ such that $\Gamma \vdash B : s$ and $(x{:}B) \in \Gamma$;*
3. *If $\Gamma \vdash (\Pi x{:}A.B) : C$ then there is $(s_1, s_2, s_3) \in \boldsymbol{R}$ such that $\Gamma \vdash A : s_1, \Gamma, x{:}A \vdash B : s_2$ and $C =_\beta s_3$;*
4. *If $\Gamma \vdash (\lambda x{:}A.b) : C$ then there is $s \in \boldsymbol{S}$ and B such that $\Gamma \vdash (\Pi x{:}A.B) : s$; $\Gamma, x{:}A \vdash b : B$; and $C =_\beta (\Pi x{:}A.B)$;*
5. *If $\Gamma \vdash Fa : C$ then there are A, B such that $\Gamma \vdash F : (\Pi x{:}A.B)$, $\Gamma \vdash a : A$ and $C =_\beta B[x{:=}a]$.* ⊠

Lemma 4.32 (Correctness of Types) *If $\Gamma \vdash A : B$ then $B \equiv s$ or $\Gamma \vdash B : s$ for some $s \in \boldsymbol{S}$.* ⊠

Lemma 4.33 (Subterm Lemma) *If A is legal and B is a subterm of A, then B is legal.* ⊠

Lemma 4.34 (Subject Reduction) *If $\Gamma \vdash A : B$ and $A \to_\beta A'$ then $\Gamma \vdash A' : B$.* ⊠

The proof of the next lemma is due to Van Benthem Jutting [9].

Lemma 4.35 (Strengthening Lemma) *Assume that $\Gamma, x{:}A$ and $\Delta \vdash B : C$. If $x \notin \mathrm{FV}(\Delta) \cup \mathrm{FV}(B) \cup \mathrm{FV}(C)$, then $\Gamma, \Delta \vdash B : C$.* ⊠

Lemma 4.36 (Unicity of Types) *If $\mathfrak{S}$ is singly sorted, $\Gamma \vdash A : B_1$ and $\Gamma \vdash A : B_2$, then $B_1 =_\beta B_2$.* ⊠

Lemma 4.37 (Strong Permutation Lemma) *If $\Gamma, x{:}A, y{:}B, \Delta \vdash C : D$ and $x \notin \mathrm{FV}(B)$, then $\Gamma, y{:}B, x{:}A, \Delta \vdash C : D$.* ⊠

Definition 4.38 (Topsort) A sort s is a *topsort* if there is no $s' \in \boldsymbol{S}$ such that $(s, s') \in \boldsymbol{A}$. ⊠

Lemma 4.39 (Topsort Lemma) *If s is a topsort and $\Gamma \vdash A : s$ then A is not of the form $A_1 A_2$ or $\lambda x{:}A_1.A_2$.* ⊠

Theorem 4.40 (Strong Normalisation for ECC) *Let A be a legal term in the Extended Calculus of Constructions. Then A is strongly normalising.* ⊠

As the systems of the Barendregt cube are subsystems of ECC, all legal terms in the systems of the Barendregt cube are strongly normalising, too.

Chapter 5

The pre-PAT RTT and STT in PAT-style

In this chapter, we describe how the two most important type systems of the pre-PAT-era, RTT and STT, can be described in a PAT style. This gives insight in the various ways in which PAT-implementations can be made.

5a RTT in PAT style

In this section we show that the system RTT of Chapter 2 can be described in a PAT style using the prop-style implementation of Curry and Howard. This will give us a better view on the various ways in which PAT can be implemented.

Before we can give a description, we must make the following observations:

- Russell and Whitehead designed their system for classical logic. As the PAT principle in prop style is based on intuitionistic logic, we need to supply extra logical axioms to obtain the classical logic of Russell and Whitehead;
- RTT is constructed with the logical connectives $\vee$, $\neg$, $\forall$, while the PAT principle is strongly based on the interpretation of $\rightarrow$ and $\forall$ as function types. In the sequel of this section we will work with the symbols $\forall$, $\rightarrow$, and an additional symbol $\bot$, representing falsum. This makes it possible to interpret a proposition $\neg f$ by $f \rightarrow \bot$, and a proposition $f \vee g$ by $(f \rightarrow \bot) \rightarrow g$;
- As RTT distinguishes between propositions of various orders, it is not enough to provide one class of types. We must distinguish between several classes of types, corresponding to the orders of RTT.[1]

[1] Something similar is done in the proof checker Nuprl, where type universes $\mathcal{U}_1, \mathcal{U}_2, \ldots$ are intro-

In Section 5a1–5a3 we present a type system λRTT for a PAT representation of RTT. The type system is (almost) a *Pure* Type System.

In Section 5a4 we make a comparison between λRTT and RTT, and in Section 5a5 we give some examples on how to "do" logic of the *Principia* in λRTT.

In Section 5a6 we discuss in which ways PAT can be implemented. The various implementations lead to different level structures within the resulting PTS.

In Section 5b we discuss STT in PAT style.

5a1 An introduction to λRTT

We now present a system λRTT that will be suitable for a PAT representation of RTT.

Definition 5.1 (Terms of λRTT) Let $\mathcal{A}$, $\mathcal{V}$ and $\mathcal{R}$ be as in Chapter 2 where $\mathcal{A}$, $\mathcal{V}$ and $\mathcal{R}$ are mutually disjoint. Let $\perp$, ι, $*_s$, $*_n$ and $\Box_n$ for $n \in \mathbb{N}^+$ be new constants (not already in $\mathcal{A}$, $\mathcal{V}$ or $\mathcal{R}$) which are all different.

Define the set $\mathcal{T}$ of terms of λRTT by:

$$\begin{array}{rcl}\mathcal{T} & ::= & *_s \mid *_{\mathbb{N}^+} \mid \Box_{\mathbb{N}^+} \mid \mathcal{A} \mid \mathcal{V} \mid \mathcal{R} \mid \iota \mid \perp \mid \\ & & \mathcal{T}\mathcal{T} \mid \lambda\mathcal{V}{:}\mathcal{T}.\mathcal{T} \mid \Pi\mathcal{V}{:}\mathcal{T}.\mathcal{T}.\end{array}$$

Define $\boldsymbol{S} = \{*_s, \Box_1, \Box_2, \ldots, *_1, *_2, \ldots\}$, $\mathbb{C} = \boldsymbol{S} \cup \{\perp\} \cup \{\iota\} \cup \mathcal{A} \cup \mathcal{R}$ and $\mathbb{V} = \mathcal{V}$. Then obviously, $\mathcal{T} = \mathcal{T}(\mathbb{V}, \mathbb{C})$.

Remark 5.2 In this definition, $*_s$, $*_n$ (for $n \in \mathbb{N}^+$) and $\Box_n$ (for $n \in \mathbb{N}^+$) are *sorts*.

- $*_s$ is the sort of object types. There will be only one object type in λRTT, namely the type of individuals ι. An individual a has type ι, this is denoted: $a : \iota$;
- $*_n$ (for $n \in \mathbb{N}^+$) is the sort containing the propositions of order $\leq n$. Notice that these propositions will be represented as *types*: We are presenting a PAT version of RTT;
- $\Box_n$ (for $n \in \mathbb{N}^+$) contains $*_n$, and (translations of) the ramified types of order n as they occur in RTT.

duced. The type universe $\mathcal{U}_n$ contains all objects of order $\leq n$. The approach below is not exactly similar to the Nuprl approach. In Nuprl, $\mathcal{U}_n$ is not only a subset of $\mathcal{U}_{n+1}$, but also an element of $\mathcal{U}_{n+1}$. See [76], [35] and the next chapter (where these type universes are denoted $*_1, *_2, \ldots$). This is the case neither in RTT nor in the system λRTT below.

Remark 5.3 A term of the form $\Pi x{:}A.B$ denotes a (dependent) function type. That is: $\Pi x{:}A.B$ is the type that contains functions f with domain A, and range $\bigcup_{x:A} B$ (it is possible that x occurs free in B), such that fa has type $B[x{:=}a]$ for all a of type A. Such function types can occur at several places:

- First of all, the translations of the propositional functions of a certain ramified type will belong to a function type. Look for instance at the ramified type $(0^0)^1$. Pfs of this type are functions that take an individual as argument, and return a proposition of order ≤ 1 as result.[2] This suggests to translate the ramified type $(0^0)^1$ by the function type $\Pi \mathtt{x}{:}\iota.*_1$ (see the forthcoming Definition 5.11);

- Secondly, certain propositions will be represented as function types. This has its origin in Heyting's description of the proof of an implication as a function. If the proof of an implication is a function, then the implication itself (a proposition, or equivalently, the type of all the proofs of a proposition) must be a function type. For example, the RTT proposition $\mathtt{R(a)} \to \mathtt{R(b)}$ will (in Definition 5.11) be translated into the type $\Pi \mathtt{x}{:}\mathsf{Ra}.\mathsf{Rb}$.

  A universal quantification will also be translated as a function type. According to Heyting, the proof of a proposition $\forall x{:}A[B]$ is a function that takes elements a of A as arguments, and returns proofs of $B(a)$. For example, the RTT proposition $\forall \mathtt{x}{:}0^0[\mathtt{R(x)}]$ will (again: In Definition 5.11) be translated by $\Pi \mathtt{x}{:}\iota.\mathsf{Rx}$. Here we see an example of a type $\Pi x{:}A.B$ where B depends on the variable x. The intuition of a function of type $\Pi \mathtt{x}{:}\iota.\mathsf{Rx}$ coincides with the intuition of a proof of the proposition $\forall \mathtt{x}{:}0^0[\mathtt{R(x)}]$: A function taking individuals a as arguments, and returning a proof of $\mathtt{R}(a)$.

The type system λRTT will have a rule to introduce terms of type $\Pi x{:}A.B$ (provided we have a term of type B): If b is a term of type B (possibly x occurs free in b), then $\lambda x{:}A.b$ is of type $\Pi x{:}A.B$. This can be understood with the above interpretations of $\Pi x{:}A.B$ in mind. For instance, we represented the RTT-proposition $\forall \mathtt{x}{:}0^0[\mathtt{R(x)}]$ by $\Pi \mathtt{x}{:}\iota.\mathsf{Rx}$. If we have a term b of type Rx (so b proves Rx for an arbitrary individual $\mathtt{x}$) then the function $\lambda \mathtt{x}{:}\iota.b$, assigning $b[\mathtt{x}{:=}a]$ to each $a \in \iota$, is indeed a proof of the proposition $\forall \mathtt{x}{:}0^0[\mathtt{R(x)}]$.

The system λRTT also has a rule that shows what can be done with a term of type $\Pi x{:}A.B$. If a is a term of type ι, and f a proof of $\Pi \mathtt{x}{:}0^0.\mathsf{Rx}$, then we can apply f to a, thus obtaining fa of type $\mathsf{Rx}[\mathtt{x}{:=}a]$, which is $\mathsf{R}a$. Indeed applying the function f to an individual a gives a proof of $\mathsf{R}a$.

[2] The proposition that is returned can be of order 1, for instance in the case we substitute the individual $\mathtt{a}$ for $\mathtt{x}$ in the pf $\forall \mathtt{y}{:}0^0[\mathtt{R(x,y)}]$, or of order 0, for instance in the case we substitute $\mathtt{a}$ for $\mathtt{x}$ in the pf $\mathtt{R(x,x)}$. However, the returned proposition can never be of order > 1, due to Corollary 2.63. See also Remark 5.5.

Remark 5.4 It is usual to write $A \to B$ for $\Pi x{:}A.B$, if x does not occur free in B. Hence, in *notation* there is hardly any difference between the RTT proposition $\mathtt{R(a)} \to \mathtt{R(b)}$ (where $\to$ denotes logical implication) and the λRTT type $\mathtt{Ra} \to \mathtt{Rb}$ (where $\to$ is used to form a function type).

Remark 5.5 One may wonder why we chose $*_n$ to be the type of all propositions of order $\leq n$ instead of the type of all propositions of order $= n$. This has a technical reason that already popped up in footnote 2, namely the translation of the ramified types of RTT into λ-terms. Consider a ramified type $(t_1^{a_1},\ldots,t_n^{a_n})^a$. If $\tau_1,\ldots,\tau_n$ are appropriate translations of $t_1^{a_1},\ldots,t_n^{a_n}$, then one would like to translate $(t_1^{a_1},\ldots,t_n^{a_n})^a$ into $\tau_1 \to \cdots \to \tau_n \to *_m$, where $*_m$ represents the type of propositions of some order m. This would suggest that a propositional function f of type $(t_1^{a_1},\ldots,t_n^{a_n})^a$ always results in a proposition of (fixed) order m as soon as values of types $t_1^{a_1},\ldots,t_n^{a_n}$ are substituted for its free variables. However, it is impossible to determine such m:

- Consider the pfs $f \equiv \mathtt{z(a)}$ and $g \equiv \mathtt{z(a)} \vee \forall \mathtt{z'}{:}(0^0)^1[\mathtt{z'(b)}]$ in a context $\Gamma = \left\{\mathtt{z}{:}(0^0)^1\right\}$. Observe: $\Gamma \vdash f : \left((0^0)^1\right)^2$ and $\Gamma \vdash g : \left((0^0)^1\right)^2$. But if we substitute $\mathtt{R(x)}$ for $\mathtt{z}$ in both f and g, we obtain $\mathtt{R(a)}$ (a proposition of order 0) and $\mathtt{R(a)} \vee \forall \mathtt{z'}{:}(0^0)^1[\mathtt{z'(b)}]$ (a proposition of order 1).

  So substituting the same propositional function $\mathtt{R(x)}$ in two different propositional functions f and g of the same type, may result in propositions of different orders;

- Let f be as above, and extend Γ with the declaration $\mathtt{x}{:}0^0$. Notice: Both $h_1 \equiv \mathtt{R(x)}$ and $h_2 \equiv \mathtt{R(x)} \vee \forall \mathtt{y}{:}0^0[\mathtt{S(y)}]$ are pfs of type $(0^0)^1$, so they can be substituted for $\mathtt{z}$ in f. As a result we obtain $\mathtt{R(a)}$, a proposition of order 0, and $\mathtt{R(a)} \vee \forall \mathtt{y}{:}0^0[\mathtt{S(y)}]$, a proposition of order 1. So substituting different propositional functions h_1, h_2 of the same type in one and the same propositional function f may result in propositions of different orders.

Therefore, we cannot interpret $*_m$ in $\tau_1 \to \cdots \to \tau_n \to *_m$ as the type of propositions of order m. But, by Corollary 2.63, the order of the proposition $f[x_1,\ldots,x_n{:=}g_1,\ldots,g_n]$ cannot be higher than the order of f. Hence, it is safe to translate the ramified type $(t_1^{a_1},\ldots,t_n^{a_n})^m$ into $\tau_1 \to \cdots \to \tau_n \to *_m$, if we interpret $*_m$ as the type of propositions of order $\leq m$.

One might wonder whether this somewhat unusual interpretation of $*_m$ makes λRTT much different from the original RTT. The idea of "propositions of order $\leq m$" however, is not very different from the idea of "propositions of order m", since all propositions of order $< m$ have a logically equivalent proposition of order m (at least in the logic that Russell and Whitehead had in mind). In other words:

If f has order $< m$ then

$$f \vee \forall \mathtt{z}{:}()^{m-1}[\mathtt{z}() \vee \neg\mathtt{z}()]$$

is of order m, and is logically equivalent to f. The system λRTT will have a special rule (Incl) stating that any proposition of order $\leq m$ is also a proposition of order $\leq m+1$.

Contexts, as in the situation of Chapter 2, contain information on the types of variables. These variables can be of different nature: First of all, we still have the variables of Chapter 2. These variables live at the level of types, as propositions are interpreted as types. But now we can also have variables that serve as assumptions: If A is a proposition, then a variable of type A refers to an arbitrary proof of A. If a context contains such a declaration $x{:}A$, then this can be read as: It is supposed that A holds (and that x is some proof of A).

In Chapter 2, contexts were sets. In particular, the order in which the various declarations were mentioned, did not matter: $\{\mathtt{x}{:}0^0, \mathtt{y}{:}()^1\}$ is not different from $\{\mathtt{y}{:}()^1, \mathtt{x}{:}0^0\}$. Now, types that occur in a context can depend on variables that are declared somewhere else in that context. We do not want a variable to occur in a context before its type has been declared. Therefore, we present contexts as lists. The order in which the variables are declared is determined by the order in the list: In a context $\langle \mathtt{x}{:}\tau, \mathtt{y}{:}\upsilon\rangle$, the variable $\mathtt{x}$ was declared before $\mathtt{y}$. Thus it is possible that υ depends on $\mathtt{x}$, i.e. $\mathtt{x}$ may occur as a free variable in υ. On the other hand, $\mathtt{y}$ must not occur in τ, as $\mathtt{y}$ is declared after τ has appeared. Therefore, the context $\langle \mathtt{x}{:}\tau, \mathtt{y}{:}\upsilon\rangle$ is different from the context $\langle \mathtt{y}{:}\upsilon, \mathtt{x}{:}\tau\rangle$.

The presented intuition leads to the type system in Definition 5.7 below, which is in fact almost a PTS. The Π-form rule determines which types (and, with the PAT principle in mind: Which propositions) can be constructed in the system. The general form of the Π-formation rule is

$$\frac{\Gamma \vdash A : s_1 \qquad \Gamma, x{:}A \vdash B : s_2}{\Gamma \vdash (\Pi x{:}A.B) : s_3},$$

where s_1, s_2, s_3 are sorts. By specifying for which combinations of triples (s_1, s_2, s_3) the Π-formation rule can be applied, one can control which Π-types can be constructed. We now informally discuss which Π-formation rules we need in a PAT description of RTT. After that, we give a formal presentation of λRTT.

The translations of the ramified types

To translate and type the ramified types, we need the rules $(*_s, \Box_n, \Box_n)$ and $(\Box_m, \Box_n, \Box_n)$ for $n \geq 1$ and $m < n$.

Assume $(t_1^{a_1}, \ldots, t_p^{a_p})^a$ is a ramified type, and that we already found proper translations $\tau_1, \ldots, \tau_p$ for the ramified types $t_1^{a_1}, \ldots, t_p^{a_p}$ such that

- $\vdash_{\lambda\text{RTT}} \tau_i : *_s$ if $t_i^{a_i} \equiv \iota$;
- $\vdash_{\lambda\text{RTT}} \tau_i : \Box_{a_i}$ if $t_i^{a_i} \not\equiv \iota$.

(we use $\vdash_{\lambda\text{RTT}}$ to denote derivability in λRTT, the PAT version of RTT). The type system λRTT will have an axiom rule that declares that $*_a$ has type $\Box_a$, therefore we have: $\mathbf{x}_p{:}\tau_p \vdash_{\lambda\text{RTT}} *_a : \Box_a$. Now distinguish:

- If $t_p^{a_p} \equiv \iota$ then we use Π-formation rule $(*_s, \Box_a, \Box_a)$:

$$\frac{\vdash_{\lambda\text{RTT}} \tau_p : *_s \qquad \mathbf{x}_p{:}\tau_p \vdash_{\lambda\text{RTT}} *_a : \Box_a}{\vdash_{\lambda\text{RTT}} (\Pi \mathbf{x}_p{:}\tau_p.*_a) : \Box_a};$$

- If $t_p^{a_p} \not\equiv \iota$ then we use Π-formation rule $(\Box_{a_p}, \Box_a, \Box_a)$ (notice that $a_p < a$):

$$\frac{\vdash_{\lambda\text{RTT}} \tau_p : \Box_{a_p} \qquad \mathbf{x}_p{:}\tau_p \vdash_{\lambda\text{RTT}} *_a : \Box_a}{\vdash_{\lambda\text{RTT}} (\Pi \mathbf{x}_p{:}\tau_p.*_a) : \Box_a}.$$

In λRTT there will be a weakening rule so that we can weaken this conclusion by adding a variable $\mathbf{x}_{p-1}$:

$$\mathbf{x}_{p-1}{:}\tau_{p-1} \vdash_{\lambda\text{RTT}} (\Pi \mathbf{x}_p{:}\tau_p.*_a) : \Box_a.$$

In a similar way, we can now deduce:

$$\vdash_{\lambda\text{RTT}} (\Pi \mathbf{x}_{p-1}{:}\tau_{p-1}.\Pi \mathbf{x}_p{:}\tau_p.*_a) : \Box_a$$

and so on until we have

$$\vdash_{\lambda\text{RTT}} (\Pi \mathbf{x}_1{:}\tau_1.\cdots.\Pi \mathbf{x}_p{:}\tau_p.*_a) : \Box_a.$$

Notice that the variables $\mathbf{x}_i$ are only dummy variables, and do not occur in any of the τ_j. We can therefore write

$$\tau_1 \to \cdots \to \tau_p \to *_a$$

instead of $(\Pi \mathbf{x}_1{:}\tau_1.\cdots.\Pi \mathbf{x}_p{:}\tau_p.*_a) : \Box_a$. This $\tau_1 \to \cdots \to \tau_p \to *_a$ is the translation of the RTT-type $(t_1^{a_1}, \ldots, t_p^{a_p})^a$.

The translation of the logical implication

The translation of propositional functions will have a lot of similarities with the mapping $\bar{\ }$ of Definition 2.7. A propositional function f with free variables $x_1 < \cdots < x_m$ will be translated into a λ-term $\lambda x_1{:}\tau_1 \ldots x_m{:}\tau_m.F$, where F is a λ-term of type $*_n$, n is the order of f, and $\tau_1, \ldots, \tau_m$ are translations of the types of the variables $x_1, \ldots, x_m$. In this way, the translation of f will have type

$\tau_1 \to \ldots \to \tau_m \to *_n$, which is exactly the translation of the type of f. In this subsection, we focus on the translation of propositions of the form $f \to g$; in the next subsection we focus on the translation of propositions of the form $\forall x{:}A[f]$.

If f and g are propositions, we must be able to form the proposition $f \to g$. According to Russell's definition in *Principia Mathematica*, $f \to g$ is only shorthand for $(\neg f) \vee g$. In a system in PAT style, the implication plays a more central role, as the logical connectives $\neg$ and $\vee$ are defined with the use of implication.

If f is a proposition of order m, and g has order n, then $f \to g$ clearly has order $\max(m, n)$. In PAT, we can obtain this effect via a rule $(*_m, *_n, *_{\max(m,n)})$. We can assume that F is a translation of f, and G is a translation of g, such that $\vdash_{\lambda\text{RTT}} F : *_m$ and $\vdash_{\lambda\text{RTT}} G : *_n$. Introducing a variable $\mathbf{x}$ of type F does not affect the fact that G is a type of sort $*_n$: $\mathbf{x}{:}F \vdash_{\lambda\text{RTT}} G : *_n$ (the system λRTT will have a so-called *weakening rule* that formalises this intuition). Applying the Π-formation rule $(*_m, *_n, *_{\max(m,n)})$ results in

$$\frac{\vdash_{\lambda\text{RTT}} F : *_m \qquad \mathbf{x}{:}F \vdash_{\lambda\text{RTT}} G : *_n}{\vdash_{\lambda\text{RTT}} (\Pi \mathbf{x}{:}F.G) : *_{\max(m,n)}}.$$

Notice that $\mathbf{x}$ is only a dummy variable here, and does not occur in G. Therefore we can write $F \to G$ instead of $\Pi\mathbf{x}{:}F.G$. This $F \to G$ is the translation of the RTT-proposition $f \to g$.

The translation of the universal quantifier

If f is a pf with x as the only free variable, then we want to translate f to some term $\lambda x{:}\tau.F$. With PAT in mind, we want to translate $\forall x{:}t[f]$ to $\Pi x{:}\tau.F$ since a proof of $\forall x{:}t[f]$ must be a function that assigns a proof of Fa to each a of type τ.

The construction of $\Pi x{:}\tau.F$ can be done in the following way. We can assume that $\lambda x{:}\tau.F$ has a type that corresponds to a propositional function with one variable: A type of the form $\tau \to *_n$, where n is the order of f. This means that F itself must be of type $*_n$, where n is the order of f. The term $\Pi x{:}\tau.F$ (a translation of $\forall x{:}t[f]$) should have type $*_n$, as it represents a proposition of order n. Now τ is a translation of a ramified type, therefore it is of sort $*_s$ (if $\tau \equiv \iota$) or sort $\Box_m$ (if $\tau \not\equiv \iota$). For the construction of $\Pi x{:}\tau.F$ we need the Π-formation rule $(*_s, *_n, *_n)$ or $(\Box_m, *_n, *_n)$. Notice that x is a free variable of f, so its order is smaller than the order of f. In other words: $m < n$.

The rules $(*_s, *_n, *_n)$ can also be used to represent universal quantification over pfs with more than one free variable. We discuss the situation for two free variables; for three or more free variables the procedure is similar.

Let g be a pf with two free variables $x_1 < x_2$, having type $(t_1^{a_1}, t_2^{a_2})^a$, where $t_i^{a_i}$ is the type of x_i. We assume that g has been translated into a term of the form $\lambda x_1{:}T(t_1^{a_1}).\lambda x_2{:}T(t_2^{a_2}).G$, of type $T(t_1^{a_1}) \to T(t_2^{a_2}) \to *_n$, where n is the order of g. With the definition of $\bar{\ }$ in mind, we translate $\forall x_1{:}t_1^{a_1}[g]$ into

$$\lambda x_2{:}T(t_2^{a_2}).\Pi x_1{:}T(t_1^{a_1}).G,$$

and $\forall x_2{:}t_2^{a_2}[g]$ by

$$\lambda x_1{:}T(t_1^{a_1}).\Pi x_2{:}T(t_2^{a_2}).G.$$

We can form $\Pi x_1{:}T(t_1^{a_1}).G$ in a similar way as $\Pi x{:}\tau.F$ above, using Π-formation rule $(*_s, *_n, *_n)$ or $(\Box_m, *_n, *_n)$ for an $n \geq 1$ and an $m < n$. The term $\Pi x_1{:}T(t_1^{a_1}).G$ has type $*_n$, and $\lambda x_2{:}T(t_2^{a_2}).\Pi x_1{:}T(t_1^{a_1}).G$ will have type $T(t_2^{a_2}) \to *_n$, which is the appropriate type for a translation of $\forall x_1{:}t_1^{a_1}[g]$. Similarly, $\Pi x_2{:}T(t_2^{a_2}).G$ has type $*_n$, and $\lambda x_1{:}T(t_1^{a_1}).\Pi x_2{:}T(t_2^{a_2}).G$ has type $T(t_1^{a_1}) \to *_n$, which is appropriate for a translation of $\forall x_2{:}t_2^{a_2}[g]$.

5a2 The system λRTT

Now that we have explained the way in which types are constructed in λRTT, we present the system in a formal way. As announced, it will (almost) have the form of a Pure Type System (PTS).

A PTS always has five fixed rules, and two "flexible" rules (axioms and Π-formation rules):

(Axioms) Axioms provide the types of certain constants that are used in the system. This rule is flexible: The axioms may vary from one PTS to another. In λRTT, we have the following axioms:

- $*_n : \Box_n$ for $n \geq 1$. See Remark 5.2;
- $\bot : *_1$. This means that we consider falsum to be a proposition of the lowest order;
- $\iota : *_s$. See Remark 5.2;
- $a : \iota$ for $a \in \mathcal{A}$. Each individual belongs to the type ι of individuals;
- $R : \underbrace{\iota \to \cdots \to \iota}_{\mathfrak{a}(R) \text{ times } \iota} \to *_1$ for $R \in \mathcal{R}$.

  This illustrates that R is an $\mathfrak{a}(R)$-ary relation on individuals.

All axioms are derivable in an empty context;

(Start) If we want to introduce a variable, we can only do this if we assign a *type* to such a variable. In the type systems that we saw before (like RTT and $\lambda_{\to\text{Church}}$), it was always clear what the types of a certain type system are. There were always two definitions: First a definition that describes which types are allowed, and then a definition that describes which terms have what type. In a PTS however, terms and types are mixed up in one derivation system. Types are recognised as follows:

- Each sort $s \in S$ is a type;
- If $\Gamma \vdash A : s$ for a $s \in S$ then A is a type (within the context Γ).

In the second case, we see that the type A has also a type itself, namely: s. For these "typable" types we make it possible to introduce variables via the start rule:

$$\frac{\Gamma \vdash A : s}{\Gamma, x{:}A \vdash x{:}A}.$$

This means that we can only introduce a variable of type A if A itself has a type $s \in S$. In λRTT, this is possible for the sort $*_n$ (because this sort has type $\Box_n$). This means that it is possible to introduce variables for propositions of order n. However, $\Box_n$ cannot be typed itself in λRTT. This is not harmful, as we do not want to introduce variables of type $\Box_n$. Such a variable would be a variable for a ramified type of order $\leq n$, and such variables do not occur in RTT, either.

As the introduced variable must be seen as a new object, we demand that x is "fresh": It must not occur anywhere in Γ or A;

(Weakening) Once we have derived $\Gamma \vdash M : N$, we want to be able to add variables to the context Γ. Compare this to the weakening rule 2.44.5 for RTT. To add such a variable to the context, we take the same precautions as in the start rule, so we have a premise $\Gamma \vdash A : s$ if we want to add a variable of type A to Γ. The weakening rule now becomes:

$$\frac{\Gamma \vdash M : N \qquad \Gamma \vdash A : s}{\Gamma, x{:}A \vdash M : N}.$$

Again, we demand that x is fresh: It must not occur in Γ, M, N, or A;

(Π-formation) This rule describes which Π-types can be constructed. It is flexible: In different PTSs, different Π-types may be allowed. We already discussed this rule in Section 5a1 where we described which Π-formation rules are needed for λRTT:

- $(*_s, \Box_n, \Box_n)$ and $(\Box_m, \Box_n, \Box_n)$ $(m < n)$ to translate the ramified types;
- $(*_m, *_n, *_{\max(m,n)})$ to translate the logical implication;
- $(*_s, *_n, *_n)$ and $(\Box_m, *_n, *_n)$ $(m < n)$ to translate the universal quantification;

(Π-introduction) This rule describes how we can form terms of type $\Pi x{:}A.B$, once we have established that the type $\Pi x{:}A.B$ can be constructed. If $\Gamma \vdash (\Pi x{:}A.B) : s$ has been derived (using the Π-formation rules), we can not only introduce variables of this type with the start and weakening rules, but we can also form terms of this type by λ-abstraction:

$$\frac{\Gamma, x{:}A \vdash b{:}B \qquad \Gamma \vdash (\Pi x{:}A.B) : s}{\Gamma \vdash (\lambda x{:}A.b) : (\Pi x{:}A.B)}.$$

This rule is a modern version of the Abstraction Principles 1.1 and 1.2, and the abstraction rules 2.44.3 and 2.44.4 for RTT.

The Π-introduction rule is also called λ-*formation*-rule;

(Π-elimination) This rule describes what can be done with a term f of type $\Pi x{:}A.B$. The intuition is clear: f is a function that takes arguments of type A, so it should be possible to apply f to any a of type A. And indeed:

$$\frac{\Gamma \vdash f : \Pi x{:}A.B \qquad \Gamma \vdash a : A}{\Gamma \vdash fa : B[x{:=}a]}.$$

The substitution in the type of the result is necessary. This can be seen if we interpret $\Pi x{:}A.B$ as a proposition $\forall x{:}A.B$. Then f is a proof of $\forall x{:}A.B$, and fa is a proof of B where x has been replaced by a, so $B[x{:=}a]$. The Π-elimination rule together with β-reduction can be seen as a modern substitution rule. We already saw in Chapter 2 that substitution in RTT can be seen as function application plus β-reduction to normal form in λ-calculus. Indeed: If we apply a term $\lambda x{:}A.b$ to a term a, we get $(\lambda x{:}A.b)a$, which β-reduces to $b[x{:=}a]$.

The Π-elimination rule is also called *application*-rule;

(Conversion) If a term A β-reduces to a term A', we consider A and A' to be equal, in some way. A property of PTSs is that if $\Gamma \vdash A : B$ and $A \rightarrow_\beta A'$, then also $\Gamma \vdash A' : B$ (the so-called "Subject Reduction" property). However, we do not have the "Type Reduction" property that $\Gamma \vdash A : B'$ whenever $\Gamma \vdash A : B$ and $B \rightarrow_\beta B'$. This property does not even hold if we demand that $\Gamma \vdash B' : s$ for some $s \in \boldsymbol{S}$. As we want to have that β-equal types have the same inhabitants, we introduce a conversion rule:

$$\frac{\Gamma \vdash A : B \qquad \Gamma \vdash B' : s \qquad B =_\beta B'}{\Gamma \vdash A : B'}.$$

As was already observed in the motivation for the start rule, there is an important difference between PTSs and the other type systems that have appeared in this book up till now. The other systems always made a clear distinction between *terms* and *types*. The definition of types is usually given first, and then a second definition indicates which terms are of what type. This distinction is not made in PTSs. There, types and terms are defined in one system. The jargon used may be confusing to the reader that is not familiar with PTSs, and therefore we give the following definition:

Definition 5.6 (Terms and Types)

Terms

- A term A is a *legal term* in a context Γ, if there is B such that $\Gamma \vdash A : B$ or $\Gamma \vdash B : A$;
- *A is a term of type B* in a context Γ, if $\Gamma \vdash A : B$;

Types

- A term A is a *type* in a context Γ if there is a sort $s \in \boldsymbol{S}$ such that $\Gamma \vdash A : s$;
- A is *the type of B* in a context Γ if $\Gamma \vdash B : A$;
- A type A is *inhabited* in a context Γ if there is a term B such that $\Gamma \vdash B : A$.

So: a type is always a term. And: a term can sometimes act as a type. Look for instance at the Π-introduction rule. In the right-hand premise, $\Pi x{:}A.B$ is a term of type s. But in the conclusion, it acts as a type: the term $\lambda x{:}A.b$ is of type $\Pi x{:}A.B$.

For λRTT we need a rule in addition to the seven rules that were stated for PTSs above. That is why we said that λRTT is *almost* a PTS. This is the so-called *inclusion*-rule, which describes the intuition behind $*_n$ as the class of propositions of order $\leq n$ (and not only the class of propositions of order n). We can formulate this rule as follows:

$$\frac{\Gamma \vdash A : *_n}{\Gamma \vdash A : *_{n+1}}.$$

We summarise the eight rules in the following definition:

Definition 5.7 (Derivation Rules for λRTT)
Let s, s_1, s_2 range over $\boldsymbol{S} = \{*_s, \Box_1, \Box_2, \ldots, *_1, *_2, \ldots\}$, and let

$$\begin{aligned}\boldsymbol{R} \;=\; & \{(*_s, \Box_n, \Box_n) \mid n \geq 1\} \cup \\ & \{(\Box_m, \Box_n, \Box_n) \mid 1 \leq m < n\} \cup \\ & \{(*_m, *_n, *_{\max(m,n)}) \mid m, n \geq 1\} \cup \\ & \{(*_s, *_n, *_n) \mid n \geq 1\} \cup \\ & \{(\Box_m, *_n, *_n) \mid 1 \leq m < n\}.\end{aligned}$$

The derivation rules for λRTT are as follows:

$$
\begin{array}{llr}
\textbf{(Axioms)} & \vdash *_n : \square_n & (n \geq 1) \\
 & \vdash \iota : *_s & \\
 & \vdash \bot : *_1 & \\
 & \vdash a : \iota & (a \in \mathcal{A}) \\
 & \vdash R : \underbrace{\iota \to \cdots \to \iota}_{\mathfrak{a}(R) \text{ times } \iota} \to *_1 & (R \in \mathcal{R})
\end{array}
$$

$$
\textbf{(Start)} \quad \frac{\Gamma \vdash A : s}{\Gamma, x{:}A \vdash x{:}A}
$$

$$
\textbf{(Weak)} \quad \frac{\Gamma \vdash M : N \qquad \Gamma \vdash A : s}{\Gamma, x{:}A \vdash M : N}
$$

$$
\textbf{(Π-form)} \quad \frac{\Gamma \vdash A : s_1 \qquad \Gamma, x{:}A \vdash B : s_2}{\Gamma \vdash (\Pi x{:}A.B) : s_3} \qquad (s_1, s_2, s_3) \in \boldsymbol{R}
$$

$$
\textbf{(Π-in)} \quad \frac{\Gamma, x{:}A \vdash b{:}B \qquad \Gamma \vdash (\Pi x{:}A.B) : s}{\Gamma \vdash (\lambda x{:}A.b) : (\Pi x{:}A.B)}
$$

$$
\textbf{(Π-el)} \quad \frac{\Gamma \vdash M : \Pi x{:}A.B \qquad \Gamma \vdash N : A}{\Gamma \vdash MN : B[x{:=}N]}
$$

$$
\textbf{(Conv)} \quad \frac{\Gamma \vdash A : B \qquad \Gamma \vdash B' : s \qquad B =_\beta B'}{\Gamma \vdash A : B'}
$$

$$
\textbf{(Incl)} \quad \frac{\Gamma \vdash A : *_n}{\Gamma \vdash A : *_{n+1}}
$$

The introduced variables x in the Start and Weakening rules are assumed to be fresh. If confusion with derivation rules of other type systems might arise, we use $\vdash_{\lambda\text{RTT}}$ instead of $\vdash$ to indicate derivability in λRTT.

5a3 Meta-properties of λRTT

We now describe some meta-properties of λRTT. Their formulation is very close to the formulation of the usual meta-properties of PTS, as described in Section 4c. However, there are a few deviations. This is due to the rule (Incl), which is not a rule in PTSs. The proofs of the meta-properties below are as in the standard literature on PTSs ([3]). In Remark 5.10 we provide some intuition behind these meta-theorems and compare them with the meta-theorems we found for RTT in Chapter 2.

Theorem 5.8 (Properties of λRTT)

1. **(Church-Rosser)** *If $A \twoheadrightarrow_\beta B_1$ and $A \twoheadrightarrow_\beta B_2$ then there is a C such that $B_1 \twoheadrightarrow_\beta C$ and $B_2 \twoheadrightarrow_\beta C$;*

2. **(Free Variables)** *Let $\Gamma \equiv x_1{:}A_1, \ldots, x_n{:}A_n$, and assume $\Gamma \vdash A : B$. Then:*

- $\text{FV}(A) \cup \text{FV}(B) \subseteq \text{DOM}(\Gamma)$;
- *For all* $1 \leq i \leq n$ *there is* $s_i \in \boldsymbol{S}$ *such that* $x_1{:}A_1, \ldots, x_{i-1}{:}A_{i-1} \vdash A_i : s_i$;

3. **(Substitution)** *Assume* $\Gamma, x{:}A, \Delta \vdash B{:}C$ *and* $\Gamma \vdash D{:}A$. *Then* $\Gamma, \Delta[x{:=}D] \vdash B[x{:=}D] : C[x{:=}D]$;

4. **(Thinning)** *Assume* Γ, Δ *are legal and* $\Gamma \subseteq \Delta$. *Then* $\Gamma \vdash A : B \Rightarrow \Delta \vdash A : B$;

5. **(Generation)** *The following hold:*

   (a) *If* $\Gamma \vdash *_n : C$ *then* $C \equiv \square_n$;
   If $\Gamma \vdash \iota : C$ *then* $C \equiv *_s$;
   If $\Gamma \vdash \bot : C$ *then* $C \equiv *_n$ *for some* $n \in \mathbb{N}$;
   For $a \in \mathcal{A}$, *if* $\Gamma \vdash a : C$ *then* $C \equiv \iota$;
   For $R \in \mathcal{R}$, *if* $\Gamma \vdash R : C$ *then* $C \equiv \underbrace{\iota \to \cdots \to \iota}_{\mathfrak{a}(R)\ \text{times}\ \iota} \to *_1$ *or* $\mathfrak{a}(R) = 0$ *and* $C \equiv *_n$ *for some* $n \in \mathbb{N}$;

   (b) *If* $\Gamma \vdash x : C$ *then there are* B, s *such that* $\Gamma \vdash B : s$, $x{:}B \in \Gamma$, *and either* $B =_\beta C$, *or there are* m, n *with* $m < n$ *and* $B \equiv *_m$, $C \equiv *_n$;

   (c) *If* $\Gamma \vdash (\Pi x{:}A.B){:}C$ *then there are* s_1, s_2, s_3 *such that* $(s_1, s_2, s_3) \in \boldsymbol{R}$, $\Gamma \vdash A : s_1$, *and* $\Gamma, x{:}A \vdash B : s_2$. *Moreover,* $C \equiv s_3$ *or there are* m, n *such that* $m < n$, $s_3 \equiv *_m$ *and* $C \equiv *_n$;

   (d) *If* $\Gamma \vdash (\lambda x{:}A.b) : C$ *then there are* B, s *such that* $\Gamma \vdash (\Pi x{:}A.B) : s$; $\Gamma, x{:}A \vdash b : B$ *and* $C =_\beta \Pi x{:}A.B$;

   (e) *If* $\Gamma \vdash (AB) : C$ *then there are* x, P, Q *such that* $\Gamma \vdash A : (\Pi x{:}P.Q)$, $\Gamma \vdash B : P$, *and either* $C =_\beta Q[x{:=}B]$ *or there are* m, n *with* $m < n$ *and* $Q[x{:=}B] \equiv *_m$, $C \equiv *_n$;

6. **(Correctness of Types)** *If* $\Gamma \vdash A : B$ *then there is* $s \in \boldsymbol{S}$ *such that* $\Gamma \vdash B : s$ *or* $B \equiv s$;

7. **(Subject Reduction)** *If* $\Gamma \vdash A : B$ *and* $A \to_\beta A'$ *then* $\Gamma \vdash A' : B$;

8. **(Permutation)** *If* $\Gamma, x{:}A, y{:}B, \Delta \vdash C : D$ *and* $x \notin \text{FV}(B)$ *then* $\Gamma, y{:}B, x{:}A, \Delta \vdash C : D$;

9. **(Topsort Lemma)** *If* s *is a topsort and* $\Gamma \vdash A : s$ *then* A *is not a variable and* A *is not of the form* $A_1 A_2$ *or* $\lambda x{:}A_1.A_2$. ⊠

Theorem 5.9 (Strong Normalisation for λRTT**)** *If* $\Gamma \vdash A : B$ *then* A *is strongly normalising.*

PROOF: We embed λRTT into system λC of the Barendregt cube by mapping $*_n$ to $*$ (for all $n \in \mathbb{N}^+$), $*_s$ to $*$, and $\Box_n$ to $\Box$. In this way, λRTT becomes a Pure Type System (as rule (Incl) disappears) that is a subsystem of λC. As all terms of λC are strongly normalising, λRTT is strongly normalising as well. ⊠

Remark 5.10 We provide some intuition for the properties in Theorems 5.8 and 5.9.

1. The Church-Rosser theorem is a basic theorem on λ-calculus. It indicates that it does not matter in which way one makes a calculation (list of β-reductions): The result will always be the same (or, more precisely: The results can be *coerced* to be the same). The various proofs that are known even present a constructive method to find a common reduct C of two β-equal terms B_1 and B_2;

2. The first part of the Free Variables Lemma is comparable to Lemma 2.56.1 of RTT.

   The second part has no counterpart in RTT. This is because in RTT the set of types is determined by a separate definition (Definition 2.36), and types do not have a type themselves. In λRTT, the derivation rules determine which types are "allowed", and which types are not: An allowed type has a sort as type. The second part of the Free Variables Lemma shows that types that occur in a context are, indeed, always typable by a sort;

3. The Substitution Lemma can be compared to the Substitution Rule of RTT (see Definition 2.44.6). Observe that the substitution $[x{:=}D]$ is now not only carried out in B (as was the case in RTT), but also in C and Δ. This is due to the fact that in λRTT, types may have free variables;

4. The Thinning Lemma is comparable to the Weakening Rule of RTT (see Definition 2.44.5);

5. The Generation Lemma is one of the most important meta-properties of a PTS. The derivation rules of λRTT are, as is the case with most usual formulations of PTSs, not syntax directed, i.e. the last rule in a derivation is not necessarily determined by the structure of the term and the context of the conclusion of the derivation. This is due to rules like (Weak), (Conv) and (Incl). If the conclusion of the derivation is $\Gamma \vdash A : C$, then the Generation Lemma provides information on the type of the subterms of A, and on the structure of C.

   The Generation Lemma for λRTT is comparable to Theorems 2.84 and 2.85 for RTT.

Case (a) of the Generation Lemma might raise the question why a relation symbol R of arity > 0 cannot have type $\iota \to \cdots \to \iota \to *_n$ for $n > 1$, while a relation symbol R' of arity 0 can have type $*_n$ for $n > 1$. This has to do with the way in which we implemented type inclusion. We only declared that $*_n$ is a subtype of $*_{n+1}$ for all n, but did not extend this to types of the form $\tau_1 \to \tau_2$. It is quite possible to work with type systems that have an extensive subtyping relation (see for instance [34]), but as we do not need an extension of subtyping in this Section, we do not introduce it here;

6. Correctness of Types shows that every term B for which there are Γ, A such that $\Gamma \vdash A : B$ is typable by a sort. Compare this to the second part of the Free Variable Lemma, that proves a similar thing for types occurring in a context;
7. Subject Reduction shows that the type of an expression does not change during a calculation. As there is no real reduction in RTT, we do not have an equivalent statement in RTT. One could see the fact that the Free Variable property 2.58 is maintained under substitution as a weak form of Subject Reduction;
8. Permutation is closely related to the Permutation Rule 2.44.7;
9. This lemma shows that the topsorts (see Definition 4.38. In λRTT, the topsorts are $*_s$ and $\Box_1, \Box_2, \ldots$.) are only inhabited by types (i.e. constants or terms of the form $\Pi x{:}A.B$).

Strong Normalisation for λRTT can be compared to the theorem on Existence of Substitution for RTT, Theorem 2.73. However, in λRTT much more reductions are possible. For instance, the proof terms of λRTT do not have an equivalent in RTT, and these proof terms are also λ-terms that might β-reduce.

5a4 Interpreting RTT in λRTT

In this section we formally prove our claim that λRTT indeed is a PAT interpretation of RTT. We translate the ramified types of RTT to types of λRTT:

Definition 5.11 We define a type $T(t^a)$ for each ramified type t^a:

- $T(0^0) \equiv \iota$;
- $T((t_1^{a_1}, \ldots, t_m^{a_m})^a) \equiv T(t_1^{a_1}) \to \cdots \to T(t_m^{a_m}) \to *_{\max(a,1)}$.

All the translations of the ramified types are typable in λRTT. As announced in Section 5a1, we use the Π-formation rules $(*_s, \Box_n, \Box_n)$ and $(\Box_m, \Box_n, \Box_n)$ for $m < n$ and $n \geq 1$.

Lemma 5.12 *In* λRTT *we can derive:*

- $\vdash T(0^0) : *_s$;
- $\vdash T((t_1^{a_1}, \ldots, t_m^{a_m})^a) : \Box_a$.

PROOF: Induction on the definition of T; use the rules $(*_s, \Box_n, \Box_n)$ and $(\Box_m, \Box_n, \Box_n)$ of λRTT as sketched in Section 5a1. ⊠

On the other hand: The inhabitants of $*_s$ and $\Box_n$ are all translations of ramified types:

Lemma 5.13

1. *If* $\Gamma \vdash A : *_s$ *then* $A \equiv \iota$;
2. *If* $\Gamma \vdash A : \Box_n$ *then there is a ramified type* t^n *with* $T(t^n) \equiv A$.

PROOF:

1. As $*_s$ is a topsort, A cannot be a variable, or of the form $A_1 A_2$ or $\lambda x{:}A_1.A_2$ (see Theorem 5.8.9). As there are no $s_1, s_2 \in \boldsymbol{S}$ for which $(s_1, s_2, *_s) \in \boldsymbol{R}$, A cannot be of the form $\Pi x{:}A_1.A_2$. Therefore, A must be a constant. By the Generation Lemma, 5.8.5(a), $A \equiv \iota$;

2. Use induction on the length of A. As $\Box_n$ is a topsort, A cannot be a variable, a λ-abstraction, or an application. Considering the Generation Lemma, 5.8.5, we conclude that $A \equiv *_n$ (and in that case we are done: Take $t^n \equiv ()^n$) or $A \equiv \Pi x{:}B_1.B_2$.

   In the last case, we use the Generation Lemma to obtain: $\Gamma \vdash B_1 : s_1$ and $\Gamma, x{:}B_1 \vdash B_2{:}s_2$, where $(s_1, s_2, \Box_n) \in \boldsymbol{R}$.

   Due to the definition of $\boldsymbol{R}$, $s_2 \equiv \Box_n$. Hence $\Gamma, x{:}B_1 \vdash B_2{:}\Box_n$. Using the induction hypothesis, we can find $u_2^{a_2}, \ldots, u_m^{a_m}$ such that $B_2 \equiv T((u_2^{a_2}, \ldots, u_m^{a_m})^n)$ (it is not the case that $B_2 \equiv T(0^0)$, otherwise we would have $\Gamma, x{:}B_1 \vdash \iota{:}\Box_n$).

   Due to the definition of $\boldsymbol{R}$ we also have $s_1 \equiv *_s$ or $s_1 \equiv \Box_{a_1}$ for some a_1. By induction, $B_1 \equiv \iota$, or $B_1 \equiv T(u_1^{a_1})$ for a ramified type $u_1^{a_1}$.

   Hence $A \equiv T((\iota, u_2^{a_2}, \ldots, u_m^{a_m})^n)$ or $A \equiv T((u_1^{a_1}, \ldots, u_m^{a_m})^n)$ (notice that $a_1 < n$, as $s_1 \equiv \Box_{a_1}$ and $(s_1, s_2, \Box_n) \in \boldsymbol{R}$). ⊠

We extend the mapping T to propositional functions:

Definition 5.14 Let Γ be a RTT-context. We define a term $T(i)$ for all $i \in \mathcal{V} \cup \mathcal{A}$, and a term $T(f)$ for each $f \in \mathcal{P}$ for which the variables of f are contained in DOM (Γ):

- $T(x) \equiv x$ for $x \in \mathcal{V}$;
- $T(a) \equiv a$ for $a \in \mathcal{A}$;
- $T(R(i_1, \ldots, i_{\mathfrak{a}(R)})) \equiv \lambda x_1{:}T(t_1) \ldots \lambda x_n{:}T(t_n).R(T(i_1)) \cdots (T(i_{\mathfrak{a}(R)}))$. Here, the free variables of $R(i_1, \ldots, i_{\mathfrak{a}(R)})$ are $x_1 < \cdots < x_n$ and $x_i{:}t_i \in \Gamma$ for $1 \leq i \leq n$;
- If $f_1, f_2 \in \mathcal{P}$ and $T(f_j) \equiv \lambda x_{j1}{:}A_{j1} \ldots \lambda x_{jm}{:}A_{jm}.F_j$, where the free variables of f_j are $x_{j1} < \cdots < x_{jm}$, then define $T(f_1 \to f_2) \equiv$

  $$\lambda x_1{:}T(t_1) \ldots \lambda x_n{:}T(t_n).(F_1 \to F_2).$$

  Here, $x_1 < \cdots < x_n$ are the free variables of $f_1 \to f_2$, and $x_i{:}t_i \in \Gamma$ for $1 \leq i \leq n$;
- If $f \in \mathcal{P}$ and $T(f) \equiv \lambda x_1{:}A_1 \ldots \lambda x_n{:}A_n.F$, where the free variables of f are $x_1 < \cdots < x_k < \cdots < x_n$, then define $T(\forall x_k{:}t_k[f]) \equiv$

  $$\lambda x_1{:}A_1 \ldots \lambda x_{k-1}{:}A_{k-1} \lambda x_{k+1}{:}A_{k+1} \ldots \lambda x_n{:}A_n.\Pi x_k{:}T(t_k).F;$$
- If $z \in \mathcal{V}$ and $k_1, \ldots, k_m \in \mathcal{A} \cup \mathcal{V} \cup \mathcal{P}$, then define $T(z(k_1, \ldots, k_m)) \equiv$

  $$\lambda x_1{:}T(t_1) \ldots \lambda x_n{:}T(t_n).z(T(k_1)) \cdots (T(k_m)),$$

  where $x_1 < \cdots < x_n$ are the free variables of $z(k_1, \ldots, k_m)$ and $x_i{:}t_i \in \Gamma$.

We show that the legal pfs of RTT are legal terms in λRTT. For one step in the proof, we need a Lemma:

Lemma 5.15 *If* $\Gamma \vdash A : \square_n$ *or* $\Gamma \vdash A : *_s$ *then* $\text{FV}(A) = \varnothing$.

PROOF: By Lemma 5.13, there is a ramified type t^a such that $A \equiv T(t^a)$. From the definition of $T(t^a)$ we conclude that $\text{FV}(T(t^a)) = \varnothing$. ⊠

Lemma 5.16 *Let* $f \in \mathcal{P}$ *and assume* $\Gamma \vdash f : t^a$ *in* RTT. *Then* $\vdash T(f) : T(t^a)$ *in* λRTT.

PROOF: Induction on the structure of f. Though the proof is rather straightforward, we treat all cases, in order to show where which rules of λRTT are used.

1. $f \equiv R(i_1, \ldots, i_{\mathfrak{a}(R)})$, and $x_1 < \cdots < x_m$ are the free variables of f. Notice: $x_i{:}0^0 \in \Gamma$ for $1 \leq i \leq m$ (Theorem 2.85). Therefore: $T(f) \equiv \lambda x_1{:}\iota \ldots \lambda x_m{:}\iota.Ri_1 \cdots i_{\mathfrak{a}(R)}$. Note that $x_1{:}\iota, \ldots, x_m{:}\iota \vdash Ri_1 \cdots i_{\mathfrak{a}(R)} : *_1$. By abstraction,

   $$\vdash T(f) : \underbrace{\iota \to \cdots \to \iota}_{\mathfrak{a}(R) \text{ times } \iota} \to *_1;$$

2. $f \equiv f_1 \to f_2$; the free variables of f are $x_1 < \cdots < x_m$, and the free variables of f_j are $x_{j1} < \cdots < x_{jm_j}$. Assume $x_i{:}t_i^{a_i} \in \Gamma$ such that $x_{ji}{:}t_{ji}^{a_{ji}}$. By Theorem 2.84, the f_j are legal, and by Theorem 2.58, $\Gamma \vdash f_j : \left(t_{j1}^{a_{j1}}, \ldots, t_{jm_j}^{a_{jm_j}}\right)^{b_j}$ for some b_j. Observe that we can write (due to Definition 5.14)

$$T(f_j) \equiv \lambda x_{j1}{:}T\left(t_{j1}^{a_{j1}}\right) \ldots \lambda x_{jm_j}{:}T\left(t_{jm_j}^{a_{jm_j}}\right).F_j,$$

and that, by the induction hypothesis,

$$\vdash T(f_j) : T\left(t_{j1}^{a_{j1}}\right) \to \cdots \to T\left(t_{jm_j}^{a_{jm_j}}\right) \to *_{b_j}.$$

This means (Generation Lemma):

$$x_{j1}{:}T\left(t_{j1}^{a_{j1}}\right), \ldots, x_{jm_j}{:}T\left(t_{jm_j}^{a_{jm_j}}\right) \vdash F_j : *_{b_j},$$

and therefore (Weakening):

$$x_1{:}T\left(t_1^{a_1}\right), \ldots, x_m{:}T\left(t_m^{a_m}\right) \vdash F_1 : *_{b_1},$$

and

$$x_1{:}T\left(t_1^{a_1}\right), \ldots, x_m{:}T\left(t_m^{a_m}\right), x{:}F_1 \vdash F_2 : *_{b_2}$$

(notice that $\{x_1, \ldots, x_m\} \supseteq \{x_{j1}, \ldots, x_{jm_j}\}$).
With rule $(*_{b_1}, *_{b_2}, *_{\max(b_1,b_2)})$,

$$x_1{:}T\left(t_1^{a_1}\right), \ldots, x_m{:}T\left(t_m^{a_m}\right) \vdash F_1 \to F_2 : *_{\max(b_1,b_2)}.$$

Now notice that $a = \max(b_1, b_2)$, and use λ-abstraction m times:

$$\vdash T(f) : T(t^a);$$

3. $f \equiv \forall x{:}u^b[f']$, $x_1 < \cdots < x_m$ are the free variables of f, and $x_i{:}t_i^{a_i} \in \Gamma$. For simplicity of notation, we will assume that $x < x_1$. As $x \in \mathrm{FV}(f')$, we can write $T(f') \equiv \lambda x{:}T(u^b).\lambda x_1{:}T(t_1^{a_1}) \ldots \lambda x_m{:}T(t_m^{a_m}).F'$. By Theorem 2.84 and Lemma 2.56, we have that f' is legal in $\Gamma \cup \{x{:}u^b\}$, and by Theorem 2.58 and Corollary 2.61: $\Gamma \cup \{x{:}u^b\} \vdash f' : (u^b, t_1^{a_1}, \ldots, t_m^{a_m})^a$. By the induction hypothesis,

$$\vdash T(f') : T(u^b) \to T(t_1^{a_1}) \to \cdots \to T(t_m^{a_m}) \to *_a.$$

Therefore (Generation Lemma),

$$x{:}T(u^b), x_1{:}T(t_1^{a_1}), \ldots, x_m{:}T(t_m^{a_m}) \vdash F' : *_a,$$

so by the permutation lemma and Lemma 5.15

$$x_1{:}T(t_1^{a_1}),\ldots,x_m{:}T(t_m^{a_m}),x{:}T(u^b)\vdash F' : *_a.$$

As $T(u^b)$ has either type $\square_b$ or type $*_s$ (Lemma 5.12), and $b < a$, we can use rule $(\square_b, *_a, *_a)$ or $(*_s, *_a, *_a)$ to derive

$$x_1{:}T(t_1^{a_1}),\ldots,x_m{:}T(t_m^{a_m})\vdash (\Pi x{:}T(u^b).F') : *_a.$$

By λ-abstraction, we find $\vdash T(f) : T(t^a)$;

4. $f \equiv z(k_1,\ldots,k_n)$, $x_1 < \cdots < x_m$ are the free variables of f, and $x_i{:}t_i^{a_i} \in \Gamma$. By Theorem 2.84, the k_j are either legal pfs of predicative type in Γ, or variables (so one of the x_is), or individuals (of type 0^0). Let $u_j^{b_j}$ be the type of k_j in Γ. By Theorem 2.85: $z{:}(u_1^{b_1},\ldots,u_n^{b_n})^{a-1} \in \Gamma$. Using the induction hypothesis for the $k_j \in \mathcal{P}$, we have that

$$x_1{:}T(t_1^{a_1}),\ldots,x_m{:}T(t_m^{a_m})\vdash T(k_j) : T(u^{b_j})$$

for $1 \le j \le n$. We also have

$$x_1{:}T(t_1^{a_1}),\ldots,x_m{:}T(t_m^{a_m})\vdash z : T(u_1^{b_1}) \to \ldots \to T(u_n^{b_n}) \to *_{a-1}.$$

Therefore,

$$x_1{:}T(t_1^{a_1}),\ldots,x_m{:}T(t_m^{a_m})\vdash z(T(k_1))\cdots(T(k_n)) : *_{a-1},$$

hence (by the (Incl) rule)

$$x_1{:}T(t_1^{a_1}),\ldots,x_m{:}T(t_m^{a_m})\vdash z(T(k_1))\cdots(T(k_n)) : *_a.$$

By λ-abstraction, $\vdash T(f) : T(t^a)$. ⊠

Remark 5.17 The use of the (Incl) rule in case 4 of the above proof is essential. There, it is shown that

$$x_1{:}T(t_1^{a_1}),\ldots,x_m{:}T(t_m^{a_m})\vdash z : T(u_1^{b_1}) \to \ldots \to T(u_n^{b_n}) \to *_{a-1},$$

and one could try to form λ-abstractions without using the (Incl) rule first. However, $z \in \mathrm{FV}(f)$, so at a certain point one has to construct a λ-abstraction over z. Let t_z be such that $z : t_z \in \Gamma$. The resulting term $\lambda z{:}T(t_z).G$ has type $T(t_z) \to \ldots \to *_{a-1}$. As z has order $a-1$, $T(t_z)$ is a term of type $\square_{a-1}$ (Lemma 5.12). One needs a rule $(\square_{a-1}, \square_p, \square_p)$ with $p > a-1$ (as for $p \le a-1$ such a rule is not present) for the construction of $T(t_z) \to \ldots \to *_{a-1}$. Hence, one has to use the rule (Incl) to replace $*_{a-1}$ by $*_a$, which has type $\square_a$. This makes it possible to use $p = a$.

5a5 Logic in RTT and λRTT

Before we can use λRTT as a system in which we can prove theorems, we must add some logical axioms to it. These axioms mainly have to do with the symbol $\bot$. For $\rightarrow$ and $\forall$ the needed derivation rules are already provided by the type theory (via the PAT principle a la Curry-Howard).

- The $\neg$-introduction rule of natural deduction systems is already incorporated in the translation of $\neg A$ to $A \rightarrow \bot$. If we have a proof T of $\bot$ under the assumption that x is a proof of A, then $\lambda x{:}A.T$ is a proof of $A \rightarrow \bot$;

- For the rule "ex falso sequitur quodlibet" the type system does not provide a natural equivalent. We therefore introduce an axiom

  $$\mathtt{ExFalso}_n : \Pi \mathtt{f}{:}*_n.\Pi \mathtt{p}{:}\bot.\mathtt{f}$$

  for each $n \in \mathbb{N}^+$. We will store these axioms in some basic context Γ_0.

  We remark that the type $\Pi\mathtt{f}{:}\ *_n\ .\Pi\mathtt{p}{:}\bot.\mathtt{f}$ is indeed a type in λRTT. It is straightforward to derive that it is a type of sort $*_{n+1}$.

  We also remark that it is necessary to introduce separate axioms

  $$\mathtt{ExFalso}_1, \mathtt{ExFalso}_2, \ldots.$$

  If we want to conclude the proposition f using the ExFalso-axiom, we must provide the type of f, and in that type the order of f is also mentioned. This is a usual thing in ramified type systems, and such constructions occur also in *Principia* (cf. [145], pp. 41–43);

- RTT is based on classical logic, and PAT on intuitionistic logic. Therefore we must add a "classical" axiom. We prefer to add the "law of double negation", and introduce axioms

  $$\mathtt{DblNeg}_n : \Pi\mathtt{f}{:}*_n.\Pi\mathtt{p}{:}(\mathtt{f}\rightarrow\bot)\rightarrow\bot.\mathtt{f}.$$

  It is easy to show that the type of this axiom is of sort $*_{n+1}$. We store the axioms $\mathtt{DblNeg}_n$ in the same context Γ_0.

We compare the obtained system with the original logical system that was proposed in *Principia Mathematica*. That system is presented in what we would now call a natural deduction style. It has one derivation rule, modus ponens (cf. *Principia*, $*1{\cdot}1$), and the following axioms:

$$(p \vee p) \to p \qquad (*1{\cdot}2);$$

$$q \to (p \vee q) \qquad (*1{\cdot}3);$$

$$(p \vee q) \to (q \vee p) \qquad (*1{\cdot}4);$$

$$(p \vee (q \vee r)) \to (q \vee (p \vee r)) \qquad (*1{\cdot}5);$$

$$(q \to r) \to ((p \vee q) \to (p \vee r)) \qquad (*1{\cdot}6);$$

$$f(x) \to \exists z[f(z)] \qquad (*9{\cdot}1);$$

$$f(x) \vee f(y) \to \exists z[f(z)] \qquad (*9{\cdot}11);$$

In any assertion containing a free variable, this free variable may be turned into an apparent variable of which all possible values are asserted to satisfy the function in question (∗9·13).

The formulation of the last axiom is not as precise as the other ones. In later formulations of logic (as the ones by Gödel [61] and Church [29]) we see that the axioms for propositional logic are mostly maintained, but that the axioms on predicate logic are replaced by two other ones:

$$\forall x[f] \to f[x{:=}a]$$

$$\forall x[f \vee g] \to f \vee \forall x[g]$$

where we assume that $x \notin \mathrm{FV}(f)$ and that a does not contain any variable that is bound in f at a place where x is free in f. These new axioms are theorems in the *Principia* (∗9·2 and ∗9·25), and Russell's axioms are proved in Church's system [29].

We must take into account that Gödel and Church both use simple type theory instead of ramified type theory. But Russell's system, by accepting the axiom of reducibility, is in fact also based on simple type theory.

Clearly, λRTT has also modus ponens (function application). We now show that all the axioms of Russell's system can be derived in λRTT as well:

Theorem 5.18 *In λRTT with the axioms* $\mathtt{ExFalso}_n$ *and* $\mathtt{DblNeg}_n$, *one can construct terms of the following types:*

$$T(\forall \mathtt{p}{:}*_k[(\mathtt{p} \vee \mathtt{p}) \to \mathtt{p}]) \qquad (*1{\cdot}2);$$

$$T(\forall \mathtt{p}{:}*_k \forall \mathtt{q}{:}*_m[\mathtt{q} \to (\mathtt{p} \vee \mathtt{q})]) \qquad (*1{\cdot}3);$$

$$T(\forall \mathtt{p}{:}*_k \forall \mathtt{q}{:}*_m[(\mathtt{p} \vee \mathtt{q}) \to (\mathtt{q} \vee \mathtt{p})]) \qquad (*1{\cdot}4);$$

$$T(\forall \mathtt{p}{:}*_k \forall \mathtt{q}{:}*_m \forall \mathtt{r}{:}*_n[(\mathtt{p} \vee (\mathtt{q} \vee \mathtt{r})) \to (\mathtt{q} \vee (\mathtt{p} \vee \mathtt{r}))]) \qquad (*1{\cdot}5);$$

$$T(\forall \mathtt{p}{:}*_k \forall \mathtt{q}{:}*_m \forall \mathtt{r}{:}*_n[(\mathtt{q} \to \mathtt{r}) \to ((\mathtt{p} \vee \mathtt{q}) \to (\mathtt{p} \vee \mathtt{r}))]) \qquad (*1{\cdot}6);$$

$$T(\forall \mathtt{f}{:}\iota \to *_m \forall \mathtt{x}{:}\iota[f(x) \to \exists z{:}\iota[f(z)]]) \qquad (*9{\cdot}1);$$

$$T(\forall \mathtt{f}{:}\iota \to *_m \forall \mathtt{x}{:}\iota \forall \mathtt{y}{:}\iota[(f(x) \vee f(y) \to \exists z{:}\iota[f(z)]]) \qquad (*9{\cdot}11).$$

PROOF: The following terms are inhabitants of the types above:

$$\lambda \mathtt{p}{:}*_k.\lambda \mathtt{x}{:}(\mathtt{p} \to \bot) \to \mathtt{p}.\mathtt{DblNeg}_k\mathtt{p}(\lambda \mathtt{y}{:}\mathtt{p} \to \bot.\mathtt{y}(\mathtt{xy})) \qquad (*1{\cdot}2);$$

$$\lambda \mathtt{p}{:}*_k.\lambda \mathtt{q}{:}*_m.\lambda \mathtt{y}{:}\mathtt{q}.\lambda \mathtt{x}{:}\mathtt{p} \to \bot.\mathtt{y} \qquad (*1{\cdot}3);$$

$$\begin{aligned}&\lambda \mathtt{p}{:}*_k.\lambda \mathtt{q}{:}*_m.\lambda \mathtt{x}{:}(\mathtt{p} \to \bot) \to \mathtt{q}.\\ &\quad \lambda \mathtt{y}{:}\mathtt{q} \to \bot.\mathtt{DblNeg}_k p(\lambda \mathtt{z}{:}\mathtt{p} \to \bot.\mathtt{y}(\mathtt{xz}))\end{aligned} \qquad (*1{\cdot}4);$$

$$\begin{aligned}&\lambda \mathtt{p}{:}*_k.\lambda \mathtt{q}{:}*_m.\lambda \mathtt{r}{:}*_n.\lambda \mathtt{x}{:}(\mathtt{p} \to \bot) \to ((\mathtt{q} \to \bot) \to \mathtt{r}).\\ &\quad \lambda \mathtt{y}{:}\mathtt{q} \to \bot.\lambda \mathtt{z}{:}\mathtt{p} \to \bot.\mathtt{xzy}\end{aligned} \qquad (*1{\cdot}5);$$

$$\begin{aligned}&\lambda \mathtt{p}{:}*_k.\lambda \mathtt{q}{:}*_m.\lambda \mathtt{r}{:}*_n.\lambda \mathtt{x}{:}\mathtt{q} \to \mathtt{r}.\\ &\quad \lambda \mathtt{y}{:}(\mathtt{p} \to \bot) \to \mathtt{q}.\lambda \mathtt{z}{:}\mathtt{p} \to \bot.\mathtt{x}(\mathtt{yz})\end{aligned} \qquad (*1{\cdot}6);$$

$$\lambda \mathtt{x}{:}\iota.\lambda \mathtt{f}{:}\iota \to *_k.\lambda \mathtt{p}{:}\mathtt{fx}.\lambda \mathtt{q}{:}(\Pi \mathtt{y}{:}\iota.\mathtt{fy} \to \bot).\mathtt{qxp} \qquad (*9{\cdot}1);$$

$$\begin{aligned}&\lambda \mathtt{x}{:}\iota.\lambda \mathtt{y}{:}\iota.\lambda \mathtt{f}{:}\iota \to *_k.\lambda \mathtt{p}{:}(\mathtt{fx} \to \iota) \to \mathtt{fy}.\\ &\quad \lambda \mathtt{q}{:}\Pi \mathtt{z}{:}\iota.\mathtt{fz} \to \bot.\mathtt{qy}(\mathtt{p}(\mathtt{qx}))\end{aligned} \qquad (*9{\cdot}11).$$

⊠

The last axiom of Russell's logical system is implemented in λRTT by the Π-elimination rule.

We conclude that the embedding T is sound with respect to the logics that are used in RTT and λRTT.

5a6 Various implementations of PAT

As was explained in Definition 5.6, types and terms are mixed up in one system λRTT. As a consequence we can spot a hierarchy of *levels* in λRTT. The hierarchies in RTT and λRTT are depicted in Figures 5.1 and 5.2. Some of the levels in these

	Ramified Types	
0^0	$()^2$	$(0^0)^1$
Objects	Propositions	Prop. functions
$\mathtt{a}$	$\forall\mathtt{x}:()^1[\mathtt{x}() \to \mathtt{x}()]$	$\mathtt{R}(\mathtt{x})$

Figure 5.1: Levels within RTT

	Topsorts	
$*_s$	$\Box_2$	$\Box_1$
	Ramified Types	
ι	$*_2$	$\iota \to *_1$
Objects	Propositions	Prop. functions
$\mathtt{a}$	$\Pi\mathtt{x}:*_1.\Pi\mathtt{y}:\mathtt{x}.\mathtt{x}$	$\lambda\mathtt{x}:\iota.\mathtt{Rx}$
	Proofs	
	$\lambda\mathtt{x}:*_1.\lambda\mathtt{y}:\mathtt{x}.\mathtt{y}$	

Figure 5.2: Levels within λRTT

figures are empty. Later, we will compare the systems RTT and λRTT with other systems. For this comparison, we will draw similar pictures, in which we need the levels that are empty in the presentation of RTT and λRTT.

We have the level of *propositional types*: This is formed by the terms that have a type of the form $\Box_n$. Below the level of propositional types we find *propositions and propositional functions*. These have a propositional type as type. Under the level of propositions and propositional functions we find the level of *proofs*. A proof has always a proposition as its type.

There is a second hierarchy of levels in λRTT. At its top we find $*_s$, the type of individual types. It has as its only inhabitant ι, the type of individuals. One could imagine situations in which $*_s$ has more inhabitants. For instance, if we would allow several sets of individuals. Or if the Π-formation rule $(*_s, *_s, *_s)$ is allowed, so that also types like $\iota \to \iota$ can be constructed. Below the type of individuals, we find the individuals themselves.

We see that the transformation of RTT to λRTT has introduced some new term levels:

- A level of *topsorts*. These topsorts are needed to type the ramified types. Such typing is needed for two reasons:

  Variable introduction If we want to introduce a variable of a certain type τ in a PTS, we have to establish that τ is an allowed type. This is done by requiring that τ itself must have a certain type;

  Type construction In order to control the construction of types with the Π-formation rules, Π-formation is only allowed with certain types. This is determined by the type of that type. The ramified types of RTT however, do not have a type in RTT. We use the topsorts $\Box_1, \Box_2, \ldots$ to type the translations of these ramified types in λRTT. This also gives us a good way to check the order of a type: A type of order n has type $\Box_n$;

- A level of *proofs*.This level was empty in RTT, as proofs are not part of the theory of RTT.

From our PTS-point of view it is remarkable that sorts of RTT which are denoted by the symbol $*$, are not all at the same level. The sort $*_s$ occurs at the level of topsorts, while the sorts $*_n$ live at the level of the ramified types. Moreover, we have already seen that each Π-formation rule of the form $(\Box_n, s_1, s_2)$ also has a variant of the form $(*_s, s_1, s_2)$, and vice versa. With this in mind it would have been more clear to write $\Box_s$ instead of $*_s$, and use $*_s$ for ι.

The reason that we chose the symbol $*_s$ (instead of $\Box_s$) has to do with traditions within the discipline of Pure Type Systems: The levels are usually partitioned in such a manner that $*_s$ and $*_n$ live at the same level. See Figure 5.3.

	Logical Topsorts $\square_2$	$\square_1$
Object Topsorts $*_s$	Logical Ramified Types $*_2$	$\iota \to *_1$
Object Types ι	Propositions $\Pi \mathtt{x}{:}*_1.\Pi \mathtt{y}{:}\mathtt{x}.\mathtt{x}$	Prop. functions $\lambda \mathtt{x}{:}\iota.\mathtt{R}\mathtt{x}$
Objects $\mathtt{a}$	Proofs $\lambda \mathtt{x}{:}*_1.\lambda \mathtt{y}{:}\mathtt{x}.\mathtt{y}$	

Figure 5.3: Levels of λRTT in PTS tradition

Obj. Tops. $*_s$	Proof Topsorts $*_2$	Logical Topsorts $\square_2$	$\square_1$
Obj. Types ι	Proof Types $\mathtt{true}(\forall \mathtt{x}{:}()^1.\mathtt{x} \to \mathtt{x})$	Logical Ramified Types $()^2$	$(0^0)^1$
Objects $\mathtt{a}$	Proofs $\lambda \mathtt{x}{:}\ *_1\ .\lambda \mathtt{y}{:}\mathtt{x}.\mathtt{y}$	Propositions $\forall \mathtt{x}{:}()^1.\mathtt{x} \to \mathtt{x}$	Prop. functions $\mathtt{R}(\mathtt{x})$

Figure 5.4: Levels of λRTT in bool-style PAT

Let's have a closer look at the traditional situation. The PAT principle within PTSs is often implemented by lifting the propositions and propositional functions from term level to type level, but leaving the individuals at term level. We see that proofs are not introduced at a new level below term level, but that the type level (as far as propositions and propositional functions are concerned) is lifted, and that the proofs are put at the term level that was originally occupied by the propositions.

The treatment of propositions at a higher level than individuals can be understood if we take a look at first-order logic. In systems for first-order logic, quantification over individuals is possible, but quantification over propositions and propositional functions is not allowed. This leads to the treatment of propositions and propositional functions at a higher level.

Contrary to the PTS tradition, we did not lift propositions from term level to type level when we constructed a PAT implementation for RTT. Instead, we built a new level below the level of propositional functions, propositions and individuals: The level of proofs. In this way the double role of propositions is more clear:

- They are terms, as they live at the same level as the individuals;
- They are types, as they can have inhabitants (their proofs).

The PAT implementation à la de Bruijn can in various ways be seen as a compromise between the two different points of view above (though it has been developed independently). A PAT implementation of RTT à la de Bruijn could be depicted as in Figure 5.4. There are three hierarchies now: The two well-known hierarchies of objects and propositions/propositional functions, plus a new hierarchy for proofs. The hierarchies of propositional functions and proofs are connected via the operator `true` (see Section 4a4), which assigns a type of proofs to each proposition. This picture:

- Respects the wish to treat propositions at term level;
- Respects the wish to treat proof classes as types;
- Can also be seen, in retrospect, as a compromise from a historical point of view. Though AUTOMATH and the PAT notion in de Bruijn style are mainly independent of other developments in logic and type theory, the bool-style PAT notion (1968) historically fits between the style of Figure 5.2 for the Ramified Theory of Types (1908-1912) and the style of Pure Type Systems in Figure 5.3 (1988).

5b STT in PAT style

From the description of RTT in PAT style it is easy to make a description λSTT of STT in PAT style: Simply remove all references to orders. This means that $*_n$ has to be replaced by $*$, and $\Box_n$ by $\Box$ (for all $n \in \mathbb{N}^+$). In fact, the same procedure is followed by Ramsey [124], Gödel [61] and Church [29] in their presentations of simple type theory.

One of the consequences is that rule (Incl) disappears. We obtain a Pure Type System with axiom $*{:}\Box$ and rules $(*_s, \Box, \Box), (\Box, \Box, \Box), (*_s, *, *), (*, *, *)$ and $(\Box, *, *)$. This looks familiar to the Calculus of Constructions λC. In λC there is the same axiom $*{:}\Box$, and rules $(*, *, *), (*, \Box, \Box), (\Box, *, *), (\Box, \Box, \Box)$. But λSTT is more restricted than λC. More specifically, rule $(*_s, \Box, \Box)$ is not as powerful as rule $(*, \Box, \Box)$ in λC. As in λC, we do have higher order logic, but we do not have the higher order functions that are present in λC. This is due to the fact that $*_s$ is a topsort in λSTT, while $*$ has type $\Box$. Therefore, λSTT is a system somewhere in between $\lambda\omega$ and λC.

Notice that we have given a PAT version of the simple theory of types, and not of the simply typed λ-calculus of Church [29]. In [29] there are more things formalised than in STT:

- In the simply typed λ-calculus there are more types. For instance, $\iota \to \iota$ is a type, and so is $* \to \iota$ (in [29] this type is denoted $o \to \iota$). More precise: For a PAT implementation of Church's theory we should add a rule $(*_s, *_s, *_s)$, and a rule $(\Box, *_s, *_s)$;

- Church has an additional logical operator ($\imath$) in his system. This operator also occurs in Russell's RTT, but only as an abbreviation and not as a new syntactical object (see [145], pp. 66–68 and pp. 173–175).

Remark 5.19 Together, the rules $(*_s, \Box, \Box)$ and $(\Box, \Box, \Box)$ form a version of the simply typed lambda calculus of Church. The identification would have been complete if we had identified $*_s$ with $\Box$.

Conclusions

In Chapter 4, we saw that there are various ways in which PAT can be implemented in type theories. There are two main streams:

Curry-Howard approach: This approach treats propositions as types, and a proof of a proposition is an inhabitant of the type that represents that proposition. The implementation is based on the Brouwer-Heyting-Kolmogorov interpretation of the logical connectives. In particular, a proof of an implication $A \to B$ is represented as a function that transforms proofs of the proposition A (terms of the type A) to proofs of B (terms of type B);

De Bruijn approach: For each proposition P we create a type $\texttt{bool}(P)$. A proof of P in this approach is not a term of type P (as in the Curry-Howard style), but a term of type $\texttt{bool}(P)$.

In Curry-Howard style implementations, logic is already part of the system. The logical connective $\rightarrow$ and the quantifier $\forall$ immediately translate to the construction of function types. Using higher-order logic, other logical connectives can be defined in terms of $\rightarrow$ and $\forall$. De Bruijn style implementations have more possibilities. One can implement the logical system independent from the type system. But it is also possible to use function types for the translation of $\rightarrow$ and/or $\forall$ as is done in the Curry-Howard style.

The various implementations lead to various levels in type systems. This was depicted in figures 5.3 and 5.4. In Curry-Howard style and the PTS tradition, propositions are at the same level as types, and therefore, proofs are at the same level as objects (terms). In de Bruijn style (bool-style), proofs, propositions, and propositional functions all live at the term level.

A third division into levels appeared when we gave a description of RTT in PAT-style. See Figure 5.2. On the one hand, λRTT uses a Curry-Howard style implementation. There is no difference between a proposition and the type of its proofs. Therefore, proofs and propositions do not live at the same level (as is the case in PAT à la de Bruijn). On the other hand, objects and propositions live at the same level.

The implementation of RTT in PAT-style not only serves as an elaborate example of PAT, but also shows that ramified types can be placed in the framework of Pure Type Systems without too many problems.

Chapter 6

A Correspondence between RTT and the system Nuprl

Recall again that in Russell's Ramified Theory of Types RTT, two hierarchical concepts dominate: *orders* and *types*. The use of orders has as a consequence that the logic part of RTT is predicative. The concept of order however, is almost dead in the type theoretic community, since Ramsey eliminated it from RTT, despite the fact that the role of orders in modern style type theory has been well understood in the proof-theory community. This is why we find Church's simple theory of types (which uses the type concept without the order one) at the bottom of the Barendregt cube rather than RTT. Despite the disappearance of orders which have a strong correlation with predicativity, predicative logic still plays an influential role in computer science. An important example is the proof checker Nuprl [35, 77], which is based on Martin-Löf's Type Theory which uses type universes. Those type universes, and also degrees of expressions in AUTOMATH, are closely related to orders. In this chapter, we show that orders are there in the background, even when they are not used explicitly. We look at orders in an explicit and fresh sense and show that orders play a crucial role in understanding the hierarchy of modern systems. In order to achieve our goal, we give a calculus of orders which we use to reason about classifications of objects and functions in Nuprl. As side results to this chapter:

- We place the historical system RTT in a context with a modern system of computer mathematics (Nuprl) and hence, another modern type theory: Martin-Löf's type theory [104].

- We present a complex type system (Nuprl) as a simple and compact PTS.

6a On the role of orders

RTT has a double hierarchy: one of types and one of orders. In Chapter 3, we have seen that the hierarchy of orders can be compared with Kripke's Hierarchy of Truths [96]. Although Church followed Ramsey's simplification in [124] of RTT into the Simple Theory of Types, he still attempted to explain orders in his book [30] and later on (as late as 1976) in [31]. Nevertheless, the hierarchy of orders remains less known than the hierarchy of types, as it became unpopular when Ramsey [124] and Hilbert and Ackermann [70] showed that one can avoid the paradoxes without this hierarchy. Furthermore, even though it became widely acknowledged that the paradoxes can be avoided without the use of orders, we believe that many logicians are (maybe unconsciously) influenced by the hierarchy of orders when constructing (non-paradoxical) theories. Moreover, orders can elegantly explain some useful hierarchies. As an example, when Kripke wanted to build a logical theory [96] which has its own truth predicate (something not straightforward according to Tarski's hierarchy of truths [141], in which the truth predicate is not definable), he used a hierarchy of languages which could elegantly be explained via the notion of orders as is shown in Chapter 3. Similarly, when Martin-Löf's impredicative type theory was shown to suffer from the paradox, he moved to the predicative version in [104] and has since, built layers of universes that again could be elegantly explained by orders (see for example, page 84 of [104]). Also, [114] provides a treatment of transfinite orders as universes, [36] introduces the "generalised" Calculus of Constructions CC^ω which includes a cumulative hierarchy of universes, [65] studies type checking and well-typedness in CC^ω and in an extended version of it with an *anonymous universe* **Type** which is intended to model Russell and Whitehead's *typical ambiguity* convention, and [133, 134] uses orders in proof theory. Moreover, orders are closely related to the degree of expression notion of AUTOMATH [112] where de Bruijn's notion of degree satisfies in the AUTOMATH systems the property that: if $E : F$ then degree(E) = degree(F) + 1. De Bruijn always assumed that degrees are finite and although he usually only had three degrees (1, 2 and 3), other finite degrees were possible in different systems of AUTOMATH. That is, although in standard formulations of AUTOMATH, de Bruijn assumed the degrees 1, 2 and 3 and took 1 to be the degree of **type**, 2 to be the degree of inhabitants of **type** and 3 to be the degree of inhabitants of inhabitants of **type**, in various other systems of AUTOMATH, other degrees were allowed. For example, in AUT-4, 4 degrees are permitted: the degrees 3, 2 and 1 are for elements, sets and the class of sets, but degrees 4, 3 and 2 are for proofs, theorems, and the class of propositions. Also, in AUT-SL, terms of any (finite) degrees are possible.

Logic based on the double hierarchy of orders and types is usually called *predicative*. The difference between predicative and impredicative logic may seem small, nevertheless, this small difference can have some drastic consequences in fundamental mathematics. When constructing the real numbers out of the rationals (with Dedekind-cuts), the Theorem of the Least Upper Bound[1], is not provable in predicative logic (see [144] and Section 3a1). The Theorem of the Least Upper Bound is, however, one of the most fundamental theorems in real analysis as is illustrated in the work of Feferman (see for instance [46]).

Many modern type systems are impredicative. For instance, the systems of the Barendregt cube [3] that have the rule "$(\square, *)$" are all impredicative. Hence, a proof checker like Coq [44], based on the Calculus of Constructions [37], is itself founded on impredicative logic.

Nevertheless, mathematics with predicative logic is possible, and from a constructive point of view it is even attractive. For instance, the proof checker Nuprl [35, 77] is based on predicative logic yet many mathematical theories can be developed using this proof checker (see [76]). Of course, we are not claiming that the motivation of Nuprl and Martin-Löf's type theory for predicativity come from mathematics or computer science. As we said above, there are parts of mathematics that need impredicativity, and this explains why for example, Chet Murthy in [109] provided an impredicative extension of Nuprl and why other research in theoretical computer science (e.g. the work on the $\lambda\mu$-calculus [115]) identifies the need for classical rather than constructive logics. Nevertheless, we are concerned here with Martin-Löf's type theory and Nuprl, and not their impredicative extensions.

Nuprl's type theory is related to type theories proposed by Martin-Löf [105, 106], used as a foundation for constructive mathematics. Nuprl's logic is related to its type theory via the well-known propositions-as-types embedding, also known as the Curry-Howard-de Bruijn isomorphism (see [75]). It is constructive on two points: it is based on intuitionistic logic (as is the Curry-Howard-de Bruijn isomorphism) and it is based on predicative logic.

In this chapter, we will try to establish the relation between predicative logic as present in modern type theory (we concentrate on a subsystem of Nuprl because Martin-Löf's type theory is one of the richest and most expressive predicative type theories) and Russell's Ramified Type Theory RTT. This has many advantages. The most important advantage is the formulation of the informal notion of universe hierarchy in these modern predicative logics using Russell's notion of order. There are however many important bonuses that result from our study:

1. We give a presentation of a subsystem of the proof checker Nuprl as a PTS. In Section 6b we give a formal description of a part of the type system of Nuprl as a Pure Type System (PTS). Nuprl in PTS style enables us to

[1]This Theorem states that any non-empty set of real numbers with an upper bound has a least upper bound. See Example 3.2

formalize the concept of order in Nuprl and to show its correctness. This order classifies types and terms of Nuprl into their relevant hierarchy.

2. In Section 6c, we present an embedding of RTT in Nuprl's type system. Note that this is very different from [65] which did not give a presentation of RTT, but instead, extended CC^{ω} with an *anonymous universe* **Type** and intended this extension to model Russell and Whitehead's *typical ambiguity* convention.

3. This chapter gives another way to connect RTT to the modern way of writing type theory as a PTS. As we present a subsystem of Nuprl within the framework of PTSs in Section 6b, and as we present an embedding of RTT in Nuprl's type system in Section 6c, we also obtain a description of RTT in PTS-style. The same remark we gave above concerning [65] applies here.

4. Our study shows that orders in the historical system RTT correspond to orders in a very powerful modern system Nuprl. Our study of orders is different from the approach of [65] whose main concerns were type checking and well-typedness in Coquand's CC^{ω}, extending CC^{ω} with anonymous universes to model Russell and Whitehead's *typical ambiguity* convention, and with definitions. [65] is another example that orders and universes play an influential role in powerful modern systems.

5. Finally, this chapter places the historical system underlying Principia Mathematica in a context with a modern system of computer mathematics (Nuprl).

Parts of this chapter are taken from Kamareddine and Laan's paper [83].

6b The Nuprl type system

Martin-Löf's type theory [104] was originally developed as a foundation of constructive mathematics. The basic idea is the interpretation of logic within type theory through the Curry-Howard-de Bruijn isomorphism where as we have seen (roughly speaking), a proposition is interpreted as a set whose elements represent the proofs of the proposition. Hence, a false proposition is interpreted as the empty set and a true proposition is interpreted as a non-empty set. In order to prove that a proposition is true, we need to show that the proposition is inhabited.

This idea has proved extremely attractive from the computational point of view and has been exploited in many theorem provers (e.g., Nuprl and Coq). This idea was already exploited in de Bruijn's AUTOMATH which played an influential role in both provers Coq and Nuprl. In this chapter, we concentrate on a subsystem of Nuprl.

6b1 A fragment of Nuprl in PTS-style

We give a description of a part of the type system on which Nuprl is based (see [76, 35]). We do not give a full presentation of all of Nuprl's type constructors, as we will only need parts of it. The description of the typing rules is given in a natural deduction style similar to that used in the Barendregt cube and Pure Type Systems.

Definition 6.1 (Terms) Let $\mathbb{V}$ be a countably infinite set of variables, $\mathbb{Z}$ be the set of integers over which $m, n, \ldots$ ranges, and let $\perp$ represent the undefined or a contradiction. We take the set of sorts $\boldsymbol{S} = \{*_1, *_2, \ldots\}$ and assume that $\mathbb{V}$, $\boldsymbol{S}$, $\mathbb{Z}$, $\{\mathbb{Z}\}$ and $\{\perp\}$ are mutually disjoint. We take $\mathbb{C} = \boldsymbol{S} \cup \{\perp\} \cup \mathbb{Z} \cup \{\mathbb{Z}\}$. Note that the sets $\mathbb{C}$ and $\mathbb{V}$ are disjoint and are countably infinite. The set of terms $\mathcal{T}$ is defined by the following abstract syntax:

$$\mathcal{T} ::= \mathbb{V} \mid \mathbb{C} \mid \mathcal{T}\mathcal{T} \mid \lambda\mathbb{V}{:}\mathcal{T}.\mathcal{T} \mid \Pi\mathbb{V}{:}\mathcal{T}.\mathcal{T} \mid \mathcal{T} \times \mathcal{T} \mid \langle \mathcal{T}, \mathcal{T} \rangle \mid \pi_1(\mathcal{T}) \mid \pi_2(\mathcal{T})$$

Again, the intuition behind the sort $*_a$ is that it represents the propositions (and, more generally, the types) of order $\leq a$. $*_a$ corresponds to the Universe of Types U_a in [76, 105]. In addition to application and abstraction (λ and Π) we take Cartesian products, pairing, and first and second projections.

We use the same notations for meta-variables, free and bound variables, etc. as in Section 4b. Moreover, when x does not occur free in B, we write $A \to B$ for $\Pi x{:}A.B$.

Definition 6.2 (Reduction) We take the usual relation $\to_\beta$ of PTSs (see Definition 4.17). In addition, we take the relation $\to_{<>}$ described by the contraction rules

$$\pi_1(\langle A, B\rangle) \to_{<>} A \text{ and } \pi_2(\langle A, B\rangle) \to_{<>} B$$

and the usual compatibility rules.

We define $\to_{\beta<>}$, $\twoheadrightarrow_{<>}$ and $\twoheadrightarrow_{\beta<>}$ in the obvious way and take $=_{\beta<>}$ to be the smallest reflexive, symmetric and transitive relation that includes $\to_{\beta<>}$.

Definition 6.3 (Specification of Nuprl) The *specification* of Nuprl is a triple $(\boldsymbol{S}, \boldsymbol{A}, \boldsymbol{R})$, such that

- $\boldsymbol{S} = \{*_1, *_2, \ldots\}$,
- $\boldsymbol{A} = \{\perp : *_1, *_n : *_{n+1}$ for $n \in \mathbb{N}^+, \mathbb{Z} : *_1, n : \mathbb{Z}$ for $n \in \mathbb{Z}\}$ and
- $\boldsymbol{R} = \{(*_n, *_n, *_n)\}$.

Contexts are defined as in Definition 4.19 and again, the Barendregt convention 4.11 is extended to contexts.

Definition 6.4 (Derivable statements in Nuprl) A statement $\Gamma \vdash A : B$ is *derivable* if it can be deduced by repeated application of the rules below:

(Axioms) $\vdash \bot : *_1 \qquad \vdash *_n : *_{n+1} \quad (n \in \mathbb{N}^+)$
$\vdash \mathbb{Z} : *_1 \qquad \vdash n : \mathbb{Z} \quad (n \in \mathbb{Z})$

(Start) $$\frac{\Gamma \vdash A : *_n}{\Gamma, x{:}A \vdash x{:}A} \quad (x \text{ is } \Gamma\text{-fresh})$$

(Weak) $$\frac{\Gamma \vdash M : N \qquad \Gamma \vdash A : *_n}{\Gamma, x{:}A \vdash M : N} \quad (x \text{ is } \Gamma\text{-fresh})$$

(Π-form) $$\frac{\Gamma \vdash A : *_n \qquad \Gamma, x{:}A \vdash B : *_n}{\Gamma \vdash (\Pi x{:}A.B) : *_n}$$

(λ) $$\frac{\Gamma, x{:}A \vdash b : B \qquad \Gamma \vdash (\Pi x{:}A.B) : *_n}{\Gamma \vdash (\lambda x{:}A.b) : (\Pi x{:}A.B)}$$

(App) $$\frac{\Gamma \vdash M : (\Pi x{:}A.B) \qquad \Gamma \vdash N : A}{\Gamma \vdash MN : B[x{:=}N]}$$

(×-form) $$\frac{\Gamma \vdash A : *_n \qquad \Gamma \vdash B : *_n}{\Gamma \vdash (A \times B) : *_n}$$

(Pairs) $$\frac{\Gamma \vdash a : A \qquad \Gamma \vdash b : B \qquad \Gamma \vdash (A \times B) : *_m}{\Gamma \vdash \langle a, b \rangle : (A \times B)}$$

(Left) $$\frac{\Gamma \vdash M : (A \times B)}{\Gamma \vdash \pi_1(M) : A}$$

(Right) $$\frac{\Gamma \vdash M : (A \times B)}{\Gamma \vdash \pi_2(M) : B}$$

(Conv) $$\frac{\Gamma \vdash M : A \qquad \Gamma \vdash B : *_n \qquad A =_{\beta<>} B}{\Gamma \vdash M : B}$$

(⊆) $$\frac{\Gamma \vdash A : *_n}{\Gamma \vdash A : *_{n+1}}$$

To those familiar with Nuprl, the above rules are straightforward. Some remarks are due however:

1. The rule (Π-form) may look restrictive. This is not the case however due to the inclusion rule (⊆). In fact, (⊆) simplifies the formulation without sacrifycing expressivity.
2. A type universe U_n of Nuprl is closed under the construction of *dependent* Cartesian products. We use non-dependent Cartesian products (×-form) .

We refrain from introducing dependent Cartesian products for two reasons: they are not needed for the purpose of the chapter and they involve many complications that will obscure our main objectives.

3. The inclusion rule ($\subseteq$) is interesting on its own. We will see below that it leads to the loss of unicity of types. However, unicity of types is valued in many PTSs but not in Nuprl or Martin-Löf's type theory. We will in any case derive a version of unicity of types that is faithful to this idea of a term having many types in Nuprl. That is, we will derive that if we collapse the orders, then a term will have only one type.

4. Nuprl itself is implicitly rather than explicitly typed. That is, Nuprl uses terms of the form $\lambda x.B$ rather than $\lambda x{:}A.B$. There is a huge literature in programming language theory and design which discusses the tradeoffs between both styles. Our reason for the explicitly typed style in Nuprl is due to the fact that PTSs deal with explicitly typed systems and only recently, have been extended to deal with the implicitly typed style ([57]).

We adopt Definition 4.22 with the extra clause:

A is called a Γ-*type* if there is n such that $\Gamma \vdash A : *_n$

We now show some PTS properties of the Nuprl type system. Omitted proofs are as in [3].

Theorem 6.5 (Church-Rosser Theorem for $\rightarrow_\beta$ and $\rightarrow_{<>}$)

1. If $A \twoheadrightarrow_\beta B_1$ and $A \twoheadrightarrow_\beta B_2$ then there is C such that $B_1 \twoheadrightarrow_\beta C$ and $B_2 \twoheadrightarrow_\beta C$.

2. If $A \twoheadrightarrow_{<>} B_1$ and $A \twoheadrightarrow_{<>} B_2$ then there is C such that

$B_1 \twoheadrightarrow_{<>} C$ and $B_2 \twoheadrightarrow_{<>} C$.

PROOF: 2: any orthogonal term rewrite system (hence $(\mathcal{T}, \rightarrow_{<>})$) is Church-Rosser (see [93]). ⊠

Theorem 6.6 (Church-Rosser Theorem for $\rightarrow_{\beta<>}$)

1. If $A \rightarrow_\beta B_1$ and $A \rightarrow_{<>} B_2$ then there is C such that $B_1 \twoheadrightarrow_{<>} C$, and either $B_2 \rightarrow_\beta C$ or $B_2 \equiv C$;

2. If $A \rightarrow_\beta B_1$ and $A \twoheadrightarrow_{<>} B_2$ then there is C such that $B_1 \twoheadrightarrow_{<>} C$, and either $B_2 \rightarrow_\beta C$ or $B_2 \equiv C$;

3. If $A \twoheadrightarrow_\beta B_1$ and $A \twoheadrightarrow_{<>} B_2$ then there is C such that $B_1 \twoheadrightarrow_{<>} C$ and $B_2 \twoheadrightarrow_\beta C$;

4. $\to_{\beta<>}$ *has the Church-Rosser property.* ⊠

PROOF: 1: induction on the structure of A. 2: use 1. 3: use 2. 4: use 3 and Theorem 6.5. ⊠

Lemma 6.7 (Free Variable Lemma) *Assume* $x_1{:}A_1, \ldots, x_n{:}A_n \vdash B : C$. *Then*

- *The* $x_1, \ldots, x_n$ *are distinct;*
- $\mathrm{FV}(B) \cup \mathrm{FV}(C) \subseteq \{x_1, \ldots, x_n\}$;
- *For each* i *there is* m *such that* $x_1{:}A_1, \ldots, x_{i-1}{:}A_{i-1} \vdash A_i : *_m$. ⊠

Lemma 6.8 (Start Lemma) *Assume* Γ *is a legal context. Then* $\Gamma \vdash \bot : *_1$, $\Gamma \vdash \mathbb{Z} : *_1$, $\Gamma \vdash n : \mathbb{Z}$ *for any* $n \in \mathbb{Z}$, *and* $\Gamma \vdash *_n{:}*_{n+1}$ *for any* $n \geq 1$. *Moreover,* $\Gamma \vdash x{:}C$ *for all* $x{:}C \in \Gamma$. ⊠

Lemma 6.9 (Transitivity Lemma) *Let* Γ, Δ *be legal contexts such that* $\Gamma \vdash x : C$ *for all* $x{:}C \in \Delta$. *Then* $\Delta \vdash A : B \Rightarrow \Gamma \vdash A : B$. ⊠

Lemma 6.10 (Substitution Lemma) *If* $\Gamma, x{:}A, \Delta \vdash B : C$ *and* $\Gamma \vdash D : A$ *then* $\Gamma, \Delta[x{:=}D] \vdash B[x{:=}D] : C[x{:=}D]$. ⊠

Lemma 6.11 (Thinning Lemma) *Let* Γ, Δ *be legal contexts such that* $\Gamma \subseteq \Delta$. *Then* $\Gamma \vdash A : B \Rightarrow \Delta \vdash A : B$. ⊠

Lemma 6.12 (Generation Lemma)

1. *If* $\Gamma \vdash *_n : C$ *then* $C =_{\beta<>} *_m$ *for a* $m > n$, *and if* $C \not\equiv *_m$ *then* $\Gamma \vdash C : *_p$ *for some* $p \geq 1$.
2. *If* $\Gamma \vdash \bot : C$ *then* $C =_{\beta<>} *_m$ *for some* $m \geq 1$, *and if* $C \not\equiv *_m$ *then* $\Gamma \vdash C : *_p$ *for some* $p \geq 1$.
3. *If* $\Gamma \vdash \mathbb{Z} : C$ *then* $C =_{\beta<>} *_m$ *for some* $m \geq 1$, *and if* $C \not\equiv *_m$ *then* $\Gamma \vdash C : *_p$ *for some* $p \geq 1$.
4. *If* $\Gamma \vdash n : C$ *then* $C =_{\beta<>} \mathbb{Z}$, *and if* $C \not\equiv \mathbb{Z}$ *then* $\Gamma \vdash C : *_p$ *for some* $p \geq 1$.
5. *If* $\Gamma \vdash x : C$ *then there is* B *such that* $x{:}B \in \Gamma$, *and either* $B =_{\beta<>} C$, *or there are* m, n *with* $m < n$ *and* $B =_{\beta<>} *_m$, $C =_{\beta<>} *_n$. *If* $C \not\equiv B$ *then* $\Gamma \vdash C : *_p$ *for some* $p \geq 1$.
6. *If* $\Gamma \vdash (\Pi x{:}A.B) : C$ *then there is* m *such that* $\Gamma \vdash A : *_m$, $\Gamma, x{:}A \vdash B : *_m$ *and* $C =_{\beta<>} *_n$ *for a* $n \geq m$. *If* $C \not\equiv *_n$ *then* $\Gamma \vdash C : *_p$ *for some* $p \geq 1$.

7. *If* $\Gamma \vdash (\lambda x{:}A.b) : C$ *then there are* m, B *such that* $\Gamma \vdash (\Pi x{:}A.B) : *_m$, $\Gamma, x{:}A \vdash b : B$ *and* $C =_{\beta<>} \Pi x{:}A.B$. *If* $C \not\equiv \Pi x{:}A.B$ *then* $\Gamma \vdash C : *_p$ *for some* $p \geq 1$.

8. *If* $\Gamma \vdash AB : C$ *then there are* x, P, Q *such that* $\Gamma \vdash A : (\Pi x{:}P.Q)$, $\Gamma \vdash B : P$ *and either* $C =_{\beta<>} Q[x{:=}B]$, *or there are* m, n *with* $m < n$ *and* $Q[x{:=}B] =_{\beta<>} *_m$ *and* $C =_{\beta<>} *_n$. *If* $C \not\equiv Q[x{:=}B]$ *then* $\Gamma \vdash C : *_p$ *for some* $p \geq 1$.

9. *If* $\Gamma \vdash (A \times B) : C$ *then there is* m *such that* $\Gamma \vdash A : *_m$, $\Gamma \vdash B : *_m$ *and* $C =_{\beta<>} *_n$ *for a* $n \geq m$. *If* $C \not\equiv *_n$ *then* $\Gamma \vdash C : *_p$ *for some* $p \geq 1$.

10. *If* $\Gamma \vdash \langle a, b\rangle : C$ *then there are* m, A, B *such that* $\Gamma \vdash (A \times B) : *_m$, $\Gamma \vdash a : A$, $\Gamma \vdash b : B$ *and* $C =_{\beta<>} A \times B$. *If* $C \not\equiv A \times B$ *then* $\Gamma \vdash C : *_p$ *for some* $p \geq 1$.

11. *If* $\Gamma \vdash \pi_i(M) : C$ *then there are* A_1, A_2 *such that* $\Gamma \vdash M : (A_1 \times A_2)$ *and either* $C =_{\beta<>} A_i$ *or there are* m, n *with* $m < n$ *and* $A_i =_{\beta<>} *_m$ *and* $C =_{\beta<>} *_n$.

PROOF: Tedious but straightforward induction on the derivation $\Gamma \vdash M : C$. We only show two cases:

(Conversion): $\Gamma \vdash M : C$ because $\Gamma \vdash C : *_p$, $\Gamma \vdash M : C'$ and $C =_{\beta<>} C'$. We treat only the case $M \equiv AB$, the others are similar or easier. With the induction hypothesis, determine x, P, Q such that $\Gamma \vdash A : (\Pi x{:}P.Q)$, $\Gamma \vdash B : P$. If $Q[x{:=}B] =_{\beta<>} C'$ then also $Q[x{:=}B] =_{\beta<>} C$; if $m < n$ such that $Q[x{:=}B] =_{\beta<>} *_m$ and $C' =_{\beta<>} *_n$ then also $C =_{\beta<>} *_n$.

($\subseteq$): $\Gamma \vdash M : *_{k+1}$ because $\Gamma \vdash M : *_k$. Notice that, by the induction hypothesis, the cases $M \equiv n$ and $M \equiv \lambda x{:}A.b$ are impossible. We treat the case $M \equiv AB$; the other cases are similar or easier. By the induction hypothesis, there are x, P, Q such that $\Gamma \vdash A : (\Pi x{:}P.Q)$, $\Gamma \vdash B : P$. If $*_k =_{\beta<>} Q[x{:=}B]$ then take $m = k$ and $n = k + 1$; if there are $m' < n'$ such that $Q[x{:=}B] =_{\beta<>} *_{m'}$ and $*_k =_{\beta<>} *_{n'}$ then notice that $k = n'$ by the Church-Rosser Theorem, and take $m = m'$ and $n = k + 1$. ⊠

Corollary 6.13 (Correctness of Types)
If $\Gamma \vdash A : B$ *then there is* $n \geq 1$ *such that* $\Gamma \vdash B : *_n$.

PROOF: Induction on $\Gamma \vdash A : B$ with the help of the Generation Lemma and the Substitution Lemma for the cases $A \equiv MN$, $A \equiv \pi_1(M)$ and $A \equiv \pi_2(M)$. ⊠

Theorem 6.14 (Subject Reduction)
If $\Gamma \vdash A : B$ *and* $A \to_{\beta<>} A'$ *then* $\Gamma \vdash A' : B$.

PROOF: As is usual in the literature, we use induction on $\Gamma \vdash A : B$ to prove simultaneously

- $\Gamma \vdash A : B, \Gamma \to_\beta \Gamma' \Rightarrow \Gamma' \vdash A : B$;
- $\Gamma \vdash A : B, A \to_\beta A' \Rightarrow \Gamma \vdash A' : B$.

⊠

Corollary 6.15 ($\twoheadrightarrow_{\beta<>}$ **preserves Γ-terms)**
If A is a Γ-term and $A \twoheadrightarrow_{\beta<>} A'$ then A' is a Γ-term.

PROOF: We only prove the case $A \to_{\beta<>} A'$. If $\Gamma \vdash A : B$ then by Subject Reduction, $\Gamma \vdash A' : B$ and A' is a Γ-term. If $\Gamma \vdash B : A$ then by correctness of types $\Gamma \vdash A : *_n$ for some n and we use Subject Reduction. ⊠

Due to ($\subseteq$), Unicity of Types doesn't hold for Nuprl. For example, $\bot : *_1$ and $\bot : *_2$. A weak version however, is possible. This version collapses the different levels of $*$'s into $*_1$:

Definition 6.16 For each term A we define a term $|A|$ as follows:

$$
\begin{array}{ll}
|*_m| = *_1 & |\Pi x{:}A.B| = \Pi x{:}A.|B| \\
|x| = x & |A \times B| = |A| \times |B| \\
|\bot| = \bot & |\langle A, B\rangle| = \langle |A|, |B|\rangle \\
|\mathbb{Z}| = \mathbb{Z} & |\pi_1(M)| = \pi_1(|M|) \\
|MN| = |M||N| & |\pi_2(M)| = \pi_2(|M|) \\
|\lambda x{:}A.b| = \lambda x{:}A.|b| &
\end{array}
$$

Theorem 6.17 (Weak Unicity of Types)
If $\Gamma \vdash A : B_1$ and $\Gamma \vdash A : B_2$ then $|B_1| =_{\beta<>} |B_2|$.

PROOF: Induction on the structure of A. We only treat $A \equiv (\lambda x{:}M.N)$. By Lemma 6.12, there are D_1, D_2 with $B_j =_{\beta<>} \Pi x{:}M.D_j$, and $\Gamma, x{:}M \vdash N : D_j$. By the induction hypothesis, $|D_1| =_{\beta<>} |D_2|$. Hence, $|B_1| =_{\beta<>} |\Pi x{:}M.D_1| \equiv \Pi x{:}M.|D_1| =_{\beta<>} \Pi x{:}M.|D_2| \equiv |\Pi x{:}M.D_2| =_{\beta<>} |B_2|$. ⊠

6b2 Orders in Nuprl

Correctness of Types makes the following lemma and definition possible:

Lemma 6.18 *If A is a Γ-term then there is a Γ-term B, there is $n \geq 1$ such that $\Gamma \vdash A : B : *_n$.*

PROOF: If A is a Γ-term then there is a Γ-term B with $\Gamma \vdash A : B$ or $\Gamma \vdash B : A$. If $\Gamma \vdash A : B$, then by Correctness of Types there is $n \geq 1$ where $\Gamma \vdash A : B : *_n$. If $\Gamma \vdash B : A$ then again by Correctness of Types there is $n \geq 1$ where $\Gamma \vdash A : *_n$ and hence by Start and Thinning, $\Gamma \vdash A : *_n : *_{n+1}$. ⊠

Note that by Corollary 6.15, if A is a Γ-term then for any A' where $A \twoheadrightarrow_{\beta<>} A'$, A' is a Γ-term. There are also $A' =_{\beta<>} A$ where $A \not\twoheadrightarrow_{\beta<>} A'$ yet A' is a Γ-term. For example, take $A = (\lambda x : \Pi z : *_1 . *_1 .xa)b$ and $A' = (\lambda y : *_1 .by)a$. For this reason, we introduce the following definition:

Definition 6.19 (Γ-terms modulo A)
We define $[A]_\Gamma = \{A' | A' \text{ is } \Gamma\text{-term and } A =_{\beta<>} A'\}$.

Now, we define the order of a term:

Definition 6.20 (Order of a Term)
Assume A is a Γ-term. We define $\mathrm{ord}_\Gamma(A)$, the *order* of A in Γ, as the smallest natural number a (i.e. $a \geq 0$) for which there are $A' \in [A]_\Gamma$ and B such that $\Gamma \vdash A' : B : *_{a+1}$.

Let us explain the intuition behind this definition. The order of a term A must be the smallest natural number n such that the type of A is of type $*_{n+1}$. By ($\subseteq$), we get that for any $m > n$, the type of A is also of type $*_m$. This captures the notion of orders à la Russell. If A itself is a type and n is the order of of A, then not only the type of A is of type $*_{n+1}$, but also $A \twoheadrightarrow_{\beta<>} A'$ for some A' of type $*_n$ (see Lemma 6.28). Moreover, $*_n$ can be regarded as the type of types of order $\leq n$ (Corollary 6.29) and a term is always of a lower order than its type (Corollary 6.30). More importantly also, is the fact that a function can never take arguments of a higher order than itself (Lemma 6.32).

Of course, we want to make sure that any element $=_{\beta<>}$ to A has the same order as A. For this reason, we defined order as above by finding one A' in $[A]_\Gamma$ which gives us the minimal n in question. Even better, there is such an A' where $A \twoheadrightarrow_{\beta<>} A'$ rather than only $A =_{\beta<>} A'$. The following lemma shows this:

Lemma 6.21 *Let A be a Γ-term and* $\mathrm{ord}_\Gamma(A) = a$. *The following holds:*

1. *If $A' \in [A]_\Gamma$ then* $\mathrm{ord}_\Gamma(A) = \mathrm{ord}_\Gamma(A')$.
2. *There are A' and B such that* $\Gamma \vdash A' : B : *_{a+1}$ *and* $A \twoheadrightarrow_{\beta<>} A'$.

PROOF: 1: easy. 2: by definition of $\mathrm{ord}_\Gamma(A)$, there are $A'' =_{\beta<>} A$ and B such that $\Gamma \vdash A'' : B : *_{a+1}$. By Church-Rosser, A, A'' have a common reduct, say A'. By Subject Reduction, $\Gamma \vdash A' : B : *_{a+1}$. ⊠

Corollary 6.22 *For a* Γ*-term* A *in* β $<>$*-normal form and* $\mathrm{ord}_\Gamma(A) = a$*, there exists a* B *such that* $\Gamma \vdash A : B : *_{a+1}$.

PROOF: Determine, with Lemma 6.21, A' and B such that $A \twoheadrightarrow_{\beta<>} A'$ and $\Gamma \vdash A' : B : *_{a+1}$. As A is in normal form, $A' \equiv A$. ⊠

In what follows, we prove some elementary properties of $\mathrm{ord}_\Gamma(A)$. The first such property states that the order of a term does not change if the context is expanded:

Lemma 6.23 (Orders are invariant under context expansion)
If $\Gamma \vdash A : B$ *and* $\Gamma, x{:}C$ *is legal, then* $\mathrm{ord}_\Gamma(A) = \mathrm{ord}_{\Gamma,x:C}(A)$.

PROOF: Let $a = \mathrm{ord}_{\Gamma,x:C}(A)$.

($\geq$) By Thinning, $\Gamma \vdash A' : P \Rightarrow \Gamma, x{:}C \vdash A' : P$ for all $A' =_{\beta<>} A$ and P, so $\mathrm{ord}_\Gamma(A) \geq a$.

($\leq$) $\exists A' =_{\beta<>} A$ and P with $\Gamma, x{:}C \vdash A' : P : *_{a+1}$. By Lemma 6.21, assume $A \twoheadrightarrow_{\beta<>} A'$. By Lemma 6.10, $\Gamma \vdash A'[x{:=}C] : P[x{:=}C] : *_{a+1}$. As $\mathrm{FV}(A') \subseteq \mathrm{FV}(A) \subseteq \mathrm{dom}(\Gamma)$, $x \notin \mathrm{FV}(A')$. Hence $A' \equiv A'[x{:=}C]$, so $\Gamma \vdash A' : P[x{:=}C] : *_{a+1}$ and $\mathrm{ord}_\Gamma(A) \leq \mathrm{ord}_{\Gamma,x:C}(A)$.

⊠

Corollary 6.24 *If* A *is a* Γ*-term and* $\Delta \supseteq \Gamma$ *is legal then* $\mathrm{ord}_\Delta(A) = \mathrm{ord}_\Gamma(A)$. ⊠

The order of a term does not increase under substitution:

Lemma 6.25 (Substitution does not increase the order) *If* $\Gamma, x{:}A, \Delta \vdash B : C$ *and* $\Gamma \vdash D : A$ *then* $\mathrm{ord}_{\Gamma,x:A,\Delta}(B) \geq \mathrm{ord}_{\Gamma,\Delta[x:=D]}(B[x{:=}D])$.

PROOF: Let $\Gamma' = \Gamma, x{:}A, \Delta$; $\Gamma'' = \Gamma, \Delta[x{:=}D]$; and $b = \mathrm{ord}_{\Gamma'}(B)$.
There are $P, B' =_{\beta<>} B$ such that $\Gamma' \vdash B' : P : *_{b+1}$.
By Lemma 6.10 $\Gamma'' \vdash B'[x{:=}D] : P[x{:=}D] : *_{b+1}$. $B[x{:=}D] =_{\beta<>} B'[x{:=}D]$.
Hence, $b \geq \mathrm{ord}_{\Gamma''}(B[x{:=}D])$. ⊠

Note here that $\mathrm{ord}_{\Gamma,x:A,\Delta}(B) = \mathrm{ord}_{\Gamma,\Delta[x:=D]}(B[x{:=}D])$ does not hold in general: take $\Gamma \equiv y{:}*_1$. Then $\Gamma, x{:}*_2 \vdash x{:}*_2$ and $\Gamma \vdash y{:}*_2$, and (by Lemma 6.31 below) $\mathrm{ord}_{\Gamma,x:*_2}(x) = 2$ and $\mathrm{ord}_\Gamma(x[x{:=}y]) = \mathrm{ord}_\Gamma(y) = 1$.

6b3 Evaluating the order of a Nuprl term

In this subsection, we attempt to provide a procedure that evaluates the order of almost any Nuprl term. We use the word almost because we are able to say how the order of almost all complex terms (like $A \times B$) is evaluated in term of the orders of the components (A and B). The only case that fails is that of an application. We cannot evaluate the order of AB precisely in terms of the orders of A and B. Rather, in the case of an application AB, we can only establish that the order of AB is $\leq$ the order of A.

We begin by evaluating the order of the first and second projections:

Lemma 6.26 (Order of Projections) *For a* Γ*-term* $\langle A, B\rangle$, $\mathrm{ord}_\Gamma(\pi_1(\langle A, B\rangle)) = \mathrm{ord}_\Gamma(A)$ *and* $\mathrm{ord}_\Gamma(\pi_2(\langle A, B\rangle)) = \mathrm{ord}_\Gamma(B)$.

PROOF: This is a direct corollary of Lemma 6.21. ⊠

The orders of constants and sorts are easy to calculate:

Lemma 6.27 (Orders of constants and sorts) *Let* Γ *be a legal context. Then* $\mathrm{ord}_\Gamma(*_a) = a + 1$, $\mathrm{ord}_\Gamma(\bot) = 1$, $\mathrm{ord}_\Gamma(\mathbb{Z}) = 1$, *and* $\mathrm{ord}_\Gamma(n) = 0$.

PROOF:

- As $\Gamma \vdash *_a : *_{a+1} : *_{a+2}$, $\mathrm{ord}_\Gamma(*_a) \leq a + 1$. Now assume $\Gamma \vdash A' : P : *_b$ for an $A' =_{\beta<>} *_a$ (hence $A' \twoheadrightarrow_{\beta<>} *_a$). By repeated Subject Reduction, $\Gamma \vdash *_a : P : *_b$. By Generation, $P =_{\beta<>} *_c$ for $c > a$ (hence $P \twoheadrightarrow_{\beta<>} *_c$). By repeated Subject Reduction, $\Gamma \vdash *_c : *_b$, so again by Generation, $\exists d > c$ where $*_b =_{\beta<>} *_d$. Hence $d = b$, so $a < c < b$, so $b \geq a + 2$, so $\mathrm{ord}_\Gamma(*_a) \geq a + 1$.
- Notice that by the Start Lemma, $\Gamma \vdash \bot : *_1 : *_2$ so $\mathrm{ord}_\Gamma(\bot) \leq 1$. Now assume $\Gamma \vdash A' : P : *_1$ for an $A' =_{\beta<>} \bot$. Notice that $\bot$ is in normal form, so $A' \twoheadrightarrow_{\beta<>} \bot$ and by repeated Subject Reduction, $\Gamma \vdash \bot : P : *_1$. By the Generation Lemma, $P =_{\beta<>} *_1$, and as $*_1$ is in normal form, $P \twoheadrightarrow_{\beta<>} *_1$. By repeated Subject Reduction, $\Gamma \vdash *_1 : *_1$, which contradicts the fact that $\mathrm{ord}_\Gamma(*_1) = 2$.
- The proof for $\mathbb{Z}$ is similar to that for $\bot$.
- By the Start Lemma, $\Gamma \vdash n : \mathbb{Z} : *_1$, so $\mathrm{ord}_\Gamma(n) \leq 0$. $\mathrm{ord}_\Gamma(n) < 0$ is not possible.

⊠

The following lemma and its corollaries are not only needed for evaluating the order of the remaining items, but they are also informative about the order of a term. This lemma says that for any Γ-type B, there is always B' of type $*_{\mathrm{ord}_\Gamma(B)}$ such that $B \twoheadrightarrow_{\beta<>} B'$. It also confirms that $*_a$ can be seen as the type of types (propositions) of order $\leq a$ (Corollary 6.29) and that a term is always of a lower order than its type (Corollary 6.30).

Lemma 6.28 (A type B reduces to a type B' of type $*_{\mathrm{ord}(B)}$) *Let B be a Γ-type and $b = \mathrm{ord}_\Gamma(B)$. There is a B' such that $\Gamma \vdash B' : *_b$ and $B \twoheadrightarrow_{\beta<>} B'$.*

PROOF: Assume $\Gamma \vdash B : *_p$. By Lemma 6.21, there are B' and P such that $\Gamma \vdash B' : P : *_{b+1}$ and $B \twoheadrightarrow_{\beta<>} B'$. By Weak Unicity of Types Theorem 6.17, $|P| =_{\beta<>} |*_p|$, say: $P =_{\beta<>} *_q$. Hence $P \twoheadrightarrow_{\beta<>} *_q$.

- By repeated Subject Reduction, $\Gamma \vdash *_q : *_{b+1} : *_{b+2}$. By Lemma 6.27, $b+1 \geq q+1$, so $b \geq q$.
- By the Conversion Rule, $\Gamma \vdash B' : *_q : *_{q+1}$, so by definition of b, $q \geq b$.

We find: $q = b$, so $P =_{\beta<>} *_b$, so $\Gamma \vdash B' : *_b$. ⊠

Corollary 6.29 ($*_a$ is the type of types of order $\leq a$)
If P is a Γ-type in β <>-normal form, then $\Gamma \vdash P : *_a \Leftrightarrow \mathrm{ord}_\Gamma(P) \leq a$.

PROOF: Let $p = \mathrm{ord}_\Gamma(P)$.

- "$\Rightarrow$" is by definition of $\mathrm{ord}_\Gamma(P)$;
- for "$\Leftarrow$", by Lemma 6.28, there is a P' where $\Gamma \vdash P' : *_p$ and $P \twoheadrightarrow_{\beta<>} P'$. As P is in normal form, $P' \equiv P$, so $\Gamma \vdash P : *_p$. Since $p \leq a$, repeated use of ($\subseteq$) derives $\Gamma \vdash P : *_a$.

⊠

Corollary 6.30 (A term is of a lower order than its type)
If $\Gamma \vdash A : B$ then $\mathrm{ord}_\Gamma(A) < \mathrm{ord}_\Gamma(B)$.

PROOF: Let $a = \mathrm{ord}_\Gamma(A)$, $b = \mathrm{ord}_\Gamma(B)$. B is a type, so by Lemma 6.28, there is a B' such that $\Gamma \vdash B' : *_b$ and $B \twoheadrightarrow_{\beta<>} B'$. $\Gamma \vdash A : B$, so by conversion, $\Gamma \vdash A : B' : *_b$. By definition of a, $b \geq a+1$, so $b > a$. ⊠

In the above corollary, $\mathrm{ord}_\Gamma(A) = \mathrm{ord}_\Gamma(B) - 1$ does not hold: take $\Gamma = \emptyset$, $A \equiv *_1$ and $B \equiv *_3$. This is as expected because, by the inclusion rule ($\subseteq$), once A is of type $*_n$, it is of type $*_m$ for any $m \geq n$.

So far, we can calculate the order of projections (Lemma 6.26) and the order of sorts and constants (Lemma 6.27). Now, we present methods to calculate the order of almost all the other terms:

Lemma 6.31 *Let C be a Γ-term. The following holds:*

1. *If* $C \equiv x$ *where* $x{:}A \in \Gamma$ *then* $\mathrm{ord}_\Gamma(x) = \mathrm{ord}_\Gamma(A) - 1$.

2. *If* $C \equiv \Pi x{:}A.B$ *then* $\mathrm{ord}_\Gamma(\Pi x{:}A.B) = \max(\mathrm{ord}_\Gamma(A), \mathrm{ord}_{\Gamma,x:A}(B))$.

3. *If* $C \equiv \lambda x{:}A.b$ *then* $\mathrm{ord}_\Gamma(\lambda x{:}A.b) = \max(\mathrm{ord}_\Gamma(A) - 1, \mathrm{ord}_{\Gamma,x:A}(b))$.

4. *If* $C \equiv A \times B$ *or* $C \equiv \langle A, B\rangle$ *then* $\mathrm{ord}_\Gamma(C) = \max(\mathrm{ord}_\Gamma(A), \mathrm{ord}_\Gamma(B))$.

PROOF:

1. Let $m = \mathrm{ord}_\Gamma(x)$. There is a B such that $\Gamma \vdash x : B : *_{m+1}$ (by Corollary 6.22). As $m+1$ is minimal, $\mathrm{ord}_\Gamma(B) = m+1$. By the Generation Lemma, $A =_{\beta<>} B$. Hence, $\mathrm{ord}_\Gamma(A) = m+1$. Note that the case $A =_{\beta<>} *_n$, $P =_{\beta<>} *_p$ with $n < p$ does not hold as m is minimal.
2. Let $a = \mathrm{ord}_\Gamma(A)$, $b = \mathrm{ord}_{\Gamma,x:A}(B)$, and $p = \mathrm{ord}_\Gamma(\Pi x{:}A.B)$. By Lemma 6.28, as $\Pi x : A.B$ is a Γ-type, there is a P such that $\Gamma \vdash P : *_p$ and $\Pi x{:}A.B \twoheadrightarrow_{\beta<>} P$. P must be of the form $\Pi x{:}A_1.B_1$, where $A \twoheadrightarrow_{\beta<>} A_1$ and $B \twoheadrightarrow_{\beta<>} B_1$. By Lemmas 6.28 and 6.12, there are A_2 and B_2 such that: $\Gamma \vdash A_2 : *_a$, $\Gamma, x{:}A \vdash B_2 : *_b$, $A \twoheadrightarrow_{\beta<>} A_2$ and $B \twoheadrightarrow_{\beta<>} B_2$.
   By Church-Rosser, A_1 and A_2 have a common reduct A_3 and also B_1 and B_2 have a common reduct B_3. By repeated Subject Reduction: $\Gamma \vdash A_3{:}*_a$; $\Gamma, x{:}A \vdash B_3{:}*_b$. As $A \twoheadrightarrow_{\beta<>} A_3$ and $B \twoheadrightarrow_{\beta<>} B_3$, Subject Reduction gives: $\Gamma \vdash (\Pi x{:}A_3.B_3) : *_p$. Now, $p = \max(a, b)$ as follows:
   - By Generation there is an $m \leq p$ such that $\Gamma, x{:}A_3 \vdash B_3 : *_m$ and $\Gamma \vdash A_3 : *_m$. By Transitivity, $\Gamma, x{:}A \vdash B_3 : *_m$. Hence $a, b \leq m \leq p$.
   - As $\Gamma \vdash A_3 : *_a$ and $\Gamma, x{:}A_3 \vdash B_3 : *_b$, so by repeated application of $(\subseteq)$, $\Gamma \vdash A_3 : *_{\max(a,b)}$ and $\Gamma, x{:}A_3 \vdash B_3 : *_{\max(a,b)}$. By ($\Pi$-form), $\Gamma \vdash (\Pi x{:}A_3.B_3) : *_{\max(a,b)}$, and so $p \leq \max(a, b)$.
3. Let $a = \mathrm{ord}_\Gamma(A)$, $m = \mathrm{ord}_\Gamma(\lambda x{:}A.b)$, $n = \mathrm{ord}_{\Gamma,x:A}(b)$. By Lemma 6.21, there are P, Q such that $\Gamma \vdash P : Q : *_{m+1}$ and $\lambda x{:}A.b \twoheadrightarrow_{\beta<>} P$. Observe that $P \equiv \lambda x{:}A'.b'$ for some A', b' with $A \twoheadrightarrow_{\beta<>} A'$ and $b \twoheadrightarrow_{\beta<>} b'$. By the Generation Lemma, there is a B such that $\Gamma, x{:}A' \vdash b' : B$ and $Q =_{\beta<>} \Pi x{:}A'.B$. Now $m + 1 = \mathrm{ord}_\Gamma(Q) = \mathrm{ord}_\Gamma(\Pi x{:}A'.B) = \mathrm{ord}_\Gamma(\Pi x{:}A.B) = \max(a, \mathrm{ord}_{\Gamma,x:A}(B))$ by 2 above. Now $m = \max(a-1, n)$ because $m+1 = \max(a, n+1)$ as is seen by the two cases:
   - $m + 1 = a$. By the Transitivity Lemma, $\Gamma, x{:}A \vdash b' : B$. By Corollary 6.30: $\mathrm{ord}_{\Gamma,x:A}(b') = n < \mathrm{ord}_{\Gamma,x:A}(B)$, so $m + 1 = \max(a, n + 1)$.
   - $m + 1 = \mathrm{ord}_{\Gamma,x:A}(B) > a$. There are B', b'' such that $b' \twoheadrightarrow_{\beta<>} b''$ and $\Gamma, x{:}A' \vdash b'' : B' : *_{n+1}$. By Transitivity, $\Gamma, x{:}A \vdash b'' : B' : *_{n+1}$. With the Π and λ rule: $\Gamma \vdash (\lambda x{:}A.b'') : (\Pi x{:}A.B') : *_{\max(a,n+1)}$. Hence, $\max(a, n + 1) \geq m + 1$, and as $a < m + 1$, $n + 1 \geq m + 1$ and

$n \geq m$. As $\Gamma, x{:}A \vdash b' : B$, $n < \text{ord}_{\Gamma,x:A}(B) = m + 1$. Hence $n = m$ and $m + 1 = \max(a, n + 1)$.

4. Case $C \equiv A \times B$ is similar to 2. Case $C \equiv \langle A, B\rangle$ is similar to 3.

⊠

As MN may be a redex, its order is harder to determine. We can, however, prove the following:

Lemma 6.32 (The order of an application)
If $\Gamma \vdash M : \Pi x{:}P.Q$ *and* $\Gamma \vdash N : P$ *then* $\text{ord}_\Gamma(N), \text{ord}_\Gamma(MN) \leq \text{ord}_\Gamma(M)$.

PROOF: Let $m = \text{ord}_\Gamma(M)$. There are M', R such that $\Gamma \vdash M' : R : *_{m+1}$ and $M \twoheadrightarrow_{\beta<>} M'$. By Subject Reduction, $\Gamma \vdash M' : \Pi x{:}P.Q$, so by Weak Unicity of Types, $|R| =_{\beta<>} |\Pi x{:}P.Q| \equiv \Pi x{:}P.|Q|$. By Church-Rosser there is an R' such that $R \twoheadrightarrow_{\beta<>} R'$ and $\Pi x{:}P.|Q| \twoheadrightarrow_{\beta<>} |R'|$. Also, R' must be of the form $\Pi x{:}P'.Q'$, where $P \twoheadrightarrow_{\beta<>} P'$ and $|Q| \twoheadrightarrow_{\beta<>} |Q'|$. By Subject Reduction and Conversion, $\Gamma \vdash M' : (\Pi x{:}P'.Q') : *_{m+1}$. As m is minimal, $\text{ord}_\Gamma(\Pi x{:}P'.Q') = m + 1$. Now, $m = \text{ord}_\Gamma(M) = \text{ord}_\Gamma(\Pi x{:}P'.Q') - 1 = \max(\text{ord}_\Gamma(P') - 1, \text{ord}_{\Gamma,x:P'}(Q') - 1) \geq \text{ord}_\Gamma(P') - 1 = \text{ord}_\Gamma(P) - 1 \geq \text{ord}_\Gamma(N)$. By conversion, $\Gamma \vdash N : P'$, so $\Gamma \vdash M'N : Q'[x{:=}N]$. As $MN =_{\beta<>} M'N$, we have $\text{ord}_\Gamma(MN) = \text{ord}_\Gamma(M'N) < \text{ord}_\Gamma(Q'[x{:=}N]) \leq \text{ord}_{\Gamma,x:P'}(Q') \leq \text{ord}_\Gamma(\Pi x{:}P'.Q') = m + 1$, so $\text{ord}_\Gamma(MN) \leq m$. ⊠

This shows that a function can never take an argument of higher order, and that the order of a term can not increase when applying an argument to that term.

6c RTT in Nuprl

We present a straightforward embedding of a slightly modified version of RTT in the type theory of Nuprl written as a PTS (Section 6b). The embedding will consist of two parts: First we give a representation of the ramified types in Nuprl (Subsection 6c1), then we represent the typable propositional functions in Nuprl.

As we will now be dealing with another typing relation (that of RTT), we will use $\vdash_N$ for the typing relation in Nuprl as defined in Definition 6.4.

6c1 Ramified Types in Nuprl

Recall that not all propositional functions should be allowed in the language. For instance, the expression $\neg x(x)$ is a perfectly legal element of $\mathcal{P}$, nevertheless, it is the propositional function that makes it possible to derive the Russell paradox. Therefore, types were introduced by Russell. We repeat here the definition of the ramified types given in Definition 2.36, but eliminating ramified types of the form $()^0$ (note in clause 2 below, that when $n = 0$ then take $a \geq 1$ instead of $a \geq 0$).

Definition 6.33 (Ramified Types)
The ramified types are defined inductively as follows:

1. 0^0 is a ramified type (0 is called the *order* of this type);
2. If $t_1^{a_1}, \ldots, t_n^{a_n}$ are ramified types, and $a \in \mathbb{N}$, $a > \max(a_1, \ldots, a_n)$, then $(t_1, \ldots, t_n)^a$ is a ramified type of order a (if $n = 0$ then take $a \geq 1$);
3. All ramified types can be constructed using the rules 1 and 2.

0^0 is the type of individuals, and $(t_1, \ldots, t_n)^a$ is the type of the propositional functions with n free variables, say $x_1, \ldots, x_n$, such that if we assign values k_1 of type t_1 to x_1, ..., k_n of type t_n to x_n, then we obtain a proposition. The type $()^a$ is the type of propositions of order a.

Recall also that Russell strictly divides his propositional functions in orders. For instance, both $\forall \mathtt{p}[\mathtt{p}() \wedge \neg \mathtt{p}()]$ and $\mathtt{R}(\mathtt{a})$ are propositions, but of different level: The first presumes a full collection of propositions, hence it cannot belong to the same collection of propositions as the propositions $\mathtt{p}$ over which it quantifies (among which $\mathtt{R}(\mathtt{a})$). This led Russell to make $\forall \mathtt{p}[\mathtt{p}() \wedge \neg \mathtt{p}()]$ belong to a type of a *higher* order (level) than the order of $\mathtt{R}(\mathtt{a})$. This can already be seen in the definition of ramified types: $(t_1, \ldots, t_n)^a$ can only be a type if a is strictly greater than each of the orders of the t_is.

The main clue to our embedding is the interpretation of $*_n$ as the sort containing all order-n-propositions for $n \in \mathbb{N}^+$. There is a small difference in that Nuprl considers any term of type $*_n$ to be of type $*_{n+1}$ as well. This means that any proposition of order n can be interpreted as a proposition of order $n+1$ as well. This inclusion is not a feature of RTT; yet it isn't a serious extension.

Another small point is that Russell doesn't specify his underlying set of "individuals" and that we want to use $\mathbb{Z}$ as translation of this underlying set. Therefore, we will assume that the set $\mathcal{A}$ of RTT-individuals is equal to the set $\mathbb{Z}$ of integers. Recall that, when $x \notin \mathrm{FV}(B)$, we write $\Pi x : A.B$ as $A \to B$.

Definition 6.34 For each ramified type t^a as given in Definition 6.33, we define a Nuprl type $T(t^a)$ as follows:

$$T(0^0) \stackrel{\text{def}}{=} \mathbb{Z} \text{ and } T((t_1^{a_1}, \ldots, t_n^{a_n})^a) \stackrel{\text{def}}{=} T(t_1^{a_1}) \to \ldots T(t_n^{a_n}) \to *_a$$

Note that $T(()^a) = *_a$ and that T does indeed interpret the type of order-a-propositions as $*_a$. Moreover, translations of ramified types are typable in Nuprl:

Lemma 6.35 *If t^a is a ramified type of order a then $\vdash_N T(t^a) : *_{a+1}$.*

PROOF: Induction on the construction of ramified types. ⊠

When we speak of a ramified type t^a of order a, we actually mean that the terms that are of type t^a have order a. $T(t^a)$ itself should, therefore, have order $a+1$ in Nuprl. Indeed, we can prove:

Lemma 6.36 *If* Γ *is a legal context then* $\mathrm{ord}_\Gamma(T(t^a)) = a + 1$.

PROOF: Induction on ramified types. $T(0^0) = \mathbb{Z}$ and $\mathrm{ord}_\Gamma(\mathbb{Z}) = 1$ by Lemma 6.27. Now assume $\mathrm{ord}_\Gamma(T(t_i^{a_i})) = a_i + 1$ for $i = 1, \ldots, n$. Notice that

$$\begin{array}{rcl} \mathrm{ord}_\Gamma(T((t_1^{a_1},\ldots,t_n^{a_n})^a)) & = & \mathrm{ord}_\Gamma(T(t_1^{a_1}) \to \ldots \to T(t_n^{a_n}) \to *_a) \\ & \stackrel{\text{lem 6.31}}{=} & \max(\mathrm{ord}_\Gamma(T(t_1^{a_1})),\ldots, \\ & & \qquad \mathrm{ord}_\Gamma(T(t_n^{a_n})), \mathrm{ord}_\Gamma(*_a)) \\ & \stackrel{\text{lem 6.27, IH}}{=} & \max(a_1+1,\ldots,a_n+1,a+1) \\ & \stackrel{a \geq a_i}{=} & a+1 \end{array}$$

⊠

We now present the typing rules for RTT. These rules are a slightly adapted version of the rules of Definition 2.44.

Definition 6.37 (Typing Rules for RTT)

- If $c \in \mathcal{A}$, then $\Gamma \vdash c : 0^0$ for any context Γ;
- If $f \in \mathcal{P}$, and $x_1 < \ldots < x_n$ are the free variables of f, and $t_1, \ldots, t_n$ are types such that $x_i{:}t_i \in \Gamma$, then $\Gamma \vdash f : (t_1, \ldots, t_n)^a$ if and only if
  - If $f \equiv R(i_1, \ldots, i_{\mathfrak{a}(R)})$ then $t_i = 0^0$ for all i, and $a = 1$;
  - If $f \equiv z(k_1, \ldots, k_m)$ then there are $u_1, \ldots, u_m$ such that $z{:}(u_1, \ldots, u_m)^{a-1} \in \Gamma$, and $\Gamma \vdash k_i{:}u_i$ for all $k_i \in \mathcal{A} \cup \mathcal{P}$, and $k_i{:}u_i \in \Gamma$ for all $k_i \in \mathcal{V}$;
  - If $f \equiv f_1 \wedge f_2$ then there are $u_1^{a_1}, u_2^{a_2}$ such that $\Gamma \vdash f_i : u_i^{a_i}$ and $a = \max(a_1, a_2)$;
    if $f \equiv \neg f'$ then $\Gamma \vdash f' : (t_1, \ldots, t_n)^a$.
  - If $f \equiv \forall x{:}t_0[f']$ then there is a j such that $\Gamma, x{:}t_0 \vdash f' : (t_1, \ldots, t_{j-1}, t_0, t_j, \ldots, t_n)^a$.

Example 6.38 $\neg\mathtt{x}(\mathtt{x})$ is not typeable in any context Γ. If $\Gamma \vdash \neg\mathtt{x}(\mathtt{x}) : t$ then t must be of the form $(u)^a$, with $\mathtt{x}{:}u \in \Gamma$, as $\neg\mathtt{x}(\mathtt{x})$ has one free variable. Hence $\Gamma \vdash \mathtt{x}(\mathtt{x}) : (u)^a$, and by Unicity of Types below, $u \equiv (u')^{a-1}$, with $\mathtt{x} : u' \in \Gamma$. As Γ is a context, $u \equiv u'$, hence $u \equiv (u)^{a-1}$. Absurd.

Recall here Corollary 2.61 which gives unicity of types.

To avoid confusion we sometimes write $\vdash_{\mathrm{N}}$ for derivability in the Nuprl type system, and $\vdash_{\mathrm{R}}$ for derivability in RTT as presented in this section.

6c2 Propositional Functions of RTT in Nuprl

We recall here the definition of Propositional functions of RTT (Definition 2.3). We depart on a small point here: we replace the connective $\vee$ by $\wedge$ simply because we will represent $f \wedge g$ using the cartesian product.

Definition 6.39 (Propositional functions)
We define a collection $\mathcal{P}$ of *propositional functions*, and for each element f of $\mathcal{P}$ we simultaneously define the collection $\text{FV}(f)$ of *free variables* of f:

1. If $R \in \mathcal{R}$ and $i_1, \ldots, i_{\mathfrak{a}(R)} \in \mathcal{A} \cup \mathcal{V}$ then $R(i_1, \ldots, i_{\mathfrak{a}(R)}) \in \mathcal{P}$. $\text{FV}\big(R(i_1, \ldots, i_{\mathfrak{a}(R)})\big) \stackrel{\text{def}}{=} \{i_1, \ldots, i_{\mathfrak{a}(R)}\} \cap \mathcal{V}$;
2. If $f, g \in \mathcal{P}$ then $f \wedge g \in \mathcal{P}$ and $\neg f \in \mathcal{P}$. $\text{FV}(f \wedge g) \stackrel{\text{def}}{=} \text{FV}(f) \cup \text{FV}(g)$; $\text{FV}(\neg f) \stackrel{\text{def}}{=} \text{FV}(f)$;
3. If $f \in \mathcal{P}$ and $x \in \text{FV}(f)$ then $\forall x[f] \in \mathcal{P}$. $\text{FV}(\forall x[f]) = \text{FV}(f) \setminus \{x\}$.
4. If $z \in \mathcal{V}$, $n \in \mathbb{N}$ and $k_1, \ldots, k_n \in \mathcal{A} \cup \mathcal{V} \cup \mathcal{P}$, then $z(k_1, \ldots, k_n) \in \mathcal{P}$. $\text{FV}(z(k_1, \ldots, k_n)) \stackrel{\text{def}}{=} \{z, k_1, \ldots, k_n\} \cap \mathcal{V}$.
   If $n = 0$, we write $z()$ so as to distinguish the propositional function $z()$ from the variable z;[2]
5. All propositional functions can be constructed by using the rules 1, 2, 3 and 4 above.

We extend the mapping T of Definition 6.34 so that a propositional function with free variables $x_1 < \ldots < x_n$ will be translated into a λ-term of the form $\lambda x_1{:}t_1 \cdots x_n{:}t_n.A$, where A itself is not of the form $\lambda x{:}t.A'$. For notational convenience, T is extended to $\mathcal{A}$ and $\mathcal{V}$ as well. Recall that we took Russell's set $\mathcal{A}$ to be $\mathbb{Z}$ in Nuprl and that all individuals of RTT are translated into elements of $\mathbb{Z}$.

Definition 6.40 Let Γ be a RTT-context. We extend T to the sets $\mathcal{A}$, $\mathcal{V}$ and $\mathcal{P}$. If $i \in \mathcal{A} \cup \mathcal{V}$ then $T(i) \stackrel{\text{def}}{=} i$. Now let $f \in \mathcal{P}$ and assume f has free variables $x_1 < \ldots < x_n$, such that $x_i{:}t_i \in \Gamma$.

- If $f = R(i_1, \ldots, i_{\mathfrak{a}(R)})$ then
  $T(f) \stackrel{\text{def}}{=} \lambda x_1{:}T(t_1) \cdots x_n{:}T(t_n).Ri_1 \cdots i_{\mathfrak{a}(R)}$
- If $f = z(k_1, \ldots, k_m)$ then
  $T(f) \stackrel{\text{def}}{=} \lambda x_1{:}T(t_1) \cdots x_n{:}T(t_n).zT(k_1) \cdots T(k_m)$;

[2] A variable is not a propositional function. See [130], Chapter VIII: "The variable", p.94 of the 7th impression.

- If $f = g_1 \wedge g_2$, and g_i has free variables $y_{i1} < \ldots < y_{im_i}$, then $T(g_i) \equiv \lambda y_{i1}{:}u_{i1} \cdots y_{im_i}{:}u_{im_i}.G_i$ for some term G_i.
  Let $T(f) \stackrel{\text{def}}{=} \lambda x_1{:}T(t_1) \cdots x_n{:}T(t_n).G_1 \times G_2$.
- If $f = \neg g$, then $T(g) \equiv \lambda x_1{:}T(t_1) \cdots x_n{:}T(t_n).G$ for some term G.
  Let $T(f) \stackrel{\text{def}}{=} \lambda x_1{:}T(t_1) \cdots x_n{:}T(t_n).G \to \bot$.
- If $f = \forall x{:}t.g$ then for some term G,
  $T(g) \equiv \lambda x_1{:}T(t_1) \cdots x_i{:}T(t_i).x{:}T(t).x_{i+1}{:}T(t_{i+1}) \cdots x_n{:}T(t_n).G$.
  Let $T(f) \stackrel{\text{def}}{=} \lambda x_1{:}T(t_1) \cdots x_n{:}T(t_n).\Pi x{:}T(t).G$.

The extension of T as defined above also depends on the context Γ. Normally it will be clear which context Γ is meant. If confusion arises, we write T_Γ to indicate the context in question.

It is important to notice that, for propositions f, $T(f)$ is exactly the interpretation of f provided by the Curry-Howard-de Bruijn isomorphism.

Finally, we define a special Nuprl-context Γ_0 which contains information on the relation and individual symbols of RTT by:

$$\Gamma_0 \stackrel{\text{def}}{=} \{R : \overbrace{\mathbb{Z} \to \ldots \to \mathbb{Z}}^{\mathfrak{a}(R) \text{ times } \mathbb{Z}} \to *_1 \mid R \in \mathcal{R}\}.$$

We assume $\mathcal{R}$ to be finite for the moment, so that Γ_0 is finite as well, and therefore is a Nuprl-context. Γ_0 is legal, as we have $\vdash_{\mathrm{N}} \mathbb{Z} \to \ldots \to \mathbb{Z} \to *_1 : *_2$.

The following theorem states that the embedding T respects the type structure of RTT. This means that we can see Nuprl as an extension of the Ramified Theory of Types.

Theorem 6.41 (Nuprl extends RTT) *If* $\Gamma \vdash_{\mathrm{R}} f : t$ *then* $\Gamma_0 \vdash_{\mathrm{N}} T(f) : T(t)$.

PROOF: Induction on the definition of $\Gamma \vdash_{\mathrm{R}} f : t$. If $\Gamma \vdash c : 0^0$ because $c \in \mathbb{Z}$ then $c{:}\mathbb{Z} \in \Gamma_0$, so $\Gamma_0 \vdash c : \mathbb{Z}$. Now assume $f \in \mathcal{P}$, f has free variables $x_1 < \ldots < x_n$, and $t_1, \ldots, t_n$ where $x_i{:}t_i \in \Gamma$ for $i = 1, \ldots, n$, and $\Gamma \vdash_{\mathrm{R}} f : (t_1, \ldots, t_n)^a$. By Lemma 6.35, $\vdash_{\mathrm{N}} T(t_i) : *_{a_i}$ for some a_i. Hence, by the Start and Weakening rules, we add $x_i{:}T(t_i)$ one by one to the context Γ_0, obtaining a legal context $\Gamma_1 = \Gamma_0, x_1{:}T(t_1), \ldots, x_n{:}T(t_n)$. We only treat the case $f = \forall x{:}t_0[g]$:

There is a j such that $\Gamma, x{:}t_0 \vdash_{\mathrm{R}} g : (t_1, \ldots, t_{j-1}, t_0, t_j, \ldots, t_n)^a$. By the induction hypothesis, and then by the Generation Lemma, we obtain respectively:
$\Gamma_0 \vdash T(g) : T(t_1) \to \cdots \to T(t_{j-1}) \to T(t_0) \to T(t_j) \to \cdots \to T(t_n) \to *_a$.
$\Gamma_0, x_1{:}T(t_1), \ldots, x_{j-1}{:}T(t_{j-1}), x{:}T(t_0), x_j{:}T(t_j), \ldots, x_n{:}T(t_n) \vdash_{\mathrm{N}} G : *_a$
where $g \equiv \lambda x_1 \cdots x_{j-1} x x_j \cdots x_n.G$. As the types of the variables in the context are independent from each other, we also have $\Gamma_1, x{:}T(t_0) \vdash_{\mathrm{N}} G : *_a$. As the

order of type t_0 is smaller than a, we have $\Gamma_1 \vdash_N T(t_0) : *_a$ (Lemma 6.35), so by (Π-form): $\Gamma_1 \vdash_N \Pi x{:}T(t_0).G : *_a$. By λ-abstracting over all the variables in $\text{FV}(f)$ we obtain $\Gamma_0 \vdash_N T(f) : T(t)$. ⊠

It would be nice if we could also prove a kind of opposite of Theorem 6.41. However, the statement "*If* $\Gamma_0 \vdash_N T(f) : T(t)$ *then there is a context* Γ *such that* $\Gamma \vdash_R f : t$" is not true. We can derive $\Gamma_0 \vdash_N T(\forall x{:}0^0[R(x)]) : *_n$ for any $n \geq 1$. Nevertheless, we have $\Gamma \vdash_R \forall x{:}0^0[R(x)] : ()^1$ for all RTT-contexts Γ, so by Unicity of Types Corollary 2.61 it is impossible that $\Gamma \vdash_R \forall x{:}0^0[R(x)] : ()^n$ for any $n > 1$. It is clear that this difference between RTT and Nuprl is caused by the type inclusion rule $\subseteq$, which is only present in Nuprl, and not in RTT. We do have a partial result, however:

Lemma 6.42 *If* $\Gamma \vdash_R f : (t_1^{a_1}, \ldots, t_n^{a_n})^a$ *and* $x_1 < \ldots < x_n$ *are the free variables of* f*, then* $\text{ord}_{\Gamma_0}(T(\forall x_1{:}t_1^{a_1}.\cdots\forall x_n{:}t_n^{a_n}[f])) = a$.

PROOF: Induction on the definition of $\Gamma \vdash_R f : (t_1^{a_1}, \ldots, t_n^{a_n})^a$.
Note that $x_i{:}t_i^{a_i} \in \Gamma$ for all i, and $\Gamma \vdash_R \forall x_1{:}t_1^{a_1} \cdots \forall x_n{:}t_n^{a_n}[f] : ()^a$.
Let $\Gamma_i \equiv \Gamma_0, x_1{:}T(t_1^{a_1}), \ldots, x_i{:}T(t_i^{a_i})$. We only treat the case $f \equiv z(k_1, \ldots, k_m)$; the other cases are similar. $z \in \text{FV}(f)$, say: $z \equiv x_p$. As $x_p{:}t_p^{a_p} \in \Gamma$, $a_p = a - 1$.
Hence $\text{ord}_{\Gamma_n}(z) = \text{ord}_{\Gamma_n}(T(t_p^{a_p})) - 1 = a - 1$.
By 6.32, $\text{ord}_{\Gamma_n}(zT(k_1)\cdots T(k_m)) \leq a - 1$. Hence

$$\begin{array}{ll}
\text{ord}_{\Gamma_0}(T(\forall x_1{:}t_1^{a_1} \cdots \forall x_n{:}t_n^{a_n}[f])) & = \\
\text{ord}_{\Gamma_0}(\Pi x_1{:}T(t_1^{a_1}).\cdots\Pi x_n{:}T(t_n^{a_n}).zT(k_1)\cdots T(k_m)) & = \\
\max(\text{ord}_{\Gamma_n}(zT(k_1)\cdots T(k_m)), \max_{i\leq n}(\text{ord}_{\Gamma_i}(T(t_i^{a_i})))) & = \\
\max(\text{ord}_{\Gamma_n}(zT(k_1)\cdots T(k_m)), \max_{i\leq n}(a_i + 1)) = a_p + 1 = a &
\end{array}$$

⊠

Corollary 6.43 *If* $\Gamma \vdash_R f : ()^a$ *then* $\text{ord}_{\Gamma_0}(T(f)) = a$.

Conclusions

In this chapter we focussed on Nuprl and described a fragment of it as a Pure Type System λN. A type universe U_n ($n \geq 1$) of Nuprl contains certain basis types, and is closed under the construction of dependent product types and Cartesian products. Moreover, U_n is an element of U_{n+1}, and all types in U_n also belong to U_{n+1}. We represented the type universe U_n by the PTS sort $*_n$. Closure under the construction of dependent products is given by rule $(*_n, *_n)$, and the fact that U_n is an element of U_{n+1} is represented by the PTS axiom $*_n : *_{n+1}$. We extended this PTS as follows:

- For Cartesian products, we introduced the rule

$$\frac{\Gamma \vdash A_1 : *_n \qquad \Gamma \vdash A_2 : *_n}{\Gamma \vdash A_1 \times A_2 : *_n}$$

  Canonical inhabitants of $A_1 \times A_2$ are terms of the form $\langle a_1, a_2 \rangle$, where $a_i : A_i$.

- We also introduced the projection functions π_i: $\frac{\Gamma \vdash a : A_1 \times A_2}{\Gamma \vdash \pi_i(a) : A_i}$ together with a reduction relation generated by the axiom $\pi_i(\langle a_1, a_2 \rangle) \to_{<>} a_i$.

- As $U_n \subseteq U_{n+1}$, we introduced an inclusion rule ($\subseteq$): $\frac{\Gamma \vdash A : *_n}{\Gamma \vdash A : *_{n+1}}$

A type universe U_n in Nuprl is closed under the construction of *dependent* Cartesian products, but as we do not need dependent Cartesian products in the chapter, we did not introduce them.

The system λN thus obtained has many properties of usual PTSs, like Church-Rosser (for $\to_{\beta<>}$), Subject Reduction and Correctness of Types. With rule ($\subseteq$), we lose Unicity of Types, but we can prove a weakened version of it.

Let Γ be a context for λN. Due to correctness of types, for each Γ-type A there is $n \geq 1$ such that $\Gamma \vdash A : *_n$. (compare this to Nuprl: each type in Nuprl belongs to some type universe U_n). We call the smallest n for which $\Gamma \vdash A : *_n$ the *order* of A (in Γ), notation $\mathrm{ord}_\Gamma(A)$. We generalize this definition to arbitrary Γ-terms A: $\mathrm{ord}_\Gamma(A)$ is the minimal n for which there is B such that $\Gamma \vdash A : B : *_n$. We prove some elementary properties of $\mathrm{ord}_\Gamma(A)$:

- $\mathrm{ord}_\Gamma(A) = \mathrm{ord}_\Delta(A)$ if Δ is legal and $\Delta \supseteq \Gamma$;
- $\mathrm{ord}_\Gamma(*_n) = n + 1$;
- If $\Gamma \vdash A : B$ then $\mathrm{ord}_\Gamma(A) < \mathrm{ord}_\Gamma(B)$;
- If $x{:}A \in \Gamma$ then $\mathrm{ord}_\Gamma(x) = \mathrm{ord}_\Gamma(A) - 1$;
- $\mathrm{ord}_\Gamma(\Pi x{:}A.B) = \max(\mathrm{ord}_\Gamma(A), \mathrm{ord}_{\Gamma,x:A}(B))$;
- $\mathrm{ord}_\Gamma(\lambda x{:}A.b) = \max(\mathrm{ord}_\Gamma(A) - 1, \mathrm{ord}_{\Gamma,x:A}(b))$;
- $\mathrm{ord}_\Gamma(\langle A_1, A_2 \rangle) = \mathrm{ord}_\Gamma(A_1 \times A_2) = \max(\mathrm{ord}_\Gamma(A_1), \mathrm{ord}_\Gamma(A_2))$.

We showed that the orders in λN (and thus the type universes in Nuprl) are closely related to orders in RTT by looking at translations of RTT propositions to λN types via a propositions-as-types embedding T: We proved that if f is an order-n proposition in RTT, then $\mathrm{ord}_{\Gamma_0}(T(f)) = n$. Here, Γ_0 is some basic context that contains only some type information of the relation symbols that are used in RTT. We conclude that our formulation of Nuprl as a PTS is faithful to the idea behind universes in Martin-Löf's type theory and our definition of order on Nuprl terms captures the hierarchy of universes in Nuprl and provides an elegant comparison between Nuprl and RTT. As a bonus, we get a description of RTT in a propositions-as-types style in which the notion of order is maintained.

There are more similarities between RTT and Nuprl. Both Nuprl and RTT have a kind of higher order substitution (see Chapter 5 of [77].) It is interesting to investigate the similarities between both notions of substitution.

Let us now recap on what conclusions one can conclude from this chapter.

At the beginning of this century, the paradoxes led to many new formulations of logical systems and an amazing variety of ideas and approaches. Later on, some of these ideas were abandoned when they shouldn't have been. Even more, some of the ideas proposed were found later to contribute nothing to the solution of the paradoxes. For example, even though ZF set theory uses the foundation axiom, it is quite clear now that it is the separation rather than the foundation axiom which was responsible for the avoidance of the paradoxes.

Our standpoint in this chapter is not to defend one line against another. Rather, we aim to clarify the different notions and philosophies assumed in the foundation of logic. In this chapter, our chosen notion is that of Russell's orders as found in the famous Ramified Theory of Types RTT. Russell, whose contribution to modern logic is historical, avoided the paradox (that he himself discovered) by adopting two layers: types and orders. Later it was found that orders contributed nothing to the avoidance of the paradox and Ramsey's work led to the abandonment of Russell's orders. Perhaps Russell did actually know that orders do not contribute to the avoidance of the paradox. We believe however that his intuition of using orders (as well as types) is a solid one and we have seen this intuition being repeated in many predicative styles logics. In Chapter 3, we showed that Russell's orders come back in Kripke's account of levels of truths. In this chapter, we showed that Russell's orders are present in Martin-Löf's type theory and the proof checker Nuprl. Of course the word "orders" is not used by Kripke, Martin-Löf and Constable. Our study however shows that formally representing (with orders) the informal hierarchies of these systems is informative about these hierarchies, about the systems themselves and about the philosophies behind them.

Not only does this chapter offer a fresh look at the "order" concept, and show its usefulness for explaining basic hierarchies and philosophies in modern systems, but also, our chapter places the historical system underlying Principia Mathematica in a context with a modern system of computer mathematics (Nuprl) and hence the modern type theory of Martin-Löf. Our main results concerning the relationship between these various systems can be summarised as follows (we take $\vdash_{\mathrm{R}}$ (resp. $\vdash_{\mathrm{N}}$) to stand for type derivation in RTT (resp. in Nuprl), and assume a translation T from types and functions in RTT into Nuprl; also Γ_0 is a basic Nuprl-context which contains information on the relation and individual symbols of RTT):

1. The system (underlying) Nuprl can be seen as a simple extension of a PTS.

2. RTT can be embedded in Nuprl.

3. Hence RTT can be regarded as a PTS.
4. Nuprl extends RTT in the sense that if $\Gamma \vdash_{R} f : t$ then $\Gamma_0 \vdash_{N} T(f) : T(t)$.

A number of questions on extending these results remain open. These questions are as follows:

1. Since Martin-Löf's type theory, Nuprl and RTT have as aim to be a foundation of mathematics, one should have an interpretation of the most basic systems of logic: predicate logic (Pred) in RTT. This would be nice and the advantages of relating RTT, PTSs and Nuprl would carry over to Pred as well. Moreover, one would get the following picture:

$$\text{Pred} < \text{RTT} < \text{Nuprl} < \text{PTSs}.$$

2. We have shown that Nuprl extends RTT (see 4 above). It would be nice to answer the question whether Nuprl is a conservative extension of RTT.

Questions 1 and 2 are very interesting and must be the subject of future research. We have thought about them and up to this stage, no clear answer has been found. Question 1 causes difficulties precisely because Russell's notion of substitution is different from substitution as is used in modern logic and type theory. We have come a long way at formalising in modern style Russell's ideas and theory. There is still work to be done in this field and we believe that this work might prove very useful to modern computer science.

Question 2 has been partially attempted in this chapter. We have said that the converse of Theorem 6.41 does not hold. We have given as a reason for this the inclusion rule ($\subseteq$) which is only present in Nuprl and not in RTT. As shown in this chapter, RTT enjoys the unicity of types property whereas Nuprl does not. Here we explain intuitively this problem caused by the difference between Nuprl and RTT and give our opinion of how future directions in establishing a form of conservativity must be followed.

We know from the fact that Nuprl extends RTT that

$$\Gamma \vdash_{R} f : t \text{ then } \Gamma_0 \vdash_{N} T(f) : T(t)$$

Now, let us take this example:
$\Gamma \vdash_{R} \forall x{:}0^0[R(x)] : ()^1 \Rightarrow$
$\Gamma_0 \vdash_{N} T(\forall x{:}0^0[R(x)]) : *_1 \Rightarrow^{(\subseteq)}$
$\Gamma_0 \vdash_{N} T(\forall x{:}0^0[R(x)]) : *_n \equiv T(()^n)$ for any $n \geq 1 \not\Rightarrow^{\text{unicity of types in RTT}}$
$\Gamma \vdash_{R} \forall x{:}0^0[R(x)] : ()^n$ for any $n > 1$.
This means that we cannot go back from Nuprl to RTT.

We can however do something about that. The idea is to establish the order of the Nuprl term A and to only go in the opposite direction of Theorem 6.41 when the type of A is $*_a$ and a is the order of A. Hence in our example above, as 1 is the order of $T(\forall x{:}0^0[R(x)])$, we can only go back with $\Gamma_0 \vdash_{\mathrm{N}} T(\forall x{:}0^0[R(x)]) : *_1$ obtaining the valid typing $\Gamma \vdash_{\mathrm{R}} \forall x{:}0^0[R(x)] : ()^1$.

We have provided a partial result related to this question (given by Lemma 6.42 and Corollary 6.43) which says that for any Russell typable propositional function f of order a, we can establish that its Nuprl order is also a and hence when we try and mimick the Nuprl typing in RTT, we should only restrict ourselves to doing this when the Nuprl type is $*_a$ and a is the order of the Nuprl term avoiding the inclusion rule as much as possible. Of course, it remains to fully work out a translation from Nuprl to RTT and show in what way it can be said that RTT extends Nuprl. This will involve a huge technicality concerning RTT's substitution and free variables. It is left as a subject for future research.

Chapter 7

Automath

The first practical use of the propositions-as-types principle sketched in Section 4a is found in the AUTOMATH project [112]. The AUTOMATH systems are the first examples of proof checkers, and in this way they are predecessors of modern proof checkers like Coq [44] and Nuprl [35].

The project was started in 1967 by N.G. de Bruijn, and

> "it was not just meant as a technical system for verification of mathematical texts, it was rather a life style with its attitudes towards understanding, developing and teaching mathematics."
>
> ([21]; see [112] p. 201)

Thus, the roots of AUTOMATH are not to be found in logic or type theory, but in mathematics and the mathematical vernacular [20]. De Bruijn had been wondering for years what a proof of a theorem in mathematics should be like, and how its correctness should be checked. The development of computers in the 60s made him wonder whether a machine could check the proof of a mathematical theorem, provided the proof was written in a very accurate way. De Bruijn developed the language AUTOMATH for this purpose. This language is not only (according to de Bruijn [19]) "*a language which we claim to be suitable for expressing very large parts of mathematics, in such a way that the correctness of the mathematical contents is guaranteed as long as the rules of grammar are obeyed*" but also "*very close to the way mathematicians have always been writing*". This is also clearly reflected in the goals of the AUTOMATH project:

> "1. The system should be able to verify entire mathematical theories.
> 2. The system should remain very general, tied as little as possible to any particular set of rules for logic and foundations of mathematics. Such basic rules should preferably belong to material that

can be presented for verification, on the same level with things like mathematical axioms that have to be explained to the reader.[1]

3. The way mathematical material is to be presented to the system should correspond to the usual way we write mathematics. The only things to be added should be details that are usually omitted in standard mathematics."

([21]; see [112] pp. 209–210)

Goal 1 was definitely achieved: Van Benthem Jutting translated and verified Landau's "Grundlagen der Analysis" [100] in AUTOMATH (see [6], [7]) and Zucker formalised classical real analysis in AUTOMATH (see [148]).

A consequence of goal 2 has already been discussed in Section 4a4. There, we saw that de Bruijn used a PAT principle that was somewhat different from Curry and Howard's. Curry and Howard identified the logical implication and the universal quantifier with function types, following Heyting's intuitionistic interpretation of logical connectives. In doing so, they do not leave a possibility for a different interpretation of implication and universal quantification. Using PAT in de Bruijn's style, the rules for manipulating the logical connectives always have to be made explicit by the user (an example of such a specification can be found in Section 12 and 13 of [8]). This makes it possible to give interpretations of logical connectives that are not based on interpreting implication and universal quantification by a function type.

De Bruijn has spent a lot of effort in achieving goal 3. He has studied the language of mathematics in great depth (see [20]), and many of his insights are reflected in AUTOMATH. We mention some AUTOMATH features that help to achieve goal 3:

- The use of books. Just like a mathematical text, AUTOMATH is written line by line, where each line may refer to definitions or results given in earlier lines;

- The use of definitions. Without definitions, expressions very soon become too long. Moreover, a definition gives a name to a certain expression, and this name makes it easier for the user to remember (or understand) what the use of the definiens is;

- The use of a parameter mechanism together with a default mechanism. We discuss the advantages of these mechanisms in Section 7a.

[1] So: the logical rules should be treated on the same level as mathematics. A logical rule can be introduced as an axiom in the same way a mathematical axiom can be introduced. Other logical rules can be derived from existing rules, like mathematical theorems can be derived from existing theorems and axioms. [remark by the authors]

As AUTOMATH was developed quite independently from other developments in the world of type theory and λ-calculus, there are many things to be explained in the relation between the various AUTOMATH languages and other type theories. In this chapter we focus on the relation between AUTOMATH and Pure Type Systems (PTSs). [3] mentions this relation in a few lines, but as far as we know a satisfactory explanation of the relation between AUTOMATH and PTSs is not available. Moreover, both works consider AUTOMATH without one of its most important mechanisms: The definition system. Even the system PAL, which roughly consists of the definition system of AUTOMATH only, is able to express some simple mathematical reasoning (see for instance Section 5 of [19]). Moreover, recent developments on the use of definitions in Pure Type Systems by Bloo, Kamareddine and Nederpelt [14] and Severi and Poll [137] justify renewed research on the relation between AUTOMATH and PTSs.

In Section 7a we give a description of AUT-68, which is one of the most elementary AUTOMATH system. In Section 7b we discuss how we can transform AUT-68 into a PTS. In doing so, we must notice that AUT-68 has some properties that are not usual for PTSs:

- AUT-68 has η-reduction;
- AUT-68 has Π-application and Π-reduction (as it does not make any difference between λ and Π);
- AUT-68 has a definition system;
- AUT-68 has a parameter mechanism.

η-reduction is the reduction relation generated by $(\lambda x.Rx) \to_\eta R$, where $x \notin \mathrm{FV}(R)$. In systems with Π-application, a term $\Pi x{:}A.B$ can be applied to a term N (of type A). This results in $(\Pi x{:}A.B)N$. The usual application rule of Pure Type Systems then changes to

$$\frac{\Gamma \vdash M : \Pi x{:}A.B \qquad \Gamma \vdash N : A}{\Gamma \vdash MN : (\Pi x{:}A.B)N}.$$

In such systems, Π behaves like λ, and as a consequence, there also is a rule of Π-reduction

$$(\Pi x{:}A.B)N \to_\Pi B[x{:=}N].$$

In AUTOMATH, one does not even make any distinction between the terms $\Pi x{:}A.B$ and $\lambda x{:}A.B$. They are both denoted $[x{:}A]B$. It is not always easy to see whether a term $[x{:}A]B$ represents (in notation of PTSs) $\lambda x{:}A.B$ or $\Pi x{:}A.B$.

We pay more attention to Π-application and Π-reduction at the end of this chapter; for more details see [90, 81] and the literature on AUTOMATH [112].

We consider η-reduction not as one of the essential features of AUTOMATH, and prefer to focus on the definition and parameter mechanisms, which are the most characteristic type-theoretical features of AUTOMATH.

In Section 7c, we present a system $\lambda 68$ that is (almost) a PTS. We show that it has the usual properties of PTSs and we prove that $\lambda 68$ can be seen as AUT-68 without η-reduction, Π-application and Π-reduction. There is no direct parameter system in $\lambda 68$ either, but this parameter system is hidden in the rules for the construction of product types. In Section 7d we discuss how our approach can be extended to other AUTOMATH systems like AUT-QE where the identification of λ and Π is more subtle than that of AUT-68 and it is not easy to tell whether $[x{:}A]B$ should stand for $\lambda x{:}A.B$ or $\Pi x{:}A.B$ in PTSs. In addition to AUT-QE, we reflect on $\Delta\Lambda$ (cf. [112], B.7) where terms are presented as lambda trees and to each AUTOMATH book, there corresponds a single lambda tree whose correctness is equivalent to that of the book.

Parts of this chapter have been taken from Kamareddine, Laan and Nederpelt: [86].

7a Description of AUTOMATH

During the AUTOMATH-project, several AUTOMATH-languages have been developed. They all have two mechanisms for describing mathematics. One of them essentially is a typed λ-calculus, with the important features of λ-abstraction, λ-application and β-reduction. The other mechanism is the use of definitions and parameters. The latter is the same for most AUTOMATH-systems, and the difference between the various systems is mainly caused by different λ-calculi that are included. In this section we will describe the system AUT-68 which not only is one of the first AUTOMATH-systems, but also a system with a relatively simple typed λ-calculus, which makes it easier to focus on the (less known) mechanism for definitions and parameters.

A more extensive description of AUT-68 on which our description below is based, can be found in [8], [18] or [42].

7a1 Books, lines and expressions

In the conception underlying the AUTOMATH-systems, a mathematical text is thought of as being a series of consecutive "clauses". Each clause is expressed in AUTOMATH as a *line*. Lines are stored in the so-called *books*. For writing lines and books in AUT-68 we need

- The symbol `type`;
- A set $\mathcal{V}$ of variables;

- A set $\mathcal{C}$ of constants;
- The symbols ( ) [ ] : — , .

We assume that $\mathcal{V}$ and $\mathcal{C}$ are infinite, or at least offer us as many different elements as needed. We also assume that $\mathcal{V} \cap \mathcal{C} = \varnothing$ and that $\texttt{type} \notin \mathcal{V} \cup \mathcal{C}$.

The elements of $\mathcal{V}$ are called *block openers*, the elements of $\mathcal{V} \cup \mathcal{C}$ are called *identifiers* in [19].

Definition 7.1 (Expressions) We define the set $\mathcal{E}$ of AUT-*68-expressions* (or, in short, *expressions*) inductively:

(variable) If $x \in \mathcal{V}$ then $x \in \mathcal{E}$;

(parameter) If $a \in \mathcal{C}$, $n \in \mathbb{N}$ ($n = 0$ is allowed) and $\Sigma_1, \ldots, \Sigma_n \in \mathcal{E}$ then $a(\Sigma_1, \ldots, \Sigma_n) \in \mathcal{E}$. We call $\Sigma_1, \ldots, \Sigma_n$ the *parameters* of $a(\Sigma_1, \ldots, \Sigma_n)$;

(abstraction) If $x \in \mathcal{V}$, $\Sigma \in \mathcal{E} \cup \{\texttt{type}\}$ and $\Omega \in \mathcal{E}$ then $[x{:}\Sigma]\Omega \in \mathcal{E}$;

(application) If $\Sigma_1, \Sigma_2 \in \mathcal{E}$ then $\langle\Sigma_2\rangle\Sigma_1 \in \mathcal{E}$.

Sometimes we will consider the set $\mathcal{E}^+ \stackrel{\text{def}}{=} \mathcal{E} \cup \{\texttt{type}\}$.

Remark 7.2 The AUT-68-expression $[x{:}\Sigma]\Omega$ is AUTOMATH-notation for abstraction terms. In PTS-notation one would write either $\lambda x{:}\Sigma.\Omega$ or $\Pi x{:}\Sigma.\Omega$. In a relatively simple AUTOMATH-system like AUT-68, it is easy to determine whether $\lambda x{:}\Sigma.\Omega$ or $\Pi x{:}\Sigma.\Omega$ is the correct interpretation for $[x{:}\Sigma]\Omega$. This is harder in AUTOMATH-systems with a more complex λ-calculus, like AUT-QE.

Remark 7.3 The AUT-68-expression $\langle\Sigma_2\rangle\Sigma_1$ is AUTOMATH-notation for the intended application of the "function" Σ_1 to the "argument" Σ_2. In PTS-notation: $\Sigma_1\Sigma_2$.

Note the unusual *order* of "function" Σ_1 and "argument" Σ_2 in $\langle\Sigma_2\rangle\Sigma_1$. An advantage of this notation with respect to the classical notation becomes clear if we assume that Σ_1 is a function $[x{:}\Omega_1]\Omega_2$. In that case, $\langle\Sigma_2\rangle\Sigma_1 \equiv \langle\Sigma_2\rangle[x{:}\Omega_1]\Omega_2$. The argument Σ_2 and the abstraction $[x{:}\Omega_1]$ belong together: As soon as the intended application of the function Σ_1 to its argument is carried out, Σ_2 is substituted for x everywhere in Ω_2. It is convenient to put expressions that belong together next to each other. In the usual, classical notation, we would write $([x{:}\Omega_1]\Omega_2)\Sigma_2$, where Σ_2 and $[x{:}\Omega_1]$ are separated from each other by the expression Ω_2. This makes the structure of the expression less clear, in particular if Ω_2 is a very long expression. The advantages of writing $\langle\Sigma_2\rangle\Sigma_1$ instead of the classical $\Sigma_1\Sigma_2$ are extensively studied by Kamareddine and Nederpelt in [91].

Definition 7.4 (Free variables)

- $\text{FV}(x) \stackrel{\text{def}}{=} \{x\}$;
- $\text{FV}(a(\Sigma_1,\ldots,\Sigma_n)) \stackrel{\text{def}}{=} \bigcup_{i=1}^{n} \text{FV}(\Sigma_i)$;
- $\text{FV}([x{:}\Sigma]\Omega) \stackrel{\text{def}}{=} \text{FV}(\Sigma) \cup (\text{FV}(\Omega) \setminus \{x\})$;
- $\text{FV}(\langle\Sigma_2\rangle\Sigma_1) \stackrel{\text{def}}{=} \text{FV}(\Sigma_1) \cup \text{FV}(\Sigma_2)$.

Convention 7.5 We adhere to the usual convention that names of bound variables in an expression differ from the free variables in that expression.

We use $\equiv$ to denote syntactical equivalence (up to renaming of bound variables) on expressions.

Definition 7.6 If $\Omega, \Sigma_1, \ldots, \Sigma_n$ are expressions (in $\mathcal{E}$), and $x_1, \ldots, x_n$ are distinct variables, then

$$\Omega[x_1, \ldots, x_n{:=}\Sigma_1, \ldots, \Sigma_n]$$

denotes the expression Ω in which all free occurrences of $x_1, \ldots, x_n$ have simultaneously been replaced by $\Sigma_1, \ldots, \Sigma_n$. This, again, is an expression in $\mathcal{E}$ (this can be proved by induction on the structure of Ω).

$\texttt{type}[x_1, \ldots, x_n{:=}\Sigma_1, \ldots, \Sigma_n]$ is defined as $\texttt{type}$.

Definition 7.7 (Books and lines) An AUT-68-*book* (or *book* if no confusion arises) is a finite list (possibly empty) of (AUT-68)-lines (to be defined next). If $l_1, \ldots, l_n$ are the lines of book $\mathfrak{B}$, we write $\mathfrak{B} \equiv l_1, \ldots, l_n$.

An AUT-68-*line* (or *line* if no confusion arises) is a 4-tuple $(\Gamma; k; \Sigma_1; \Sigma_2)$. Here,

- Γ is a context, i.e. a finite (possibly empty) list $x_1{:}\alpha_1, \ldots, x_n{:}\alpha_n$, where the x_is are different elements of $\mathcal{V}$ and the α_is are elements of $\mathcal{E}^+$;
- k is an element of $\mathcal{V} \cup \mathcal{C}$;
- Σ_1 can be (only):
  - The symbol — (if $k \in \mathcal{V}$);
  - The symbol PN (if $k \in \mathcal{C}$) (PN stands for "primitive notion");
  - An element of $\mathcal{E}$ (if $k \in \mathcal{C}$);
- Σ_2 is an element of $\mathcal{E}^+$.

Remark 7.8 As regards the intended meaning of an AUTOMATH-line, we note the following. There are three sorts of lines:

1. $(\Gamma; k; —; \Sigma_2)$ with $k \in \mathcal{V}$. This is a *variable declaration* of the variable k having type Σ_2. This does not really add a new statement to the book, but these declarations are needed to form contexts.

   Variables can play two roles. First of all they can represent an unspecified object of a certain type (compare this to the mathematical way of speaking: "let x be a natural number"). Secondly, a variable can act as a logical assumption. This happens if the variable has as type the proof of a certain proposition A. The usual mathematical way of speaking in such a situation is not "let x be a proof of A", but: "assume A";

2. $(\Gamma; k; \text{PN}; \Sigma_2)$ with $k \in \mathcal{C}$. This line introduces a *primitive notion*: A constant k of type Σ_2. This constant can act as a primitive notion (for instance introducing the type of natural numbers, or introducing the number 0), or as an axiom (to be precise, a postulated inhabitant of the set of proofs of the proposition expressing the axiom).

   The introduction of k is *parametrised* by the context Γ. For instance, if we want to introduce the primitive notion of "logical conjunction", we do not want to have a separate primitive notion for each possible conjunction $\mathtt{and}(A, B)$.[2] Instead, we want to have one primitive notion $\mathtt{and}$, to which we can add two propositions A and B as parameters when we want to form the proposition $\mathtt{and}(A, B)$. Therefore, we introduce $\mathtt{and}$ in a context $\Gamma \equiv$ $\mathtt{x{:}prop}, \mathtt{y{:}prop}$. Given certain propositions A, B this enables us to form the AUT-68-expression $\mathtt{and}(A, B)$;

3. $(\Gamma; k; \Sigma_1; \Sigma_2)$ with $k \in \mathcal{C}$ and $\Sigma_1 \in \mathcal{E}$. This line introduces a *definition*. The *definiendum* k is defined by the *definiens* Σ_1 and has *type* Σ_2. Definitions can be parametrised in a similar way as primitive definitions. Definitions have two important applications:

   - They make it possible to abbreviate long expressions, thus keeping the structure of a book clear, and making manipulations with expressions more efficient;

   - They make it possible to give a name to an expression. For instance, we can abbreviate S(S(S(S(S(S(S(0))))))) by 7.

Example 7.9 In Figure 7.1 we give an example of an AUTOMATH-book that introduces some elementary notions of propositional logic. We have numbered each

[2]Contrary to the habit in mathematics to use only one character (possibly indexed) for a variable, AUTOMATH adopts the convention of computer science to use variables that consist of more than one character. So $\mathtt{and}$ represents only one variable, and not the application of $\mathtt{a}$ to $\mathtt{n}$ and $\mathtt{d}$.

line in the example, and use these line numbers for reference in our comments below. To keep things clear, we have omitted the types of the variables in the context. The book consists of three parts:

- In lines 1–5 we introduce some basic material:

    1. We take the type `prop` as a primitive notion. This type can be interpreted as the type of propositions;
    2. We declare a variable `x` of type `prop`. This variable will be used in the sequel of the book;
    3. We similarly define a variable `y` of type `prop`. We do this within the context `x:prop`. For reasons of space, we do not explicitly mention the type of `x` in the context; if necessary we can find that type in line 2;
    4. Given propositions `x` and `y`, we introduce a new primitive notion, the conjunction `and(x,y)` of `x` and `y`;
    5. Given a proposition `x` we introduce the type `proof(x)` of the proofs of `x` as a primitive notion. In this way, we can use the PAT principle à la de Bruijn (cf. Section 4a4);

- In lines 6–11 we show how we can construct proofs of propositions of the form `and(x,y)`, and how we can use proofs of such propositions:

    6. Given propositions `x` and `y`, we assume that we have a expression `px` $\in \mathcal{V}$ of type `proof(x)`. In other words, the variable `px` represents an arbitrary proof of the proposition `x`;
    7. We also assume a proof `py` of `y`;
    8. Given the propositions `x` and `y`, and proofs `px` and `py` of `x` and `y`, we want to conclude that `and(x,y)` holds. This is an axiom of natural deduction, and we call this axiom `and-I` (and-introduction) in our book. An expression `and-I(x,y,px,py)` is a proof of `and(x,y)`, so of type `proof(and(x,y))`.

       In line 8, we see `proof(and)` instead of `proof(and(x,y))` as the type of `and-I`. This is usual notation in AUTOMATH, and keeps lines short. To be precise, this "default mechanism" works as follows. From line 4, we conclude that `and` should always carry two parameters. This is because the context of line 4 has two variables `x` and `y`. In the expression `proof(and)` in line 8, no parameters are provided for `and`. It is then implicitly assumed that the first two variables of the context of line 8 are used as "default parameters". The first two variables of the

∅	prop	PN	type	(1)
∅	x	—	prop	(2)
x	y	—	prop	(3)
x,y	and	PN	prop	(4)
x	proof	PN	type	(5)
x,y	px	—	proof(x)	(6)
x,y,px	py	—	proof(y)	(7)
x,y,px,py	and-I	PN	proof(and)	(8)
x,y	pxy	—	proof(and)	(9)
x,y,pxy	and-01	PN	proof(x)	(10)
x,y,pxy	and-02	PN	proof(y)	(11)
x	prx	—	proof(x)	(12)
x,prx	and-R	and-I(x,x,prx,prx)	proof(and(x,x))	(13)
x,y,pxy	and-S	and-I(y,x,and-02,and-01)	proof(and(y,x))	(14)

Figure 7.1: Example of an AUTOMATH-book

context of line 8 are x and y. Therefore, proof(and) in line 8 should be read as proof(and(x,y)).

In a similar way, we could write proof instead of proof(x) in line 6. From line 5 (where proof is introduced) we find that proof carries one parameter. Writing just proof in line 6 means that we must use the first variable of the context of line 6, x, as a default parameter. We must write proof(y) in line 7. Writing just proof would give proof(x);

9. We also want to express how we can use a proof of and(x,y). Therefore we introduce a variable pxy that represents an arbitrary proof of and(x,y);
10. First of all, we want that x holds whenever and(x,y) holds. Therefore we introduce an axiom and-01 (and-out, first and-elimination). Given propositions x,y and a proof pxy of the proposition and(x,y), and-01(x,y,pxy) is a proof of x;
11. Similarly, we introduce an axiom and-02 that represents a proof of y;

- We can now derive some elementary theorems:

12. We want to prove that we can derive and(x,x) from x. That is: Whenever we have a proof of x, we can construct a proof of and(x,x). In line 6, we already introduced a variable for a proof of x: px. However, we declared this variable in the context x,y. As we do not want a second proposition y to occur in this theorem, we declare a new proof variable prx, in the context x;
13. We derive our first small theorem: The reflexivity of the logical conjunction. Given a proposition x, and a proof prx of x, we can use the axiom and-I to find a proof of and(x,x): we can use the expression and-I(x,x,prx,prx) thanks to line 8. We give a name to this proof: and-R. If, anywhere in the sequel of the book, Σ is a proposition, and Ω is a proof of Σ, we can write and-R(Σ, Ω) for a proof of and(Σ, Σ). This is shorter, and more expressive, than the original expression and-I($\Sigma, \Sigma, \Omega, \Omega$);
14. We can also show that and is symmetric. That is: Whenever and(x,y) holds, we also have and(y,x). The idea is as follows. Given propositions x,y and a proof pxy of and(x,y), we can form proofs and-01(x,y,pxy) of x and and-02(x,y,pxy) of y. We can feed these proofs "in reverse order" to the axiom and-I: The expression and-I(y,x,and-02,and-01) represents a proof of and(y,x). The expression and-02 should be read as and-02(x,y,pxy) due to the

"default parameter" mechanism. Similarly, `and-01` must be read as `and-01(x,y,pxy)`.

7a2 Correct books

Not all books are good books. If $(\Gamma; k; \Sigma_1; \Sigma_2)$ is a line of a book $\mathfrak{B}$, the expressions Σ_1 and Σ_2 (as long as Σ_1 is not PN or —, and Σ_2 is not `type`) must be well-defined, i.e. the elements of $\mathcal{V} \cup \mathcal{C}$ occurring in them must have been established (as variables, primitive notions, or defined constants) in previous parts of $\mathfrak{B}$. The same holds for the type assignments $x_i{:}\alpha_i$ that occur in Γ. Moreover, if Σ_1 is not PN or —, then Σ_1 must be of the same type as k, hence Σ_1 must be of type Σ_2 (within the context Γ). Finally, there should be only one definition of any object in a book, so k should not occur in the preceding lines of the book.

Hence we need notions of correctness (with respect to a book and/or a context) and a definition of the notion "Σ_1 is of type Σ_2" (within a book and a context).

We write $\mathfrak{B}; \varnothing \vdash$ OK to indicate that a book $\mathfrak{B}$ is correct, and $\mathfrak{B}; \Gamma \vdash$ OK to indicate that the context Γ is correct with respect to the (correct) book $\mathfrak{B}$. As the empty context will be correct with respect to any correct book, this does not lead to misunderstandings.

We write $\mathfrak{B}; \Gamma \vdash \Sigma$ (or $\mathfrak{B}; \Gamma \vdash_{\mathrm{AUT-68}} \Sigma$ if confusion with other derivation systems might arise) to indicate that Σ is a correct expression with respect to $\mathfrak{B}$ and Γ. We write $\mathfrak{B}; \Gamma \vdash \Sigma_1 : \Sigma_2$ (or $\mathfrak{B}; \Gamma \vdash_{\mathrm{AUT-68}} \Sigma_1 : \Sigma_2$) to indicate that Σ_1 is a correct expression of type Σ_2 with respect to $\mathfrak{B}$ and Γ. We also say: $\Sigma_1 : \Sigma_2$ is a correct *statement* with respect to $\mathfrak{B}$ and Γ.

The following two interrelated definitions are based on [42].

Definition 7.10 (Correct books and contexts) A book $\mathfrak{B}$ and a context Γ are *correct* if $\mathfrak{B}; \Gamma \vdash$ OK can be derived with the following rules (The relation $=_{\beta\mathrm{d}}$ ("definitional equality") will be explained in Section 7a3. The rules use the notion of *correct statement* as given in Definition 7.11).

(axiom) $$\varnothing; \varnothing \vdash \mathrm{OK}$$

(context ext.) $$\frac{\mathfrak{B}_1, (\Gamma; x; —; \alpha), \mathfrak{B}_2; \Gamma \vdash \mathrm{OK}}{\mathfrak{B}_1, (\Gamma; x; —; \alpha), \mathfrak{B}_2; \Gamma, x{:}\alpha \vdash \mathrm{OK}}$$

(book ext.: var1) $$\frac{\mathfrak{B}; \Gamma \vdash \mathrm{OK}}{\mathfrak{B}, (\Gamma; x; —; \mathtt{type}); \varnothing \vdash \mathrm{OK}}$$

(book ext.: var2) $$\frac{\mathfrak{B}; \Gamma \vdash \Sigma_2 : \mathtt{type}}{\mathfrak{B}, (\Gamma; x; —; \Sigma_2); \varnothing \vdash \mathrm{OK}}$$

(book ext.: pn1)

$$\frac{\mathfrak{B};\Gamma\vdash\text{OK}}{\mathfrak{B},(\Gamma;k;\text{PN};\texttt{type});\varnothing\vdash\text{OK}}$$

(book ext.: pn2)

$$\frac{\mathfrak{B};\Gamma\vdash\Sigma_2:\texttt{type}}{\mathfrak{B},(\Gamma;k;\text{PN};\Sigma_2);\varnothing\vdash\text{OK}}$$

(book ext.: def1)

$$\frac{\mathfrak{B};\Gamma\vdash\Sigma_1:\texttt{type}}{\mathfrak{B},(\Gamma;k;\Sigma_1;\texttt{type});\varnothing\vdash\text{OK}}$$

(book ext.: def2)

$$\frac{\mathfrak{B};\Gamma\vdash\Sigma_2:\texttt{type}\qquad\mathfrak{B};\Gamma\vdash\Sigma_1:\Sigma_2'\qquad\mathfrak{B};\Gamma\vdash\Sigma_2=_{\beta\text{d}}\Sigma_2'}{\mathfrak{B},(\Gamma;k;\Sigma_1;\Sigma_2);\varnothing\vdash\text{OK}}$$

For the (book ext.) rules, we assume that the introduced identifiers $x\in\mathcal{V}$ and $k\in\mathcal{C}$ do not occur anywhere in $\mathfrak{B}$ and Γ.

Definition 7.11 (Correct statements) A statement $\mathfrak{B};\Gamma\vdash\Sigma:\Omega$ is *correct* if it can be derived with the rules below (the start rule uses the notions of correct context and correct book as given in Definition 7.10).

(start)

$$\frac{\mathfrak{B};\Gamma_1,x{:}\alpha,\Gamma_2\vdash\text{OK}}{\mathfrak{B};\Gamma_1,x{:}\alpha,\Gamma_2\vdash x{:}\alpha}$$

(parameters)

$$\frac{\begin{array}{c}\mathfrak{B}\equiv\mathfrak{B}_1,(x_1{:}\alpha_1,\ldots,x_n{:}\alpha_n;b;\Omega_1;\Omega_2),\mathfrak{B}_2\\ \mathfrak{B};\Gamma\vdash\Sigma_i{:}\alpha_i[x_1,\ldots,x_{i-1}{:=}\Sigma_1,\ldots,\Sigma_{i-1}](i=1,\ldots,n)\end{array}}{\mathfrak{B};\Gamma\vdash b(\Sigma_1,\ldots,\Sigma_n):\Omega_2[x_1,\ldots,x_n{:=}\Sigma_1,\ldots,\Sigma_n]}$$

(abstr.1)

$$\frac{\mathfrak{B};\Gamma\vdash\Sigma_1{:}\texttt{type}\qquad\mathfrak{B};\Gamma,x{:}\Sigma_1\vdash\Omega_1{:}\texttt{type}}{\mathfrak{B};\Gamma\vdash[x{:}\Sigma_1]\Omega_1:\texttt{type}}$$

(abstr.2)

$$\frac{\mathfrak{B};\Gamma\vdash\Sigma_1{:}\texttt{type}\qquad\mathfrak{B};\Gamma,x{:}\Sigma_1\vdash\Omega_1{:}\texttt{type}\qquad\mathfrak{B};\Gamma,x{:}\Sigma_1\vdash\Sigma_2{:}\Omega_1}{\mathfrak{B};\Gamma\vdash[x{:}\Sigma_1]\Sigma_2:[x{:}\Sigma_1]\Omega_1}$$

(application)

$$\frac{\mathfrak{B};\Gamma\vdash\Sigma_1:[x{:}\Omega_1]\Omega_2\qquad\mathfrak{B};\Gamma\vdash\Sigma_2:\Omega_1}{\mathfrak{B};\Gamma\vdash\langle\Sigma_2\rangle\Sigma_1:\Omega_2[x{:=}\Sigma_2]}$$

(conversion)

$$\frac{\mathfrak{B};\Gamma\vdash\Sigma:\Omega_1\qquad\mathfrak{B};\Gamma\vdash\Omega_2{:}\texttt{type}\qquad\mathfrak{B};\Gamma\vdash\Omega_1=_{\beta\text{d}}\Omega_2}{\mathfrak{B};\Gamma\vdash\Sigma:\Omega_2}$$

When using the parameter rule, we assume that $\mathfrak{B};\Gamma\vdash\text{OK}$, even if $n=0$.

Example 7.12 The book of Example 7.9 (see Figure 7.1) is correct. We prove this line by line for the first four lines (the reader is invited to check lines 5–14 for himself). We write $(m\text{–}n)$ to denote the book that consists of lines m to n of Example 7.9.

1. By (axiom), $\varnothing;\varnothing \vdash$ OK, so by (book ext.: pn1),

$$(\varnothing;\mathtt{prop};\text{PN};\mathtt{type});\varnothing \vdash \text{OK}.$$

2. By (parameters), (1–1); $\varnothing \vdash \mathtt{prop} : \mathtt{type}$. Therefore by (book ext.: var1), (1–1), $(\varnothing,\mathtt{x},\text{—},\mathtt{prop});\varnothing \vdash$ OK.

3. By (context ext.), (1–2); $\mathtt{x{:}prop} \vdash$ OK. Therefore by (book ext.: var1), (1–2), $(\mathtt{x{:}prop};\mathtt{y};\text{—};\mathtt{prop}) \vdash$ OK.

4. By two applications of (context ext.), (1–3); $\mathtt{x{:}prop},\mathtt{y{:}prop} \vdash$ OK. By (parameters), (1–3); $\mathtt{x{:}prop},\mathtt{y{:}prop} \vdash \mathtt{prop{:}type}$.
Therefore by (book ext.: pn2), (1–4); $\varnothing \vdash$ OK.

7a3 Definitional equality

We still need to describe the relation $=_{\beta\mathrm{d}}$ ("definitional equality"). This notion is based on both the definition mechanism and the abstraction/application mechanism of AUT-68. The abstraction/application mechanism provides the well-known notion of β-equality, originating from the rule of β-conversion:

$$\langle\Sigma\rangle[x{:}\Omega_2]\Omega_1 \quad \to_\beta \quad \Omega_1[x{:=}\Sigma].$$

We will use notations like $\twoheadrightarrow_\beta$, $=_\beta$, and $\to_\beta^+$ as usual (see 4.14).

We now describe the definition mechanism of AUT-68 via the notion of d-*equality*. This definition depends on the definition of derivability, and the definition of derivability given in the previous subsection depends on the definition of definitional equality. In fact, the definitions of correct book, correct line, correct context, correct expression and definitional equality should be given within one definition, using induction on the length of the book. This would lead to a correct but very long definition, and that is probably the reason why the definitions are split into smaller parts (in this book as well as in [42]).

Definition 7.13 (d-equality) Assume, $\mathfrak{B};\Gamma \vdash \Sigma : \Sigma'$. We define the d-*normal form* $\mathrm{nf_d}(\Sigma)$ of Σ with respect to $\mathfrak{B}$ by induction on the length of the book $\mathfrak{B}$. So, assume $\mathrm{nf_d}(\Sigma)$ has already been defined for all books $\mathfrak{B}'$ with less lines than $\mathfrak{B}$, and all expressions Σ that are correct with respect to $\mathfrak{B}'$ and a context Γ. Use induction on the structure of Σ:

- If Σ is a variable x, then $\mathrm{nf_d}(\Sigma) \stackrel{\text{def}}{=} x$;

- Now assume $\Sigma \equiv b(\Omega_1,\ldots,\Omega_n)$, and assume that the normal forms of the Ω_is have already been defined.

Determine a line $(\Delta; b; \Xi_1; \Xi_2)$ in the book $\mathfrak{B}$ (there is exactly one such line, and this line is determined by b).

Write $\Delta \equiv x_1{:}\alpha_1, \ldots, x_n{:}\alpha_n$. Distinguish:

- $\Xi_1 \equiv$ —. This case doesn't occur, as $b \in \mathcal{C}$;
- $\Xi_1 \equiv$ PN. Then define $\mathrm{nf_d}(\Sigma) \stackrel{\text{def}}{=} b(\mathrm{nf_d}(\Omega_1), \ldots, \mathrm{nf_d}(\Omega_n))$;
- Ξ_1 is an expression. Then Ξ_1 is correct with respect to a book $\mathfrak{B}'$ that contains less lines than $\mathfrak{B}$ ($\mathfrak{B}'$ doesn't contain the line $(\Delta; b; \Xi_1; \Xi_2)$, and all lines of $\mathfrak{B}'$ are also lines of $\mathfrak{B}$), hence we can assume that $\mathrm{nf_d}(\Xi_1)$ has already been defined. Now define

$$\mathrm{nf_d}(\Sigma) \stackrel{\text{def}}{=} \mathrm{nf_d}(\Xi_1)[x_1, \ldots, x_n{:=}\mathrm{nf_d}(\Omega_1), \ldots, \mathrm{nf_d}(\Omega_n)];$$

- If $\Sigma \equiv [x{:}\Omega_1]\Omega_2$ then $\mathrm{nf_d}(\Sigma) \stackrel{\text{def}}{=} [x{:}\mathrm{nf_d}(\Omega_1)]\mathrm{nf_d}(\Omega_2)$;
- If $\Sigma \equiv \langle\Omega_2\rangle\Omega_1$ then $\mathrm{nf_d}(\Sigma) \stackrel{\text{def}}{=} \langle\mathrm{nf_d}(\Omega_2)\rangle\mathrm{nf_d}(\Omega_1)$.

We write $\Sigma_1 =_\mathrm{d} \Sigma_2$ when $\mathrm{nf_d}(\Sigma_1) \equiv \mathrm{nf_d}(\Sigma_2)$.

As we see, the d-normal form $\mathrm{nf_d}(\Sigma)$ of a correct expression Σ depends on the book $\mathfrak{B}$, and in order to be completely correct we should write $\mathrm{nf}_{\mathrm{d}\mathfrak{B}}(\Sigma)$ instead of only $\mathrm{nf_d}(\Sigma)$. We will, however, omit the subscript $\mathfrak{B}$ as long as no confusion arises. We write $=_{\beta\mathrm{d}}$ for the smallest equivalence relation that contains both $=_\beta$ and $=_\mathrm{d}$.

Definition 7.14 (Definitional equality) Σ_1 and Σ_2 are called *definitionally equal* (with respect to a book $\mathfrak{B}$) if $\Sigma_1 =_{\beta\mathrm{d}} \Sigma_2$.

This definition completes the description of AUT-68. Again, definitional equality of expressions Σ_1 and Σ_2 depends on the book $\mathfrak{B}$, so we should write $=_{\beta\mathrm{d}\mathfrak{B}}$ instead of $=_{\beta\mathrm{d}}$. Also in this case we leave out the subscript $\mathfrak{B}$ as long as no confusion arises.

As an alternative to Definition 7.13, we describe the notion of d-equality via a reduction relation.

Definition 7.15 (δ-reduction) Let $\mathfrak{B}$ be a book, Γ a correct context with respect to $\mathfrak{B}$, and Σ a correct expression with respect to $\mathfrak{B}; \Gamma$. We define $\Sigma \to_\delta \Omega$ by the usual compatibility rules, and

(δ) If $\Sigma = b(\Sigma_1, \ldots, \Sigma_n)$, and $\mathfrak{B}$ contains a line $(x_1{:}\alpha_1, \ldots, x_n{:}\alpha_n; b; \Xi_1; \Xi_2)$ where $\Xi_1 \in \mathcal{E}$, then

$$\Sigma \to_\delta \Xi_1[x_1, \ldots, x_n{:=}\Sigma_1, \ldots, \Sigma_n].$$

We say that Σ is in δ-normal form if for no expression Ω, $\Sigma \rightarrow_\delta \Omega$, and use notations like $\twoheadrightarrow_\delta$, $\twoheadrightarrow_\delta^+$ and $=_\delta$ as usual. $\rightarrow_\delta$ depends on $\mathfrak{B}$, but as we did before with $\mathrm{nf_d}$ and $=_\mathrm{d}$, we only explicitly mention this if it is not clear in relation to which book $\mathfrak{B}$ $\rightarrow_\delta$ is considered.

The relations $=_\mathrm{d}$ and $=_\delta$ are the same:

Lemma 7.16

1. *(Church-Rosser) If* $A_1 =_\delta A_2$ *then there is* B *such that* $A_1 \twoheadrightarrow_\delta B$ *and* $A_2 \twoheadrightarrow_\delta B$*;*
2. $\mathrm{nf_d}(\Sigma)$ *is the (unique)* δ*-normal form of* Σ*;*
3. $\Sigma =_\delta \Omega$ *if and only if* $\Sigma =_\mathrm{d} \Omega$.

PROOF: AUT-68 with $\rightarrow_\delta$ can be seen as an orthogonal term rewrite system (see [93]).

1. Such a term rewrite system has the Church-Rosser property (see [93]);
2. It is not hard to show that $\Sigma \twoheadrightarrow_\delta \mathrm{nf_d}(\Sigma)$. By induction on the definition of $\mathrm{nf_d}(\Sigma)$ one shows that $\mathrm{nf_d}(\Sigma)$ is in δ-normal form. The uniqueness of this normal form follows from the Church-Rosser property;
3. If $\Sigma =_\delta \Omega$ then by (1) there is Ψ such that $\Sigma \twoheadrightarrow_\delta \Psi$ and $\Omega \twoheadrightarrow_\delta \Psi$. This means that the δ-normal forms of Σ and Ω are equal, so by (2), $\mathrm{nf_d}(\Sigma) \equiv \mathrm{nf_d}(\Omega)$. On the other hand, if $\mathrm{nf_d}(\Sigma) \equiv \mathrm{nf_d}(\Omega)$, then Σ and Ω have the same δ-normal forms (by (2)), so $\Sigma =_\delta \Omega$.

⊠

Lemma 7.17 *The relation* $\rightarrow_\delta$ *is strongly normalising.*

PROOF: We already know that $\rightarrow_\delta$ is weakly normalising (by 2). Moreover, the definition of $\mathrm{nf_d}(\Sigma)$ in 7.13 induces an innermost reduction strategy. By a theorem of O'Donnell (see [113], or pp. 75–76 of [93]), $\rightarrow_\delta$ is strongly normalising. ⊠

7a4 Some elementary properties

Although we do not want to give a complete overview of all the meta-theoretical properties of AUTOMATH (these are studied in [110] and [42]), we do present some properties that we will need at a later stage.

Definition 7.18 A book $\mathfrak{B}$ is part of another book $\mathfrak{B}'$, denoted as $\mathfrak{B} \subseteq \mathfrak{B}'$, if all lines of $\mathfrak{B}$ are lines of $\mathfrak{B}'$ as well. Similarly, a context Γ is part of another context Γ', notation $\Gamma \subseteq \Gamma'$, if all declarations $x{:}\alpha$ of Γ are declarations in Γ' as well.

Lemma 7.19 (Weakening for AUT-68) *If* $\mathfrak{B};\Gamma \vdash \Sigma : \Omega$, $\mathfrak{B} \subseteq \mathfrak{B}'$, $\Gamma \subseteq \Gamma'$ *and* $\mathfrak{B}';\Gamma' \vdash \mathrm{OK}$ *then* $\mathfrak{B}';\Gamma' \vdash \Sigma : \Omega$.

PROOF: By induction on the derivation of $\mathfrak{B};\Gamma \vdash \Sigma : \Omega$. ⊠

7b From AUT-68 towards a PTS

We want to give a description of AUT-68 within the framework of the Pure Type Systems. There are several ways to do this. One of the most important choices to be made is whether or not to maintain the parameter mechanism (that is: To allow expressions with parameters, as in the second clause of Definition 7.1). On the one hand, the parameter mechanism is an important feature of AUTOMATH. On the other hand PTSs do not have a parameter mechanism, and the parameter mechanism can be easily imitated by function application (cf. the second clause of the forthcoming Definition 7.21). Moreover, the description by van Benthem Jutting in [3] of the systems AUT-68 and AUT-QE in a PTS style does not use parameters.

In this chapter, we provide a translation to PTSs without parameters. In doing so, we can explain van Benthem Jutting's description of AUT-68 and AUT-QE.

We will see, however, that the way in which we must handle parameters in the resulting PTS is a bit artificial. Moreover, we think that parameters play an important role in the AUTOMATH systems, and that they could play a similar role in other PTSs. Therefore, we will present an extension of PTSs with parameters in Chapter 10. This extension is based on the way in which parameters are handled in AUTOMATH, and it will be shown that AUTOMATH can be described very well within these PTSs with parameters.

For a description of AUT-68 in PTSs without parameters, we must first make a translation of the expressions in AUT-68 to typed λ-terms of PTSs. This translation is very straightforward. First, we recall the definition of the terms of PTSs (Definition 4.16):

Definition 7.20 (Terms of PTSs) Let $\mathbb{V}$ be a set of variables and $\mathbb{C}$ a set of constants. Assume that $\mathbb{V}$ and $\mathbb{C}$ are countably infinite and are disjoint. The set $\mathcal{T}(\mathbb{V},\mathbb{C})$ (shorthand: $\mathcal{T}$) is given by:

$$\mathcal{T} ::= \mathbb{V} \mid \mathbb{C} \mid \mathcal{T}\mathcal{T} \mid \lambda\mathbb{V}{:}\mathcal{T}.\mathcal{T} \mid \Pi\mathbb{V}{:}\mathcal{T}.\mathcal{T}.$$

Definition 7.21 We define a mapping $\overline{[\ldots]}$ from the correct expressions in $\mathcal{E}$ (relative to a book $\mathfrak{B}$ and a context Γ) to $\mathcal{T}$, the set of terms for PTSs. We assume that $\mathcal{C} \cup \mathcal{V} \subseteq \mathbb{V}$ ($\mathbb{V}$ is the set of variables for PTS-terms).

- $\overline{x} \stackrel{\text{def}}{=} x$ for $x \in \mathcal{V}$;

- $\overline{b(\Sigma_1,\ldots,\Sigma_n)} \stackrel{\text{def}}{=} b\overline{\Sigma_1}\cdots\overline{\Sigma_n}$;
- $\overline{[x{:}\Sigma]\Omega} \stackrel{\text{def}}{=} \Pi x{:}\overline{\Sigma}.\overline{\Omega}$ if $[x{:}\Sigma]\Omega$ has type $\mathtt{type}$, otherwise $\overline{[x{:}\Sigma]\Omega} \stackrel{\text{def}}{=} \lambda x{:}\overline{\Sigma}.\overline{\Omega}$;
- $\overline{\langle\Omega\rangle\Sigma} \stackrel{\text{def}}{=} \overline{\Sigma}\,\overline{\Omega}$.

Moreover, we define: $\overline{\mathtt{type}} \stackrel{\text{def}}{=} *$.

In the second clause of this definition we see that the parameter mechanism of Definition 7.1 is replaced by repeated function application in PTSs.

With this translation in mind, we want to find a type system λ68 that "suits" AUT68, i.e. if Σ is a correct expression of type Ω with respect to a book $\mathfrak{B}$ and a context Γ, then we want $\mathfrak{B}', \Gamma' \vdash \overline{\Sigma} : \overline{\Omega}$ to be derivable in λ68, and vice versa. Here, $\mathfrak{B}'$ and Γ′ are some suitable translations of $\mathfrak{B}$ and Γ. The search for a suitable λ68 will concentrate on three points, which we first discuss informally. In the next section we give a formal definition of λ68, and prove that it has the desired property.

7b1 The choice of the correct formation (Π) rules and the parameter types ¶$x{:}A.B$

When we keep in mind that $\overline{\mathtt{type}} \equiv *$, the definition of correct expressions 7.11 gives a clear answer to the question of which Π-rules are implied by the abstraction mechanism of AUT-68. The rule

$$\frac{\mathfrak{B};\Gamma \vdash \Sigma_1{:}\mathtt{type} \qquad \mathfrak{B};\Gamma, x{:}\Sigma_1 \vdash \Omega_1{:}\mathtt{type}}{\mathfrak{B};\Gamma \vdash [x{:}\Sigma_1]\Omega_1 : \mathtt{type}}$$

immediately translates into Π-rule $(*,*,*)$ for PTSs:

$$\frac{\overline{\mathfrak{B}},\overline{\Gamma} \vdash \overline{\Sigma_1}{:}\, * \qquad \overline{\mathfrak{B}},\overline{\Gamma}, x{:}\overline{\Sigma_1} \vdash \overline{\Omega_1}{:}*}{\overline{\mathfrak{B}},\overline{\Gamma} \vdash (\Pi x{:}\overline{\Sigma_1}.\overline{\Omega_1}) : *},$$

where $\overline{\mathfrak{B}}$ and $\overline{\Gamma}$ are suitable translations of $\mathfrak{B}$ and Γ.

It is, however, not immediately clear which Π-rules are induced by the parameter mechanism of AUT-68.

Let $\Sigma \equiv b(\Sigma_1,\ldots,\Sigma_n)$ be a correct expression of type Ω with respect to a book $\mathfrak{B}$ and a context Γ. Here is a line (see Definition 7.10)

$$(x_1{:}\alpha_1,\ldots,x_n{:}\alpha_n; b; \Xi_1; \Xi_2)$$

in $\mathfrak{B}$ such that each Σ_i is a correct expression with respect to $\mathfrak{B}$ and Γ, and has a type that is definitionally equal to $\alpha_i[x_1,\ldots,x_{i-1}{:=}\Sigma_1,\ldots,\Sigma_{i-1}]$. We also know that $\Omega =_{\beta\mathrm{d}} \Xi_2[x_1,\ldots,x_n{:=}\Sigma_1,\ldots\Sigma_n]$.

Now $\overline{\Sigma} \equiv b\overline{\Sigma_1}\cdots\overline{\Sigma_n}$, and, assuming that we can derive in $\lambda 68$ that $\overline{\Sigma_i}$ has type

$$\overline{\alpha_i}[x_1,\ldots,x_{i-1}{:=}\overline{\Sigma_1},\ldots,\overline{\Sigma_{i-1}}],$$

it is not unreasonable to assign the type $\Pi x_1{:}\overline{\alpha_1}\cdots\Pi x_n{:}\overline{\alpha_n}to\mathrm{b}.\overline{\Xi_2}$. We will abbreviate this last term by $\prod_{i=1}^{n} x_i{:}\overline{\alpha_i}.\overline{\Xi_2}$. Then we can derive (using n times the application rule that we will introduce for $\lambda 68$) that $\overline{\Sigma}$ has type $\overline{\Omega}$ in $\lambda 68$.

It is important to notice that the type of b, $\prod_{i=1}^{n} x_i{:}\overline{\alpha_i}.\overline{\Xi_2}$, does not necessarily have an equivalent in AUT-68, as in AUT-68 abstractions over `type` are not allowed (only abstractions over expressions Σ that have `type` as type are possible — cf. Definition 7.11). In other words, the type of b, $\prod_{i=1}^{n} x_i{:}\overline{\alpha_i}.\overline{\Xi_2}$, is not necessarily a first-class citizen of AUT-68 and should therefore have special treatment in $\lambda 68$. This is the reason to create a special sort Δ, in which these types of AUT-68 constants and definitions are stored. This idea originates from Van Benthem Jutting, and was firstly presented in [3].

If we construct $\Pi x_n{:}\overline{\alpha_n}.\overline{\Xi_2}$ from $\overline{\Xi_2}$, we must use a rule (s_1, s_2, s_3), where s_1, s_2, s_3 are sorts. Sort s_1 must be the type of $\overline{\alpha_n}$. As $\alpha_n \equiv$ `type` or α_n has type `type`, we must allow the possibilities $s_1 \equiv *$ and $s_1 \equiv \Box$. Similarly, $\Xi_2 \equiv$ `type` or Ξ_2 has type `type`, so we also allow $s_2 \equiv *$ and $s_2 \equiv \Box$. As we intended to store the new type in sort Δ, we take $s_3 \equiv \Delta$.

For similar reasons, we introduce rules $(*, \Delta, \Delta)$ and $(\Box, \Delta, \Delta)$ to construct $\prod_{i=1}^{n} x_i{:}\overline{\alpha_i}.\overline{\Xi_2}$ from $\Pi x_n{:}\overline{\alpha_n}.\overline{\Xi_2}$ for $n > 1$.

As a result, we have the following Π-rules:

$$\begin{array}{cc} & (*,*,*); \\ (*,*,\Delta); & (\Box,*,\Delta); \\ (*,\Box,\Delta); & (\Box,\Box,\Delta); \\ (*,\Delta,\Delta); & (\Box,\Delta,\Delta). \end{array}$$

We do not have rules of the form (Δ, s_2, s_3) or (s_1, Δ, s_3) with $s_3 \equiv *$ or $s_3 \equiv \Box$. So types of sort Δ cannot be used to construct types of other sorts. In this way, we can keep the types of the λ-calculus part of AUT-68 separated from the types of the parameter mechanism: The last ones are stored in Δ.

In Example 5.2.4.8 of [3], there is no rule $(*, *, \Delta)$. In principle, this rule is superfluous, as each application of rule $(*, *, \Delta)$ can be replaced by an application of rule $(*, *, *)$. Nevertheless we want to maintain this rule:

- First of all, the presence of both $(*, *, *)$ and $(*, *, \Delta)$ in the system stresses the fact that AUT-68 has two type mechanisms: One provided by the parameter mechanism and one by the λ-abstraction mechanism;

- Secondly, there are technical arguments to make a distinction between types formed by the abstraction mechanism and types that appear via the parameter mechanism. In this book, we will denote product types constructed by the abstraction mechanism in the usual way (so: $\Pi x{:}A.B$), whilst we will (from now on) use the notation $¶x{:}A.B$ for a type constructed by the parameter mechanism. Hence, we have for the constant b given above that $b : ¶_{i=1}^{n}\, x_i{:}\overline{\alpha_i}.\overline{\Xi_2}$[3]. As an additional advantage, the resulting system will maintain Unicity of Types.[4] This would have been lost if we had introduced rules $(*,*,*)$ and $(*,*,\triangle)$ without making this difference, as we can then derive both

$$\frac{\alpha{:}* \vdash \alpha : * \qquad \alpha{:}*, x{:}\alpha \vdash \alpha{:}*}{\alpha{:}* \vdash (\Pi x{:}\alpha.\alpha) : *}\ (\text{rule } (*,*,*))$$

and

$$\frac{\alpha{:}* \vdash \alpha : * \qquad \alpha{:}*, x{:}\alpha \vdash \alpha{:}*}{\alpha{:}* \vdash (\Pi x{:}\alpha.\alpha) : \triangle}\ (\text{rule } (*,*,\triangle));$$

- There is another reason to make a distinction between types formed by the abstraction mechanism and types that appear in the translation via the definition mechanism. For the moment, we consider AUT-68 *without* the so-called Π-application. In AUT-68 with Π-application (call this system AUT-68Π for the moment; see also Section 8a1) the application rule of Definition 7.11

$$\frac{\mathfrak{B};\Gamma \vdash \Sigma_1{:}[x{:}\Omega_1]\Omega_2 \qquad \mathfrak{B};\Gamma \vdash \Sigma_2{:}\Omega_1}{\mathfrak{B};\Gamma \vdash \langle\Sigma_2\rangle\Sigma_1 : \Omega_2[x{:=}\Sigma_2]}$$

is replaced by

$$\frac{\mathfrak{B};\Gamma \vdash \Sigma_1{:}[x{:}\Omega_1]\Omega_2 \qquad \mathfrak{B};\Gamma \vdash \Sigma_2{:}\Omega_1}{\mathfrak{B};\Gamma \vdash \langle\Sigma_2\rangle\Sigma_1 : \langle\Sigma_2\rangle\Omega_2},$$

but the rule describing the type of $b(\Sigma_1,\ldots,\Sigma_n)$ is the same as the rule in Definition 7.11 (parameters).

So if we want to make a translation of AUT-68Π, the application rule for Π-terms has to be different from the application rule for ¶-terms. Without distinction between Π-terms and ¶-terms, it would be impossible to amend the system to represent AUT-68Π. Distinguishing between Π-terms and ¶-terms makes it possible to obtain a translation of AUT-68Π from the translation of AUT-68 in a simple way.

[3]we use $¶_{i=1}^{n}\, x_i{:}\overline{\alpha_i}.\overline{\Xi_2}$ as an abbreviation for $¶x_1{:}\overline{\alpha_1}\cdots ¶x_n{:}\overline{\alpha_n}.\overline{\Xi_2}$

[4]The system as presented in [3] has Unicity of Types as well, because it does not have the Π-formation rule $(*,*,\triangle)$ and is therefore singly sorted.

7b2 The different treatment of constants and variables

When we seek for a translation in $\lambda68$ of the AUT-68 judgement $\mathfrak{B};\Gamma\vdash\Sigma:\Omega$, we must pay extra attention to the translation of $\mathfrak{B}$, as there is no equivalent of books in PTSs. Our solution is to store the information on identifiers of $\mathfrak{B}$ in a PTS-context. Therefore, contexts of $\lambda68$ will have the form $\Delta;\Gamma$. The left part Δ contains type information on primitive notions and definitions, and can be seen as the translation of the information on primitive notions and definitions in $\mathfrak{B}$. In the right part Γ we find the usual type information on variables.

The idea to store the constant information of $\mathfrak{B}$ in the left part of the context arises in a natural way. Let $\mathfrak{B}$ be a correct AUT-68 book, to which we add a line $(\Gamma;b;\text{PN};\Xi_2)$. Then $\Gamma\equiv x_1{:}\alpha_1,\dots,x_n{:}\alpha_n$ is a correct context with respect to $\mathfrak{B}$, and $\mathfrak{B};\Gamma\vdash\Xi_2{:}\mathtt{type}$ or $\Xi_2\equiv\mathtt{type}$. In $\lambda68$ we can work as follows. Assume the information on constants in $\mathfrak{B}$ has been translated into the left part Δ of a $\lambda68$ context. We have (assuming that $\lambda68$ is a type system that behaves like AUT-68, and writing $\overline{\Gamma}$ for the translation $x_1{:}\overline{\alpha_1},\dots,x_n{:}\overline{\alpha_n}$ of Γ):

$$\Delta;\overline{\Gamma}\vdash\overline{\Xi_2}{:}s$$

($s\equiv *$ if $\mathfrak{B};\Gamma\vdash\Xi_2{:}\mathtt{type}$; $s\equiv\Box$ if $\Xi_2\equiv\mathtt{type}$). Applying the ¶-formation rule n times, we obtain

$$\Delta;\varnothing\vdash\P\overline{\Gamma}.\overline{\Xi_2}:\triangle$$

(If Γ is the empty context, then $\P\overline{\Gamma}.\overline{\Xi_2}\equiv\overline{\Xi_2}$, and $\overline{\Xi_2}$ has type $*$ or $\Box$ instead of $\triangle$. We write $\P\overline{\Gamma}$ for $\P_{i=1}^{n}x_i{:}\overline{\alpha_i}$). As $\P\overline{\Gamma}.\overline{\Xi_2}$ is exactly the type that we want to give to b (see the discussion in Subsection 7b1), we use this statement as premise for the start rule that introduces b. As the right part $\overline{\Gamma}$ of the original context has disappeared when we applied the ¶-formation rules, $b{:}\,\P\overline{\Gamma}.\overline{\Xi_2}$ is automatically placed at the righthand end of Δ: The conclusion of the start rule is

$$\Delta,b{:}\,\P\overline{\Gamma}.\overline{\Xi_2}\vdash b{:}\,\P\overline{\Gamma}.\overline{\Xi_2}.$$

Adding $b{:}\,\P\overline{\Gamma}.\overline{\Xi_2}$ at the end of Δ can be compared with adding the line $(\Gamma;b;\text{PN};\Xi_2)$ at the end of $\mathfrak{B}$.

The process above can be captured in one rule:

$$\frac{\Delta;\overline{\Gamma}\vdash\overline{\Xi_2}{:}s_1 \qquad \Delta;\vdash\P\overline{\Gamma}.\overline{\Xi_2}{:}s_2}{\Delta,b{:}\,\P\overline{\Gamma}.\overline{\Xi_2};\vdash b{:}\,\P\overline{\Gamma}.\overline{\Xi_2}}.$$

Here $s_1\in\{*,\Box\}$ (compare: $\Xi_2{:}\mathtt{type}$ or $\Xi_2\equiv\mathtt{type}$) and $s_2\in\{*,\Box,\triangle\}$ (usually, $s_2\equiv\triangle$; the cases $s_2\equiv *,\Box$ only occur if Γ is empty).

7b3 The definition system and the translation using §

A line $(x_1{:}\alpha_1,\dots,x_n{:}\alpha_n;b;\Xi_1;\Xi_2)$, in which b is a constant and $\Xi_1\in\mathcal{E}$, represents a definition. Such a line should be read as:

For all expressions $\Omega_1, \ldots, \Omega_n$ (obeying certain type conditions), $b(\Omega_1, \ldots, \Omega_n)$ is an abbreviation for $\Xi_1[x_1, \ldots, x_n{:=}\Omega_1, \ldots, \Omega_n]$, and has type

$$\Xi_2[x_1, \ldots, x_n{:=}\Omega_1, \ldots, \Omega_n].$$

So in $\lambda 68$, the context should also mention that for all terms $X_1, \ldots, X_n$, we have that $bX_1 \cdots X_n$ "is equal to" $\Xi_1[x_1, \ldots, x_n{:=}X_1, \ldots, X_n]$. The most straightforward way to do this, is to write

$$b{:=}\left(\overset{n}{\underset{i=1}{\lambda}} x_i{:}\overline{\alpha_i}.\overline{\Xi_1}\right) : \left(\P_{i=1}^{n} x_i{:}\overline{\alpha_i}.\overline{\Xi_2}\right)$$

in the context instead of only $b{:}\ \P_{i=1}^{n} x_i{:}\overline{\alpha_i}.\overline{\Xi_2}$, and adding a δ-reduction rule that allows to unfold the definition of b:

$$\Delta \vdash b \to_\delta \overset{n}{\underset{i=1}{\lambda}} x_i{:}\overline{\alpha_i}.\overline{\Xi_1}$$

whenever $b{:=}\left(\lambda_{i=1}^{n} x_i{:}\overline{\alpha_i}.\overline{\Xi_1}\right) : \left(\P_{i=1}^{n} x_i{:}\overline{\alpha_i}.\overline{\Xi_2}\right) \in \Delta$.

Unfolding the definition of b in a term $b\overline{\Sigma_1} \cdots \overline{\Sigma_n}$ and applying β-reduction n times results in $\overline{\Xi_1}[x_1{:=}\overline{\Sigma_1}] \cdots [x_n{:=}\overline{\Sigma_n}]$. This procedure corresponds exactly to the δ-reduction

$$\Delta \vdash b(\Sigma_1, \ldots, \Sigma_n) \to_\delta \Xi_1[x_1, \ldots, x_n{:=}\Sigma_1, \ldots, \Sigma_n]$$

in AUT-68[5].

This method, however, has some disadvantages.

- Look again at a line $(x_1{:}\alpha_1, \ldots, x_n{:}\alpha_n; b; \Xi_1; \Xi_2)$ in an AUT-68 book. Then $b(\Sigma_1, \ldots, \Sigma_n)$ has $b\overline{\Sigma_1} \cdots \overline{\Sigma_n}$ as its equivalent in $\lambda 68$. If $n > 0$, the latter $\lambda 68$-term has $B \equiv b\overline{\Sigma_1} \cdots \overline{\Sigma_m}$ as a subterm for any $m < n$. But B has no equivalent in AUT-68: Only after B has been applied to suitable terms $\overline{\Sigma_{m+1}}, \ldots, \overline{\Sigma_n}$ the resulting term $B\overline{\Sigma_{m+1}} \cdots \overline{\Sigma_n}$ has $b(\Sigma_1, \ldots, \Sigma_n)$ as its equivalent in AUT-68. Hence B must not be seen as a term directly translatable into AUTOMATH, but only as an intermediate result that is necessary to construct the equivalent of the expression $b(\Sigma_1, \ldots, \Sigma_n)$. B is recognisable as an intermediate result via its type $\P_{i=m+1}^{n} x_i{:}\overline{\alpha_i}.\overline{\Xi_2}$, which has sort Δ (instead of $*$ or $\Box$).

  The method above allows to unfold the definition of b already in B, because $b\overline{\Sigma_1} \cdots \overline{\Sigma_m}$ can reduce to $\left(\lambda_{i=1}^{n} x_i{:}\overline{\alpha_i}.\overline{\Xi_1}\right) \overline{\Sigma_1} \cdots \overline{\Sigma_m}$, and we can β-reduce this term m times to $\left(\lambda_{i=m+1}^{n} x_i{:}\overline{\alpha_i}.\overline{\Xi_1}\right) [x_j{:=}\overline{\Sigma_j}]_{j=1}^{m}$. It is more in line with AUT-68 to make such unfolding not possible before *all* n arguments $\overline{\Sigma_1}, \ldots, \overline{\Sigma_n}$ have been applied to b, so only when the construction of the equivalent of $b(\Sigma_1, \ldots, \Sigma_n)$ has been completed;

[5] We can assume that the x_i do not occur in the Σ_j, so the simultaneous substitution $\Xi_1[x_1, \ldots, x_n{:=}\Sigma_1, \ldots, \Sigma_n]$ is equal to $\Xi_1[x_1{:=}\Sigma_1] \cdots [x_n{:=}\Sigma_n]$.

- Moreover, $\lambda_{i=1}^{n}\ x_i{:}\overline{\alpha_i}.\overline{\Xi_1}$ not necessarily has an equivalent in AUT-68. Consider for instance the constant b in the line

$$(\alpha{:}\texttt{type}; b; [x{:}\alpha]x; [x{:}\alpha]\alpha).$$

  In this case, $\lambda_{i=1}^{n}\ x_i{:}\overline{\alpha_i}.\overline{\Xi_1} \equiv \lambda\alpha{:}{*}.\lambda x{:}\alpha.x$. Its equivalent in AUT-68 would be $[\alpha{:}\texttt{type}][x{:}\alpha]x$, but an abstraction $[\alpha{:}\texttt{type}]$ cannot be made in AUT-68.[6] This is the reason why we do not incorporate $\lambda_{i=1}^{n}\ x_i{:}\overline{\alpha_i}.\overline{\Xi_1}$ as a citizen of $\lambda 68$; we feel that this is better than making it a (first-class or second-class) citizen of $\lambda 68$.

Therefore we choose a different translation. The line

$$(x_1{:}\alpha_1, \ldots, x_n{:}\alpha_n; b; \Xi_1; \Xi_2),$$

where $\Xi_1 \in \mathcal{E}$, will be translated by putting

$$b{:=}\left(\mathop{\S}_{i=1}^{n}\ x_i{:}\overline{\alpha_i}.\overline{\Xi_1}\right) : \left(\mathop{\P}_{i=1}^{n}\ x_i{:}\overline{\alpha_i}.\overline{\Xi_2}\right)$$

instead of

$$b{:=}\left(\mathop{\lambda}_{i=1}^{n}\ x_i{:}\overline{\alpha_i}.\overline{\Xi_1}\right) : \left(\mathop{\P}_{i=1}^{n}\ x_i{:}\overline{\alpha_i}.\overline{\Xi_2}\right)$$

in the left part of the translated context Δ. And a reduction rule

$$bX_1 \cdots X_n \to_\delta \overline{\Xi_1}[x_1, \ldots, x_n{:=}X_1, \ldots, X_n]$$

is added for all terms $X_1, \ldots, X_n$. The symbol § is used instead of λ. This is to emphasise that, though both $\S x{:}A$ and $\lambda x{:}A$ are abstractions, they are not the same kind of abstraction.

7c $\lambda 68$

Here, we give $\lambda 68$, show that it has the desirable properties of PTSs and that it is the PTS version of AUT-68.

7c1 Definition and elementary properties

We give the formal definition of $\lambda 68$, based on the motivation in Section 7b.

[6] This situation can be compared to the situation in Section 7b1, where we found that the type of b is not necessarily a first-class citizen of AUT-68. There, we could not avoid that the type of b became a citizen of $\lambda 68$ (though we made it a second-class citizen by storing it in the sort Δ).

Definition 7.22 (λ68)

1. The terms of λ68 form a set $\mathcal{T}$ defined by

$$\mathcal{T} ::= \mathcal{V} \mid \mathcal{C} \mid \boldsymbol{S} \mid \mathcal{T}\mathcal{T} \mid \lambda\mathcal{V}{:}\mathcal{T}.\mathcal{T} \mid \S\mathcal{V}{:}\mathcal{T}.\mathcal{T} \mid \Pi\mathcal{V}{:}\mathcal{T}.\mathcal{T} \mid \P\mathcal{V}{:}\mathcal{T}.\mathcal{T},$$

   where $\boldsymbol{S}$ is the set of sorts $\{*, \Box, \triangle\}$.

   Let $\mathbb{V} = \mathcal{C} \cup \mathcal{V}$ and assume that $\boldsymbol{S} \cap \mathbb{V} = \varnothing$.

   (Recall that $\mathcal{V} \cap \mathcal{C} = \varnothing$ from the start of Section 7a1).

   We also define the sets of free variables $\text{FV}(T)$ and ("free")[7] constants $\text{FC}(T)$ of a term T in the straightforward way;

2. We define the notion of context inductively:

   - $\varnothing; \varnothing$ is a context; $\text{DOM}(\varnothing; \varnothing) = \varnothing$;
   - If $\Delta; \Gamma$ is a context, $x \in \mathcal{V}$, x does not occur in $\Delta; \Gamma$ and $A \in \mathcal{T}$, then $\Delta; \Gamma, x{:}A$ is a context (x is a newly introduced variable); $\text{DOM}(\Delta; \Gamma) = \text{DOM}(\Delta; \Gamma) \cup \{x\}$;
   - If $\Delta; \Gamma$ is a context, $b \in \mathcal{C}$, b does not occur in $\Delta; \Gamma$ and $A \in \mathcal{T}$ then $\Delta, b{:}A; \Gamma$ is a context (in this case b is a *primitive* constant; cf. the primitive notions of AUTOMATH in Section 7a1); $\text{DOM}(\Delta, b{:}A; \Gamma) = \text{DOM}(\Delta; \Gamma) \cup \{b\}$;
   - If $\Delta; \Gamma$ is a context, $b \in \mathcal{C}$, b does not occur in $\Delta; \Gamma$, $A \in \mathcal{T}$, and $T \in \mathcal{T}$, then $\Delta, b{:=}T{:}A; \Gamma$ is a context (in this case b is a *defined* constant; cf. the definitions of AUTOMATH in Section 7a1); $\text{DOM}(\Delta, b := T : A; \Gamma) = \text{DOM}(\Delta; \Gamma) \cup \{b\}$.

   Observe that a semicolon is used as the separation mark between the two parts of the context, and that a comma is used to separate the different expressions within each of these parts.

   We define

$$\begin{aligned} \text{PRIMCONS}(\Delta; \Gamma) &= \{b \in \text{DOM}(\Delta; \Gamma) \mid b \text{ is a primitive constant}\}; \\ \text{DEFCONS}(\Delta; \Gamma) &= \{b \in \text{DOM}(\Delta; \Gamma) \mid b \text{ is a defined constant}\}; \\ \text{FV}(\Delta; \Gamma) &= \text{DOM}(; \Gamma); \end{aligned}$$

3. We define the notion of δ-reduction on terms. Let Δ be the left part of a context. If $(b{:=}(\S_{i=1}^{n} x_i{:}A_i.T) : (\P_{i=1}^{n} x_i{:}A_i.B)) \in \Delta$, where B is not of the form $\P y{:}B_1.B_2$, then

$$\Delta \vdash bX_1 \cdots X_n \to_\delta T[x_1, \ldots, x_n{:=}X_1, \ldots, X_n]$$

[7] Of course, to call a constant "free" is a bit peculiar, since there are no *bound* constants.

for all $X_1, \ldots X_n \in \mathcal{T}$.

We also have the usual compatibility rules on δ-reduction. We use notations like $\twoheadrightarrow_\delta, \twoheadrightarrow_\delta^+, =_\delta$ as usual. When there is no confusion about which Δ is considered, we simply write

$$bX_1 \cdots X_n \rightarrow_\delta T[x_1, \ldots, x_n{:=}X_1, \ldots, X_n];$$

4. We use the usual notion of β-reduction;

5. Judgements in $\lambda 68$ have the form $\Delta; \Gamma \vdash A : B$, where $\Delta; \Gamma$ is a context and A and B are terms. In the case that a judgement $\Delta; \Gamma \vdash A : B$ is derivable according to the rules below, $\Delta; \Gamma$ is a *legal* context and A and B are *legal* terms. We write $\Delta; \Gamma \vdash A : B : C$ if both $\Delta; \Gamma \vdash A : B$ and $\Delta; \Gamma \vdash B : C$ are derivable in $\lambda 68$.

   Here are the rules for $\lambda 68$ (v, pc, and dc are shorthand for variable, primitive constant, and defined constant, respectively. † stands for: *where* $s_1 \equiv *, \Box$):

(Axiom) $\quad ;\vdash * : \Box$

(Start: v) $$\frac{\Delta; \Gamma \vdash A : s_1}{\Delta; \Gamma, x{:}A \vdash x : A} \quad \dagger$$

(Start: pc) $$\frac{\Delta; \Gamma \vdash B : s_1 \qquad \Delta; \vdash \P\Gamma.B : s_2}{\Delta, b{:}\P\Gamma.B; \vdash b : \P\Gamma.B} \quad \dagger$$

(Start: dc) $$\frac{\Delta; \Gamma \vdash T : B : s_1 \qquad \Delta; \vdash \P\Gamma.B : s_2}{\Delta, b{:=}(\S\Gamma.T){:}(\P\Gamma.B); \vdash b : \P\Gamma.B} \quad \dagger$$

(Weak: v) $$\frac{\Delta; \Gamma \vdash M : N \qquad \Delta; \Gamma \vdash A : s_1}{\Delta; \Gamma, x{:}A \vdash M : N} \quad \dagger$$

(Weak: pc) $$\frac{\Delta; \vdash M : N \qquad \Delta; \Gamma \vdash B : s_1 \qquad \Delta; \vdash \P\Gamma.B : s_2}{\Delta, b{:}\P\Gamma.B; \vdash M : N} \quad \dagger$$

(Weak: dc)
$$\frac{\Delta; \vdash M : N \qquad \Delta; \Gamma \vdash T : B : s_1 \qquad \Delta; \vdash \P\Gamma.B : s_2}{\Delta, b{:=}(\S\Gamma.T){:}(\P\Gamma.B); \vdash M : N} \quad \dagger$$

(Π-form) $\dfrac{\Delta;\Gamma \vdash A : * \qquad \Delta;\Gamma, x{:}A \vdash B : *}{\Delta;\Gamma \vdash (\Pi x{:}A.B) : *}$

(¶-form) $\dfrac{\Delta;\Gamma \vdash A : s_1 \qquad \Delta;\Gamma, x{:}A \vdash B : s_2}{\Delta;\Gamma \vdash (\P x{:}A.B) : \Delta}$ †

(λ) $\dfrac{\Delta;\Gamma \vdash \Pi x{:}A.B : * \qquad \Delta;\Gamma, x{:}A \vdash F : B}{\Delta;\Gamma \vdash (\lambda x{:}A.F) : (\Pi x{:}A.B)}$

(App₁) $\dfrac{\Delta;\Gamma \vdash M : \Pi x{:}A.B \qquad \Delta;\Gamma \vdash N : A}{\Delta;\Gamma \vdash MN : B[x{:=}N]}$

(App₂) $\dfrac{\Delta;\Gamma \vdash M : \P x{:}A.B \qquad \Delta;\Gamma \vdash N : A}{\Delta;\Gamma \vdash MN : B[x{:=}N]}$

(Conv) $\dfrac{\Delta;\Gamma \vdash M : A \qquad \Delta;\Gamma \vdash B : s \qquad \Delta \vdash A =_{\beta\delta} B}{\Delta;\Gamma \vdash M : B}$

The newly introduced variables in the Start-rules and Weakening-rules are assumed to be fresh. Moreover, when introducing a variable b with a "pc"-rule or a "dc"-rule, we assume $b \in \mathcal{C}$, and when introducing x via a "v"-rule, we assume $x \in \mathcal{V}$.

We write $\Delta;\Gamma \vdash_{\lambda 68} A : B$ instead of $\Delta;\Gamma \vdash A : B$ if the latter gives rise to confusion with other derivation systems.

Notice that there is no rule (§). This is because we do not want that terms of the form $\S\, x{:}A.B$ are first-class citizens of λ68: they do not have an equivalent in AUTOMATH.

Many basic properties for Pure Type Systems also hold for λ68 and can be proved by the same methods as in the standard literature on PTSs. Due to the split of contexts and the different treatment of constants and variables, these properties are on some points differently formulated than usual (see Section 4c2). The proofs of the lemmas of this section follow [3].

Lemma 7.23 (Free Variable Lemma) *Assume* $\Delta;\Gamma \vdash M : N$.
Write $\Delta \equiv b_1{:}B_1, \ldots, b_m{:}B_m$*;* $\Gamma \equiv x_1{:}A_1, \ldots, x_n{:}A_n$ *(in* Δ*, also expressions* $b_i{:=}T_i{:}B_i$ *may occur, but for uniformity of notation we leave out the* $:=T_i$*-part). Then:*

- *The* $b_1, \ldots, b_m \in \mathcal{C}$ *and* $x_1, \ldots, x_n \in \mathcal{V}$ *are all distinct;*
- $\text{FC}(M), \text{FC}(N) \subseteq \{b_1, \ldots, b_m\}$*;* $\text{FV}(M), \text{FV}(N) \subseteq \{x_1, \ldots, x_n\}$*;*
- $b_1{:}B_1, \ldots, b_{i-1}{:}B_{i-1}; \vdash B_i{:}s_i$ *for some* $s_i \in \{*, \square, \Delta\}$*;*
  $\Delta; x_1{:}A_1, \ldots, x_{j-1}{:}A_{j-1} \vdash A_j{:}t_j$ *for some* $t_j \in \{*, \square\}$. ⊠

Lemma 7.24 (Start Lemma) *Let* $\Delta;\Gamma$ *be a legal context. Then* $\Delta;\Gamma \vdash * : \Box$, *and if* $b{:}A \in \Delta;\Gamma$, *or* $c{:=}T{:}A \in \Delta$, *then* $\Delta;\Gamma \vdash c : A$. ⊠

The following lemma is not a basic PTS-property. However, it can be seen as an extension of the Start Lemma.

Lemma 7.25 (Definition Lemma) *Assume*

$$\Delta_1, b{:=}\left(\mathop{\S}_{i=1}^{n} x_i{:}A_i.T\right) : \left(\mathop{\P}_{i=1}^{n} x_i{:}A_i.B\right), \Delta_2;\Gamma \vdash M : N,$$

where B *is not of the form* $\P y{:}B_1.B_2$. *Then* $\Delta_1; x_1{:}A_1,\ldots,x_n{:}A_n \vdash T : B : s$ *for an* $s \in \{*,\Box\}$. ⊠

The Transitivity Lemma must be formulated in a somewhat different way than usual (cf. 4.28). This has to do with the fact that contexts may contain definitions. To the usual formulation

> *"Let* $\Delta_1;\Gamma_1$ *and* $\Delta_2;\Gamma_2$ *be contexts, of which* $\Delta_1;\Gamma_1$ *is legal. Assume that for all* $b{:}A \in \Delta_2;\Gamma_2$ *and for all* $b{:=}T{:}A \in \Delta_2;\Gamma_2$, $\Delta_1;\Gamma_1 \vdash b{:}A$. *Then* $\Delta_2;\Gamma_2 \vdash B : C \Rightarrow \Delta_1;\Gamma_1 \vdash B : C$."

we must add an extra clause that b is defined in $\Delta_1;\Gamma_1$ in a similar way as it has been defined in $\Delta_2;\Gamma_2$. In the following example we show that things go wrong otherwise:

Example 7.26 Let

$$\Delta_1 \equiv \mathtt{b}_1{:}*, \mathtt{b}_2{:}*, \mathtt{b}_3{:=}\mathtt{b}_1{:}*;$$
$$\Delta_2 \equiv \mathtt{b}_1{:}*, \mathtt{b}_2{:}*, \mathtt{b}_3{:=}\mathtt{b}_2{:}*$$

and $\Gamma_1 \equiv \Gamma_2 \equiv \mathtt{x}_3{:}\mathtt{b}_3$. Notice that all the assumptions of the traditional formulation of the Transitivity Lemma (see above) hold for $\Delta_1;\Gamma_1$ and $\Delta_2;\Gamma_2$. Nevertheless, we can derive

$$\Delta_2;\Gamma_2 \vdash \mathtt{x}_3 : \mathtt{b}_2$$

(because $\Delta_2;\Gamma_2 \vdash \mathtt{x}{:}\mathtt{b}_3$ and according to Δ_2, $\mathtt{b}_3 =_{\beta\mathrm{d}} \mathtt{b}_2$, so we can use the conversion rule). But we cannot derive

$$\Delta_1;\Gamma_1 \vdash \mathtt{x}_3 : \mathtt{b}_2$$

(because $\mathtt{b}_3$ and $\mathtt{b}_2$ are *not* definitionally equal according to Δ_1).

The following formulation of the Transitivity Lemma is correct:

Definition 7.27 We define: $\Delta_1;\Gamma_1 \vdash \Delta_2;\Gamma_2$ if and only if

- If $b{:}A \in \Delta_2;\Gamma_2$ then $\Delta_1;\Gamma_1 \vdash b{:}A$;
- If $b{:=}T{:}A \in \Delta_2$ then $\Delta_1;\Gamma_1 \vdash b{:}A$;
- If $b{:=}(\S_{i=1}^n\, x_i : A_i.U){:}B \in \Delta_2$ and $U \not\equiv \S\, y{:}B.A'$ then $\Delta_1 \vdash bx_1 \cdots x_n =_{\beta\delta} U$.

Lemma 7.28 (Transitivity Lemma)
Assume $\Delta_1;\Gamma_1 \vdash \Delta_2;\Gamma_2$ and $\Delta_2;\Gamma_2 \vdash B : C$. Then $\Delta_1;\Gamma_1 \vdash B : C$. ⊠

Lemma 7.29 (Substitution Lemma)
Assume $\Delta;\Gamma_1, x{:}A, \Gamma_2 \vdash B : C$ and $\Delta;\Gamma_1 \vdash D : A$.
Then $\Delta;\Gamma_1, \Gamma_2[x{:=}D] \vdash B[x{:=}D] : C[x{:=}D]$. ⊠

Lemma 7.30 (Thinning Lemma) *Let $\Delta_1;\Gamma_1$ be a legal context, and let $\Delta_2;\Gamma_2$ be a legal context such that $\Delta_1 \subseteq \Delta_2$ and $\Gamma_1 \subseteq \Gamma_2$.*
Then $\Delta_1;\Gamma_1 \vdash A : B \Rightarrow \Delta_2;\Gamma_2 \vdash A : B$. ⊠

Lemma 7.31 (Generation Lemma)

- *If $x \in \mathcal{V}$ and $\Delta;\Gamma \vdash x{:}C$ then there is $s \in \{*, \Box\}$ and $B =_{\beta\delta} C$ such that $\Delta;\Gamma \vdash B : s$ and $x{:}B \in \Gamma$;*
- *If $b \in \mathcal{C}$ and $\Delta;\Gamma \vdash b{:}C$ then there is $s \in \boldsymbol{S}$ and $B =_{\beta\delta} C$ such that $\Delta;\Gamma \vdash B : s$, and either $b{:}B \in \Delta$ or there is T such that $b{:=}T{:}B \in \Delta$;*
- *If $s \in \boldsymbol{S}$ and $\Delta;\Gamma \vdash s{:}C$ then $s \equiv *$ and $C =_{\beta\delta} \Box$;*
- *If $\Delta;\Gamma \vdash MN : C$ then there are A, B such that $\Delta;\Gamma \vdash M : (\Pi x{:}A.B)$ or $\Delta;\Gamma \vdash M : (\P x{:}A.B)$, and $\Delta;\Gamma \vdash N{:}A$ and $C =_{\beta\delta} B[x{:=}N]$;*
- *If $\Delta;\Gamma \vdash (\lambda x{:}A.b) : C$ then there is B such that $\Delta;\Gamma \vdash (\Pi x{:}A.B) : *$, $\Delta;\Gamma, x{:}A \vdash b : B$ and $C =_{\beta\delta} \Pi x{:}A.B$;*
- *Assume $\Delta;\Gamma \vdash (\Pi x{:}A.B) : C$.*
  Then $C =_{\beta\delta} *$, $\Delta;\Gamma \vdash A{:}*$ and $\Delta;\Gamma, x{:}A \vdash B{:}*$;
- *If $\Delta;\Gamma \vdash (\P x{:}A.B) : C$ then $C =_{\beta\delta} \Delta$, $\Delta;\Gamma \vdash A{:}s_1$ for some $s_1 \in \{*, \Box\}$, and $\Delta;\Gamma, x{:}A \vdash B{:}s_2$ for some $s_2 \in \{*, \Box, \triangle\}$.* ⊠

Lemma 7.32 (Unicity of Types) *If $\Delta;\Gamma \vdash A : B_1$ and $\Delta;\Gamma \vdash A : B_2$ then $B_1 =_{\beta\delta} B_2$.* ⊠

Lemma 7.33 (Correctness of Types) *If $\Delta;\Gamma \vdash A : B$ then there is $s \in \boldsymbol{S}$ such that $B \equiv s$ or $\Delta;\Gamma \vdash B : s$.* ⊠

From Correctness of Types and the Generation Lemma we conclude:

Lemma 7.34 *If* $\Delta;\Gamma \vdash A : (\Pi x{:}B_1.B_2)$ *then*

- $\Delta;\Gamma \vdash B_1 : *$;
- $\Delta;\Gamma, x{:}B_1 \vdash B_2 : *$. ⊠

Lemma 7.35 *If* $\Delta;\Gamma \vdash A : (\P x{:}B_1.B_2)$ *then*

- $\Delta;\Gamma \vdash B_1 : s_1$ *for some* $s_1 \in \{*, \Box\}$;
- $\Delta;\Gamma, x{:}B_1 \vdash B_2{:}s_2$ *for some sort* s_2. ⊠

7c2 Reduction and conversion

In this section we show some properties of the reduction relations $\to_\beta$, $\to_\delta$ and $\to_{\beta\delta}$. As δ-reduction also depends on books, we first have to give a translation of AUT-68 books and AUT-contexts to λ68-contexts:

Definition 7.36 Let Γ be a AUT-68-context $x_1{:}\alpha_1, \ldots, x_n{:}\alpha_n$.
Then $\overline{\Gamma} \stackrel{\text{def}}{=} x_1{:}\overline{\alpha_1}, \ldots, x_n{:}\overline{\alpha_n}$.

Definition 7.37 Let $\mathfrak{B}$ be a book. We define the left part $\overline{\mathfrak{B}}$ of a context in λ68:

- $\overline{\varnothing} \stackrel{\text{def}}{=} \varnothing$;
- $\overline{\mathfrak{B}, (\Gamma; b; \text{PN}; \Omega)} \stackrel{\text{def}}{=} \overline{\mathfrak{B}}, b{:}\,\P\overline{\Gamma}.\overline{\Omega}$;
- $\overline{\mathfrak{B}, (\Gamma; x; —; \Omega)} \stackrel{\text{def}}{=} \overline{\mathfrak{B}}$;
- $\overline{\mathfrak{B}, (\Gamma; b; \Sigma; \Omega)} \stackrel{\text{def}}{=} \overline{\mathfrak{B}}, b{:=}\,\S\,\overline{\Gamma}.\overline{\Sigma}{:}\,\P\overline{\Gamma}.\overline{\Omega}$.

Example 7.38 The translation of the book of Example 7.9 is given in Figure 7.2 (because of the habit in computer science to use more than one digit for a variable, we have to write some additional brackets around subterms like `proof` to keep things unambiguous). We see that all variable declarations of the original book have disappeared in the translation. In the original book, they do not add any new knowledge but are only used to construct contexts. In our translation, this happens in the right part of the context, instead of the left part.

Lemma 7.39 *Assume,* Σ *is a correct expression with respect to a book* $\mathfrak{B}$.

1. $\Sigma \to_\beta \Sigma'$ *if and only if* $\overline{\Sigma} \to_\beta \overline{\Sigma'}$;
2. $\mathfrak{B} \vdash \Sigma \to_\delta \Sigma'$ *if and only if* $\overline{\mathfrak{B}} \vdash \overline{\Sigma} \to_\delta \overline{\Sigma'}$.

PROOF: An easy induction on the structure of Σ. ⊠

```
prop    :   *,
and     :   ¶x:prop.¶y:prop.prop,
proof   :   ¶x:prop.*,
and-I   :   ¶x:prop.¶y:prop.¶px:(proof)x.¶py:(proof)y.(proof)((and)xy),
and-O1  :   ¶x:prop.¶y:prop.¶pxy:(proof)((and)xy).(proof)x,
and-O2  :   ¶x:prop.¶y:prop.¶pxy:(proof)((and)xy).(proof)y,
and-R   :=  §x:prop.§prx:(proof)x.(and-I)xx(prx)(prx)
        :   ¶x:prop.¶prx:(proof)x.(proof)((and)xx),
and-S   :=  §x:prop.§y:prop.§pxy:(proof)((and)xy).
            (and-I)yx((and-O2)xy(pxy))((and-O1)xy(pxy))
        :   ¶x:prop.¶y:prop.¶pxy:(proof)((and)xy).(proof)((and)yx)
```

Figure 7.2: Translation of Example 7.9

The Church-Rosser property of $\rightarrow_{\beta\delta}$ will be proved by the method of Parallel Reduction, invented by Martin-Löf and Tait (see Section 3.2 of [2]).

Definition 7.40 Let Δ be the left part of a context. We define a reduction relation $\Rightarrow_{\beta\delta}$ ("parallel reduction") on the set of terms $\mathcal{T}$:

- For $x \in \mathcal{V}$, $\Delta \vdash x \Rightarrow_{\beta\delta} x$;
- For $b \in \mathcal{C}$, $\Delta \vdash b \Rightarrow_{\beta\delta} b$;
- For $s \in \boldsymbol{S}$, $\Delta \vdash s \Rightarrow_{\beta\delta} s$;
- If $\Delta \vdash P \Rightarrow_{\beta\delta} P'$ and $\Delta \vdash Q \Rightarrow_{\beta\delta} Q'$, then
  - $\Delta \vdash \lambda x{:}P.Q \Rightarrow_{\beta\delta} \lambda x{:}P'.Q'$;
  - $\Delta \vdash \Pi x{:}P.Q \Rightarrow_{\beta\delta} \Pi x{:}P'.Q'$;
  - $\Delta \vdash \P x{:}P.Q \Rightarrow_{\beta\delta} \P x{:}P'.Q'$;
  - $\Delta \vdash PQ \Rightarrow_{\beta\delta} P'Q'$;
- If $\Delta \vdash Q \Rightarrow_{\beta\delta} Q'$ and $\Delta \vdash R \Rightarrow_{\beta\delta} R'$, then $\Delta \vdash (\lambda x{:}P.Q)R \Rightarrow_{\beta\delta} Q'[x{:=}R']$;
- If $b{:=}(\S_{i=1}^{n}\, x_i{:}A_i.T){:}(\P_{i=1}^{n}\, x_i{:}A_i.U) \in \Delta$, the term T is not of the form $\S y{:}T_1.T_2$, $\Delta \vdash T \Rightarrow_{\beta\delta} T'$ and $\Delta \vdash M_i \Rightarrow_{\beta\delta} M_i'$ for $i = 1, \ldots, n$, then $\Delta \vdash bM_1 \cdots M_n \Rightarrow_{\beta\delta} T'[x_1, \ldots, x_n{:=}M_1', \ldots, M_n']$.

Some elementary properties of $\Rightarrow_{\beta\delta}$ are:

Lemma 7.41 (Properties of $\Rightarrow_{\beta\delta}$) *Let Δ be the left part of a context. For all terms M, N:*

1. $\Delta \vdash M \Rightarrow_{\beta\delta} M$;
2. *If* $\Delta \vdash M \rightarrow_{\beta\delta} M'$ *then* $\Delta \vdash M \Rightarrow_{\beta\delta} M'$;
3. *If* $\Delta \vdash M \Rightarrow_{\beta\delta} M'$ *then* $\Delta \vdash M \twoheadrightarrow_{\beta\delta} M'$.

PROOF: All proofs can be given by induction on the structure of M. ⊠

We conclude from this lemma that $\twoheadrightarrow_{\beta\delta}$ (the reflexive and transitive closure of $\rightarrow_{\beta\delta}$) in the context Δ is the same relation as the reflexive and transitive closure of $\Rightarrow_{\beta\delta}$ in Δ. Therefore, if we want to prove the Church-Rosser theorem for $\twoheadrightarrow_{\beta\delta}$, it suffices to prove the Diamond Property for $\Rightarrow_{\beta\delta}$. We first make some preliminary definitions and remarks:

Lemma 7.42 (Substitution and $\Rightarrow_{\beta\delta}$) *If* $\Delta \vdash M \Rightarrow_{\beta\delta} M'$ *and* $\Delta \vdash N \Rightarrow_{\beta\delta} N'$ *then* $\Delta \vdash M[y{:=}N] \Rightarrow_{\beta\delta} M'[y{:=}N']$.

PROOF: Induction on the structure of M. ⊠

Lemma 7.43 *Assume, Δ and Δ, Δ' are left parts of legal contexts, and* $\text{FC}(M) \subseteq \text{DOM}(\Delta)$. *Then $\Delta \vdash M \Rightarrow_{\beta\delta} N$ if and only if $\Delta, \Delta' \vdash M \Rightarrow_{\beta\delta} N$.*

PROOF: By induction on the length of Δ and by induction on the definition of $\Delta \vdash M \Rightarrow_{\beta\delta} N$. All cases in the definition of $\Delta \vdash M \Rightarrow_{\beta\delta} N$ follow directly from the induction hypothesis for $\Delta \vdash M \Rightarrow_{\beta\delta} N$, except for the case $bM_1 \cdots M_n \Rightarrow_{\beta\delta} T'[x_1, \ldots, x_n := M'_1, \ldots, M'_n]$.

As $\text{FC}(M) \subseteq \text{DOM}(\Delta)$, we have $b \in \text{DOM}(\Delta)$.

Write $\Delta \equiv \Delta_1, b := (\S_{i=1}^n x_i{:}A_i.T) : (\P_{i=1}^n x_i{:}A_i.U), \Delta_2$.

- Notice that T is typable in $\Delta_1; x_1{:}A_1, \ldots, x_n{:}A_n$ (Definition Lemma). By the Free Variable Lemma: $\text{FC}(T) \subseteq \text{DOM}(\Delta_1)$. By the induction hypothesis on the length of Δ we have $\Delta_1 \vdash T \Rightarrow_{\beta\delta} T'$ iff $\Delta \vdash T \Rightarrow_{\beta\delta} T'$, and $\Delta_1 \vdash T \Rightarrow_{\beta\delta} T'$ iff $\Delta, \Delta' \vdash T \Rightarrow_{\beta\delta} T'$;
- We conclude: $\Delta \vdash T \Rightarrow_{\beta\delta} T'$ iff $\Delta, \Delta' \vdash T \Rightarrow_{\beta\delta} T'$;
- By the induction hypothesis on the definition of $\Delta \vdash M \Rightarrow_{\beta\delta} N$, we have $\Delta \vdash M_i \Rightarrow_{\beta\delta} M'_i$ iff $\Delta, \Delta' \vdash M_i \Rightarrow_{\beta\delta} M'_i$;
- Notice that $b := (\S_{i=1}^n x_i{:}A_i.T) : (\P_{i=1}^n x_i{:}A_i.U)$ is an element of both Δ and Δ, Δ'. Moreover, $b \notin \text{DOM}(\Delta')$ (because Δ, Δ' is the left part of a *legal* context). Therefore we have that $\Delta \vdash bM_1 \cdots M_n \Rightarrow_{\beta\delta} N$ if and only if $\Delta, \Delta' \vdash bM_1 \cdots M_n \Rightarrow_{\beta\delta} N$.

⊠

For left parts Δ of contexts and for $M \in \mathcal{T}$ with $\text{FC}(M) \subseteq \text{DOM}(\Delta)$, we define a term M^Δ. In M^Δ, all β-redexes that exist in M are contracted simultaneously (this is a usual step in a proof of Church-Rosser by Parallel Reduction), but also all δ-redexes are contracted. We will show that $\Delta \vdash N \Rightarrow_{\beta\delta} M^\Delta$ for any N with $\Delta \vdash M \Rightarrow_{\beta\delta} N$; so M^Δ helps us to show the Diamond Property for $\Rightarrow_{\beta\delta}$.

Definition 7.44 We define, for any left part Δ of a context and any $M \in \mathcal{T}$ such that $\text{FC}(M) \subseteq \text{DOM}(\Delta)$, M^Δ. The definition of M^Δ is by induction on the length of Δ. So assume $M^{\Delta'}$ has been defined for contexts Δ' that are shorter than Δ. We use induction on the structure of M:

- $x^\Delta \stackrel{\text{def}}{=} x$ for any $x \in \mathcal{V}$;
- $M \equiv b$. Distinguish:
  - $b^\Delta \stackrel{\text{def}}{=} b$ for any $b \in \text{PRIMCONS}(\Delta;)$;

- $b^{\Delta} \stackrel{\text{def}}{=} b$ for any $b \in$ DEFCONS $(\Delta;)$ that is not a δ-redex;
- If $b \in$ DEFCONS $(\Delta;)$ is a δ-redex, then $\Delta \equiv \Delta_1, b{:=}T{:}U, \Delta_2$, where $T \not\equiv \S y{:}T_1.T_2$. By the Definition Lemma, $\Delta_1; \vdash T : U$, so we can assume that T^{Δ_1} has already been defined. Then $b^{\Delta} \stackrel{\text{def}}{=} T^{\Delta_1}$;

- $s^{\Delta} \stackrel{\text{def}}{=} s$ for any $s \in \boldsymbol{S}$;
- $(\lambda x{:}P.Q)^{\Delta} \stackrel{\text{def}}{=} \lambda x{:}P^{\Delta}.Q^{\Delta}$;

  $(\Pi x{:}P.Q)^{\Delta} \stackrel{\text{def}}{=} \Pi x{:}P^{\Delta}.Q^{\Delta}$;

  $(\P x{:}P.Q)^{\Delta} \stackrel{\text{def}}{=} \P x{:}P^{\Delta}.Q^{\Delta}$;
- M is an application term. We distinguish three possibilities:
  - $M \equiv PQ$ is not a $\beta\delta$-redex. Then we define $M^{\Delta} \stackrel{\text{def}}{=} P^{\Delta}Q^{\Delta}$;
  - M is a β-redex $(\lambda x{:}P.Q)R$. We define $M^{\Delta} \stackrel{\text{def}}{=} Q^{\Delta}[x{:=}R^{\Delta}]$;
  - M is a δ-redex $bM_1 \cdots M_n$, and

$$\Delta \equiv \Delta_1, b{:=}\left(\mathop{\S}_{i=1}^{n} x_i{:}A_i.T\right) : \left(\mathop{\P}_{i=1}^{n} x_i{:}A_i.U\right), \Delta_2,$$

  where T is not of the form $\S y{:}T_1.T_2$. In that case

$$\Delta_1; x_1{:}A_1, \ldots, x_n{:}A_n \vdash T : U$$

  (by the Definition Lemma), so we can assume that T^{Δ_1} has already been defined. Then $M^{\Delta} \stackrel{\text{def}}{=} T^{\Delta_1}[x_1, \ldots, x_n{:=}M_1^{\Delta}, \ldots, M_n^{\Delta}]$.

Lemma 7.45 *Let Δ be the left part of a legal context. $\Delta \vdash M \Rightarrow_{\beta\delta} M^{\Delta}$ for all M with* FC$(M) \subseteq$ DOM (Δ).

PROOF: By induction on the definition of M^{Δ}.
We only treat the case $\Delta \vdash bM_1 \cdots M_n \Rightarrow_{\beta\delta} (bM_1 \cdots M_n)^{\Delta}$ where $bM_1 \cdots M_n$ is a δ-redex. As in the definition of $(bM_1 \cdots M_n)^{\Delta}$, write

$$\Delta \equiv \Delta_1, b{:=}\left(\mathop{\S}_{i=1}^{n} x_i{:}A_i.T\right) : \left(\mathop{\P}_{i=1}^{n} x_i{:}A_i.U\right), \Delta_2.$$

By induction, we may assume that $\Delta_1 \vdash T \Rightarrow_{\beta\delta} T^{\Delta_1}$ and $\Delta \vdash M_i \Rightarrow_{\beta\delta} M_i^{\Delta}$. By the Definition Lemma, T is typable in $\Delta_1; x_1{:}A_1, \ldots, x_n{:}A_n$, so by the Free Variable Lemma, FC$(T) \subseteq$ DOM (Δ_1). By Lemma 7.43, $\Delta \vdash T \Rightarrow_{\beta\delta} T^{\Delta_1}$. So $\Delta \vdash bM_1 \cdots M_n \Rightarrow_{\beta\delta} T^{\Delta_1}[x_1, \ldots, x_n{:=}M_1^{\Delta}, \ldots, M_n^{\Delta}]$. ⊠

Theorem 7.46 *Let Δ be the left part of a legal context. Assume* $\text{FC}(M) \subseteq \text{DOM}(\Delta)$. *If* $\Delta \vdash M \Rightarrow_{\beta\delta} N$ *then* $\Delta \vdash N \Rightarrow_{\beta\delta} M^\Delta$.

PROOF: Induction on the the definition of M^Δ.

- $M \equiv x$. Then $N \equiv x$ and $M^\Delta \equiv x$;
- $M \equiv b$. Distinguish:
  - $b \in \text{PRIMCONS}(\Delta;)$. Then $N \equiv b$ and $M^\Delta \equiv b$;
  - $b \in \text{DEFCONS}(\Delta;)$, but b is not a δ-redex. Then $N \equiv b$ and $M^\Delta \equiv b$;
  - $b \in \text{DEFCONS}(\Delta;)$, and $\Delta \equiv \Delta_1, b{:=}T{:}U, \Delta_2$, and $T \not\equiv \S y{:}T_1.T_2$. Then either $N \equiv b$ or $N \equiv T'$ where $T \Rightarrow_{\beta\delta} T'$. If $N \equiv b$ then $M \equiv N$ and we can use Lemma 7.45. If $N \equiv T$ then observe that by the induction hypothesis, $\Delta_1 \vdash T \Rightarrow_{\beta\delta} T^{\Delta_1}$, that by Lemma 7.43 $\Delta \vdash T \Rightarrow_{\beta\delta} T^{\Delta_1}$, and that $M^\Delta \equiv T^{\Delta_1}$;
- $M \equiv s$. Then $N \equiv s$ and $M^\Delta \equiv s$;
- $M \equiv \lambda x{:}P.Q$. Then $N \equiv \lambda x{:}P'.Q'$ for some P', Q' with $\Delta \vdash P \Rightarrow_{\beta\delta} P'$ and $\Delta \vdash Q \Rightarrow_{\beta\delta} Q'$. By the induction hypothesis on P and Q we find $\Delta \vdash P' \Rightarrow_{\beta\delta} P^\Delta$ and $\Delta \vdash Q' \Rightarrow_{\beta\delta} Q^\Delta$.
  Therefore $\Delta \vdash \lambda x{:}P'.Q' \Rightarrow_{\beta\delta} \lambda x{:}P^\Delta.Q^\Delta$.

  The cases $M \equiv \Pi x{:}P.Q$, $M \equiv \P x{:}P.Q$, and $M \equiv PQ$ where PQ is not a $\beta\delta$-redex, are proved similarly;
- M is an application term (and is either a β or a δ-redex). Distinguish:
  - M is a β-redex, $M \equiv (\lambda x{:}P.Q)R$. Distinguish:
    * $N \equiv (\lambda x{:}P'.Q')R'$ where $\Delta \vdash P \Rightarrow_{\beta\delta} P'$, $\Delta \vdash Q \Rightarrow_{\beta\delta} Q'$ and $\Delta \vdash R \Rightarrow_{\beta\delta} R'$. By induction, we get that $\Delta \vdash Q' \Rightarrow_{\beta\delta} Q^\Delta$ and $\Delta \vdash R' \Rightarrow_{\beta\delta} R^\Delta$. Therefore $\Delta \vdash N \Rightarrow_{\beta\delta} Q^\Delta[x{:=}R^\Delta]$;
    * $N \equiv Q'[x{:=}R']$ where $\Delta \vdash Q \Rightarrow_{\beta\delta} Q'$ and $\Delta \vdash R \Rightarrow_{\beta\delta} R'$. By induction, $\Delta \vdash Q' \Rightarrow_{\beta\delta} Q^\Delta$ and $\Delta \vdash R' \Rightarrow_{\beta\delta} R^\Delta$. By Lemma 7.42, $\Delta \vdash Q'[x{:=}R'] \Rightarrow_{\beta\delta} Q^\Delta[x{:=}R^\Delta]$;
  - M is a δ-redex, $M \equiv bM_1 \cdots M_n$,
  $$\Delta \equiv \Delta_1, b{:=}\left(\mathop{\S}_{i=1}^{n} x_i{:}A_i.T\right) : \left(\mathop{\P}_{i=1}^{n} x_i{:}A_i.U\right), \Delta_2,$$
  where $T \not\equiv \S y{:}T_1.T_2$. Distinguish:
    * $N \equiv bM_1' \cdots M_n'$ for M_i' where $\Delta \vdash M_i \Rightarrow_{\beta\delta} M_i'$. By induction, $\Delta \vdash M_i' \Rightarrow_{\beta\delta} M_i^\Delta$. By the Definition Lemma, T is typable in a context $\Delta_1; x_1{:}A_1, \ldots, x_n{:}A_n$, so by the Free Variable Lemma, $\text{FC}(T) \subseteq \text{DOM}(\Delta_1)$. By Lemma 7.45, $\Delta_1 \vdash T \Rightarrow_{\beta\delta} T^{\Delta_1}$. By Lemma 7.43, $\Delta \vdash T \Rightarrow_{\beta\delta} T^{\Delta_1}$.
    Hence $\Delta \vdash N \Rightarrow_{\beta\delta} T^{\Delta_1}[x_1, \ldots, x_n{:=}M_1^\Delta, \ldots, M_n^\Delta]$;

* $N \equiv T'[x_1,\ldots,x_n{:=}M'_1,\ldots,M'_n]$ where $\Delta \vdash T \Rightarrow_{\beta\delta} T'$ and $\Delta \vdash M_i \Rightarrow_{\beta\delta} M'_i$. By the Definition Lemma, T is typable in $\Delta_1; x_1{:}A_1,\ldots,x_n{:}A_n$, so by the Free Variable Lemma, $\text{FC}(T) \subseteq \text{DOM}(\Delta_1)$. By Lemma 7.43, $\Delta_1 \vdash T \Rightarrow_{\beta\delta} T'$. By the induction hypothesis on T, $\Delta_1 \vdash T' \Rightarrow_{\beta\delta} T^{\Delta_1}$. As $\Delta_1 \vdash T \Rightarrow_{\beta\delta} T'$, $\text{FC}(T') \subseteq \text{DOM}(\Delta_1)$, so by Lemma 7.43, $\Delta \vdash T' \Rightarrow_{\beta\delta} T^{\Delta_1}$. By the induction hypothesis, also $\Delta \vdash M'_i \Rightarrow_{\beta\delta} M_i^{\Delta}$. Repeatedly applying Lemma 7.42, we find[8]

$$\Delta \vdash T'[x_1,\ldots,x_n{:=}M'_1,\ldots M'_n] \Rightarrow_{\beta\delta} T^{\Delta_1}[x_1,\ldots,x_n{:=}M_1^{\Delta},\ldots,M_n^{\Delta}].$$

⊠

Corollary 7.47 (Diamond Property for $\Rightarrow_{\beta\delta}$) *Let Δ be the left part of a context in which M is typable. Assume $\Delta \vdash M \Rightarrow_{\beta\delta} N_1$ and $\Delta \vdash M \Rightarrow_{\beta\delta} N_2$. Then there is P such that $\Delta \vdash N_1 \Rightarrow_{\beta\delta} P$ and $\Delta \vdash N_2 \Rightarrow_{\beta\delta} P$.*

PROOF: Immediately from the theorem above: Take $P \equiv M^{\Delta}$. ⊠

Corollary 7.48 (Church-Rosser property for $\rightarrow_{\beta\delta}$) *Let Δ be the left part of a context in which M is typable. If $\Delta \vdash M \twoheadrightarrow_{\beta\delta} N_1$ and $\Delta \vdash M \twoheadrightarrow_{\beta\delta} N_2$ then there is P such that $\Delta \vdash N_1 \twoheadrightarrow_{\beta\delta} P$ and $\Delta \vdash N_2 \twoheadrightarrow_{\beta\delta} P$.*

PROOF: Directly from Lemma 7.41.2, Lemma 7.41.3 and Corollary 7.47. ⊠

7c3 Subject reduction

Lemma 7.49 (Subject Reduction for $\rightarrow_{\beta}$)
If $\Delta;\Gamma \vdash A : B$ and $A \rightarrow_{\beta} A'$ then $\Delta;\Gamma \vdash A' : B$.

PROOF: The proof is as in [3]. ⊠

Subject Reduction also holds for the reduction relation $\rightarrow_{\delta}$:

Lemma 7.50 (Subject Reduction for $\rightarrow_{\delta}$)
If $\Delta;\Gamma \vdash A : B$ and $A \rightarrow_{\delta} A'$ then $\Delta;\Gamma \vdash A' : B$.

---

[8] We must remark that

$$T'[x_1,\ldots,x_n{:=}M'_1,\ldots,M'_n] \equiv T'[x_1{:=}M'_1]\cdots[x_n{:=}M'_n]$$

and

$$T^{\Delta_1}[x_1,\ldots,x_n{:=}M_1^{\Delta},\ldots,M_n^{\Delta}] \equiv T^{\Delta_1}[x_1{:=}M_1^{\Delta}]\cdots[x_n{:=}M_n^{\Delta}].$$

This is correct as we can assume that the x_i do not occur in the M'_j and M_j^{Δ}.

PROOF: Following the line of [3], we define $\Delta;\Gamma \to_\delta \Delta;\Gamma'$ if $\Gamma \equiv \Gamma_1, x{:}A, \Gamma_2$, and $\Gamma' \equiv \Gamma_1, x{:}A', \Gamma_2$, and $\Delta \vdash A \to_\delta A'$. We define $\Delta;\Gamma \to_\delta \Delta';\Gamma$ similarly, and we simultaneously prove

$$\begin{array}{lcl} \Delta;\Gamma \vdash A{:}B \text{ and } \Delta \vdash A \to_\delta A' & \Rightarrow & \Delta;\Gamma \vdash A'{:}B \\ \Delta;\Gamma \vdash A{:}B \text{ and } \Delta;\Gamma \to_\delta \Delta';\Gamma & \Rightarrow & \Delta';\Gamma \vdash A{:}B \\ \Delta;\Gamma \vdash A{:}B \text{ and } \Delta;\Gamma \to_\delta \Delta;\Gamma' & \Rightarrow & \Delta;\Gamma' \vdash A{:}B, \end{array}$$

using induction on the derivation of $\Delta;\Gamma \vdash A{:}B$. We only treat the case in which the last applied rule is the 2nd application rule, and we only prove the first of the three statements for this case.

We write $A[x_i{:=}B_i]_{i=m}^n$ for $A[x_m{:=}B_m][x_{m+1}{:=}B_{m+1}]\cdots[x_n{:=}B_n]$. We assume that

$$\Delta \equiv \Delta_1, b{:=}\left(\mathop{\S}_{i=1}^{n} x_i{:}A_i.T\right) : \left(\mathop{\P}_{i=1}^{n} x_i{:}A_i.B\right), \Delta_2 \tag{7.1}$$

with $B \not\equiv \P y{:}B_1.B_2$, and that the conclusion of the 2nd application rule is

$$\Delta;\Gamma \vdash bM_1 \cdots M_n : K_n \tag{7.2}$$

for some K_n, and therefore

$$\Delta \vdash bM_1 \cdots M_n \to_\delta T[x_i{:=}M_i]_{i=1}^n.$$

We must prove: $\Delta;\Gamma \vdash T[x_i{:=}M_i]_{i=1}^n : K_n$. We do this in two steps.

1. We analyse the structure of K_n, and derive that
$$\Delta \vdash K_n =_{\beta\delta} B[x_i{:=}M_i]_{i=1}^n;$$
2. We show that $\Delta;\Gamma \vdash T[x_i{:=}M_i]_{i=1}^n : B[x_i{:=}M_i]_{i=1}^n$.

Ad 1.

We repeatedly apply the Generation Lemma, starting with (7.2), thus obtaining $K_n, K_{n-1}, \ldots, K_1, K'_n, K'_{n-1}, \ldots, K'_1, L_n, L_{n-1}, \ldots, L_1$ such that

$$\Delta;\Gamma \vdash bM_1 \cdots M_{i-1} : (\P x_i{:}L_i.K'_i); \tag{7.3}$$

$$\Delta;\Gamma \vdash M_i : L_i; \tag{7.4}$$

$$\Delta \vdash K_i =_{\beta\delta} K'_i[x_i{:=}M_i]; \tag{7.5}$$

$$\Delta \vdash K_{i-1} =_{\beta\delta} \P x_i{:}L_i.K'_i. \tag{7.6}$$

We end with $\Delta;\Gamma \vdash b : (\P x_1{:}L_1.K'_1)$. By (7.1) and the Generation Lemma:

$$\Delta \vdash \P x_1{:}L_1.K'_1 =_{\beta\delta} \mathop{\P}_{j=1}^{n} x_j{:}A_j.B.$$

By the Church-Rosser Theorem we have $L_1 =_{\beta\delta} A_1$ and

$$\Delta \vdash K_1' =_{\beta\delta} \mathop{\P}_{j=2}^{n} x_j{:}A_j.B. \tag{7.7}$$

Hence

$$\begin{aligned} \Delta \vdash \P x_2{:}L_2.K_2' &\overset{(7.6)}{=}_{\beta\delta} K_1 \\ &\overset{(7.5,7.7)}{=}_{\beta\delta} \left(\mathop{\P}_{j=2}^{n} x_j{:}A_j.B\right)[x_1{:=}M_1] \\ &\equiv \mathop{\P}_{i=2}^{n} x_i{:}A_i[x_1{:=}M_1].B[x_1{:=}M_1], \end{aligned}$$

so by the Church-Rosser Theorem $L_2 =_{\beta\delta} A_2[x_1{:=}M_1]$. Proceeding in this way, we obtain for $i = 1, \ldots, n$:

$$\begin{aligned} \Delta \vdash L_i &=_{\beta\delta} A_i[x_j{:=}M_j]_{j=1}^{i-1}; \\ \Delta \vdash K_i' &=_{\beta\delta} \mathop{\P}_{j=i+1}^{n} x_j{:}A_j[x_k{:=}M_k]_{k=1}^{i-1}.B[x_k{:=}M_k]_{k=1}^{i-1}; \\ \Delta \vdash K_i &=_{\beta\delta} \mathop{\P}_{j=i+1}^{n} x_j{:}A_j[x_k{:=}M_k]_{k=1}^{i}.B[x_k{:=}M_k]_{k=1}^{i}. \end{aligned} \tag{7.8}$$

In particular,

$$\Delta \vdash K_n =_{\beta\delta} B[x_i{:=}M_i]_{i=1}^{n}. \tag{7.9}$$

Ad 2.

Now we calculate the type of $T[x_i{:=}M_i]_{i=1}^{n}$. By the Definition Lemma on (7.1) we also have

$$\Delta_1; x_1{:}A_1, \ldots, x_n{:}A_n \vdash T : B, \tag{7.10}$$

so by the Start Lemma: $\Delta_1; x_1{:}A_1, \ldots, x_{i-1}{:}A_{i-1} \vdash A_i{:}s_i$ for sorts $s_i \in \boldsymbol{S}$. This yields:

$$\begin{aligned} \Delta; \Gamma \vdash A_1 : s_1 &\quad \text{(Thinning Lemma);} \\ \Delta; \Gamma, x_1{:}A_1 \text{ is legal} &\quad \text{(Start Rule);} \\ \Delta; \Gamma, x_1{:}A_1 \vdash A_2 : s_2 &\quad \text{(Thinning Lemma);} \\ \Delta; \Gamma, x_1{:}A_1, x_2{:}A_2 \text{ is legal} &\quad \text{(Start Rule);} \\ &\ \ \vdots \\ \Delta; \Gamma, x_1{:}A_1, \ldots, x_n{:}A_n \text{ is legal.} &\quad \text{(Start Rule).} \end{aligned}$$

Therefore, we can apply the Thinning Lemma to (7.10), and we find:

$$\Delta; \Gamma, x_1{:}A_1, \ldots, x_n{:}A_n \vdash T : B.$$

As $\Delta;\Gamma \vdash M_1 : L_1$ (7.4) and $\Delta;\Gamma \vdash A_1 : s_1$, we have $\Delta;\Gamma \vdash M_1 : A_1$ by the Conversion rule and (7.8), so by the Substitution Lemma:

$$\begin{array}{rcl} \Delta;\Gamma,x_2{:}A_2[x_1{:=}M_1],\ldots,x_n{:}A_n[x_1{:=}M_1] & \vdash & T[x_1{:=}M_1] : B[x_1{:=}M_1]; \\ \Delta;\Gamma & \vdash & A_2[x_1{:=}M_1] : s_2. \end{array}$$

As $\Delta;\Gamma \vdash M_2 : L_2$ (7.4) and $\Delta \vdash A_2[x_1{:=}M_1] =_{\beta\delta} L_2$ (7.8) we have by conversion $\Delta;\Gamma \vdash M_2 : A_2[x_1{:=}M_1]$, and again by the Substitution Lemma:

$$\begin{array}{rcl} \Delta;\Gamma,x_3{:}A_3[x_i{:=}M_i]_{i=1}^2,\ldots,x_n{:}A_n[x_i{:=}M_i]_{i=1}^2 & & \\ & \vdash & T[x_i{:=}M_i]_{i=1}^2 : B[x_i{:=}M_i]_{i=1}^2; \\ \Delta;\Gamma & \vdash & A_3[x_1{:=}M_1][x_2{:=}M_2] : s_3. \end{array}$$

Proceeding in this way we eventually find

$$\Delta;\Gamma \vdash T[x_i{:=}M_i]_{i=1}^n : B[x_i{:=}M_i]_{i=1}^n. \tag{7.11}$$

Applying Lemma 7.33 to (7.9) we have $\Delta;\Gamma \vdash K_n : s$. Now use the Conversion Rule, (7.11), and the fact that $\Delta \vdash K_n =_{\beta\delta} B[x_i{:=}M_i]_{i=1}^n$. ⊠

Corollary 7.51 (Subject Reduction for $\to_{\beta\delta}$) *If $\Delta;\Gamma \vdash A : B$ and $A \twoheadrightarrow_{\beta\delta} A'$ then $\Delta;\Gamma \vdash A' : B$.* ⊠

The Subject Reduction Theorem for $\to_\delta$ is used to prove:

Lemma 7.52 *Assume $s \in \boldsymbol{S}$ and M legal.*
Then $(\Delta \vdash M =_{\beta\delta} s) \Rightarrow M \equiv s$.

PROOF: First assume $s \in \{\Box, \Delta\}$. If $\Delta;\Gamma \vdash M : N$ for some Γ and N, and $\Delta \vdash M =_{\beta\delta} s$ then by Church-Rosser $\Delta \vdash M \twoheadrightarrow_{\beta\delta} s$, so by Subject Reduction $\Delta;\Gamma \vdash s : N$, contradicting the Generation Lemma. If $\Delta;\Gamma \vdash N : M$ and $\Delta \vdash M =_{\beta\delta} s$ and $M \not\equiv s$ then we have by Lemma 7.33 that $\Delta;\Gamma \vdash M : P$ for some P, so again $\Delta;\Gamma \vdash s : P$, in contradiction with the Generation Lemma.

Now assume $s \equiv *$, $\Delta;\Gamma \vdash M : N$, and $\Delta \vdash M =_{\beta\delta} s$. Again by Church-Rosser, $\Delta \vdash M \twoheadrightarrow_{\beta\delta} *$, say $\Delta \vdash M \to_{\beta\delta} \ldots \to_{\beta\delta} M' \to_{\beta\delta} *$. By Subject Reduction, $\Delta;\Gamma \vdash M' : N$ and $\Delta;\Gamma \vdash * : N$. By the Generation Lemma $\Delta \vdash N =_{\beta\delta} \Box$, so $N \equiv \Box$. Distinguish:

- $M' \equiv (\lambda x{:}A.B)C$ and $* \equiv B[x{:=}C]$. By the Generation Lemma there is B' such that $\Delta \vdash B'[x{:=}C] =_{\beta\delta} \Box$, $\Delta;\Gamma \vdash (\lambda x{:}A.B) : (\Pi x{:}A.B')$ and $\Delta;\Gamma \vdash C : A$. $C \equiv \Box$ contradicts $\Delta;\Gamma \vdash C : A$, so $B' \equiv \Box$. By Lemma 7.34 $\Delta;\Gamma \vdash (\Pi x{:}A.\Box) : *$, so by the Generation Lemma $\Delta;\Gamma, x{:}A \vdash \Box : *$, contradiction;

- $M' \equiv bM_1 \cdots M_n$ and $\Delta \vdash bM_1 \cdots M_n \rightarrow_\delta T[x_i{:=}M_i]_{i=1}^n \equiv *$. The argument is similar as in the case $M' \equiv (\lambda x{:}A.B)C$.

If $s \equiv *$, $\Delta;\Gamma \vdash N : M$, and $\Delta \vdash M =_{\beta\delta} s$ then by Lemma 7.33 $M \equiv s$ (and we are done) or $\Delta;\Gamma \vdash M : s'$ (which implies $M \equiv s$ by the above argument). ⊠

7c4 Strong normalisation

We prove Strong Normalisation for $\beta\delta$-reduction in $\lambda 68$ by mapping a typable term M (in a context $\Delta;\Gamma$) of $\lambda 68$ to a term $|M|_\Delta$ that is typable in a strongly normalising PTS. The mapping is constructed in such a way that if $M \rightarrow_\beta N$, $|M|_\Delta \twoheadrightarrow_\beta^+ |N|_\Delta$, and that if $\Delta \vdash M \rightarrow_\delta N$, $|M|_\Delta \twoheadrightarrow_\beta |N|_\Delta$.

Definition 7.53 Let Δ be the left part of a legal context and let $M \in \mathcal{T}$. We define $|M|_\Delta$ by induction on the length of Δ and the structure of M.

- $|x|_\Delta \stackrel{\text{def}}{=} x$ for $x \in \mathcal{V}$;
- $|b|_\Delta \stackrel{\text{def}}{=} b$ for all $b \in \mathcal{C} \setminus \text{DEFCONS}\,(\Delta;)$;
- $|b|_\Delta \stackrel{\text{def}}{=} \lambda_{i=1}^n\, x_i{:}\, |A_i|_{\Delta_1} . |T|_{\Delta_1}$

  $$\text{if } \Delta \equiv \Delta_1, b{:=}(\S_{i=1}^n\, x_i{:}A_i.T){:}(\P_{i=1}^n\, x_i{:}A_i.U), \Delta_2;$$
- $|s|_\Delta \stackrel{\text{def}}{=} s$ for $s \in \boldsymbol{S}$;
- $|\lambda x{:}P.Q|_\Delta \stackrel{\text{def}}{=} \lambda x{:}\, |P|_\Delta . |Q|_\Delta$;
- $|\Pi x{:}P.Q|_\Delta \stackrel{\text{def}}{=} \Pi x{:}\, |P|_\Delta . |Q|_\Delta$;
- $|\P x{:}P.Q|_\Delta \stackrel{\text{def}}{=} \Pi x{:}\, |P|_\Delta . |Q|_\Delta$;
- $|PQ|_\Delta \stackrel{\text{def}}{=} |P|_\Delta\, |Q|_\Delta$.

The following lemmas are useful:

Lemma 7.54 *Let Δ be the left part of a legal context and $M \in \mathcal{T}$. Then* $\text{FV}(|M|_\Delta) = \text{FV}(M)$.

PROOF: The proof is by induction on the definition of $|M|_\Delta$ and is trivial for all cases except for the case $M \equiv b$ and $\Delta \equiv \Delta_1, b{:=}(\S\,\Gamma.T){:}(\P\,\Gamma.U), \Delta_2$ (where $T \not\equiv \S\, y{:}T_1.T_2$). By the Definition Lemma, T is typable in $\Delta_1;\Gamma$; therefore $\text{FV}(T) \subseteq \text{DOM}\,(\Gamma)$ (Free Variable Lemma). By the induction hypothesis, we get $\text{FV}\big(|T|_{\Delta_1}\big) \subseteq \text{DOM}\,(\Gamma)$ and therefore $\text{FV}(|b|_\Delta) = \varnothing$. ⊠

Lemma 7.55 *If Δ_1 and Δ_2 are left parts of legal contexts and $\Delta_2 \equiv \Delta_1, \Delta'$ then $|M|_{\Delta_2} \equiv |M|_{\Delta_1}$ for all $M \in \mathcal{T}$ with $\mathrm{FC}(M) \subseteq \mathrm{DOM}(\Delta_1)$.*

PROOF: An easy induction on the definition of $|M|_{\Delta_1}$. ⊠

Lemma 7.56 *Let Δ be the left part of a legal context. For all M, N:*

$$|M[x{:=}N]|_\Delta \equiv |M|_\Delta\,[x{:=}\,|N|_\Delta].$$

PROOF: By induction on the definition of $|M|_\Delta$. In the case $M \equiv b$ and $b{:=}T{:}U \in \Delta$, use the fact that $\mathrm{FV}(|M|_\Delta) = \mathrm{FV}(M) = \varnothing$ (Lemma 7.54) and therefore $|M|_\Delta\,[x{:=}\,|N|_\Delta] \equiv |M|_\Delta \equiv |M[x{:=}N]|_\Delta$. ⊠

The purpose of the definition of $|M|_\Delta$ is explained in the following two lemmas:

Lemma 7.57 *If $M \to_\beta N$ then $|M|_\Delta \twoheadrightarrow_\beta^+ |N|_\Delta$.*

PROOF: Induction on the structure of M. We only treat the case $M \equiv (\lambda x{:}P.Q)R$ and $N \equiv Q[x{:=}R]$.

$$\begin{array}{rcl} |M|_\Delta & \equiv & (\lambda x{:}\,|P|_\Delta\,.\,|Q|_\Delta)\,|R|_\Delta \\ & \to_\beta & |Q|_\Delta\,[x{:=}\,|R|_\Delta] \\ & \stackrel{7.56}{\equiv} & |Q[x{:=}R]|_\Delta\,. \end{array}$$

⊠

Lemma 7.58 *If $\Delta \vdash M \to_\delta N$, then $|M|_\Delta \twoheadrightarrow_\beta |N|_\Delta$.*

PROOF: Induction on the structure of M. We only treat the case in which

$$\begin{array}{l} M \equiv bM_1 \cdots M_n; \\ \Delta \equiv \Delta_1, b{:=}\left(\S_{i=1}^{n}\, x_i{:}A_i.T\right) : \left(\P_{i=1}^{n}\, x_i{:}A_i.U\right), \Delta_2; \\ N \equiv T[x_1,\ldots,x_n{:=}M_1,\ldots,M_n]. \end{array}$$

Notice that

$$\begin{array}{rcl} |M|_\Delta & \equiv & \left(\lambda_{i=1}^{n}\, x_i{:}\,|A_i|_{\Delta_1}\,.\,|T|_{\Delta_1}\right)|M_1|_\Delta \cdots |M_n|_\Delta \\ & \twoheadrightarrow_\beta & |T|_{\Delta_1}\,[x_i{:=}\,|M_i|_\Delta]_{i=1}^n \\ & \stackrel{7.55}{\equiv} & |T|_\Delta\,[x_i{:=}\,|M_i|_\Delta]_{i=1}^n \\ & \stackrel{7.56}{\equiv} & |T[x_i{:=}M_i]_{i=1}^n|_\Delta \\ & \equiv & |T[x_1,\ldots,x_n{:=}M_1,\ldots,M_n]|_\Delta\,. \end{array}$$

At the last equivalence, we must make a remark similar to footnote 8 on page 212. ⊠

Let λSN be the PTS over λ-terms with variables from $\mathcal{V} \cup \mathcal{C}$ and sorts from $\boldsymbol{S}$, and the following rules (we choose the name λSN because this system will help us in showing that $\lambda 68$ is SN):

$$\begin{array}{cc} & (*,*,*); \\ (*,*,\Delta); & (\Box,*,\Delta); \\ (*,\Box,\Delta); & (\Box,\Box,\Delta); \\ (*,\Delta,\Delta); & (\Box,\Delta,\Delta). \end{array}$$

This is in fact the Pure Type System that is based on the Π-formation rules that were proposed in Section 7b1. λSN is contained in the system **ECC** (see [3]). As **ECC** is β-strongly normalising, also λSN is β-strongly normalising.

We present a translation of $\lambda 68$-contexts to λSN-contexts:

Definition 7.59 Let $\Delta;\Gamma$ be a legal $\lambda 68$-context.

- We define $|\Delta|$ by induction on the length of Δ:
  - $|\varnothing| \stackrel{\text{def}}{=} \varnothing$;
  - $|\Delta, b{:}U| \stackrel{\text{def}}{=} |\Delta|, b{:}\,|U|_\Delta$;
  - $|\Delta, b{:=}T{:}U| \stackrel{\text{def}}{=} |\Delta|$;
- If $\Gamma \equiv x_1{:}A_1, \ldots, x_n{:}A_n$ then $|\Delta;\Gamma| \stackrel{\text{def}}{=} |\Delta|, x_1{:}\,|A_1|_\Delta, \ldots, x_n{:}\,|A_n|_\Delta$.

We see that the definitions $b{:=}T{:}U$ in Δ are not translated into $|\Delta|$. This corresponds to the fact that all these definitions are unfolded (replaced by their definiendum) in $|b|_\Delta$.

Now we are able to prove the most important lemma of this subsection:

Lemma 7.60 *If* $\Delta;\Gamma \vdash_{\lambda 68} M : N$ *then* $|\Delta;\Gamma| \vdash_{\lambda \mathrm{SN}} |M|_\Delta : |N|_\Delta$.

PROOF: The proof is by induction on the derivation of $\Delta;\Gamma \vdash M : N$. We treat a few cases:

(Start: Primitive Constants)

$$\frac{\Delta;\Gamma \vdash_{\lambda 68} B : s_1 \qquad \Delta; \vdash_{\lambda 68} \P\,\Gamma.B : s_2}{\Delta, b{:}\P\,\Gamma.B; \vdash_{\lambda 68} b : \P\,\Gamma.B}\,(s_1 = *, \Box).$$

By the induction hypothesis, $|\Delta| \vdash_{\lambda \mathrm{SN}} |\P\,\Gamma.B|_\Delta : s_2$, so by the Start rule:

$$|\Delta|, b{:}\,|\P\,\Gamma.B|_\Delta \vdash b{:}\,|\P\,\Gamma.B|_\Delta\,.$$

Observe that $|\Delta, b{:}\,\P\,\Gamma.B| \equiv |\Delta|, b{:}\,|\P\,\Gamma.B|_\Delta$, that $|b|_{\Delta, b:\,\P\,\Gamma.B} \equiv b$ and that (by Lemma 7.55) $|\P\,\Gamma.B|_\Delta \equiv |\P\,\Gamma.B|_{\Delta, b:\,\P\,\Gamma.B}$;

(Start: Defined Constants)

$$\frac{\Delta;\Gamma \vdash_{\lambda 68} T : B : s_1 \qquad \Delta; \vdash_{\lambda 68} \P\Gamma.B : s_2}{\Delta, b{:=}(\Gamma.T){:}(\P\Gamma.B); \vdash_{\lambda 68} b : \P\Gamma.B}(s_1 = *, \Box).$$

By induction we have

$$|\Delta;| \vdash_{\lambda\mathrm{SN}} |\P\Gamma.B|_\Delta : s_2,$$

so (write $\Gamma \equiv x_1{:}A_1, \ldots, x_n{:}A_n$):

$$|\Delta;| \vdash_{\lambda\mathrm{SN}} \prod_{i=1}^{n} x_i{:}\,|A_i|_\Delta \,.\, |B|_\Delta : s_2. \tag{7.12}$$

By induction, we also have $|\Delta;\Gamma| \vdash_{\lambda\mathrm{SN}} |T|_\Delta : |B|_\Delta$, so:

$$|\Delta|, x_1{:}\,|A_1|_\Delta, \ldots, x_n{:}\,|A_n|_\Delta \vdash_{\lambda\mathrm{SN}} |T|_\Delta : |B|_\Delta, \tag{7.13}$$

and by repeatedly applying the λ-rule on (7.13) and using the fact that, by the Induction Hypothesis, the types $\prod_{j=i}^{n} x_j{:}\,|A_j|_\Delta \,.\, |B|_\Delta$ are all typable, we find:

$$|\Delta;| \vdash_{\lambda\mathrm{SN}} \left(\mathop{\lambda}_{i=1}^{n} x_i{:}\,|A_i|_\Delta \,.\, |T|_\Delta\right) : \left(\prod_{i=1}^{n} x_i{:}\,|A_i|_\Delta \,.\, |B|_\Delta\right); \tag{7.14}$$

(Application 1) (the Application 2-case is similar)

$$\frac{\Delta;\Gamma \vdash_{\lambda 68} M : (\Pi x{:}A.B) \qquad \Delta;\Gamma \vdash_{\lambda 68} N : A}{\Delta;\Gamma \vdash_{\lambda 68} MN : B[x{:=}N]}$$

By the induction hypothesis, we have

$$|\Delta;\Gamma| \vdash_{\lambda\mathrm{SN}} |M|_\Delta : (\Pi x{:}\,|A|_\Delta \,.\, |B|_\Delta),$$

and $|\Delta;\Gamma| \vdash_{\lambda\mathrm{SN}} |N|_\Delta : |A|_\Delta$. The application rule gives

$$|\Delta;\Gamma| \vdash_{\lambda\mathrm{SN}} |M|_\Delta\, |N|_\Delta : |B|_\Delta\, [x{:=}\, |A|_\Delta].$$

Use the definition of $|MN|_\Delta$ and Lemma 7.56 to obtain

$$|\Delta;\Gamma| \vdash_{\lambda\mathrm{SN}} |MN|_\Delta : |B[x{:=}A]|_\Delta\,.$$

⊠

Corollary 7.61 (Strong Normalisation) λ68 *is* $\beta\delta$*-strongly normalising.*

PROOF: Assume, we have an infinite $\beta\delta$-reduction path in $\lambda 68$:

$$M_1 \to_{\beta\delta} M_2 \to_{\beta\delta} M_3 \to_{\beta\delta} \cdots \tag{7.15}$$

As δ-reduction is strongly normalising (7.17 and 7.39.2), there must be infinitely many β-reductions in this reduction path, so we have a path

$$N_1 \to_\beta N_1' \twoheadrightarrow_\delta N_2 \to_\beta N_2' \twoheadrightarrow_\delta N_3 \to_\beta N_3' \twoheadrightarrow_\delta \cdots$$

By Lemmas 7.57 and 7.58, this gives us a reduction path

$$|N_1|_\Delta \twoheadrightarrow_\beta^+ |N_1'|_\Delta \twoheadrightarrow_\beta |N_2|_\Delta \twoheadrightarrow_\beta^+ |N_2'|_\Delta \twoheadrightarrow_\beta |N_3|_\Delta \twoheadrightarrow_\beta^+ |N_3'|_\Delta \twoheadrightarrow_\beta \cdots$$

which is an infinite β-reduction path in λSN. By Lemma 7.60, $|N_1|_\Delta$ is a legal term in λSN. But as λSN is strongly normalising, the above infinite β-reduction path cannot exist. Hence, the infinite $\beta\delta$-reduction path (7.15) does not exist, either. ⊠

7c5 The formal relation between AUT-68 and $\lambda 68$

Theorem 7.62 *Let $\mathfrak{B}$ be an* AUTOMATH *book and Γ an* AUTOMATH *context.*

- *If $\mathfrak{B};\Gamma \vdash_{AUT-68}$ OK then $\overline{\mathfrak{B}};\overline{\Gamma}$ is legal;*
- *If $\mathfrak{B};\Gamma \vdash_{AUT-68} \Sigma : \Omega$ then $\overline{\mathfrak{B}};\overline{\Gamma} \vdash \overline{\Sigma} : \overline{\Omega}$.*

PROOF: We prove both statements simultaneously, using induction on the derivation of $\mathfrak{B};\Gamma \vdash_{\text{AUT}-68}$ OK and $\mathfrak{B};\Gamma \vdash \Sigma : \Omega$ of Definition 7.10 and Definition 7.11. We only treat one case; the other cases are similar or trivial. Assume, the last step of the derivation has been an application of the book extension rule def2:

$$\frac{\mathfrak{B};\Gamma \vdash_{\text{AUT}-68} \Sigma_2{:}\texttt{type} \quad \mathfrak{B};\Gamma \vdash_{\text{AUT}-68} \Sigma_1{:}\Sigma_2' \quad \mathfrak{B};\Gamma \vdash_{\text{AUT}-68} \Sigma_2 =_{\beta\text{d}} \Sigma_2'}{\mathfrak{B},(\Gamma;k;\Sigma_1;\Sigma_2);\varnothing \vdash_{\text{AUT}-68} \text{OK}}.$$

By the induction hypothesis, we have

$$\overline{\mathfrak{B}};\overline{\Gamma} \vdash_{\lambda 68} \overline{\Sigma_2} : * \tag{7.16}$$

and

$$\overline{\mathfrak{B}};\overline{\Gamma} \vdash_{\lambda 68} \overline{\Sigma_1} : \overline{\Sigma_2'}. \tag{7.17}$$

By Lemma 7.39, we have

$$\overline{\mathfrak{B}} \vdash_{\lambda 68} \overline{\Sigma_2} =_{\beta\delta} \overline{\Sigma_2'}. \tag{7.18}$$

Applying the conversion rule of $\lambda 68$ to (7.16), (7.17) and (7.18) yields

$$\overline{\mathfrak{B}};\overline{\Gamma} \vdash_{\lambda 68} \overline{\Sigma_1} : \overline{\Sigma_2}. \tag{7.19}$$

Notice that $\overline{\mathfrak{B}};\overline{\Gamma}$ is legal, so for each $x{:}\alpha \in \overline{\Gamma}$ (say: $\overline{\Gamma} \equiv \Gamma_1, x{:}\alpha, \Gamma_2$) we have $\overline{\mathfrak{B}};\Gamma_1 \vdash \alpha : s$ for an $s \in \{*, \Box\}$, by the Free Variable Lemma 7.23. Thus we can repeatedly apply the ¶-formation rule (starting with (7.16)) to obtain:

$$\overline{\mathfrak{B}};\vdash_{\lambda 68} \P\,\overline{\Gamma}.\overline{\Sigma_2} : \Delta \tag{7.20}$$

(If $\Gamma \equiv \varnothing$ then we apply the ¶-formation rule zero times, and the type of $\P\,\overline{\Gamma}.\overline{\Sigma_2}$ is $*$ instead of Δ). Now we can apply the (Start: dc) rule on (7.19), (7.16) and (7.20) to obtain:

$$\overline{\mathfrak{B}}; k{:=}(\S\,\overline{\Gamma}.\overline{\Sigma_1}){:}(\P\,\overline{\Gamma}.\overline{\Sigma_2});\vdash_{\lambda 68} k : \P\,\overline{\Gamma}.\overline{\Sigma_2},$$

so $\overline{\mathfrak{B}, (\Gamma; k; \Sigma_1; \Sigma_2);} \equiv \overline{\mathfrak{B}}, k{:=}(\S\,\overline{\Gamma}.\overline{\Sigma_1}){:}(\P\,\overline{\Gamma}.\overline{\Sigma_2});$ is legal. ⊠

It is possible to prove a conservativity theorem (in the style: If $\overline{\mathfrak{B}};\overline{\Gamma} \vdash_{\lambda 68} \overline{\Sigma} : \overline{\Omega}$, then $\mathfrak{B};\Gamma \vdash_{\text{AUT-68}} \Sigma : \Omega$), but we want to prove that all the typable terms of λ68 have some interpretation in AUT-68, and not only the terms that have an equivalent in AUT-68. We have to distinguish six different cases, and the interpretation of these six cases is given after the proof of the next theorem.

Theorem 7.63 *Assume $\Delta;\Gamma \vdash_{\lambda 68} M : N$. Then there is an* AUTOMATH *book* $\mathfrak{B}$ *and an* AUTOMATH *context* Γ' *such that* $\mathfrak{B};\Gamma' \vdash_{AUT-68}$ OK, *and* $\overline{\mathfrak{B}}, \overline{\Gamma'} \equiv \Delta;\Gamma$. *Moreover,*

1. *If $N \equiv \Box$ then $M \equiv *$;*

2. *If $\Delta;\Gamma \vdash_{\lambda 68} N : \Box$ then $N \equiv *$ and there is $\Omega \in \mathcal{E}$ such that $\overline{\Omega} \equiv M$ and $\mathfrak{B};\Gamma' \vdash_{AUT-68} \Omega : \mathtt{type}$;*

3. *If $N \equiv \Delta$ then there is $\Gamma'' \equiv x_1{:}\Sigma_1, \ldots, x_n{:}\Sigma_n$ and $\Omega \in \mathcal{E}^+$ such that*
   - *Γ', Γ'' is correct with respect to $\mathfrak{B}$;*
   - *$M \equiv \P\,\overline{\Gamma''}.\overline{\Omega}$;*
   - *$\Omega \equiv \mathtt{type}$ or $\mathfrak{B};\Gamma' \vdash_{AUT-68} \Omega : \mathtt{type}$;*

4. *If $\Delta;\Gamma \vdash_{\lambda 68} N : \Delta$ then there are $b \in \mathcal{C}$ and $\Sigma_1, \ldots, \Sigma_n \in \mathcal{E}$ such that $M \equiv b\overline{\Sigma_1} \cdots \overline{\Sigma_n}$. Moreover, $\mathfrak{B}$ contains a line*

   $$(x_1{:}\Omega_1, \ldots, x_m{:}\Omega_m; b; \Xi_1; \Xi_2)$$

   such that

   - *$m > n$;*
   - *$\mathfrak{B};\Gamma' \vdash_{AUT-68} \Sigma_i{:}\Omega_i[x_1, \ldots, x_{i-1}{:=}\Sigma_1, \ldots, \Sigma_{i-1}]$ $(1 \le i \le n)$;*
   - *$N \equiv \left(\P_{i=n+1}^{m}\, x_i{:}\overline{\Omega_i}.\overline{\Xi_2}\right)[x_1, \ldots, x_n{:=}\overline{\Sigma_1}, \ldots, \overline{\Sigma_n}]$;*

5. *If* $N \equiv *$ *then there is* $\Omega \in \mathcal{E}$ *such that* $\overline{\Omega} \equiv M$ *and* $\mathfrak{B}; \Gamma' \vdash_{AUT-68} \Omega : \mathtt{type}$;
6. *If* $\Delta; \Gamma \vdash_{\lambda 68} N : *$ *then there are* $\Sigma, \Omega \in \mathcal{E}$ *such that* $\overline{\Sigma} \equiv M$ *and* $\overline{\Omega} \equiv N$, *and* $\mathfrak{B}; \Gamma' \vdash_{AUT-68} \Sigma : \Omega$, *and* $\mathfrak{B}; \Gamma' \vdash_{AUT-68} \Omega : \mathtt{type}$.

PROOF: We use induction on the derivation of $\Delta; \Gamma \vdash_{\lambda 68} M : N$. We only treat a few cases:

Weakening: definitions The last step in the derivation has been

$$\frac{\Delta; \vdash_{\lambda 68} M : N \qquad \Delta; \Gamma \vdash_{\lambda 68} T : B : s_1 \qquad \Delta; \vdash_{\lambda 68} \P\Gamma.B : s_2}{\Delta, b{:=}(\S\,\Gamma.T){:}(\P\,\Gamma.B); \vdash_{\lambda 68} M : N}$$

where $s_1 \equiv *$ or $s_1 \equiv \Box$. Use the induction hypothesis and determine $\mathfrak{B}$, Γ', Σ_1, Σ_2, Ω_1, and Ω_2 such that $\overline{\mathfrak{B}} \equiv \Delta$, $\overline{\Gamma'} \equiv \Gamma$, $\overline{\Sigma_1} \equiv T$, $\overline{\Sigma_2} \equiv B$, $\overline{\Omega_1} \equiv M$ and $\overline{\Omega_2} \equiv N$. By induction, $\mathfrak{B}; \Gamma' \vdash_{\mathrm{AUT-68}} \Sigma_2 : \mathtt{type}$ (if $s_1 \equiv *$) or $\Sigma_2 \equiv *$ (if $s_2 \equiv \Box$). Also, $\mathfrak{B}; \Gamma' \vdash_{\mathrm{AUT-68}} \Sigma_1 : \Sigma_2$. This makes it possible to extend $\mathfrak{B}$ with a new line, thus obtaining a legal book $\mathfrak{B}, (\Gamma'; b; \Sigma_1; \Sigma_2)$. Using Weakening for AUT-68 (Lemma 7.19) and the induction hypothesis on $\Delta; \vdash_{\lambda 68} M : N$, it is not hard to verify the cases 1–6 for $\Delta, b{:=}(\S\,\Gamma.T){:}(\P\,\Gamma.B); \vdash_{\lambda 68} M : N$;

Application 2 The last step in the derivation has been

$$\frac{\Delta; \Gamma \vdash_{\lambda 68} M_1 : (\P x{:}A.B) \qquad \Delta; \Gamma \vdash_{\lambda 68} M_2 : A}{\Delta; \Gamma \vdash_{\lambda 68} M_1 M_2 : B[x{:=}M_2]}.$$

Determine $\mathfrak{B}$, Γ' such that $\overline{\mathfrak{B}} \equiv \Delta$ and $\overline{\Gamma'} \equiv \Gamma$. By Correctness of Types 7.33 and the Generation Lemma 7.31, we have $\Delta; \Gamma \vdash_{\lambda 68} (\P x{:}A.B) : \Delta$, so by the induction hypothesis (case 4), there are $b, \Sigma_1, \ldots, \Sigma_n$ such that $M_1 \equiv b\overline{\Sigma_1} \cdots \overline{\Sigma_n}$, and there is a line

$$(x_1{:}\Omega_1, \ldots, x_m{:}\Omega_m; b; \Xi_1; \Xi_2)$$

in $\mathfrak{B}$ such that $m > n$, $\mathfrak{B}; \Gamma' \vdash_{\mathrm{AUT-68}} \Sigma_i{:}\Omega_i[x_j{:=}\Sigma_j]_{j=1}^{i-1}$ for $i = 1, \ldots, n$, and

$$\P x{:}A.B \equiv \left(\P_{i=n+1}^{m} x_i{:}\overline{\Omega_i}.\overline{\Xi_2}\right)[x_j{:=}\overline{\Sigma_j}]_{j=1}^{n}.$$

Observe that $A \equiv \overline{\Omega_{n+1}[x_j{:=}\Sigma_j]_{j=1}^{n}}$. As $\mathfrak{B}; \Gamma' \vdash_{\mathrm{AUT-68}} \Omega_{n+1} : \mathtt{type}$ or $\Omega_{n+1} \equiv \mathtt{type}$, we have $\Delta; \Gamma \vdash_{\lambda 68} \overline{\Omega_{n+1}} : s$ for an $s \in \{*, \Box\}$, and by the Substitution and Transitivity Lemmas we have
$\Delta; \Gamma \vdash_{\lambda 68} \overline{\Omega_{n+1}}[x_j{:=}\overline{\Sigma_j}]_{j=1}^{n} : s$, hence $\Delta; \Gamma \vdash_{\lambda 68} A : s$.
With the induction hypothesis we determine $\Sigma \in \mathcal{E}$ such that

$$\mathfrak{B}; \Gamma' \vdash_{\mathrm{AUT-68}} \Sigma : \Omega_{n+1}[x_j{:=}\Sigma_j]_{j=1}^{n},$$

and $M_2 \equiv \overline{\Sigma}$. We now treat the most important ones of the cases 1–6:

4. The only thing that does not directly follow from the results above is $m > n+1$. Assume, for the sake of the argument, $m = n+1$. Then $B[x{:=}M_2] \equiv \overline{\Xi_2}[x_j{:=}\overline{\Sigma_j}]_{j=1}^{n+1}$. As $\Delta;\Gamma \vdash_{\lambda 68} B[x{:=}M_2] : \Delta$, $\overline{\Xi_2}[x_j{:=}\overline{\Sigma_j}]_{j=1}^{n+1}$ is of the form $\P x{:}P.Q$, which is impossible;
6. Notice: $B[x{:=}M_2] \equiv \left(\P_{j=n+2}^{m} x_i{:}\overline{\Omega_i}.\overline{\Xi_2}\right)[x_j{:=}\overline{\Sigma_j}]_{j=1}^{n+1}$. We have $\Delta;\Gamma \vdash_{\lambda 68} B[x{:=}M_2] : *$. Therefore $B[x{:=}M_2]$ cannot be of the form $\P y{:}P.Q$, and therefore $m = n+1$.
   Therefore, $\mathfrak{B};\Gamma' \vdash_{\mathrm{AUT-68}} b(\Sigma_1,\ldots,\Sigma_{n+1}) : \Xi_2[x_i{:=}\Sigma_i]_{i=1}^{n+1}$. ⊠

Remark 7.64 We give some explanation to the different cases mentioned in the formulation of Theorem 7.63.

- The cases $N \equiv \Box$ and $\Delta;\Gamma \vdash N : \Box$ imply that there are no other terms in $\lambda 68$ than $*$ itself at the same level as $*$. This corresponds to the fact that `type` is the only "top-expression" in AUT-68;

- The cases $N \equiv *$ and $\Delta;\Gamma \vdash N : *$ give a precise correspondence between expressions of AUT-68 and terms of $\lambda 68$: If $M : N$ in $\lambda 68$ then there are expressions Σ, Ω in AUT-68 such that $\Sigma : \Omega$ in AUT-68 and $\overline{\Sigma} \equiv M$ and $\overline{\Omega} \equiv N$;

- The cases $N \equiv \Delta$ and $\Delta;\Gamma \vdash N : \Delta$ cover terms that do not have an equivalent in AUT-68 but are necessary in $\lambda 68$ to form terms that have equivalents in AUT-68. More specific, this concerns terms of the form $\P_{i=1}^{n} x_i{:}A_i.B$ (which are needed to introduce constants) and terms of the form $bM_1 \cdots M_n$, where b is a constant of type $\P_{i=1}^{m} x_i{:}A_i.B$ for certain $m > n$ (which are needed to construct $\lambda 68$-equivalents of expressions of the form $b(\Sigma_1,\ldots,\Sigma_m)$).

We conclude that $\lambda 68$ and AUT-68 coincide as much as possible, and that the terms in $\lambda 68$ that do not have an equivalent in AUT-68 can be traced easily (these are the terms of type Δ and the terms of a type $N : \Delta$, and the sorts $\Box$ and Δ, which are needed to give a type to $*$ and to the $\P$-types).

Notice that the alternative definition of δ-reduction in $\lambda 68$, discussed at the end of Subsection 7a3, would introduce more terms in $\lambda 68$ without an equivalent in AUT-68, namely terms of the form $\lambda_{i=1}^{n} x_i{:}A_i.B$.

7d More Suitable Pure Type Systems for AUTOMATH

Recall that we related the system AUT-68 to a PTS $\lambda 68$ ignoring the AUTOMATH features: parameters, and identifying λs and Πs or at least, providing both Π-reduction and Π-application. In particular, in Definition 7.21, we gave $\overline{b(\Sigma_1,\ldots,\Sigma_n)} \stackrel{\text{def}}{=} b\overline{\Sigma_1}\cdots\overline{\Sigma_n}$ as λ68 does not have direct parameters. Also, although we had λs and Πs in λ68, unlike AUTOMATH which used expressions of the form $[x{:}\Sigma]\Omega$ for both abstractions, we did not allow:

- Π-reduction where the reduction rule $\to_\Pi$ works like β-reduction:

  Π-reduction $\quad (\Pi x{:}A.B)N \to_\Pi B[x{:=}N]$

- Π-application where the λ68 rule (App_1) is changed into

  Π-application $\quad \dfrac{\Delta;\Gamma \vdash M : \Pi x{:}A.B \qquad \Delta;\Gamma \vdash N : A}{\Delta;\Gamma \vdash MN : (\Pi x{:}A.B)N}$

There are good reasons to use parameters (cf. [84, 87] and Part III of this book), Π-reduction and Π-application (cf. [81, 90]). In Section 7d1 we discuss how we might remedy the above shorcomings to create more faithful interpretations of AUT-68 as PTSs. More details can be found in Part III.

The system AUT-68 is one of several AUTOMATH-systems that have been proposed. Another frequently used system is AUT-QE. In Section 7d2 we compare AUT-68 to AUT-QE and describe how we can easily adapt λ68 to a system λQE. In Section 7d3 we reflect on the system ΔΛ which is claimed by de Bruijn to embrace all the essential aspects of AUTOMATH apart from type inclusion.

7d1 λ68 with parameters, Π-reduction and Π-application

PTSs don't usually follow AUTOMATH in identifying λs and Πs. PTSs don't even follow AUTOMATH in allowing Π-reduction and Π-application. We have the following results in the area:

- [80] showed that as long as the usual application rule of PTSs is used, a PTS system remains unchanged whether Π-reduction is included or not. As a result, if the usual application rule of PTSs is used, a PTS system remains unchanged whether λs and Πs are unified or not. [80] concluded that a PTS system where λs and Πs are unified and where the application is changed to Π-application faces the same problem (and inherits the same solution) as that of the PTSs where λs and Πs are not unified but where Π-application and Π-reduction are used.

- [90] showed that PTSs with Π-reduction and Π-application lose Subject Reduction. For instance, one can derive $\alpha{:}*, x{:}\alpha \vdash (\lambda y{:}\alpha.y)x : (\Pi y{:}\alpha.\alpha)x$, but it is not possible to derive $\alpha{:}*, x{:}\alpha \vdash x : (\Pi y{:}\alpha.\alpha)x$.

- [81] showed that PTSs with Π-reduction and Π-application have all the desirable properties if a definition system is used. Let us call the PTS with Π-reduction and Π-application and definitions as in [81], λβΠd.

Though our system λ68 does not have Π-reduction and Π-application, it is easy to extend it to a system λΠ68 by adding these rules:

- Changing rule (App_1) into $\dfrac{\Delta;\Gamma \vdash M : \Pi x{:}A.B \qquad \Delta;\Gamma \vdash N : A}{\Delta;\Gamma \vdash MN : (\Pi x{:}A.B)N}$

  (Rule (App_2) remains unchanged — see the discussion in Section 7b1);

- Adding the new reduction rule $\to_\Pi$ by $(\Pi x{:}A.B)N \to_\Pi B[x{:=}N]$.

The system λΠ68 is actually much closer to AUT-68 than λ68.

In λΠ68 we do not have Subject Reduction, either: we can derive

$$;\alpha{:}*, x{:}\alpha \vdash (\lambda y{:}\alpha.y)x : (\Pi y{:}\alpha.\alpha)x$$

Nevertheless, we can not derive in λΠ68

$$;\alpha{:}*, x{:}\alpha \vdash x : (\Pi y{:}\alpha.\alpha)x$$

The "restoration" of Subject Reduction in λβΠd is only because of the special way in which definitions are introduced and removed from the context. In λΠ68, once definitions have been introduced, they cannot be removed from the left part of the context any more. So, we need to investigate whether the method of [81] can be extended to λΠ68 in order to restore Subject Reduction in λΠ68.

As for parameters, [84] gives a formulation of the cube with parameters, [87] formulates PTSs with parameters, Π-reduction, Π-application, definitions à la [81] and explicit substitutions, [98, 13] formulate PTSs with parameters and definitions as in AUTOMATH and [80] gives a formulation of PTSs where λs and Πs are unified, and with parameters, Π-application, explicit substitutions and definitions à la [81]. All these formulations satisfy the good properties of PTSs.

We ignore Π-reductions in this book. Instead, in Chapters 9 and 10, we give the extensions with parameters and show how the Barendregt cube of Figure 4.2 can be refined into the eight smaller cubes of Figure 9.1, where the AUTOMATH systems AUT-68 and AUT-QE, as well as the Edinburgh LF and Milner's ML find a more accurate placing in this refined cube as on the picture on the right in Figure 10.2.

7d2 AUT-QE

The system AUT-QE has many similarities with AUT-68. There are a few extensions:

1. We can also build abstraction expressions of the form $[x{:}\Sigma]\mathtt{type}$ (thus extending Definition 7.1);

2. Inhabitants of types of the form $[x{:}\Sigma]\mathtt{type}$ are introduced by extending the abstraction rules 1 and 2 of Definition 7.11 with the following rule for AUT-QE:
$$\frac{\mathfrak{B};\Gamma\vdash\Sigma_1{:}\mathtt{type}\qquad\mathfrak{B};\Gamma,x{:}\Sigma_1\vdash\Sigma_2{:}\mathtt{type}}{\mathfrak{B};\Gamma\vdash[x{:}\Sigma_1]\Sigma_2:[x{:}\Sigma_1]\mathtt{type}}.$$
Notice that the expression $[x{:}\Sigma_1]\mathtt{type}$ is not typable, just as $\mathtt{type}$ is not typable. In a translation to a PTS, these expressions should get type $\square$;

3. There is a new reduction relation on expressions, which is specific for AUT-QE and therefore will be called $\rightarrow_{\mathrm{QE}}$ in the sequel. The relation is described by the rule
$$[x_1{:}\Sigma_1]\cdots[x_n{:}\Sigma_n][y{:}\Omega]\mathtt{type}\rightarrow_{\mathrm{QE}}[x_1{:}\Sigma_1]\cdots[x_n{:}\Sigma_n]\mathtt{type}$$
(for $n\geq 0$).

The first two rules are rather straightforward. They correspond to an extension of $\lambda_{\rightarrow}$ to λP in Pure Type Systems. It is also easy to extend λ68 with similar rules: We just add the Π-formation rule $(*,\square,\square)$:
$$\frac{\Delta;\Gamma\vdash A:*\qquad\Delta;\Gamma,x{:}A\vdash B:\square}{\Delta;\Gamma\vdash(\Pi x{:}A.B):\square}.$$

In AUT-68 PAT is implemented in de Bruijn-style (see Section 4a4 and Example 7.9). An implementation of predicate logic in Howard-style is not possible in AUT-68, but due to the extension with types of the form $[x{:}\Sigma]\mathtt{type}$, such an implementation becomes possible in AUT-QE. See [41].

The third rule deserves some extra attention, as it is very unusual. It is needed in AUT-QE because that system does not distinguish between λs and Πs. In AUT-68 this did not matter, as from the context it could always be derived whether an expression $[x{:}\Sigma]\Omega$ should be interpreted as $\lambda x{:}\Sigma.\Omega$ or as $\Pi x{:}\Sigma.\Omega$. The latter should have type $\mathtt{type}$, and the first should not have type $\mathtt{type}$.

In AUT-QE the situation is more complicated. A expression $[x{:}\Sigma]\Omega$ may have more than one type:

Example 7.65 Let $\mathfrak{B}$ consist of two lines:
$$(\varnothing,\alpha,\text{—},\mathtt{type}),$$
$$(\alpha{:}\mathtt{type},x,\text{—},\alpha).$$

Notice that, using rule (abstr.1) of Definition 7.11, we can derive that
$$\mathfrak{B};\alpha{:}\mathtt{type}\vdash_{\mathrm{QE}}[x{:}\alpha]\alpha:\mathtt{type}. \tag{7.21}$$

But using the new abstraction rule of AUT-QE we can also derive
$$\mathfrak{B};\alpha{:}\mathtt{type}\vdash_{\mathrm{QE}}[x{:}\alpha]\alpha:[x{:}\alpha]\mathtt{type}. \tag{7.22}$$

More generally, we can prove that the two statements below are equivalent (that is: if either of them is derivable then they are both derivable) in AUT-QE:

$$\mathfrak{B};\Gamma \;\vdash_{\mathrm{QE}}\; [x_1{:}\Sigma_1]\cdots[x_n{:}\Sigma_n]\Omega : [x_1{:}\Sigma_1]\cdots[x_n{:}\Sigma_n]\mathtt{type}; \quad (7.23)$$
$$\mathfrak{B};\Gamma \;\vdash_{\mathrm{QE}}\; [x_1{:}\Sigma_1]\cdots[x_n{:}\Sigma_n]\Omega : [x_1{:}\Sigma_1]\cdots[x_m{:}\Sigma_m]\mathtt{type} \quad (7.24)$$

(for $m < n$). In (7.23), the expression $[x_1{:}\Sigma_1]\cdots[x_n{:}\Sigma_n]\Omega$ should be read as $\lambda_{i=1}^n\, x_i{:}\Sigma_i.\Omega$; in (7.24) it should be read as $\lambda_{i=1}^m\, x_i{:}\Sigma_i.\prod_{j=m+1}^n x_j{:}\Sigma_j.\Omega$.

But this equivalence holds only for expressions of the form

$$[x_1{:}\Sigma_1]\cdots[x_n{:}\Sigma_n]\Omega$$

and not for general expressions Σ (take, for instance, Σ a variable). In order that the equivalence holds for general expressions Σ, de Bruijn introduced a rule for type inclusion:

$$\frac{\mathfrak{B};\Gamma \vdash_{\mathrm{QE}} \Sigma : [x_1{:}\Sigma_1]\cdots[x_n{:}\Sigma_n]\mathtt{type}}{\mathfrak{B};\Gamma \vdash_{\mathrm{QE}} \Sigma : [x_1{:}\Sigma_1]\cdots[x_{n-1}{:}\Sigma_{n-1}]\mathtt{type}}.$$

Lists of abstractions $[x_1{:}\Sigma_1]\cdots[x_n{:}\Sigma_n]$ were also called *telescopes* by de Bruijn. In the rule for type inclusion, we see that one part of the telescope "collapses".

7d3 $\Delta\Lambda$

As we saw above, de Bruijn departed from the classical notation of the λ-calculus and wrote the argument before the function and used $[x : A]$ instead of $\lambda x : A$ or $\Pi x : A$. For example, de Bruijn wrote $(\lambda x : *.\lambda y : x.y)z$ as $\langle z\rangle[x : *][y : x]y$.

De Bruijn called items of the form $\langle B\rangle$ and $[x : C]$, A- (for application) resp. T- (for typing) *wagons*. De Bruijn called $\langle B\rangle[x : C]$, an AT-pair.

In de Bruijn's notation, the β-rule $(\lambda x : C.A)B \to_\beta A[x := B]$ becomes:

$$\langle B\rangle[x : C]A \to_\beta [x := B]A$$

Note that the A-wagon $\langle B\rangle$ and the T-wagon $[x : C]$ occur NEXT to each other.

Here is an example which compares β-reduction in both the classical and the de Bruijn notation. Wagons that have the same symbol on top, are matched (we ignore types for the sake of simplicity):

Classical Notation	De Bruijn's Notation
$\underline{(\overset{\circ}{\lambda x}\,.(\overset{+}{\lambda y}\,.\,\overset{-}{\lambda z}\,.zD)\,\overset{+}{C})\,\overset{\circ}{B})}\,\overset{-}{A}$	$\overset{-}{\langle A\rangle}\ \underline{\overset{\circ}{\langle B\rangle}\overset{\circ}{[x]}}\ \overset{+}{\langle C\rangle}\overset{+}{[y]}\overset{-}{[z]}\ \langle D\rangle z$
$\downarrow_\beta$	$\downarrow_\beta$
$(\underline{(\overset{+}{\lambda y}\,.\,\overset{-}{\lambda z}\,.zD)\,\overset{+}{C})}\,\overset{-}{A}$	$\overset{-}{\langle A\rangle}\ \underline{\overset{+}{\langle C\rangle}\overset{+}{[y]}}\ \overset{-}{[z]}\ \langle D\rangle z$
$\downarrow_\beta$	$\downarrow_\beta$
$\underline{(\overset{-}{\lambda z}\,.zD)\,\overset{-}{A}}$	$\underline{\overset{-}{\langle A\rangle}\overset{-}{[z]}}\langle D\rangle z$
$\downarrow_\beta$	$\downarrow_\beta$
AD	$\langle D\rangle A$

The *bracketing structure* of $((\overset{\circ}{\lambda x}\,.(\overset{+}{\lambda y}\,.\,\overset{-}{\lambda z}\,.zD)\,\overset{+}{C})\,\overset{\circ}{B})\,\overset{-}{A})$, in classical notation is $\overset{\circ}{[_1}\ \overset{+}{[_2}\ \overset{-}{[_3}\ \overset{+}{]_2}\ \overset{\circ}{]_1}\ \overset{-}{]_3}$, where $[_i$ and $]_i$ match. Whereas $\overset{-}{\langle A\rangle}\overset{\circ}{\langle B\rangle}\overset{\circ}{[x]}\overset{+}{\langle C\rangle}\overset{+}{[y]}\overset{-}{[z]}\ \langle D\rangle z$ has the simpler bracketing structure $\overset{-}{[}\,\overset{\circ}{[}\,\overset{\circ}{]}\,\overset{+}{[}\,\overset{+}{]}\,\overset{-}{]}$ or even better: $[\,[\,][\,]\,]$ in de Bruijn's notation. An A-wagon $\langle B\rangle$ and a T-wagon $[x : C]$ are *partners* when they match. Non-partnered wagons are *bachelors*. A sequence of wagons is called a *segment*. A segment is *well balanced* when it contains only partnered wagons.

Moreover, de Bruijn defined local β-reduction, which keeps the AT-pair and does β-reduction at one instance (instead of all the instances). For example (we take a simpler example than above and again ignore types for simplicity):

$\langle y\rangle[x]\langle x\rangle x$ β-reduces locally to $\langle y\rangle[x]\langle x\rangle y$ and to $\langle y\rangle[x]\langle y\rangle x$. Doing a further local β-reduction gives $\langle y\rangle[x]\langle y\rangle y$. Now that the $[x]$ does not bind any variable any more, we can remove the AT-pair $\langle y\rangle[x]$ obtaining $\langle y\rangle y$.

Furthermore, de Bruijn generalised the AT-pair to the AT-couple where for example, in $\langle A\rangle\langle B\rangle[x]\langle C\rangle[y][z]\langle D\rangle z$, we have the AT-pairs: $\langle B\rangle[x]$ and $\langle C\rangle[y]$ and the AT-couple $\langle A\rangle[z]$. This definition of AT-couples leads to a natural generalisation of β-reduction as follows:

$$\langle B\rangle\overline{s}[x : C]A \rightsquigarrow_\beta \overline{s}[x := B]A \qquad \text{where } \overline{s} \text{ is a well balanced segment.}$$

So for example, $\langle A\rangle\langle B\rangle[x]\langle C\rangle[y][z]\langle D\rangle z \rightsquigarrow_\beta \langle B\rangle[x]\langle C\rangle[y][z := A]\langle D\rangle z$.

The λ-calculus à la de Bruijn has many advantages over the classical λ-calculus. Some of these advantages are summarised in [91].

In AUT-SL (cf. B.2 of [112]), de Bruijn described how a complete AUTOMATH book can be written as a single lambda calculus formula. The disadvantage of AUT-SL was that in order to put the book into the lambda calculus framework, it was necessary to first eliminate all definitional lines of the book. De Bruijn did not like this idea as without definitions, formulae can exponentially grow.

For this reason, de Bruijn developed the $\Delta\Lambda$ calculus (cf. B.7 of [112]), with which he attempts to embrace all essential aspects of AUTOMATH apart from type inclusion. $\Delta\Lambda$ is the lambda calculus written in his notation (as above)[9] but where β-reduction[10] is presented as the result of local β-reductions and AT-removals. The reason for this is that the delta reductions of AUTOMATH can be considered as local β-reductions, and not as ordinary β-reductions.

We have fully investigated PTSs and the type free lambda calculus in de Bruijn's notation [89, 91, 14]. We have also shown that $\rightsquigarrow_\beta$ satisfies nice properties in the type free lambda calculus [79] and that it loses subject reduction in PTSs but that subject reduction can be regained if definitions are added in the contexts [14]. We have not yet studied PTSs with local β-reductions and AT-removal, although we have studied the type free lambda calculus with local β-reduction, AT-removal and explicit substitution [88]. We leave the study of PTSs with de Bruijn's local β-reduction and AT-removal for future work.

Conclusions

In this chapter we described the most basic AUTOMATH-system, AUT-68, in a PTS style. Though such descriptions have been given in [3] we feel that our description is more accurate. Moreover, our description pays attention to the definition system, which is a crucial item in AUTOMATH. The descriptions mentioned above do not.

We gave a PTS called $\lambda 68$ which is closely related to AUT-68. Although $\lambda 68$ does not include Π-conversion (while AUTOMATH does), one can adapt it to include Π-conversion following the lines of [81].

The adaptation of $\lambda 68$ to a system λQE, representing the AUTOMATH-system AUT-QE is not hard, either: it requires adaptation of the Π-formation rule to include not only the rule $(*, *, *)$ but also $(*, \Box, \Box)$ and the introduction of the additional reduction rule of type inclusion. We leave this as a future work. We also leave as a future work the extension of PTSs with local β-reduction and AT-removal à la de Bruijn and hence the connection between de Bruijn's $\Delta\Lambda$ and PTSs with definitions.

Of course, the properties of $\lambda 68$ presented in Section 7c have to be reviewed for these new systems.

There is no doubt that AUTOMATH has had an amazing influence in theorem proving, type theory and logical frameworks. AUTOMATH however, was developed independently from other developments in type theory and uses a λ-calculus and type-theoretical style that is unique to AUTOMATH. Writing AUTOMATH in the modern style of type theory will enable useful comparisons between type sys-

[9] In $\Delta\Lambda$, de Bruijn favours trees over character strings and does not make use of AT-couples.
[10] Recall this is now both β- and Π-reduction as he unifies λs and Πs.

tems to take place. There are still many lessons to learn from AUTOMATH and writing it in modern style is a useful step in this direction.

Part III

Extensions of Pure Type Systems

Chapter 8

Pure Type Systems with definitions

In many type theories and lambda calculi, there is no formal possibility to use abbreviations, i.e., to introduce names for large expressions which can be used several times in a program or a proof. This possibility is essential for practical use, and indeed implementations of Pure Type Systems such as Coq [44] and Nuprl [35] do provide this possibility. Moreover, most implementations of programming languages (Haskell, ML, CAML, etc.) use names for large expressions via a well-known programming language concept: *let expressions.*

Example 8.1 $\mathtt{Let}\ id : A \to A\ \mathtt{be}\ (\lambda_{x:A}.x)\ \mathtt{in}\ (\lambda_{y:A\to A}.id)id$ abbreviates the complex expression $(\lambda_{x:A}.x)$ as id in a more complex expression in which id occurs two times.

The intended meaning of "$\mathtt{let}\ x : A\ \mathtt{be}\ a\ \mathtt{in}\ b$" is that a can be substituted for x in the expression b. In a sense, the expression $\mathtt{let}\ x : A\ \mathtt{be}\ a\ \mathtt{in}\ b$ is similar to $(\lambda_{x:A}.b)a$ which β-reduces to $b[x := a]$, i.e., b with all free occurrences of x replaced by a. In the `let`-expression, however, it is not intended to necessarily replace all the occurrences of x in b by a. Nor is it intended that such a `let`-expression is a part of our term. Rather, the `let`-expression will live in the environment (or context) in which we evaluate or reason about the expression.

One of the advantages of the expression $\mathtt{let}\ x : A\ \mathtt{be}\ a\ \mathtt{in}\ b$ over the redex $(\lambda_{x:A}.b)a$ is that it is convenient to have the freedom of substituting only some of the occurrences of an expression in a given formula. Another advantage is efficiency; one evaluates a in $\mathtt{let}\ x : A\ \mathtt{be}\ a\ \mathtt{in}\ b$ only once, even in lazy languages.[1]

[1]Note that *smart* lazy languages will use explicit substitution or sharing techniques to evaluate a in $b[x := a]$ only once. Nevertheless, our extension with abbreviations as is considered here, is a

A further advantage is that using x to be a in b can be used to type b efficiently, since the type A of a has to be calculated only once.[2] Furthermore, practical experiences with type systems show that `let`-expressions are absolutely indispensable for any realistic application. Without `let`-expressions, terms soon become forbiddingly complicated. By using `let`-expressions one can avoid such an explosion in complexity. This is, by the way, a very natural thing to do: the apparatus of mathematics, for instance, is unimaginable without a form of `let`-expressions (viz. definitions).

There exist already three formal studies of `let`-expressions in PTSs [14, 81, 137] where those `let`-expressions are called *definitions*. In this chapter we will present both the version of [81] and that of [137]. The account of [81] differs from [14] in that it avoids nested definitions, which were needed for generalised reduction in [14]. Moreover, this account differs from [137] in that it does not introduce new terms (`let`-terms) into the syntax and does not extend β-reduction to deal with those new terms. This simple account of [81] helps illustrate the usefulness of definitions. However, for various systems (including historical ones like AUTOMATH and modern ones like Milner's ML), we need to extend the terms with definitions and to unfold definitions inside the terms. For this reason, we will also present definitions inside the terms as is done in [137].

The presentation of AUT-68 in the PTS-like system $\lambda 68$ makes a good comparison between these systems and the definition system in AUT-68 possible. This will be done in Sections 8a1 and 8b1.

8a Definitions in contexts

We assume the various notational conventions set out in Sections 4b and 4c. In particular, we start from a set of typed lambda terms $\mathcal{T}(\mathbb{V}, \mathbb{C})$ as given in Definition 4.3. We adopt Definition 4.18 as it is, but we change the definition of contexts to include now not only declarations but also definitions. Contexts Γ (see Definition 4.19) are now a list of declarations of the form $x : A$ or of *definitions* of the form $x = B : A$. These latter definitions define x to be B and to have the type A.

---

straightforward extension of the λ-calculus and hence has less machinery than is involved with sharing or explicit substitution. Moreover, it may be that this simple concept of abbreviations as we introduce it can be used to formalise the notion of sharing.

[2]Here, in $b[x := a]$, the type A of a is calculated many times. Of course with the presence of Subject Reduction (SR), we do not need to calculate the type of $b[x := a]$. Instead, we calculate the type of $(\lambda x : A.b)a$ (where the type of a is calculated only once) and we use subject reduction to derive the type of $b[x := a]$. However, many programming languages (PLs) do not have a clear notion of SR. Similarly, although in many PLs, sharing is used in order to calculate the type A of a only once in $b[x := a]$, there can be no escape from showing that this sharing technique of PLs is correct. Our system of definitions can be used to formalize this sharing technique of PLs.

Definition 8.2 (declarations, definitions, contexts, $\subseteq'$)

1. A *declaration* d is as in Definition 4.19.

2. A *definition* d is of the form $x = B : A$ and defines x of type A to be B. We define $\mathtt{var}(d)$, $\mathtt{type}(d)$ and $\mathtt{ab}(d)$ to be x, A, and B respectively. We define $\text{FV}(d) \equiv \text{FV}(A) \cup \text{FV}(B)$.

3. We use $d, d', d_1, \ldots$ to range over declarations and definitions.

4. A *context* Γ is a (possibly empty) concatenation of declarations and definitions $d_1, d_2, \cdots, d_n$ such that if $i \neq j$, then $\mathtt{var}(d_i) \not\equiv \mathtt{var}(d_j)$. We define $\text{DOM}(\Gamma) = \{\mathtt{var}(d) \mid d \in \Gamma\}$, $\Gamma\text{-}\mathtt{decl} = \{d \in \Gamma \mid d \text{ is a declaration }\}$ and $\Gamma\text{-}\mathtt{abb} = \{d \in \Gamma \mid d \text{ is a definition }\}$.
   Note that $\text{DOM}(\Gamma) = \{\mathtt{var}(d) \mid d \in \Gamma\text{-}\mathtt{decl} \cup \Gamma\text{-}\mathtt{abb}\}$.
   We use $\Gamma, \Delta, \Gamma', \Gamma_1, \Gamma_2, \ldots$ to range over contexts.

5. Define $\subseteq'$ between contexts as the least reflexive transitive relation satisfying:

   - $\Gamma, \Delta \subseteq' \Gamma, d, \Delta$ for d a declaration or a definition.
   - $\Gamma, x : A, \Delta \subseteq' \Gamma, x = B : A, \Delta$

Substitution on contexts is now defined by extending Definition 4.20 with the clause:

$$(\Gamma', y = C{:}B)[x{:=}A] \equiv \begin{cases} \Gamma', y = C{:}B & \text{if } x \equiv y; \\ \Gamma'[x{:=}A], y = C{:}B[x{:=}A] & \text{if } x \not\equiv y. \end{cases}$$

Definition 8.3 The new typing relation $\vdash$ is obtained by adding three new rules to the typing rules of Definition 4.21: (start-def), (weak-def), and (def) below, and by replacing the (conv) rule by (new-conv) as follows:

$$\text{(start-def)} \qquad \frac{\Gamma \vdash A : s \qquad \Gamma \vdash B : A}{\Gamma, x = B{:}A \vdash x : A} \qquad x \notin \text{DOM}(\Gamma)$$

$$\text{(weak-def)} \qquad \frac{\Gamma \vdash A : B \qquad \Gamma \vdash C : s \qquad \Gamma \vdash D : C}{\Gamma, x = D{:}C \vdash A : B} \qquad x \notin \text{DOM}(\Gamma)$$

(def) $$\frac{\Gamma, x = B{:}A \vdash C : D}{\Gamma \vdash (\pi x : A.C)B : D[x := B]} \qquad \text{for } \pi \in \{\lambda, \Pi\}$$

(new-conv) $$\frac{\Gamma \vdash A : B \qquad \Gamma \vdash B' : s \qquad \Gamma \vdash B \stackrel{\text{def}}{=} B'}{\Gamma \vdash A : B'}$$

In (new-conv), $\Gamma \vdash B \stackrel{\text{def}}{=} B'$ is defined on $\mathcal{T}$ as the smallest equivalence relation closed under:

- If $B =_\beta B'$ then $\Gamma \vdash B \stackrel{\text{def}}{=} B'$
- If $x = D : C \in \Gamma$ and B' arises from B by substituting one particular free occurrence of x in B by D then $\Gamma \vdash B \stackrel{\text{def}}{=} B'$.

In Definition 8.3, (start-def) and (weak-def) are the start and weakening rules that deal with definitions in the context. The (def) rule types λ- and Π-redexes using definitions in the context. Finally, (new-conv) accommodates the new reductions.

Note that the (def) rule makes type derivation more efficient, because it permits avoiding the (Π) rule. This can be seen by the following example (taken from [87]):

Example 8.4 Without (def) we have the following type derivation:

0.		$\vdash$	$* : \Box$	axiom
1.	$\alpha : *$	$\vdash$	$\alpha : *$	0, (start)
2.	$\alpha : *, x : \alpha$	$\vdash$	$\alpha : *$	1, 1, (weak)
3.	$\alpha : *$	$\vdash$	$\Pi x : \alpha.\alpha : *$	1, 2, (Π), (*,*)
4.	$\alpha : *, x : \alpha$	$\vdash$	$x : \alpha$ 1, (start)	
5.	$\alpha : *$	$\vdash$	$\lambda x : \alpha.x : \Pi x : \alpha.\alpha$	4, 3, (λ)
6.	$\alpha : *, y : \alpha$	$\vdash$	$\lambda x : \alpha.x : \Pi x : \alpha.\alpha$	5, 1, (weak)
7.	$\alpha : *, y : \alpha$	$\vdash$	$y : \alpha$	1, (start)
8.	$\alpha : *, y : \alpha$	$\vdash$	$(\lambda x : \alpha.x)y : (\Pi x : \alpha.\alpha)y$	6, 7, (app)
9.	$\alpha : *, y : \alpha$	$\vdash$	$(\Pi x : \alpha.\alpha)y =_{def} \alpha$	
10.	$\alpha : *, y : \alpha$	$\vdash$	$(\lambda x : \alpha.x)y : \alpha$	

Using the (def) rule this type derivation can be shortened as follows:

0.		$\vdash * : \Box$	axiom
1.	$\alpha : *$	$\vdash \alpha : *$	0, (start)
2.	$\alpha : *, y : \alpha$	$\vdash \alpha : *$	1, 1, (weak)
3.	$\alpha : *, y : \alpha$	$\vdash y : \alpha$	1, (start)
4.	$\alpha : *, y : \alpha, x = y : \alpha$	$\vdash x : \alpha$	2, 3, (start-def)
5.	$\alpha : *, y : \alpha$	$\vdash (\lambda x : \alpha.x)y : \alpha[x := y] \equiv \alpha$	4, (def)

8a1 Comparison with the definitions of AUTOMATH

In [81], Kamareddine, Bloo and Nederpelt extend the usual PTSs with both Π-conversion and definitions. [81] starts with PTSs extended with Π-reduction, but *without* definitions (see [90]). This system (which we will call $\lambda\beta\Pi$ for the moment) does not have the Subject Reduction property. For instance, one can derive

$$\alpha{:}*, x{:}\alpha \vdash (\lambda y{:}\alpha.y)x : (\Pi y{:}\alpha.\alpha)x,$$

but it is not possible to derive

$$\alpha{:}*, x{:}\alpha \vdash x : (\Pi y{:}\alpha.\alpha)x.$$

Adding a definition mechanism results in a system that we will call $\lambda\beta\Pi$d and is the main point of interest in [81]. As a sort of "side effect" of adding this definition mechanism, $\lambda\beta\Pi$d has Subject Reduction.

It will be clear that it is useful to take Π-conversion into consideration when comparing AUTOMATH with $\lambda\beta\Pi$. Though our system λ68 does not have Π-conversion, it is very easy to extend it to a system λΠ68 by:

- Changing rule (App_1) into

$$\frac{\Delta;\Gamma \vdash M : \Pi x{:}A.B \qquad \Delta;\Gamma \vdash N : A}{\Delta;\Gamma \vdash MN : (\Pi x{:}A.B)N}$$

  (Rule (App_2) remains unchanged — see also the discussion in Section 7b1);

- Adding a new reduction rule $\to_\Pi$ by

$$(\Pi x{:}A.B)N \to_\Pi B[x{:=}N].$$

The system λΠ68 is actually much closer to AUT-68 than λ68 as AUT-68 has Π-conversion as well.

In λΠ68 we do not have Subject Reduction, either: It is not hard to derive

$$;\alpha{:}*, x{:}\alpha \vdash (\lambda y{:}\alpha.y)x : (\Pi y{:}\alpha.\alpha)x$$

in λΠ68. Nevertheless, we can not derive

$$;\alpha{:}*, x{:}\alpha \vdash x : (\Pi y{:}\alpha.\alpha)x$$

(In such a derivation, no definitions can occur: Definitions, once they have been introduced, cannot be removed from the left part of the context any more; when we are not allowed to use any definition rules, λΠ68 has not more rules than the system $\lambda\beta\Pi$ of Kamareddine, Bloo and Nederpelt).

The "restoration" of Subject Reduction in $\lambda\beta\Pi$d is only because of the special way in which definitions are introduced and removed from the context. We do not go into details on this; the interested reader can consult [81].

8b Definitions in the terms and the contexts

[137] presents an extension of PTSs with definitions, thus obtaining Pure Type Systems with definitions (DPTSs). Again here, we assume the various notational conventions set out in Sections 4b and 4c. [137] extends the syntax of terms of Definition 4.3 to

$$\mathcal{T} ::= \mathbb{V} \mid \mathbb{C} \mid \mathcal{T}\mathcal{T} \mid \lambda\mathbb{V}{:}\mathcal{T}.\mathcal{T} \mid \Pi\mathbb{V}{:}\mathcal{T}.\mathcal{T} \mid \mathbb{V}{=}\mathcal{T}{:}\mathcal{T} \text{ in } \mathcal{T}$$

The expression $x{=}a{:}A$ in B means, let x by equal to a of type A in term B.

[137] also extends the contexts as we did in Definition 8.2 and then extend the usual PTS-rules of Definition 4.21 by adding the rules (D-start), (D-weak), (D-form) and (D-intro) and (D-conv) as follows:

(D-start) $$\frac{\Gamma \vdash a : A}{\Gamma, x{=}a{:}A \vdash x : A}$$

(D-weak) $$\frac{\Gamma \vdash b : B \qquad \Gamma \vdash a : A}{\Gamma, x{=}a{:}A \vdash b : B}$$

(D-form) $$\frac{\Gamma, x{=}a{:}A \vdash B : s}{\Gamma \vdash (x{=}a{:}A \text{ in } B) : s}$$

(D-intro) $$\frac{\Gamma, x{=}a{:}A \vdash b : B \qquad \Gamma \vdash (x{=}a{:}A \text{ in } B) : s}{\Gamma \vdash (x{=}a{:}A \text{ in } b) : (x{=}a{:}A \text{ in } B)}$$

(D-conv) $$\frac{\Gamma \vdash b : B \qquad \Gamma \vdash B' : s \qquad \Gamma \vdash B =_{\mathrm{D}} B'}{\Gamma \vdash b : B'}$$

where D-reduction is defined by the following rules:

$$\Gamma_1, x{=}a{:}A, \Gamma_2 \vdash x \to_{\mathrm{D}} a;$$

$$\Gamma \vdash (x{=}a{:}A \text{ in } b) \to_{\mathrm{D}} b \qquad (x \notin \mathrm{FV}(b));$$

$$\frac{\Gamma, x{=}a{:}A \vdash b \to_{\mathrm{D}} b'}{\Gamma \vdash (x{=}a{:}A \text{ in } b) \to_{\mathrm{D}} (x{=}a{:}A \text{ in } b')}$$

and the usual compatibility rules. As we see, there is an extra class of terms in DPTSs, namely those of the form $(x{=}a{:}A$ in $b)$.

8b1 Comparison with the definitions of AUTOMATH

When regarding both systems we find that:

- In DPTSs, definitions do not only occur in a context, but may also occur in terms. Moreover, definitions may disappear from contexts when they are introduced in terms (e.g. the D-form and the D-intro rules, and the last of the three D-reduction rules), and definitions may disappear from terms when the definiendum does not occur in that term (the middle D-reduction rule).

This gives definitions a more temporary character: We can use them as long as needed, and when we do not need them any more, we can remove them from the context.

Definitions can also play a more local role: A definition that is needed in only one term can be imported into that term while it is not necessary to carry it around in the (global) context, as well.

This temporary and local behaviour of definitions is not present in AUTOMATH;

- Due to the fact that definitions can also play a local role, D-reduction can also unfold definitions which are not present in the (global) context, but which are given within the term. For example, though there is no definition of *id* in the context $\alpha{:}*$, we have $\alpha{:}* \vdash (id{=}\lambda x{:}\alpha.x \text{ in } id) \twoheadrightarrow_{\mathrm{D}} \lambda x{:}\alpha.x$.

  Again, this is not possible in AUTOMATH;

- The start rule for definitions in DPTSs,

$$\frac{\Gamma \vdash T : B}{\Gamma, x{=}T{:}B \vdash x : B}$$

  does not require $\Gamma \vdash B : s$ for a sort s. In $\lambda 68$ we have the rule (Start: dc):

$$\frac{\Delta;\Gamma \vdash T : B : s_1 \qquad \Delta;\vdash \P\,\Gamma.B : s_2}{\Delta, x{:=}\,\S\,\Gamma.T{:}\P\,\Gamma.B; \vdash x : \P\,\Gamma.B}(s_1 = *, \Box)$$

  where we see that both B and $\P\,\Gamma.B$ need to be of a certain sort (and B must be of sort $*$ or $\Box$);

- The start rules for definitions in DPTSs and in $\lambda 68$ also differ in another respect, namely the type of definiens and definiendum. In DPTSs they have the same type (in the notation of the previous paragraph: B), while in $\lambda 68$ the definiens T has type B and the definiendum x has type $\P\,\Gamma.B$. This topic has already been discussed when we introduced the definition mechanism of $\lambda 68$ in Section 7b3. This difference also holds for the system of Kamareddine, Bloo and Nederpelt of [81].

- D-reduction differs from δ-reduction, also when only global definitions are taken into account. For instance, δ-reduction is *substitutive*. That is, if $\Delta \vdash A \rightarrow_\delta A'$ then $\Delta \vdash A[x{:=}b] \rightarrow_\delta A'[x{:=}b]$ (proof: Induction on the structure of A). D-reduction is not substitutive: take $\Gamma \equiv \alpha{:}*, y{=}\alpha{:}*$. Then $\Gamma \vdash y \rightarrow_{\mathrm{D}} \alpha$, but $\Gamma \not\vdash y[\alpha{:=}M] \rightarrow_{\mathrm{D}} \alpha[\alpha{:=}M]$ for arbitrary M.

  In $\lambda 68$, this example would look as follows. Take $\Delta \equiv y{:=}\alpha{:}\P\alpha{:}*.*$. Then $\Delta \vdash y\alpha \rightarrow_\delta \alpha$ and $\Delta \vdash y\alpha[\alpha{:=}M] \rightarrow_\delta \alpha[\alpha{:=}M]$.

Substitutivity for $\to_D$ is lost, because unfolding a definition by D-reduction may introduce new free variables in the term. In AUTOMATH, all free variables in the definiens must be added as parameters to the definiendum. In $\lambda 68$ this is visible in the Start and Weakening rules for defined constants: The right part Γ of the context $\Delta; \Gamma$ that is used to type the definiens T in these rules, serves as list of parameters in the definiendum. When an AUTOMATH-definition is unfolded, the free variables occurring in the definiens are replaced by the parameters;

- We see that the definition of y in $\lambda 68$ in the example above is more general than in the corresponding DPTS situation. In the DPTS-example, y D-reduces to one, fixed term α. In the $\lambda 68$ version, yM is defined for any (typable) term M. To do something similar in DPTSs, one needs to define y as $\lambda\alpha{:}{*}.\alpha$. In particular, one needs to type the term $\lambda\alpha{:}{*}.\alpha$, which involves the use of Π-formation rule $(\Box, \Box)$, so the use of a higher type system. One could say that AUTOMATH and $\lambda 68$ use an implicit λ-abstraction where DPTSs need an explicit λ-abstraction. On this point, AUTOMATH and $\lambda 68$ are more flexible than DPTSs. This is due to the parameter mechanism of AUTOMATH. It is possible to extend DPTSs with a parameter mechanism as well. This will be the main topic of Chapter 10.

We summarise the differences between DPTSs and AUTOMATH:

- DPTSs have global and local definitions. AUTOMATH has only global definitions;
- In DPTSs, the type B of a definition $x{=}T{:}B$ does not have to be typable itself. In AUTOMATH, B has to be typable;
- The D-reduction of DPTSs is not substitutive; δ-reduction of AUTOMATH is substitutive;
- AUTOMATH has a parameter mechanism, DPTSs do not have such a mechanism.

Conclusions

When comparing $\lambda 68$ to other type systems with definitions, we find an important difference. In $\lambda 68$, the correspondence between types of definiendum and definiens differs from the similar correspondence in the systems in [137] and [14].

The reason why $\lambda 68$ differs from other theories in this respect has been discussed in Section 7b3: The definition system in AUTOMATH allows *parameters* to occur in the definiens, and there is no parameter mechanism in PTSs. In Chapters 9 and 10, we extend PTSs with a parameter mechanism. This extension has AUT-68 as a subsystem. Moreover, we show that a parameter mechanism has also other advantages.

Chapter 9

The Barendregt cube with parameters

There are many motivations for parameters. Here we repeat two:

First-order logic First-order predicate logic has no λ-calculus. It only has parametric constructs. In [99] it is shown that parametric constructs make it possible to give a description of first-order predicate logic in type theory that is much more accurate than the traditional approach in typed λ-calculus. Moreover, implementations of first-order logic in a PTS in PAT-style usually use a PTS that is related to λP. λP has sorts $*, \square$, axiom $*{:}\square$, and two Π-formation rules, $(*, *, *)$ and $(*, \square, \square)$. In this PTS it is possible to construct types (that is: terms of type $*$ or $\square$) that are not in β-normal form. Hence, a derivation in λP can have non-trivial applications of the conversion rule

$$\frac{\Gamma \vdash A : B_1 \qquad \Gamma \vdash B_2 : s \qquad B_1 =_\beta B_2}{\Gamma \vdash A : B_2}.$$

This can be problematic in implementations. In theory, it is always decidable whether two terms B_1, B_2 are β-equal or not (simply: check whether their β-normal forms are syntactically equal or not). In practice, such a calculation may take quite some time and memory. Therefore, it would be better to use a PTS in which applications of the conversion rule are only possible when $B_1 \equiv B_2$. This is the case if all types in such a PTS are in β-normal form. As all types in $\lambda_\to$ (that is: λP without Π-formation rule $(*, \square, \square)$) are in β-normal form, it would be a good candidate for an implementation of first-order predicate logic. Unfortunately, first-order predicate logic cannot be described in PAT-style in $\lambda_\to$. The introduction of the relation symbols in a first order language involves the Π-formation rule $(*, \square, \square)$.

But in a first-order language, a relation symbol R always has a fixed arity $\mathfrak{a}(R)$. This means that R itself is not a proposition. It can only be used to *construct* a proposition: if $t_1, \ldots, t_{\mathfrak{a}(R)}$ are terms, then $R(t_1, \ldots, t_{\mathfrak{a}(R)})$ is a proposition. With the use of parameters in PTSs, it is possible to introduce the relation symbols without Π-formation rule $(*, \Box, \Box)$. This results in a system in which the conversion rule is superfluous, and therefore easier to handle in implementations. See Section 10f;

A different form of abstraction and application In λ-calculus without parameters there is one mechanism for abstraction and application. For abstraction, we use λ-abstraction, and application is implemented via function application. Abstraction and application form the basis for a type system. A parameter mechanism is a different abstraction-and-application mechanism. The parametric scheme for induction could only be used when parameters were supplied. In other words: abstraction is allowed, but has to be followed immediately by application. In the perspective of our study of the various ways in which application and abstraction are present in type theory, we conclude that this mechanism for combined abstraction and application, being different from the λ-calculus mechanism, deserves our attention.

This chapter gives the simplest extension of the Barendregt cube with parameters. In the next chapter, we will extend PTSs in general, and not only with parameters, but also with definitions. This chapter serves as a preparation for the larger extension in the next chapter.

9a On parameters in the Barendregt cube

There are several ways in which the Barendregt cube can be extended with parameters. For instance, when working in the systems of the Barendregt cube, we may want to add only parametric terms $t(p_1, \ldots, p_n)$ for which the parameters $p_1, \ldots, p_n$ have types $A_1, \ldots, A_n$ that are of sort $*$. But we could also decide to add parametric terms $t(p_1, \ldots, p_n)$ without this restriction to the types of the $p_1, \ldots, p_n$.

There is a method to classify these various parametric extensions that corresponds to the classification of type systems that is used in the framework of Pure Type Systems.

In the Barendregt cube, there are two sorts $*$ and $\Box$, and the various PTSs in the cube are determined by the various ways in which type abstractions can be made. If all constructions of Π-types are allowed, we obtain the Calculus of Constructions, with rules $(*, *, *)$, $(*, \Box, \Box)$, $(\Box, *, *)$ and $(\Box, \Box, \Box)$. If we do not allow all Π-type constructions, we get one of the subsystems of the Calculus of Constructions in the Barendregt cube.

Something similar can be done with the parameter mechanism. One option is to provide one, general way of parametric abstraction and parametric application. We then allow all kinds of parameters. On the other hand, there are several ways in which a parameter mechanism may be restricted. We mention two ways:

- Assume, we are working in one of the systems of the Barendregt cube, extended with parameters, and we have that $t(p_1,\ldots,p_m)$ has type A. By Correctness of Types, A has either type $*$ or type $\Box$. One can imagine that we only allow $t(p_1,\ldots,p_m)$ if it has type A of type $*$ (so we only allow parametric *terms*);

- Still working in systems of the Barendregt cube extended with parameters, we will show that the parameters $p_1,\ldots,p_m$ in a term $t(p_1,\ldots,p_m)$ are typable themselves. Again, a parameter p_i can have a type P_i of type $*$ (so p_i is at *term level*), or a type P_i' of type $\Box$ (so p_i is at *type level*), and there are systems in which one would only allow parameters p_i that have a type P_i of type $*$ (or of type $\Box$).

These two possibilities for restriction are orthogonal in the sense that they can be combined. In many Pascal versions, for instance, parametric *terms* can only have parameters at *term level*. It is, for instance, not possible in Pascal to write a function `CartProd` that takes two types A and B as parameters, and returns a type that represents the Cartesian product $A \times B$ of A and B.

It is possible to incorporate such restrictions in our system in a similar way as the restrictions on the formation of Π-types in PTSs. We then obtain rules for parameter constructions. These rules have the form (s_1, s_2). The sort s_1 indicates that the parameters $p_1,\ldots,p_m$ have to have types $P_1,\ldots,P_m$ of sort s_1. The sort s_2 indicates that the resulting parametric term must have a type P of sort s_2. The combination of the rules for parameter constructions with the well-known rules for the construction of Π-types in the Barendregt cube leads to a division of the Barendregt cube into eight sub-cubes (we illustrate this in Figure 9.1 on page 251). As in the Barendregt cube, one dimension in the cube still corresponds with one of the rules $(*, \Box)$, $(\Box, *)$ or $(\Box, \Box)$. Following an edge of the cube in dimension (s_1, s_2) can now be done in two ways:

- As was already possible, we can follow the edge to the end. This still corresponds to accepting the Π-formation rule (s_1, s_2, s_2);

- We can also follow the edge only half-way. This means that we do not accept the Π-formation rule (s_1, s_2, s_2), but that we do accept the parameter construction rule (s_1, s_2).

This viewpoint suggests that allowing the Π-formation rule (s_1, s_2, s_2) also allows the parameter construction rule (s_1, s_2). Formally, one can work with systems in which we do allow the Π-formation rule, but do not allow the parameter construction rule. We can prove, however, that if the Π-construction rule (s_1, s_2, s_2) is allowed, a parameter construction involving rule (s_1, s_2) can be imitated by λ-abstractions (Theorem 10.79).

We extend the eight systems of the Barendregt cube with parametric constructs. Parametric constructs are of the form $c(b_1, \ldots, b_n)$ where $b_1, \ldots, b_n$ are terms of certain prescribed types. Just as we can allow several kinds of Π-constructs (via the set $\boldsymbol{R}$) in the Barendregt cube, we can also allow several kinds of parametric constructs. This is indicated by a set $\boldsymbol{P}$, consisting of tuples (s_1, s_2) where $s_1, s_2 \in \{*, \Box\}$. We understand by $(s_1, s_2) \in \boldsymbol{P}$ to mean that we allow parametric constructs $c(b_1, \ldots, b_n) : A$ where $b_1, \ldots, b_n$ have types $B_1, \ldots, B_n$ of sort s_1, and A is of type s_2. However, if both $(*, s_2) \in \boldsymbol{P}$ and $(\Box, s_2) \in \boldsymbol{P}$ then combinations of parameters are possible. For example, it is allowed that B_1 has type $*$, whilst B_2 has type $\Box$.

First we describe the extended syntax.

Definition 9.1 The set $\mathcal{T}_P$ of *parametric terms* is defined together with the set $\mathcal{L}_T$ of *lists of terms* as follows:

$$\begin{array}{rcl} \mathcal{T}_P & ::= & \mathcal{V} \mid \boldsymbol{S} \mid \mathcal{C}(\mathcal{L}_T) \mid \mathcal{T}_P\mathcal{T}_P \mid \lambda\mathcal{V}{:}\mathcal{T}_P.\mathcal{T}_P \mid \Pi\mathcal{V}{:}\mathcal{T}_P.\mathcal{T}_P; \\ \mathcal{L}_T & ::= & \varnothing \mid \langle \mathcal{L}_T, \mathcal{T}_P \rangle. \end{array}$$

where, as usual, $\mathcal{V}$ is a set of variables, $\mathcal{C}$ is a set of constants, and $\boldsymbol{S} = \{*, \Box\}$ is a set of sorts. We assume that $\mathcal{V}$, $\mathcal{C}$ and $\boldsymbol{S}$ are mutually disjoint and that $\mathcal{V}$ and $\mathcal{C}$ are countable (possibly infinite). Formally, lists of terms are of the form $\langle \ldots \langle\langle \varnothing, A_1\rangle, A_2\rangle \ldots A_n\rangle$. We usually write $\langle A_1, \ldots, A_n\rangle$ or even $A_1, \ldots, A_n$. In a parametric term of the form $c(b_1, \ldots, b_n)$, the subterms $b_1, \ldots, b_n$ are called the *parameters* of the term.

Definition 9.2 (Free Variables and Constants)

- We extend the usual definition of $\text{FV}(A)$ (cf. Definition 4.8), the set of *free variables* of a term A, to parametric terms as follows:

$$\text{FV}(c(a_1, \ldots, a_n)) \quad = \quad \textstyle\bigcup_{i=1}^{n} \text{FV}(a_i);$$

- We define $\text{CONS}\,(A)$, the set of *constants* of A as follows:

$$\begin{array}{l} \text{CONS}\,(s) = \text{CONS}\,(x) = \emptyset; \\ \text{CONS}\,(c(a_1, \ldots, a_n)) = \{c\} \cup \bigcup_{i=1}^{n} \text{CONS}\,(a_i)\,; \\ \text{CONS}\,(AB) = \text{CONS}\,(A) \cup \text{CONS}\,(B)\,; \\ \text{CONS}\,(\lambda x{:}A.B) = \text{CONS}\,(\Pi x{:}A.B) = \text{CONS}\,(A) \cup \text{CONS}\,(B)\,; \end{array}$$

- $\text{FV}(A) \cup \text{CONS}\,(A)$ forms the *domain* $\text{DOM}\,(A)$ of A.

We also adopt the Barendregt convention of Definition 4.11 extended for parametric terms.

We extend the Definition 4.12 of substitution of a term a for a variable x in a term b, $b[x{:=}a]$, to parametric terms, assuming that x is not a bound variable of either b or a:

$$c(b_1,\ldots,b_n)[x{:=}a] \quad \equiv \quad c(b_1[x{:=}a],\ldots,b_n[x{:=}a]);$$

Compatibility of β-reduction is extended to parametric terms in the obvious way:

if $b_i \to_\beta b_i'$ then $c(b_1,\ldots,b_i,\ldots,b_n) \to_\beta c(b_1,\ldots,b_i',\ldots,b_n)$.

This is the only way in which β-reduction on $\mathcal{T}_P$ differs from β-reduction on $\mathcal{T}$.

Definition 9.3 (Contexts) Given the set of parametric terms, we define the set $\mathcal{C}_P$ of *parametric contexts* (which we denote by $\Gamma,\Gamma',\ldots$) and the set $\mathcal{L}_V$ of *lists of variable declarations* as follows:

$$\begin{array}{lcl} \mathcal{C}_P & ::= & \varnothing \mid \langle \mathcal{C}_P, \mathcal{V}{:}\mathcal{T}_P\rangle \mid \langle \mathcal{C}_P, \mathcal{C}(\mathcal{L}_V){:}\mathcal{T}_P\rangle \\ \mathcal{L}_V & ::= & \varnothing \mid \langle \mathcal{L}_V, \mathcal{V}{:}\mathcal{T}_P\rangle. \end{array}$$

Notice that $\mathcal{L}_V \subseteq \mathcal{C}_P$: all lists of variable declarations are contexts, as well.

Definition 9.4 (Specification) Let (s_1,s_2) denote (s_1,s_2,s_2). A *parametric cube specification* is a quadruple $(\boldsymbol{S},\boldsymbol{A},\boldsymbol{R},\boldsymbol{P})$, such that:

- $\boldsymbol{S} = \{*,\Box\}$,
- $\boldsymbol{A} = \{* : \Box\} \subseteq \boldsymbol{S} \times \boldsymbol{S}$,
- $\boldsymbol{R} \subseteq \{(*,*),(*,\Box),(\Box,*),(\Box,\Box)\} \subseteq \boldsymbol{S} \times \boldsymbol{S} \times \boldsymbol{S}$ such that $(*,*) \in \boldsymbol{R}$,
- $\boldsymbol{P} \subseteq \{(*,*),(*,\Box),(\Box,*),(\Box,\Box)\}$, such that $(*,*) \in \boldsymbol{P}$.

We use the same notations as Definition 4.18. Note that the difference between one parametric cube specification and another is in the set of rules $\boldsymbol{R}$ and in the set of parametric rules $\boldsymbol{P}$.

Definition 9.5 (The Barendregt cube with parametric constants)
Let $(\boldsymbol{S},\boldsymbol{A},\boldsymbol{R},\boldsymbol{P})$ be a parametric cube specification as in Definition 9.4. The judgments that are derivable in $\lambda\boldsymbol{RP}$ are determined by the rules for $\lambda\boldsymbol{R}$ of Definition 4.24 and the following two rules where $\Delta \equiv x_1{:}B_1,\ldots,x_n{:}B_n$ and $\Delta_i \equiv x_1{:}B_1,\ldots,x_{i-1}{:}B_{i-1}$:

$$(\vec{\mathrm{C}}\textbf{-weak}) \quad \frac{\Gamma \vdash b : B \qquad \Gamma,\Delta_i \vdash B_i : s_i \qquad \Gamma,\Delta \vdash A : s}{\Gamma, c(\Delta) : A \vdash b : B} \quad (s_i,s) \in \boldsymbol{P}$$

$$(\vec{\mathrm{C}}\textbf{-app}) \quad \frac{\begin{array}{lcll} \Gamma_1, c(\Delta){:}A, \Gamma_2 & \vdash & b_i{:}B_i[x_j{:=}b_j]_{j=1}^{i-1} & (i=1,\ldots,n) \\ \Gamma_1, c(\Delta){:}A, \Gamma_2 & \vdash & A : s & (\text{if } n=0) \end{array}}{\Gamma_1, c(\Delta){:}A, \Gamma_2 \vdash c(b_1,\ldots,b_n) : A[x_j{:=}b_j]_{j=1}^{n}}$$

where the c that is introduced in the $\vec{\mathrm{C}}$-weakening rule is assumed to be Γ-fresh.

At first sight one might miss a $\vec{C}$-introduction rule. Such a rule, however, is not necessary, as c (on its own) is not a term. c can only be (part of) a term in the form $c(b_1, \ldots, b_n)$, and such terms can be typed by the ($\vec{C}$-app) rule.

Constant weakening ($\vec{C}$-weak) explains how we can introduce a declaration of a parametric constant in the context. The context Δ indicates the arity of the parametric constants (the number of declarations in Δ), and of which type each parameter must be ($x_j : B_j$ in Δ means the j-th parameter must be of type B_j).

The extra condition $\Gamma_1, c(\Delta){:}A, \Gamma_2 \vdash A : s$ in the ($\vec{C}$-app) for $n = 0$ is necessary to prevent an empty list of premises. Such an empty list of premises would make it possible to have almost arbitrary contexts in the conclusion. The extra condition is needed to assure that the context in the conclusion is legal (see Definition 4.22).

Now we illustrate the difference between the cube without and with parameters.

Example 9.6

- In the cube system $\lambda_{\rightarrow}$ (with one Π-formation rule $(*, *)$) we could introduce a type variable $N : *$ and a variable $o : N$ when we want to work with natural numbers. N represents the type of natural numbers and o represents the natural number zero;

- Though the representation of objects like the type of natural numbers and the natural number zero as a variable works fine in practice, there is a philosophical problem with such a representation. We do not consider the set $\mathbb{N}$ and the number $0 \in \mathbb{N}$ to be variables, because these objects 'do not vary'. If we have a derivation of $N{:}*, o{:}N \vdash t : N$ for some term t, it is technically possible to make a λ-abstraction over the variable o and obtain $N{:}* \vdash \lambda o{:}N.t : N \rightarrow N$. This is permitted since o is introduced as a variable, but it is probably not what we had in mind. In systems with parameters, we can distinguish between constants and variables. If o is introduced as a constant, it is not possible to form a λ-abstraction $\lambda o{:}N.t$;

- In some cases, we may need to introduce for each proposition Σ the type $\texttt{proof}(\Sigma)$ of proofs of Σ. This cannot be done in the cube system $\lambda_{\rightarrow}$ extended with (unparametrised) constants: such a constant `proof` should be of type $\texttt{prop} \rightarrow \texttt{type}$ and this type cannot be constructed in $\lambda_{\rightarrow}$ (notice that $\texttt{type} \equiv *$, so the construction of $\texttt{prop} \rightarrow \texttt{type}$ would involve the Π-formation rule $(*, \Box)$).

  However, the term `proof` will hardly ever be used on its own. It is usually used when applied to a proposition Σ. With parameters, it is possible to introduce a parametric version of `proof` by the following context declaration: $\texttt{proof(p:prop)} : \texttt{type}$.

This does not involve the construction of a type $\texttt{prop} \to \texttt{type}$. Nevertheless it is possible to construct the term $\texttt{proof}(P)$ for any term $P : \texttt{prop}$. We obtain a form of polymorphism without using the polymorphism of λ-calculus.

A disadvantage may be that we cannot speak about the term `proof` 'as it is'. When using `proof` in the syntax, it must always be applied to a parameter $T : \texttt{prop}$.

However, an advantage is that we can restrict ourselves to a much more simple type system. In the situation above we remain within the types of the system $\lambda_{\to}$. We do not need to use types of the system λP. This may have advantages in implementations of type systems. For instance, the system $\lambda_{\to}$ does not involve the conversion rule

$$\frac{\Gamma \vdash A : B \qquad \Gamma \vdash B : s \qquad B =_\beta B'}{\Gamma \vdash A : B'}$$

while λP does involve such a rule. The conversion rule involves β-equality of terms, and though it is decidable whether two λ-terms of λP are β-equal or not, it may take a lot of time and/or memory to establish such a fact. This may cause serious problems when implementing certain type systems. Using parameters whenever possible may therefore simplify implementations.

The parametric type system of Definition 9.5 has similar meta-theoretical properties as the systems of the Barendregt cube. We list them below. The proofs are similar to those of the Barendregt cube.

Lemma 9.7 *Assume* $\Gamma \vdash b : B$. *Then* $\textsc{dom}(b), \textsc{dom}(B) \subseteq \textsc{dom}(\Gamma)$.

PROOF: By induction on the derivation $\Gamma \vdash b : B$. ⊠

Lemma 9.8 (Generation Lemma)

1. *If* $\Gamma \vdash s : C$ *then* $s \equiv *$, $C =_\beta \square$ *and if* $C \not\equiv \square$ *then* $\Gamma \vdash C : s'$ *for some sort* s'.
2. *If* $\Gamma \vdash x : C$ *then there is* $s \in \boldsymbol{S}$ *and* $B =_\beta C$ *such that* $\Gamma \vdash B : s$ *and* $(x{:}B) \in \Gamma$;
3. *If* $\Gamma \vdash (\Pi x{:}A.B) : C$ *then there is* $(s_1, s_2) \in \boldsymbol{R}$ *such that* $\Gamma \vdash A : s_1$, $\Gamma, x{:}A \vdash B : s_2$ *and* $C =_\beta s_2$;
4. *If* $\Gamma \vdash (\lambda x{:}A.b) : C$ *then there is* $s \in \boldsymbol{S}$ *and* B *such that* $\Gamma \vdash (\Pi x{:}A.B) : s$; $\Gamma, x{:}A \vdash b : B$; *and* $C =_\beta (\Pi x{:}A.B)$;
5. *If* $\Gamma \vdash Fa : C$ *then there are* A, B *such that* $\Gamma \vdash F : (\Pi x{:}A.B)$, $\Gamma \vdash a : A$ *and* $C =_\beta B[x{:=}a]$.

6. *If* $\Gamma \vdash c(b_1, \ldots, b_n) : D$ *then there exist* s, $\Delta \equiv x_1 : B_1, \ldots, x_n : B_n$ *and* A *such that* $\Gamma \vdash D =_\beta A[x_j{:=}b_j]_{j=1}^{n}$, *and* $\Gamma \vdash b_i{:}B_i[x_j{:=}b_j]_{j=1}^{i-1}$. *Moreover,* $\Gamma \equiv \Gamma_1, c(\Delta) : A, \Gamma_2$ *and* $\Gamma_1, \Delta \vdash A : s$. *Finally, there are* $s_i \in S$ *such that* $\Gamma, \Delta_i \vdash B_i : s_i$ *and* $(s_i, s) \in \boldsymbol{P}$.

PROOF: By induction on the derivation. ⊠

Lemma 9.9 (Correctness of Types) *If* $\Gamma \vdash A : B$ *then* $B \equiv \Box$ *or* $\Gamma \vdash B : s$ *for some* $s \in \boldsymbol{S}$.

PROOF: By induction on the derivation rules. ⊠

Lemma 9.10 (Subterm Lemma) *If* A *is legal and* B *is a subterm of* A, *then* B *is legal.*

PROOF: By induction on the structure of A. ⊠

Lemma 9.11 (Subject Reduction) *If* $\Gamma \vdash A : B$ *and* $A \to_\beta A'$ *then* $\Gamma \vdash A' : B$.

PROOF: By simultaneous induction on the derivation rules:

- If $\Gamma \vdash A : B$ and $A \to_\beta A'$ then $\Gamma \vdash A' : B$.
- If $\Gamma \vdash A : B$ and $\Gamma \to_\beta \Gamma'$ then $\Gamma' \vdash A : B$ where $\Gamma \to_\beta \Gamma'$ iff $\Gamma = d_1, \ldots, d_i, \ldots d_n$, $\Gamma' = d_1, \ldots, d_i', \ldots d_n$ and
  - $d_i = x : A$, $d_i' = x : A'$ and $A \to_\beta A'$ or
  - $d_i = c(\Delta)$, $d_i' = c(\Delta)$, and $\Delta \to_\beta \Delta'$.

⊠

Lemma 9.12 (Unicity of Types) *If* $\Gamma \vdash A : B_1$ *and* $\Gamma \vdash A : B_2$, *then* $B_1 =_\beta B_2$.

PROOF: By induction on the structure of A. ⊠

Theorem 9.13 (Strong Normalisation) *If* $\Gamma \vdash A : B$ *then* A *and* B *are* β-*strongly normalising, that is: any* β-*reduction path of* A *or* B *is finite.*

PROOF: By the usual technique of [3]. See [98]. ⊠

9b The Barendregt cube refined with parameters

The systems of Definition 9.5 have six degrees of freedom: three for the possible choices of $(*, \Box)$, $(\Box, *)$ and $(\Box, \Box) \in \boldsymbol{R}$ and three for the possible choices of $(*, \Box)$, $(\Box, *)$, and $(\Box, \Box) \in \boldsymbol{P}$. However, these choices are not independent

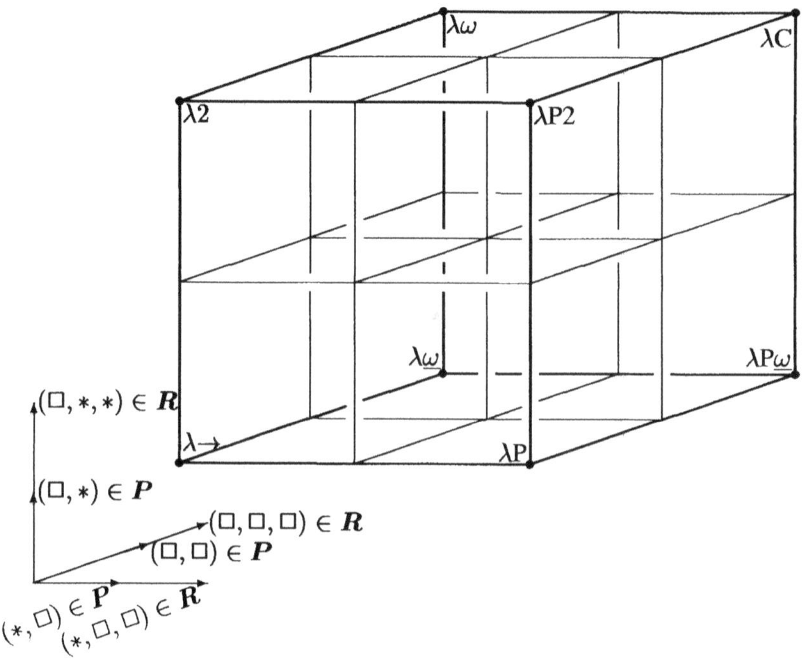

Figure 9.1: The Barendregt cube refined with parameters

since constructs that can be made with $\boldsymbol{P}$-rule (s_1, s_2) can be imitated in a typed λ-calculus with $\boldsymbol{R}$-rule (s_1, s_2). This means that the parameter-free type system with $\boldsymbol{R} = \{(*, *), (*, \Box)\}$ is at least as strong as the type system with parameters with the same set $\boldsymbol{R}$, but with $\boldsymbol{P} = \{(*, *), (*, \Box)\}$. We make this precise in Theorem 9.19.

The insight of Theorem 9.19 can be expressed by depicting the systems with parameters of Definition 9.5 as a *refinement* of the Barendregt cube. As in the Barendregt cube, we start with the system $\lambda_{\rightarrow}$, which has $\boldsymbol{R} = \{(*, *)\}$ and $\boldsymbol{P} = \{(*, *)\}$. Adding an extra element (s_1, s_2) to $\boldsymbol{R}$ still corresponds to moving in one dimension in the cube. Now we add the possibility of moving in one dimension in the cube but stopping half-way. We let this movement correspond to extending $\boldsymbol{P}$ with (s_1, s_2). This "going only half-way" is in line with the intuition that Π-formation with (s_1, s_2) can imitate the construction of a parametric construct with (s_1, s_2). In other words, the system obtained by "going all the way" is at least as strong as the system obtained by "going only half-way".

The refinement of the Barendregt cube with parameters is depicted in Figure 9.1.

We now make the above intuition that "$\boldsymbol{R}$ can imitate $\boldsymbol{P}$" precise.

Definition 9.14 Consider the system $\lambda \boldsymbol{RP}$. We call this system *parametrically conservative* if $(s_1, s_2) \in \boldsymbol{P}$ implies $(s_1, s_2) \in \boldsymbol{R}$.

Let $\lambda \boldsymbol{RP}$ be parametrically conservative. In order to show that the parameter-free system $\lambda \boldsymbol{R}$ is at least as powerful as $\lambda \boldsymbol{RP}$, we need to remove the parameters from the syntax of $\lambda \boldsymbol{RP}$. To do so, we replace the parametric application in a term $c(b_1, \ldots, b_n)$ by function application $cb_1, \ldots, b_n$:

Definition 9.15 Define the parameter-free translation $\{t\}$ of a term $t \in \mathcal{T}_P$ by:

$$\begin{array}{rcll} \{a\} & \equiv & a & \text{if } a \equiv x \text{ or } a \equiv s; \\ \{c(b_1, \ldots, b_n)\} & \equiv & c\,\{b_1\} \cdots \{b_n\}; & \\ \{ab\} & \equiv & \{a\}\,\{b\}; & \\ \{\pi x{:}A.B\} & \equiv & \pi x{:}\,\{A\}\,.\,\{B\} & \text{if } \pi \text{ is } \lambda \text{ or } \Pi \end{array}$$

Definition 9.16 We extend the definition of $\{\_\}$ to contexts:

$$\begin{array}{rcl} \{\langle\rangle\} & \equiv & \langle\rangle; \\ \{\Gamma, x{:}A\} & \equiv & \{\Gamma\}, x{:}\,\{A\}; \\ \{\Gamma, c(\Delta){:}A\} & \equiv & \{\Gamma\}, c(){:}\,\{\prod \Delta.A\}\,. \end{array}$$

Here, $\Delta \equiv x_1 : B_1, \ldots, x_n : B_n$, and $\prod \Delta.A$ is shorthand for $\prod_{i=1}^{n} x_n : B_i.A$.

To demonstrate the behaviour of $\{\_\}$ under β-reduction, we need a lemma that shows how to manipulate with substitutions and $\{\_\}$. The proof is straightforward, using induction on the structure of a.

Lemma 9.17 *For $a, b \in \mathcal{T}_P$: $\{a[x{:=}b]\} \equiv \{a\}\,[x{:=}\,\{b\}]$.* ⊠

The mapping $\{\_\}$ maintains β-reduction:

Lemma 9.18 *$a \to_\beta a'$ if and only if $\{a\} \to_\beta \{a'\}$.*

PROOF: Follows easily by induction on the structure of a, and Lemma 9.17. ⊠

Now we show that $\{\_\}$ embeds the parametrically conservative $\lambda \boldsymbol{RP}$ in the parameter-free $\lambda \boldsymbol{R}$:

Theorem 9.19 *Let $\lambda \boldsymbol{RP}$ be parametrically conservative. If $\Gamma \vdash_{\boldsymbol{RP}} a : A$ then $\{\Gamma\} \vdash_{\boldsymbol{R}} \{a\} : \{A\}$.*

PROOF: Induction on the derivation of $\Gamma \vdash_{\boldsymbol{RP}} a : A$. By Lemma 9.17, all cases are easy except for ($\vec{\mathrm{C}}$-weak). So: assume the last step of the derivation was

$$\frac{\Gamma \vdash_{\boldsymbol{RP}} b : B \qquad \Gamma, \Delta_i \vdash_{\boldsymbol{RP}} B_i : s_i \qquad \Gamma, \Delta \vdash_{\boldsymbol{RP}} A : s}{\Gamma, c(\Delta){:}A \vdash_{\boldsymbol{RP}} b : B} \quad (s_i, s) \in \boldsymbol{P}.$$

By the induction hypothesis, we have:

$$\{\Gamma\} \vdash_{\boldsymbol{R}} \{b\} : \{B\}; \tag{9.1}$$

$$\{\Gamma, \Delta_i\} \vdash_{\boldsymbol{R}} \{B_i\} : s_i; \tag{9.2}$$

$$\{\Gamma, \Delta\} \vdash_{\boldsymbol{R}} \{A\} : s. \tag{9.3}$$

$\lambda \boldsymbol{RP}$ is parametrically conservative, so $(s_i, s) \in \boldsymbol{R}$ for $i = 1, \ldots, n$. Therefore, we can repeatedly use the Π-formation rule, starting with (9.3) and (9.2), obtaining

$$\{\Gamma\} \vdash_{\boldsymbol{R}} \textstyle\prod_{i=1}^{n} x_i : \{B_i\} . \{A\} : s. \tag{9.4}$$

Notice that $\prod_{i=1}^{n} x_i : \{B_i\} . \{A\} \equiv \{\prod \Delta.A\}$. Using ($\vec{C}$-weak) on (9.1) and (9.4) gives

$$\{\Gamma\}, c() : \{\textstyle\prod \Delta.A\} \vdash_{\boldsymbol{R}} \{b\} : \{B\}. \qquad \boxtimes$$

Conclusions

Many of the existing type systems do not fit exactly in the Barendregt cube. For example, AUTOMATH uses parameters heavily and these parameters do not exist in the cube. Moreover, there are some types that are only used in special situations by systems like the Edinburgh Logical Framework LF, and the programming language ML, but those types and situations could be covered by parameters which would allow one to only cover those special situations instead of using a more powerful type system just to cover a couple of situations. In fact, ML (as well as LF and AUTOMATH) allows Π-types, but not all of them. But, in any corner of the cube, as soon as an abstraction of a sort is allowed, all abstractions of that sort are allowed too. So, if we place systems like ML, LF and AUTOMATH at the corners of the cube, we will be allowing far more abstractions in these systems than they accept.

For this reason, we studied a refinement of the cube where not only the eight corners can be inhabited, but also points half way between these corners. We described an extension of the Barendregt cube with parameters. This is more a *refinement* than an *extension*, as new systems that are introduced can be depicted by dividing the traditional Barendregt cube into eight sub-cubes. This is due to the fact that parametric constructs can be imitated by constructions of typed λ-calculus (see Theorem 9.19) but not the other way around.

In Chapter 10 we will show that our refinement makes it possible to:

- Give a better description of practical type systems like LF and ML than the systems in the usual cube.

- Position systems that could not be placed in the usual cube (several AUTOMATH-systems).

This will make it possible to give a more detailed comparison between the expressiveness of several type systems.

Chapter 10

Pure Type Systems with parameters and definitions

This chapter is devoted to the description of Pure Type Systems with parameters and definitions. One reason to study this extension of PTSs is to give a better description of AUTOMATH than in Chapter 7, where we had to work with the sort Δ to store terms and types that did not have a counterpart in AUTOMATH (cf. Subsection 7b1). Such terms and types were needed for the description of the system because no parameters or definitions were used. But as we have seen in the previous three chapters, there are many more arguments why type systems with parameters and definitions deserve to be studied.

This chapter is organised as follows. In Section 10a, we give definitions of PTSs extended with parametric constants and definitions. This definition includes the unfolding of definitions (parametric or not) via the so-called δ-reduction. In Section 10b we show that the δ-reduction and $\beta\delta$-reductions have the Church-Rosser property, and that δ-reduction (under some reasonable conditions) is strongly normalising. In Section 10c, we show some elementary properties of the system introduced in Section 10a, like a Generation Lemma, and the Subject Reduction theorem for $\beta\delta$-reduction. We also prove that $\beta\delta$-reduction is strongly normalising if a slightly stronger PTS is β-strongly normalising.

Section 10d is devoted to the various ways in which parameters can be added to a PTS in a more restricted way, with the refined Barendregt cube of Figure 9.1 as a result.

In Section 10e, we compare our system with some other type systems, like AUTOMATH. We place various AUTOMATH systems in the refined Barendregt cube of Figure 9.1.

In Section 10f we see that the use of parameters can sometimes result in simpler and more realistic implementations of type systems.

10a Parametric constants and definitions

In this section, we combine the techniques developed in Chapters 8 and 9 and we extend PTSs with parametric constants and definitions. This extension will also contain the DPTSs presented in Chapter 8 (definitions in DPTSs can be interpreted as parametric definitions with zero parameters) and the extension of the cube with parameters as presented in Chapter 9. In Section 10e, we show that AUT-68 can be seen as a (on some points somewhat restricted version of a) PTS with parameters and definitions.

Definition 10.1 The set $\mathcal{T}_P$ of *parametric terms* is defined together with the set $\mathcal{L}_V$ of *lists of variables* and the set $\mathcal{L}_T$ of *lists of terms*:

$$\begin{array}{lcl}\mathcal{T}_P & ::= & \mathcal{V} \mid \boldsymbol{S} \mid \mathcal{C}(\mathcal{L}_T) \mid \mathcal{T}_P\mathcal{T}_P \mid \lambda\mathcal{V}{:}\mathcal{T}_P.\mathcal{T}_P \mid \\ & & \Pi\mathcal{V}{:}\mathcal{T}_P.\mathcal{T}_P \mid \mathcal{C}(\mathcal{L}_V){=}\mathcal{T}_P{:}\mathcal{T}_P \text{ IN } \mathcal{T}_P; \\ \mathcal{L}_V & ::= & \varnothing \mid \langle \mathcal{L}_V, \mathcal{V}{:}\mathcal{T}_P\rangle; \\ \mathcal{L}_T & ::= & \varnothing \mid \langle \mathcal{L}_T, \mathcal{T}_P\rangle.\end{array}$$

where, as usual, $\mathcal{V}$ is a set of variables, $\mathcal{C}$ is a set of constants, and $\boldsymbol{S}$ is a set of sorts. We assume that $\mathcal{V}$, $\mathcal{C}$ and $\boldsymbol{S}$ are mutually disjoint and that $\mathcal{V}$ and $\mathcal{C}$ are countable (possibly infinite). Formally, lists of variables are of the form

$$\langle \ldots \langle\langle \varnothing, x_1{:}A_1\rangle, x_2{:}A_2\rangle \ldots x_n{:}A_n\rangle.$$

We usually write $\langle x_1{:}A_1, \ldots, x_n{:}A_n\rangle$ or even $x_1{:}A_1, \ldots, x_n{:}A_n$. A similar convention is adopted for lists of terms. In a parametric term of the form $c(b_1, \ldots, b_n)$, the subterms $b_1, \ldots, b_n$ are called the *parameters* of the term.

Terms of the form $\mathcal{C}(\mathcal{L}_V){=}\mathcal{T}_P{:}\mathcal{T}_P$ in $\mathcal{T}_P$ represent parametric local definitions. An example of such a term is $\texttt{double}(\texttt{x}{:}\mathbb{N}){=}(\texttt{x}{+}\texttt{x}){:}\mathbb{N} \text{ IN } A$. The term indicates that a subterm of A of the form $\texttt{double}(P)$ is to be interpreted as $P + P$, and has type $\mathbb{N}$. The definition is local, that is: the scope of the definition is the term A. Local definitions stand in contrast to global definitions. Global definitions are given in a context Γ, and refer to any term that is considered within Γ (see the forthcoming Definition 10.7). The definition system in AUTOMATH can be compared to the system of global definitions in this chapter. However, there are no local definitions in AUTOMATH.

Definition 10.2

- We extend the definition of $\text{FV}(A)$, the set of *free variables* of a term A, to parametric terms:

$$\begin{array}{rcl}\text{FV}(c(a_1,\ldots,a_n)) & = & \bigcup_{i=1}^n \text{FV}(a_i);\\ \text{FV}\Big(c(\vec{x}{:}\vec{A}){=}A{:}B \text{ IN } C\Big) & = & \bigcup_{i=1}^n (\text{FV}(A_i)\setminus\{x_1,\ldots,x_{i-1}\})\\ & & \cup\,(\text{FV}(A)\cup\text{FV}(B))\setminus\{x_1,\ldots,x_n\}\\ & & \cup\,\text{FV}(C)\end{array}$$

  where $\vec{x}{:}\vec{A}$ denotes $x_1{:}A_1,\ldots,x_n{:}A_n$;

- We similarly define $\text{CONS}\,(A)$, the set of *constants and global definitions* of A:

$$\begin{array}{rcl}\text{CONS}\,(s) = \text{CONS}\,(x) & = & \emptyset;\\ \text{CONS}\,(c(a_1,\ldots,a_n)) & = & \{c\}\cup\bigcup_{i=1}^n \text{CONS}\,(a_i);\\ \text{CONS}\,(AB) & = & \text{CONS}\,(A)\cup\text{CONS}\,(B);\\ \text{CONS}\,(\lambda x{:}A.B) & = & \text{CONS}\,(A)\cup\text{CONS}\,(B);\\ \text{CONS}\,(\Pi x{:}A.B) & = & \text{CONS}\,(A)\cup\text{CONS}\,(B);\\ \text{CONS}\,\Big(c(\vec{x}{:}\vec{A}){=}A{:}B \text{ IN } C\Big) & = & \bigcup_{i=1}^n \text{CONS}\,(A_i)\\ & & \cup\,\text{CONS}\,(A)\cup\text{CONS}\,(B)\\ & & \cup\,(\text{CONS}\,(C)\setminus\{c\}).\end{array}$$

$\text{FV}(A)\cup\text{CONS}\,(A)$ forms the *domain* $\text{DOM}\,(A)$ of A.

Remark 10.3 The definition of

$$\text{FV}\Big(c(\vec{x}{:}\vec{A}){=}A{:}B \text{ IN } C\Big)$$

and

$$\text{CONS}\,\Big(c(\vec{x}{:}\vec{A}){=}A{:}B \text{ IN } C\Big)$$

make clear what the binding structure in a term $c(\vec{x}{:}\vec{A}){=}A{:}B \text{ IN } C$ is.

- A variable declaration $x_i{:}A_i$ in the parameter list $\vec{x}{:}\vec{A}$ binds all the occurrences of x_i in A_j, for $j \geq i$. That is: the type of a parameter x_j may depend on earlier declared parameters;

- Moreover, the declaration $x_i{:}A_i$ binds all the occurrences of x_i in A and B. This corresponds to the intuitive idea of a parametric definition: x_i can serve as a parameter in the definiens A and in the type B of the definiens;

- However, the variable declaration $x_i{:}A_i$ does *not* bind any occurrence of x_i in C. The definiendum c will occur in C only with a list of parameters $a_1,\ldots,a_n$ behind it, so in the form $c(a_1,\ldots,a_n)$. The variables $x_1,\ldots,x_n$ in the definition of c only serve to indicate what the type of the a_is must be (below, we will see that a_i must have type $A_i[x_j{:=}a_j]_{j=1}^{i-1}$), and what the type of the term $c(a_1,\ldots,a_n)$ is (this appears to be $B[x_j{:=}a_j]_{j=1}^{n}$);

- Moreover, we see that c is not included in the constants of

$$c(\vec{x}{:}\vec{A}){=}A{:}B \text{ IN } C.$$

This is because c is a *local* definition, and acts as a binder for the occurrences of c in C.

Remark 10.4 There are several reasons for including the type B in a local definition $c(\vec{x}{:}\vec{A}){=}A{:}B$ IN C:

- We want to remain consistent with other binders, such as λ and Π. In a term $\lambda x{:}A.B$ or $\Pi x{:}A.B$ we mention the type of the binder x, therefore we also mention the type of the binder c in a local definition $c(\vec{x}{:}\vec{A}){=}A{:}B$ IN C;

- Sometimes $A : B$ indicates that the term A is a proof of a theorem B (using PAT). If we want to use B in the proof of a new theorem B', we must use the proof term A of B in the proof A' of B'. In that case it is attractive to abbreviate A by introducing a definition $c(\vec{x}{:}\vec{A}){=}A{:}B$ IN A'. It is important to remember that c is (an abbreviation of) a proof of B, and that is a reason to mention B, the type of A, in the definition declaration;

- For practical purposes like proof assistants or proof checkers, it may seem to be problematic to have B in the definition declaration. However, the program does not always have to ask the user to explicitly mention the type of the abbreviation. Often it can find this type itself via a type checking algorithm. Of course, this also depends on whether type checking is decidable in the underlying type system.

  Sometimes, the user may wish to manually enter the type, because he/she may prefer a certain formulation of the type to a β-equivalent formulation that the program automatically offers.

Again, as usual in PTSs, we do not make difference between terms that are equal up to renaming of bound variables: we consider these terms to be syntactically equal. Moreover, we assume the Barendregt variable convention of Definition 4.11 taking into account the new terms. Hence, we do not write $(\lambda \mathtt{x}{:}A.\mathtt{x})\mathtt{x}$ but $(\lambda \mathtt{y}{:}A.\mathtt{y})\mathtt{x}$. Similarly, we write $\mathtt{c}(\mathtt{x}'{:}A){=}\mathtt{x}'{:}A$ IN $\mathtt{c}(\mathtt{x})$ instead of $\mathtt{c}(\mathtt{x}{:}A){=}\mathtt{x}{:}A$ IN $\mathtt{c}(\mathtt{x})$.

Definition 10.5 We extend the definition of substitution of a term a for a variable x in a term b, $b[x{:=}a]$, to parametric terms, assuming that x is not a bound variable of either b or a:

$$\begin{array}{rcl} c(b_1,\ldots,b_n)[x{:=}a] & \equiv & c(b_1[x{:=}a],\ldots,b_n[x{:=}a]); \\ (c(\vec{x}{:}\vec{A}) = A{:}B \text{ IN } C)[x{:=}a] & \equiv & c(x_1{:}A_1[x{:=}a],\ldots,x_n{:}A_n[x{:=}a])= \\ & & A[x{:=}a]{:}B[x{:=}a] \text{ IN } C[x{:=}a]. \end{array}$$

We now define contexts for type systems with parameters and definitions.

Definition 10.6 The set of *contexts* is given by

$$\mathcal{C}_P \quad ::= \quad \varnothing \mid \langle \mathcal{C}_P, \mathcal{V}{:}\mathcal{T}_P \rangle \mid \langle \mathcal{C}_P, \mathcal{C}(\mathcal{L}_V){=}\mathcal{T}_P{:}\mathcal{T}_P \rangle \mid \langle \mathcal{C}_P, \mathcal{C}(\mathcal{L}_V){:}\mathcal{T}_P \rangle.$$

Notice that $\mathcal{L}_V \subseteq \mathcal{C}_P$: all lists of variable declarations are contexts, as well. We denote contexts by $\Gamma, \Gamma', \ldots$.

Definition 10.7 Let Γ be a context.
Elements $x{:}A$, $c(x_1{:}B_1,\ldots,x_n{:}B_n){:}A$, $c(x_1{:}B_1,\ldots,x_n{:}B_n){=}a{:}A$ of Γ are called *declarations.*

- $x{:}A$ is a *variable declaration.*
  - The variable x is the *subject* of the declaration;
  - A is the *type* or *predicate* of the declaration;
- A declaration of the form $c(x_1{:}B_1,\ldots,x_n{:}B_n){:}A$ is a *constant declaration.*
  - The constant c is the *subject* of the declaration. As c is introduced without further definition, c is called a *primitive constant* (cf. the primitive notions in AUTOMATH);
  - $x_1,\ldots,x_n$ are the *parameters* of the declaration;
  - A is the *type* (*predicate*) of the declaration;
- A declaration $c(x_1{:}B_1,\ldots,x_n{:}B_n){=}a{:}A$ is called a *global definition declaration* or shorthand *global definition* or *definition.*
  - The constant c is the *subject* or *definiendum* of the declaration. c is called a *(globally) defined constant*;
  - $x_1,\ldots,x_n$ are the *parameters* of the declaration;
  - a is the *definiens* of the declaration;
  - A is the *type* (*predicate*) of the declaration.

The reasons for including the type of a global definition or a parametric constant in its declaration are the same as for local definitions. See Remark 10.4.

In the rest of this chapter, Δ denotes a context $x_1{:}B_1, \ldots, x_n{:}B_n$ consisting of variable declarations only. Such a context is typically used as a list of parameters in a definition $c(\Delta){=}a{:}A$. We write

$$\Delta_i \equiv x_1{:}B_1, \ldots, x_{i-1}{:}B_{i-1}$$

for $i \leq n$.

We extend the definition of substitution to contexts:

Definition 10.8 Let $\Gamma \in \mathcal{C}_P$, $M \in \mathcal{T}_P$. We define $\Gamma[x{:=}M]$ as follows:

$$\begin{array}{rcll} \varnothing[x{:=}M] & \equiv & \varnothing; & \\ \langle\Gamma, x{:}A\rangle[x{:=}M] & \equiv & \Gamma[x{:=}M]; & \\ \langle\Gamma, x'{:}A\rangle[x{:=}M] & \equiv & \langle\Gamma[x{:=}M], x'{:}A[x{:=}M]\rangle & \text{if } x \not\equiv x'; \\ \langle\Gamma, c(\Delta){:}A\rangle[x{:=}M] & \equiv & \langle\Gamma[x{:=}M], c(\Delta[x{:=}M]){:}A[x{:=}M]\rangle; & \\ \langle\Gamma, c(\Delta){=}a{:}A\rangle[x{:=}M] & \equiv & \langle\Gamma[x{:=}M], c(\Delta[x{:=}M]){=}a[x{:=}M]{:}A[x{:=}M]\rangle. & \end{array}$$

For a term A we defined $\text{FV}(A)$ and $\text{CONS}\,(A)$. For a context Γ we do not form one set $\text{CONS}\,(\Gamma)$, but we split this set into a set $\text{PRIMCONS}\,(\Gamma)$, containing the primitive constants of Γ, and a set $\text{DEFCONS}\,(\Gamma)$, containing the defined constants of Γ.

Definition 10.9 Let Γ be a context. We define the free variables, constants and definitions of Γ:

Γ	$\text{FV}(\Gamma)$	$\text{PRIMCONS}\,(\Gamma)$	$\text{DEFCONS}\,(\Gamma)$
$\varnothing$	$\varnothing$	$\varnothing$	$\varnothing$
$\Gamma, x{:}A$	$\text{FV}(\Gamma) \cup \{x\}$	$\text{PRIMCONS}\,(\Gamma)$	$\text{DEFCONS}\,(\Gamma)$
$\Gamma, c(\Delta){:}A$	$\text{FV}(\Gamma)$	$\text{PRIMCONS}\,(\Gamma) \cup \{c\}$	$\text{DEFCONS}\,(\Gamma)$
$\Gamma, c(\Delta){=}a{:}A$	$\text{FV}(\Gamma)$	$\text{PRIMCONS}\,(\Gamma)$	$\text{DEFCONS}\,(\Gamma) \cup \{c\}$

Finally we define the *domain* of Γ, $\text{DOM}\,(\Gamma)$, by

$$\text{FV}(\Gamma) \cup \text{PRIMCONS}\,(\Gamma) \cup \text{DEFCONS}\,(\Gamma)\,.$$

In ordinary Pure Type Systems we have that, for a legal term A in a legal context Γ, $\text{FV}(A) \subseteq \text{FV}(\Gamma)$. The type of a free variable in A, therefore, can always be determined via Γ. In our Pure Type Systems with definitions and parameters we will have: $\text{FV}(A) \subseteq \text{FV}(\Gamma)$ and $\text{CONS}\,(A) \subseteq \text{PRIMCONS}\,(\Gamma) \cup \text{DEFCONS}\,(\Gamma)$. This has not only as an effect that the type of a free variable or a constant can be determined via Γ, but also that Γ determines whether a constant in A that is not serving as a *local* definition within A, is a defined constant or a primitive constant. We therefore define:

Definition 10.10 For a context Γ and a term A with $\text{DOM}(A) \subseteq \text{DOM}(\Gamma)$ we define

$$\begin{aligned}\text{DEFCONS}_\Gamma(A) &= \text{CONS}(A) \cap \text{DEFCONS}(\Gamma);\\ \text{PRIMCONS}_\Gamma(A) &= \text{CONS}(A) \cap \text{PRIMCONS}(\Gamma).\end{aligned}$$

We see that a constant $c \in C$ can play three roles in a term A, with respect to a context Γ:

- If c occurs in a subterm $(c(\Delta){=}b{:}B \text{ IN } a)$ of A, then c is a *locally defined constant*;
- If $c \in \text{DEFCONS}_\Gamma(A)$, then c is a *globally defined constant*;
- If $c \in \text{PRIMCONS}_\Gamma(A)$ (or $c \notin \text{DOM}(\Gamma)$), then c is a *primitive constant*.

Example 10.11 It is possible that $c \in \text{CONS}(A)$ is a globally defined constant with respect to a context Γ, but a primitive constant with respect to a context Γ'. Take for example $A \equiv \texttt{id}$, $\Gamma \equiv \alpha{:}*, \texttt{id}()=(\lambda \texttt{x}{:}\alpha.\texttt{x}){:}(\alpha \to \alpha)$, and $\Gamma' \equiv \alpha{:}*, \texttt{id}(){:}(\alpha \to \alpha)$.

A natural condition on a context $\Gamma_1, c(\Delta){=}a{:}A, \Gamma_2$ is that all the free variables and constants of a and A are declared in either Γ_1 or Δ, and that all free variables and constants in a declaration $x_i{:}B_i \in \Delta$ are declared in Γ_1, Δ_i (recall that Δ is a standard context $x_1{:}B_1, \ldots, x_n{:}B_n$ and $\Delta_i \equiv x_1{:}B_1, \ldots, x_{i-1}{:}B_{i-1}$). We call such a context *sound*:

Definition 10.12 $\Gamma \in \mathcal{C}_P$ is *sound* if $\Gamma \equiv \Gamma_1, c(\Delta){=}a{:}A, \Gamma_2$ implies

$$\text{DOM}(a) \cup \text{DOM}(A) \subseteq \text{DOM}(\Gamma_1) \cup \text{DOM}(\Delta)$$

and

$$\text{DOM}(B_i) \subseteq \text{DOM}(\Gamma_1, \Delta_i).$$

The contexts occurring in the type systems proposed in this chapter are all sound (see Lemma 10.23). This fact will be useful when proving properties of these systems.

We will consider some extensions of Pure Type Systems (PTSs).

- An extension that includes globally and locally defined constants (cf. [137]) presented in Chapter 8: "PTSs with definitions" (D-PTSs);
- Orthogonally, we can extend PTSs with parameter-free primitive constants. Then we obtain C-PTSs. C-PTSs are not very interesting, as the role of

parameter-free primitive constants can usually be imitated by variables.[1] One could agree that a parameter-free primitive constant is a special sort of variable, and promise not to make any (λ or Π) abstraction over such a variable;

- Our first real extension describes PTSs with *parametric* primitive constants, but without definitions ($\vec{\text{C}}$-PTSs). The $\vec{\text{C}}$-PTSs include the C-PTSs, as a parameter-free primitive constant can be seen as a parametric primitive constant with zero parameters;
- Another extension includes parametric defined constants, and can be seen as a generalisation of D-PTSs: $\vec{\text{D}}$-PTSs. This is the extension of [84] presented in Chapter 9.
- We can combine the extensions with primitive constants and defined constants, choosing between parametrised or parameter-free variants. For instance, we can make an extension that includes parameter-free defined constants, and parametric primitive constants. We call this extension $\vec{\text{C}}$D-PTSs.

Combining the various extensions, we obtain a hierarchy that can be depicted as in Figure 10.1.

Example 10.13 We give some examples of the possibilities of parameters and definitions.

- We illustrate the difference between PTSs, C-PTSs and $\vec{\text{C}}$-PTSs.
  - In the PTS $\lambda_{\to}$ (with only one axiom $* : \Box$ and one Π-formation rule $(*, *, *)$) we could introduce a type variable $N : *$ and a variable $o : N$ when we want to work with natural numbers. N represents the type of natural numbers and o represents the natural number zero;
  - Though the representation of objects like the type of natural numbers and the natural number zero as a variable works fine in practice, there is a philosophical problem with such a representation. We do not consider the set $\mathbb{N}$ and the number $0 \in \mathbb{N}$ to be variables, because these objects "do not vary". If we have a derivation of $N{:}*, o{:}N \vdash t : N$ for some term t, it is technically possible to make a λ-abstraction over the variable o and obtain $N{:}* \vdash \lambda o{:}N.t : N \to N$. In this judgement, o acts as a variable, while it was initially introduced as a constant.

    In C-PTSs we can distinguish between constants and variables. If o is introduced as a constant, it is not possible to form a λ-abstraction $\lambda o{:}N.t$;

[1] There are, however, extensions of PTSs in which constants play an essential role. See for instance the Modal PTSs in the thesis of Borghuis [16], p. 28–29

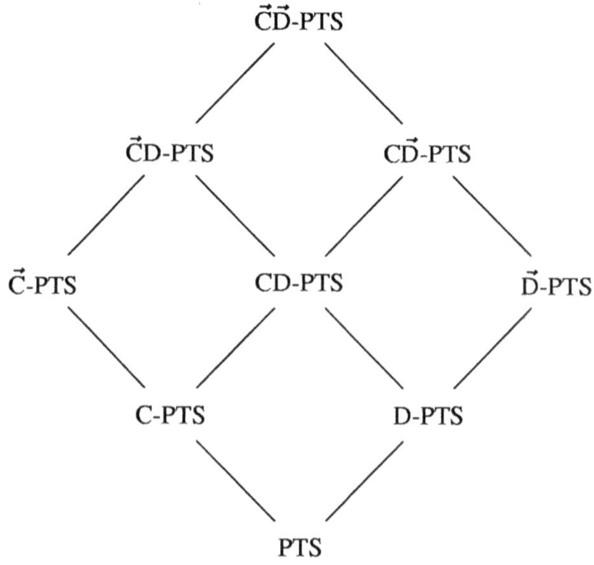

Figure 10.1: The hierarchy of parameters and definitions

- In Example 7.9, we introduced for each proposition Σ the type $\texttt{proof}(\Sigma)$ of proofs of Σ. This cannot be done in the PTS $\lambda_{\to}$ extended with (unparametrised) constants: such a constant `proof` should be of type $\texttt{prop} \to \texttt{type}$ and this type cannot be constructed in $\lambda_{\to}$ (notice that $\texttt{type} \equiv *$, so the construction of $\texttt{prop} \to \texttt{type}$ would involve the Π-formation rule $(*, \square, \square)$).

  However, the term `proof` will hardly ever be used on its own. It is usually used when applied to a proposition Σ. In $\vec{C}$-PTSs it is possible to introduce a parametric version of `proof` by the following context declaration:

$$\texttt{proof}(\texttt{p:prop}) : \texttt{type}.$$

  This does not involve the construction of a type $\texttt{prop} \to \texttt{type}$. Nevertheless it is possible to construct the term $\texttt{prop}(P)$ for any term $P : \texttt{prop}$. We obtain a form of polymorphism without using the polymorphism of λ-calculus.

  A disadvantage may be that we cannot speak about the term `proof` "as it is". When using `proof` in the syntax, it must always be applied to a parameter $T : \texttt{prop}$.

  However, an advantage is that we can restrict ourselves to a much more simple type system. In the situation above we remain within the types of the system $\lambda_{\rightarrow}$. We do not need to use types of the system λP. This may have advantages in implementations of type systems. For instance, the system $\lambda_{\rightarrow}$ does not involve the conversion rule

$$\frac{\Gamma \vdash A : B \qquad \Gamma \vdash B : s \qquad B =_{\beta} B'}{\Gamma \vdash A : B'}$$

while λP does involve such a rule. The conversion rule involves β-equality of terms, and though it is decidable whether two λ-terms of λP are β-equal or not, it may take a lot of time and/or memory to establish such a fact. This may cause serious problems when implementing certain type systems. Using parameters whenever possible may therefore simplify implementations. We give an example in Section 10f;

- We illustrate the difference between PTSs, D-PTSs and $\vec{\mathrm{D}}$-PTSs.
  - In a simple PTS like $\lambda_{\rightarrow}$ one can derive the following statement for an identity function:
  $$\alpha{:}* \vdash (\lambda \mathtt{x}{:}\alpha.\mathtt{x}) : \alpha \rightarrow \alpha;$$
  - The same derivation can be made in the corresponding D-PTS, but in that D-PTS we have the possibility of abbreviating the term $\lambda\mathtt{x}{:}\alpha.\mathtt{x}$. We can do this in two ways. First of all, we can introduce this definition in the context:
  $$\alpha{:}*, \mathtt{id}{=}(\lambda \mathtt{x}{:}\alpha.\mathtt{x}){:}(\alpha{\rightarrow}\alpha) \vdash \mathtt{id} : \alpha{\rightarrow}\alpha.$$
  But we can also decide to make a local definition:
  $$\alpha{:}* \vdash (\mathtt{id}{=}(\lambda \mathtt{x}{:}\alpha.\mathtt{x}){:}(\alpha{\rightarrow}\alpha) \text{ IN } \mathtt{id}) : \\ \mathtt{id}{=}(\lambda \mathtt{x}{:}\alpha.\mathtt{x}){:}(\alpha{\rightarrow}\alpha) \text{ IN } \alpha{\rightarrow}\alpha.$$
  We see that the definition of `id` appears both in the term and in the type of the term, but not in the context.
  The advantages of definitions are:
    - We can abbreviate long expressions. This makes terms more *surveyable*: `id` is shorter than $\lambda\mathtt{x}{:}\alpha.\mathtt{x}$;
    - We can give names to important expressions. This makes terms more *understandable*: `id` expresses that we have to do with the `identity` function, whilst $\lambda\mathtt{x}{:}\alpha.\mathtt{x}$ does not express this fact;
  - In a $\vec{\mathrm{D}}$-PTS we have more options for abbreviating the identity function.

- First, we can make the same derivation as in the D-PTS. Formally, there is a small difference: we cannot use id but must work with id(), a parametric term with zero parameters (as in $\vec{\mathrm{D}}$-PTSs we can only work with parametric definitions). We obtain (in the case of the global definition):

$$\alpha{:}*, \mathtt{id}() {=} (\lambda \mathtt{x}{:}\alpha.\mathtt{x}){:}(\alpha{\to}\alpha) \vdash \mathtt{id}() : \alpha{\to}\alpha;$$

- But we could also decide to use one or more parameters in the definition of id. For instance, we could parametrise the variable α. This results in the declaration

$$\mathtt{id}(\alpha{:}*) {=} (\lambda \mathtt{x}{:}\alpha.\mathtt{x}){:}(\alpha{\to}\alpha).$$

  If we want to use this declaration, we must have a term T of type $*$. Assuming that we have such a term T, we can derive:

$$\mathtt{id}(\alpha{:}*) {=} (\lambda \mathtt{x}{:}\alpha.\mathtt{x}){:}(\alpha{\to}\alpha) \vdash \mathtt{id}(T) : T \to T.$$

  We see that we obtain a restricted form of polymorphism in this way. The type system may not allow the construction of $\lambda\alpha{:}*.\lambda \mathtt{x}{:}\alpha.\mathtt{x}$; nevertheless the parameter mechanism makes it possible to express id(T) for any type $T : *$;
- We could also decide to parametrise the variable x, and leave the variable α unparametrised. This yields a context

$$\alpha{:}*, \mathtt{id}(\mathtt{x}{:}\alpha) {=} \mathtt{x}{:}\alpha.$$

  We see that the λ-abstraction $\lambda \mathtt{x}{:}\alpha.\mathtt{x}$ is parametrised now. The definition declaration means: For any term t of type α, the term id(t) of type α is defined by t. If we have such a term t, then we can derive

$$\alpha{:}*, \mathtt{id}(\mathtt{x}{:}\alpha) {=} \mathtt{x}{:}\alpha \vdash \mathtt{id}(t) : \alpha.$$

  Observe that id(t) does not have type $\alpha \to \alpha$ (as was the case with id) but type α (which would also be the type of id$t$ if we had used the identity id$=\lambda \mathtt{x}{:}\alpha.\mathtt{x}$ from λ-calculus);
- Finally, one could parametrise both α and x. This results in a declaration

$$\mathtt{id}(\alpha{:}*, \mathtt{x}{:}\alpha) {=} \mathtt{x}{:}\alpha$$

  in the context. If we have a term T of type $*$ and a term t of type T, we can derive

$$\mathtt{id}(\alpha{:}*, \mathtt{x}{:}\alpha) {=} \mathtt{x}{:}\alpha \vdash \mathtt{id}(t) : T.$$

The global definitions given in the $\vec{\mathrm{D}}$-PTS case could also be made local, as was done in the D-PTS case.

We now start a more detailed description of the various extensions of PTSs with definitions and parameters.

We define two reduction relations, namely the δ- and β-reduction. β-reduction is defined as usual. As far as global definitions are concerned, δ-reduction is comparable to δ-reduction in AUTOMATH. This is reflected in rule (δ1) in the definition below. But now, a δ-reduction step can also unfold *local* definitions. Therefore, two new reduction steps are introduced. Rule (δ2) below removes the declaration of a local definition if there is no position within its scope where it can be unfolded ("removal of void local definitions"). Rule (δ3) shows how one can treat a local definition as a global definition, and thus how the problem of unfolding local definitions can be reduced to unfolding global definitions ("localisation of global definitions").

Remember that $\Delta \equiv x_1{:}B_1, \ldots, x_n{:}B_n$.

Definition 10.14 We define the following three reduction rules:

$$\Gamma_1, c(\Delta){=}a{:}A, \Gamma_2 \vdash c(b_1, \ldots, b_n) \to_\delta a[x_i{:=}b_i]_{i=1}^n \qquad (\delta 1)$$

$$\Gamma \vdash c(\Delta){=}a{:}A \text{ IN } b \to_\delta b \quad \text{if } c \notin \text{CONS}\,(b) \qquad (\delta 2)$$

$$\frac{\Gamma, c(\Delta){=}a{:}A \vdash b \to_\delta b'}{\Gamma \vdash c(\Delta){=}a{:}A \text{ IN } b \to_\delta c(\Delta){=}a{:}A \text{ IN } b'} \qquad (\delta 3)$$

Furthermore, we have some compatibility rules. These rules are not very complicated, there are only quite a lot of them.

Definition 10.15 We define the following compatibility rules:

$$\frac{\Gamma, \Delta \vdash a \to_\delta a'}{\Gamma \vdash c(\Delta){=}a{:}A \text{ IN } b \to_\delta c(\Delta){=}a'{:}A \text{ IN } b}$$

$$\frac{\Gamma, \Delta \vdash A \to_\delta A'}{\Gamma \vdash c(\Delta){=}a{:}A \text{ IN } b \to_\delta c(\Delta){=}a{:}A' \text{ IN } b}$$

$$\frac{\Gamma, \Delta_i \vdash B_i \to_\delta B_i'}{\Gamma \vdash c(\Delta){=}a{:}A \text{ IN } b \to_\delta c(x_1{:}B_1, \ldots, x_i{:}B_i', \ldots, x_n{:}B_n){=}a{:}A \text{ IN } b}$$

$$\frac{\Gamma \vdash a \to_\delta a'}{\Gamma \vdash ab \to_\delta a'b} \qquad \frac{\Gamma \vdash b \to_\delta b'}{\Gamma \vdash ab \to_\delta ab'}$$

$$\frac{\Gamma, x{:}A \vdash a \to_\delta a'}{\Gamma \vdash \lambda x{:}A.a \to_\delta \lambda x{:}A.a'} \qquad \frac{\Gamma \vdash A \to_\delta A'}{\Gamma \vdash \lambda x{:}A.a \to_\delta \lambda x{:}A'.a}$$

$$\frac{\Gamma, x{:}A \vdash a \to_\delta a'}{\Gamma \vdash \Pi x{:}A.a \to_\delta \Pi x{:}A.a'} \qquad \frac{\Gamma \vdash A \to_\delta A'}{\Gamma \vdash \Pi x{:}A.a \to_\delta \Pi x{:}A'.a}$$

$$\frac{\Gamma \vdash a_j \to_\delta a_j'}{\Gamma \vdash c(a_1,\ldots,a_n) \to_\delta c(a_1,\ldots,a_j',\ldots,a_n)}$$

Remark 10.16 One might also expect a compatibility rule

$$\frac{\Gamma \vdash b \to_\delta b'}{\Gamma \vdash c(\Delta){=}a{:}A \text{ IN } b \to_\delta c(\Delta){=}a{:}A \text{ IN } b'}.$$

However, this rule is a derived rule (see the forthcoming Lemma 10.26).

Now we can give a formal definition of our reduction:

Definition 10.17
- β-reduction is defined as before and we use $\to_\beta$, $\twoheadrightarrow_\beta$, $\twoheadrightarrow_\beta^+$, and $=_\beta$ as usual.
- δ-reduction is defined as the smallest relation $\to_\delta$ on $\mathcal{C}_P \times \mathcal{T}_P \times \mathcal{T}_P$ closed under the rules ($\delta 1$), ($\delta 2$) and ($\delta 3$) of Definition 10.14 and under the compatibility rules of Definition 10.15.

When Γ is the empty context, we write $a \to_\delta a'$ instead of $\Gamma \vdash a \to_\delta a'$.
We extend $\to_\delta$ to contexts:

Definition 10.18 δ-reduction between contexts is the smallest relation $\to_\delta$ on $\mathcal{C}_P \times \mathcal{C}_P$ closed under the following rules:

$$\frac{\Gamma_1 \vdash A \to_\delta A'}{\Gamma_1, x{:}A, \Gamma_2 \to_\delta \Gamma_1, x{:}A', \Gamma_2}$$

$$\frac{\Gamma_1, \Delta \to_\delta \Gamma_1, \Delta'}{\Gamma_1, c(\Delta){:}A, \Gamma_2 \to_\delta \Gamma_1, c(\Delta'){:}A, \Gamma_2}$$

$$\frac{\Gamma_1, \Delta \vdash A \to_\delta A'}{\Gamma_1, c(\Delta){:}A, \Gamma_2 \to_\delta \Gamma_1, c(\Delta){:}A', \Gamma_2}$$

$$\frac{\Gamma_1, \Delta \vdash a \to_\delta a'}{\Gamma_1, c(\Delta){=}a{:}A, \Gamma_2 \to_\delta \Gamma_1, c(\Delta){=}a'{:}A, \Gamma_2}$$

$$\frac{\Gamma_1, \Delta \to_\delta \Gamma_1, \Delta'}{\Gamma_1, c(\Delta){=}a{:}A, \Gamma_2 \to_\delta \Gamma_1, c(\Delta'){=}a{:}A, \Gamma_2}$$

$$\frac{\Gamma_1, \Delta \vdash A \to_\delta A'}{\Gamma_1, c(\Delta){=}a{:}A, \Gamma_2 \to_\delta \Gamma_1, c(\Delta){=}a{:}A', \Gamma_2}$$

We now describe the extensions to PTSs that are needed to obtain $\vec{C}$-PTSs and $\vec{D}$-PTSs. We don't discuss D-PTSs and $\vec{C}$D-PTSs: D-PTSs are introduced in [137] and presented in Chapter 8, $\vec{C}$D-PTSs can be constructed by extending D-PTSs with the additional rules for $\vec{C}$-PTSs.

Definition 10.19 ($\vec{C}$-PTS: Pure Type Systems with parametric constants) The *typing relation* $\vdash^{\vec{C}}$ is the smallest relation on $\mathcal{C}_P \times \mathcal{T}_P \times \mathcal{T}_P$ closed under the rules in Definition 4.21 and the following ones (we still write $\Delta \equiv x_1{:}B_1, \ldots, x_n{:}B_n$):

$$(\vec{C}\textbf{-weak}) \qquad \frac{\Gamma \vdash^{\vec{C}} b : B \qquad \Gamma, \Delta \vdash^{\vec{C}} A : s}{\Gamma, c(\Delta) : A \vdash^{\vec{C}} b : B}$$

$$(\vec{C}\textbf{-app}) \qquad \frac{\begin{array}{lll} \Gamma_1, c(\Delta){:}A, \Gamma_2 & \vdash^{\vec{C}} \; b_i{:}B_i[x_j{:=}b_j]_{j=1}^{i-1} & (i = 1, \ldots, n) \\ \Gamma_1, c(\Delta){:}A, \Gamma_2 & \vdash^{\vec{C}} \; A : s & (\text{if } n = 0) \end{array}}{\Gamma_1, c(\Delta){:}A, \Gamma_2 \vdash^{\vec{C}} c(b_1, \ldots, b_n) : A[x_j{:=}b_j]_{j=1}^{n}}$$

where $s \in S$ and the c that is introduced in the $\vec{C}$-weakening rule is assumed to be Γ-fresh.

Note that $\vec{C}$-PTSs exts PTSs with parametric constants exactly as we did for the cube in Chapter 9. The following remark is familiar from that chapter:

Remark 10.20 At first sight one might miss a $\vec{C}$-introduction rule. Such a rule, however, is not necessary, as c (on its own) is not a term. c can only be (part of) a term in the form $c(b_1, \ldots, b_n)$, and such terms can be typed by the $\vec{C}$-application rule.

The extra condition $\Gamma_1, c(\Delta){:}A, \Gamma_2 \vdash^{\vec{C}} A : s$ in the $\vec{C}$-application rule for $n = 0$ is necessary to prevent an empty list of premises. Such an empty list of premises would make it possible to have almost arbitrary contexts in the conclusion. The extra condition is only needed to assure that the context in the conclusion is a legal context.

Adapting these rules for $\vdash^{\vec{C}}$ and the rules for definitions of [137] results in rules for *parametric definitions*:

Definition 10.21 ($\vec{D}$-PTS: Pure Type Systems with parametric definitions) The *typing relation* $\vdash^{\vec{D}}$ is the smallest relation on $\mathcal{C}_P \times \mathcal{T}_P \times \mathcal{T}_P$ closed under the rules in Definition 4.21 and the following ones:

($\vec{\mathrm{D}}$-weak)
$$\frac{\Gamma \vdash^{\vec{D}} b : B \qquad \Gamma, \Delta \vdash^{\vec{D}} a : A}{\Gamma, c(\Delta){=}a{:}A \vdash^{\vec{D}} b : B}$$

($\vec{\mathrm{D}}$-app)
$$\frac{\begin{array}{ll}\Gamma_1, c(\Delta){=}a{:}A, \Gamma_2 \vdash^{\vec{D}} b_i : B_i[x_j{:=}b_j]_{j=1}^{i-1} & (i = 1, \ldots, n) \\ \Gamma_1, c(\Delta){=}a{:}A, \Gamma_2 \vdash^{\vec{D}} a : A & (\text{if } n = 0)\end{array}}{\Gamma_1, c(\Delta){=}a{:}A, \Gamma_2 \vdash^{\vec{D}} c(b_1, \ldots, b_n) : A[x_j{:=}b_j]_{j=1}^{n}}$$

($\vec{\mathrm{D}}$-form)
$$\frac{\Gamma, c(\Delta){=}a{:}A \vdash^{\vec{D}} B : s}{\Gamma \vdash^{\vec{D}} c(\Delta){=}a{:}A \text{ IN } B : s}$$

($\vec{\mathrm{D}}$-intro)
$$\frac{\Gamma, c(\Delta){=}a{:}A \vdash^{\vec{D}} b : B \qquad \Gamma \vdash^{\vec{D}} c(\Delta){=}a{:}A \text{ IN } B : s}{\Gamma \vdash^{\vec{D}} c(\Delta){=}a{:}A \text{ IN } b : c(\Delta){=}a{:}A \text{ IN } B}$$

($\vec{\mathrm{D}}$-conv)
$$\frac{\Gamma \vdash^{\vec{D}} b : B \qquad \Gamma \vdash^{\vec{D}} B' : s \qquad \Gamma \vdash B =_\delta B'}{\Gamma \vdash^{\vec{D}} b : B'}$$

where $s \in \boldsymbol{S}$, and the c that is introduced in the $\vec{\mathrm{D}}$-weakening rule is assumed to be Γ-fresh.

$\vdash^{\vec{D}}$ includes the definition system of [137]: The $\vec{\mathrm{D}}$-application rule for $n = 0$ can be seen as the δ-start rule of D-PTSs.

Definition 10.22 (Pure Type Systems with (parametric) constants and (parametric) definitions) Let $\mathfrak{S}$ be a specification (see 4.18).

- A *Pure Type System with (parametric) constants* $\vec{\mathrm{C}}$-PTS is denoted as $\lambda^{\vec{C}}(\mathfrak{S})$ and consists of a set of terms $\mathcal{T}_P$, a set of contexts $\mathcal{C}_P$, the β-reduction rule and the typing relation $\vdash^{\vec{C}}$;

- A *Pure Type System with (parametric) definitions* $\vec{\mathrm{D}}$-PTS is denoted as $\lambda^{\vec{D}}(\mathfrak{S})$ and consists of a set of terms $\mathcal{T}_P$, a set of contexts $\mathcal{C}_P$, the β- and δ-reduction rules and the typing relation $\vdash^{\vec{D}}$;

- A *Pure Type System with (parametric) constants and (parametric) definitions* $\vec{\mathrm{C}}\vec{\mathrm{D}}$-PTS is denoted as $\lambda^{\vec{C}\vec{D}}(\mathfrak{S})$ and consists of a set of terms $\mathcal{T}_P$, a set of contexts $\mathcal{C}_P$, the β- and δ-reduction rules and the typing relation $\vdash^{\vec{C}\vec{D}}$, which is the smallest relation on $\mathcal{C}_P \times \mathcal{T}_P \times \mathcal{T}_P$ that is closed under the rules of Definition 4.21 and the rules of $\vdash^{\vec{C}}$ and $\vdash^{\vec{D}}$.

A term a is *legal* (with respect to a certain type system) if there are Γ, b such that either $\Gamma \vdash a : b$ or $\Gamma \vdash b : a$ is derivable (in that type system). Similarly, a context Γ is *legal* if there are a, b such that $\Gamma \vdash a : b$.

All contexts occurring in $\vec{C}\vec{D}$-PTSs are sound (see Definition 10.12). Moreover, since $\vec{C}\vec{D}$-PTSs are clearly extensions of PTSs, $\vec{C}$-PTSs and $\vec{D}$-PTSs, this implies that all contexts occurring in PTSs, $\vec{C}$-PTSs and $\vec{D}$-PTSs are sound. We need this fact in many proofs in the next sections. The proof of the lemma below is by induction on the derivation of $\Gamma \vdash^{\vec{C}\vec{D}} a : A$.

Lemma 10.23 *Assume* $\Gamma \vdash^{\vec{C}\vec{D}} b : B$.

1. $\text{DOM}(b), \text{DOM}(B) \subseteq \text{DOM}(\Gamma)$;
2. Γ *is sound.*

PROOF: We prove the statements (1) and (2) simultaneously by induction on the derivation of $\Gamma \vdash^{\vec{C}\vec{D}} b : B$. We treat the two most important cases:

- ($\vec{D}$-weakening) $\Gamma, c(\Delta){=}a{:}A \vdash b{:}B$ because $\Gamma \vdash b{:}B$ and $\Gamma, \Delta \vdash a{:}A$. (1) is trivial; (2) follows from the induction hypothesis for (1);
- ($\vec{D}$-formation) $\Gamma \vdash (c(\Delta){=}a{:}A \text{ IN } B) : s$ because $\Gamma, c(\Delta){=}a{:}A \vdash B : s$. (1) follows from the induction hypothesis for (2); (2) is trivial.

⊠

10b Properties of terms

In this section, we prove properties of terms without wondering whether these terms are legal or not. In Section 10b1 we discuss some basic properties, such as a Substitution Lemma, and substitutivity. Section 10b2 is devoted to the Church-Rosser property for $\beta\delta$-reduction, and in Section 10b3 we prove strong normalisation for δ-reduction.

Though we do not restrict ourselves to legal terms in this section, we often demand that the free variables and constants of a term are contained in the domain of a sound context.

10b1 Basic properties

In the following lemma we show that a δ-reduction step remains invariant if we enlarge the context. The proof is done by induction on the definition of $\rightarrow_\delta$.

Lemma 10.24 **($\rightarrow_\delta$-weakening)** *Let* $\langle \Gamma_1, \Gamma_2, \Gamma_3 \rangle \in \mathcal{C}_P$ *be such that*

$$\Gamma_1, \Gamma_3 \vdash b \rightarrow_\delta b'.$$

Then

$$\Gamma_1, \Gamma_2, \Gamma_3 \vdash b \rightarrow_\delta b'.$$

⊠

The implications from left to right of the following lemma are a particular case of Lemma 10.24.

The implications from right to left allow to make the context shorter. The first two parts state that declarations of the form $c(\Delta){:}A$ and $x{:}A$ in a context do not have any influence on the reduction relation $\rightarrow_{\beta\delta}$. The last part states that declarations of the form $c(\Delta){=}a{:}A$ in a context do not have any influence on the $\rightarrow_{\beta\delta}$-reduction behaviour of terms $b \in \mathcal{T}_P$ with $c \notin \textsc{cons}(b)$. This allows to remove definition declarations, as rule ($\delta 2$) of the definition of δ-reduction does for local definitions.

The lemma is proved by induction on the definition of $\rightarrow_\beta$ and $\rightarrow_\delta$.

Lemma 10.25

1. *Let* $\langle \Gamma_1, x{:}A, \Gamma_2 \rangle \in \mathcal{C}_P$ *and* $b \in \mathcal{T}_P$.
   $\Gamma_1, \Gamma_2 \vdash b \rightarrow_{\beta\delta} b'$ *if and only if* $\Gamma_1, x{:}A, \Gamma_2 \vdash b \rightarrow_{\beta\delta} b'$;

2. *Let* $\langle \Gamma_1, c(\Delta){:}A, \Gamma_2 \rangle \in \mathcal{C}_P$ *and* $b \in \mathcal{T}_P$.
   $\Gamma_1, \Gamma_2 \vdash b \rightarrow_{\beta\delta} b'$ *if and only if* $\Gamma_1, c(\Delta){:}A, \Gamma_2 \vdash b \rightarrow_{\beta\delta} b'$;

3. *Let* $\langle \Gamma_1, c(\Delta){=}a{:}A, \Gamma_2 \rangle \in \mathcal{C}_P$ *and* $b \in \mathcal{T}_P$ *be such that* $c \notin \textsc{cons}(b)$.
   $\Gamma_1, \Gamma_2 \vdash b \rightarrow_{\beta\delta} b'$ *if and only if* $\Gamma_1, c(\Delta){=}a{:}A, \Gamma_2 \vdash b \rightarrow_{\beta\delta} b'$.

⊠

Now we show that the compatibility rule for $c(\Delta){=}a{:}A$ IN b when we reduce inside b is a derived rule (and therefore not included in the list of compatibility rules in Definition 10.15).

Lemma 10.26 *The following rule is derivable from the ones in the definition of* $\rightarrow_\delta$:

$$\frac{\Gamma \vdash b \rightarrow_\delta b'}{\Gamma \vdash c(\Delta){=}a{:}A \text{ IN } b \rightarrow_\delta c(\Delta){=}a{:}A \text{ IN } b'}.$$

PROOF: Suppose $\Gamma \vdash b \rightarrow_\delta b'$. By Lemma 10.24, $\Gamma, c(\Delta){=}a{:}A \vdash b \rightarrow_\delta b'$. By definition of $\rightarrow_\delta$, it follows that $\Gamma \vdash c(\Delta){=}a{:}A$ IN $b \rightarrow_\delta c(\Delta){=}a{:}A$ IN b'. ⊠

The following lemma is proved by induction on the structure of a.

Lemma 10.27 (Substitution Lemma) *Suppose* $x \not\equiv y$ *and* $x \notin \textsc{fv}(d)$. *Then*

$$a[x{:=}b][y{:=}d] \equiv a[y{:=}d][x{:=}b[y{:=}d]].$$

⊠

The following lemma shows that $\rightarrow_\beta$ is substitutive. It is proved by induction on the generation of $\rightarrow_\beta$ and by the Substitution Lemma.

Lemma 10.28 (Substitutivity for $\to_\beta$)
If $a \to_\beta a'$ *then* $a[x{:=}b] \to_\beta a'[x{:=}b]$. ⊠

The relation $\to_\delta$ is not substitutive. For example, let

$$\Gamma \equiv \mathtt{x}{:}\alpha, \mathtt{x}'{:}\alpha, \mathsf{c}(){=}\mathtt{x}{:}\alpha.$$

We have

$$\Gamma \vdash^{\vec{D}} \mathsf{c}() : \alpha$$

and

$$\Gamma \vdash \mathsf{c}() \to_\delta \mathtt{x},$$

but not

$$\Gamma \vdash \mathsf{c}()[\mathtt{x}{:=}\mathtt{x}'] \to_\delta \mathtt{x}[\mathtt{x}{:=}\mathtt{x}'].$$

The reason for this is to be found in the δ-weakening rule. When we introduce a new parametric definition $c(\Delta){=}a{:}A$, the term a may contain free variables that are not in the domain of Δ but in the domain of Γ. When unfolding the definition c, these new variables can appear, thus destroying substitutivity.

However, we do have the following version of substitutivity. It is adapted so that the substitution now occurs in the context as well. The proof is by induction on the derivation of $\Gamma \vdash a \to_\delta a'$.

Lemma 10.29 (Weak substitutivity for $\to_\delta$) *If* $\Gamma \vdash a \to_\delta a'$ *then*

$$\Gamma[x{:=}b] \vdash a[x{:=}b] \to_\delta a'[x{:=}b].$$

PROOF: Induction on the derivation of $\Gamma \vdash a \to_\delta a'$. We only consider the two most interesting cases:

- $\Gamma_1, c(\Delta){=}d{:}A, \Gamma_2 \vdash c(b_1, \ldots, b_n) \to_\delta d[x_i{:=}b_i]_{i=1}^n$. ($\delta 1$)
  Now

  $$\langle \Gamma_1, c(\Delta){=}d{:}A, \Gamma_2\rangle[x{:=}b] \equiv \langle \Gamma_1[x{:=}b], c(\Delta[x{:=}b]){=}d[x{:=}b]{:}A[x{:=}b], \Gamma_2[x{:=}b]\rangle$$

  so

  $$\langle \Gamma_1, c(\Delta){=}d{:}A, \Gamma_2\rangle[x{:=}b] \vdash c(b_1[x{:=}b], \ldots, b_n[x{:=}b]) \to_\delta d[x{:=}b][x_i{:=}b_i[x{:=}b]]_{i=1}^n$$

  and as the x_i are bound in $\langle \Gamma_1, c(\Delta){=}d{:}A, \Gamma_2\rangle$, $x_i \notin \mathrm{FV}(b)$ by the variable convention, so by the Substitution Lemma

  $$d[x{:=}b][x_i{:=}b_i[x{:=}b]]_{i=1}^n \equiv \left(d[x_i{:=}b_i]_{i=1}^n\right)[x{:=}b];$$

- $c \notin$ CONS (a) and $\Gamma \vdash c(\Delta){=}d{:}A$ IN $a \to_\delta a$ ($\delta 2$). We have that c is bound in $c(\Delta){=}d{:}A$ IN a, so by the variable convention $c \notin$ CONS (b), so $c \notin$ CONS $(a[x{:=}b])$. Hence

$$\Gamma[x{:=}b] \vdash (c(\Delta){=}d{:}A \text{ IN } a)[x{:=}b] \to_\delta a[x{:=}b].$$

⊠

In the following lemma we reduce inside the term b of $a[x{:=}b]$. The proof is by induction on the structure of a.

Lemma 10.30 *If* $\Gamma \vdash b \to_{\beta\delta} b'$ *then* $\Gamma \vdash a[x{:=}b] \twoheadrightarrow_{\beta\delta} a[x{:=}b']$. ⊠

10b2 Church-Rosser for $\to_{\beta\delta}$

In this section we prove the Church-Rosser theorem for $\twoheadrightarrow_\beta$, $\twoheadrightarrow_\delta$ and $\twoheadrightarrow_{\beta\delta}$. As for ordinary λ-terms, we have:

Theorem 10.31 (Church-Rosser theorem for β-reduction) *If* $a \twoheadrightarrow_\beta a_1$ *and* $a \twoheadrightarrow_\beta a_2$ *then there exists a term* a_3 *such that* $a_1 \twoheadrightarrow_\beta a_3$ *and* $a_2 \twoheadrightarrow_\beta a_3$. ⊠

The proof is similar to the proof for λ-terms without definitions and parameters.

For a context Γ and a term b we define $|b|_\Gamma$, which is, intuitively, b in which all definitions are unfolded. That is: both the local definitions inside b, and the global definitions given in Γ. The definition is by induction on the total number of symbols occurring in Γ and b.

Definition 10.32 For $a \in \mathcal{T}_P$ and $\Gamma \in \mathcal{C}_P$ we define a term $|a|_\Gamma \in \mathcal{T}_P$ as follows:

$$\begin{array}{rcl}
|x|_\Gamma & \equiv & x \text{ (for } x \in \mathcal{V});\\
|s|_\Gamma & \equiv & s \text{ (for } s \in \boldsymbol{S});\\
|c(b_1,\ldots,b_n)|_\Gamma & \equiv & \begin{cases} |a|_{\Gamma_1,\Delta}[x_i{:=}|b_i|_\Gamma]_{i=1}^n & \text{if } \Gamma \equiv \\ & \langle \Gamma_1, c(\Delta){=}a{:}A, \Gamma_2\rangle; \\ c(|b_1|_\Gamma,\ldots,|b_n|_\Gamma) & \text{if } c \notin \text{DEFCONS}(\Gamma); \end{cases}\\
|ab|_\Gamma & \equiv & |a|_\Gamma |b|_\Gamma;\\
|\lambda x{:}A.a|_\Gamma & \equiv & \lambda x{:}|A|_\Gamma.|a|_{\Gamma,x:A};\\
|\Pi x{:}A.B|_\Gamma & \equiv & \Pi x{:}|A|_\Gamma.|B|_{\Gamma,x:A};\\
|c(\Delta){=}a{:}A \text{ in } b|_\Gamma & \equiv & |b|_{\Gamma,c(\Delta)=a:A} \text{ (where } c \text{ is } \Gamma\text{-fresh)}.
\end{array}$$

The following lemma shows that $|b|_\Gamma$ is independent from variable declarations $x{:}A$ and constant declarations $c(\Delta){:}A$ in Γ. The proof is by induction on the definition of $|b|_{\Gamma_1,\Gamma_2}$.

Lemma 10.33

- $|b|_{\Gamma_1,\Gamma_2} \equiv |b|_{\Gamma_1,x:A,\Gamma_2}$;
- $|b|_{\Gamma_1,\Gamma_2} \equiv |b|_{\Gamma_1,c(\Delta):A,\Gamma_2}$. ⊠

By induction on the definition of $|b|_\Gamma$ one shows that $|b|_\Gamma$ does not contain any local definitions.

Lemma 10.34 *For all $b \in \mathcal{T}_P$ and $\Gamma \in \mathcal{C}_P$, $|b|_\Gamma$ has no subterms of the form $(c(\Delta){=}a{:}A \text{ IN } d)$.* ⊠

The intuition on $|b|_\Gamma$ suggests that all definitions of b are unfolded in $|b|_\Gamma$. However, there may be global definitions in Γ that have not been unfolded in $|b|_\Gamma$. Take, for example, $\Gamma \equiv \langle \mathsf{c}(){=}\mathsf{c}'(){:}*, \mathsf{c}'(){=}\mathsf{c}''(){:}*\rangle$. Then $|\mathsf{c}()|_\Gamma \equiv |\mathsf{c}'()|_\varnothing \equiv \mathsf{c}'()$, but $\mathsf{c}'()$ is not in δ-normal form with respect to Γ. This is due to the fact that Γ is not a sound context (see 10.12).

By induction on the definition of $|b|_\Gamma$, we show that if Γ is sound, $|b|_\Gamma$ is equal to b with all the definitions in b and Γ unfolded. It is no serious restriction to consider only sound contexts, as all contexts that appear in $\vec{C}\vec{D}$-PTSs are sound (Lemma 10.23).

Lemma 10.35 *Let Γ be a sound context such that* $\text{DOM}(b) \subseteq \text{DOM}(\Gamma)$. *Then* $\text{DOM}(|b|_\Gamma) \subseteq \text{DOM}(\Gamma) \setminus \text{DEFCONS}(\Gamma)$.

PROOF: Induction on the definition of $|b|_\Gamma$. We treat the two most interesting cases (at (IH) we use the induction hypothesis):

- $b \equiv c(b_1, \dots, b_n)$ and $\Gamma \equiv \langle \Gamma_1, c(\Delta){=}a{:}A, \Gamma_2\rangle$.

$$\begin{aligned}
\text{DOM}(|b|_\Gamma) &= \text{DOM}\left(|a|_{\Gamma_1,\Delta}[x_i{:=}|b_i|_\Gamma]_{i=1}^n\right)\\
&\subseteq \left(\text{DOM}\left(|a|_{\Gamma_1,\Delta}\right) \setminus \{x_1,\dots,x_n\}\right) \cup \bigcup_{i=1}^n \text{DOM}(|b_i|_\Gamma)\\
&\overset{\text{(IH)}}{\subseteq} (\text{DOM}(\Gamma_1,\Delta) \setminus \text{DEFCONS}(\Gamma_1)) \setminus \{x_1,\dots,x_n\}\\
&\qquad \cup (\text{DOM}(\Gamma) \setminus \text{DEFCONS}(\Gamma))\\
&= \text{DOM}(\Gamma) \setminus \text{DEFCONS}(\Gamma).
\end{aligned}$$

We can use the induction hypothesis at (IH) because Γ is sound, and therefore $\text{DOM}(a) \subseteq \text{DOM}(\Gamma_1, \Delta)$;

- $b \equiv c(\Delta){=}a{:}A \text{ IN } b'$.

$$\begin{aligned}\text{DOM}\,(|b|_\Gamma) &= \text{DOM}\left(|b'|_{\Gamma,c(\Delta)=a:A}\right)\\ &\stackrel{\text{(IH)}}{\subseteq} \text{DOM}\,(\Gamma, c(\Delta){=}a{:}A) \setminus \text{DEFCONS}\,(\Gamma, c(\Delta){=}a{:}A)\\ &= \text{DOM}\,(\Gamma) \setminus \text{DEFCONS}\,(\Gamma)\,.\end{aligned}$$

⊠

With the above we can show:

Lemma 10.36 *If* Γ *is sound and* DOM $(d) \subseteq$ DOM (Γ), *then*

$$\Gamma \vdash d \twoheadrightarrow_\delta |d|_\Gamma.$$

PROOF: Induction on the total number of symbols occurring in Γ and d. We treat the two most important cases:

- $\Gamma \equiv \langle \Gamma_1, c(\Delta){=}a{:}A, \Gamma_2 \rangle$ and $d \equiv c(b_1, \ldots, b_n)$.
  Notice that $\Gamma \vdash d \to_\delta a[x_i{:=}b_i]_{i=1}^n$.
  By induction, $\Gamma_1, \Delta \vdash a \twoheadrightarrow_\delta |a|_{\Gamma_1,\Delta}$, so $\Gamma \vdash a \twoheadrightarrow_\delta |a|_{\Gamma_1,\Delta}$ (Lemma 10.24), so by Lemma 10.29,

$$\Gamma[x_i{:=}b_i]_{i=1}^n \vdash a[x_i{:=}b_i]_{i=1}^n \twoheadrightarrow_\delta |a|_{\Gamma_1,\Delta}[x_i{:=}b_i]_{i=1}^n.$$

  As the x_i are bound in $c(\Delta){=}a{:}A$, they do not occur free in Γ, so $\Gamma[x_i{:=}b_i]_{i=1}^n \equiv \Gamma$. Therefore

$$\begin{aligned}\Gamma \vdash a[x_i{:=}b_i]_{i=1}^n &\twoheadrightarrow_\delta |a|_{\Gamma_1,\Delta}[x_i{:=}b_i]_{i=1}^n\\ &\stackrel{\text{(IH, 10.30)}}{\twoheadrightarrow_\delta} |a|_{\Gamma_1,\Delta}[x_i{:=}|b_i|_\Gamma]_{i=1}^n\\ &\equiv |c(b_1, \ldots, b_n)|_\Gamma;\end{aligned}$$

- $d \equiv c(\Delta){=}a{:}A \text{ IN } b$. By the induction hypothesis,

$$\Gamma, c(\Delta){=}a{:}A \vdash b \twoheadrightarrow_\delta |b|_{\Gamma,c(\Delta)=a:A}$$

  hence

$$\Gamma \vdash c(\Delta){=}a{:}A \text{ IN } b \twoheadrightarrow_\delta c(\Delta){=}a{:}A \text{ IN } |b|_{\Gamma,c(\Delta)=a:A}.$$

  Now DOM $(d) \subseteq$ DOM (Γ), so DOM $(b) \subseteq$ DOM $(\Gamma, c(\Delta){=}a{:}A)$,
  so by Lemma 10.35,

$$\text{DOM}\left(|b|_{\Gamma,c(\Delta)=a:A}\right) \subseteq \text{DOM}\,(\Gamma, c(\Delta){=}a{:}A) \setminus \text{DEFCONS}\,(\Gamma, c(\Delta){=}a{:}A)\,,$$

  so $c \notin \text{CONS}\left(|b|_{\Gamma,c(\Delta)=a:A}\right)$, so

$$\Gamma \vdash d \twoheadrightarrow_\delta c(\Delta){=}a{:}A \text{ IN } |b|_{\Gamma,c(\Delta)=a:A} \to_\delta |b|_{\Gamma,c(\Delta)=a:A}.$$

⊠

Corollary 10.37 *In any $\vec{C}\vec{D}$-PTS, the relation $\rightarrow_\delta$ is weakly normalising, i.e. each legal term has a β-normal form.*

PROOF: By Lemma 10.34 and Lemma 10.35, $|b|_\Gamma$ is in δ-normal form; by Lemma 10.36, $|b|_\Gamma$ is a δ-normal form of b. ⊠

The mapping $|-|_\_$ also helps us to show that $\rightarrow_{\beta\delta}$ is confluent (Theorem 10.42). For the proof we use some lemmas:

Lemma 10.38 *Assume $\langle\Gamma_1, \Gamma_3\rangle$ is sound and* $\text{DOM}(b) \subseteq \text{DOM}(\Gamma_1, \Gamma_3)$. *Then* $|b|_{\Gamma_1,\Gamma_2,\Gamma_3} \equiv |b|_{\Gamma_1,\Gamma_3}$.

PROOF: Induction on the definition of $|b|_{\Gamma_1,\Gamma_3}$. We consider only a few non-trivial cases:

- $b \equiv c(b_1, \ldots, b_n)$ and $\Gamma_3 \equiv \langle\Gamma_{31}, c(\Delta){=}a{:}A, \Gamma_{32}\rangle$.

$$\begin{aligned} |c(b_1,\ldots,b_n)|_{\Gamma_1,\Gamma_3} &\equiv |a|_{\Gamma_1,\Gamma_{31},\Delta}[x_i{:=}|b_i|_{\Gamma_1,\Gamma_3}] \\ &\overset{(\text{IH},10.30)}{\equiv} |a|_{\Gamma_1,\Gamma_2,\Gamma_{31},\Delta}[x_i{:=}|b_i|_{\Gamma_1,\Gamma_2,\Gamma_3}] \\ &\equiv |c(b_1,\ldots,b_n)|_{\Gamma_1,\Gamma_2,\Gamma_3}; \end{aligned}$$

- $b \equiv c(\Delta){=}a{:}A \text{ IN } b$. Notice that

$$\begin{aligned} |c(\Delta){=}a{:}A \text{ IN } b|_{\Gamma_1,\Gamma_3} &\equiv |b|_{\Gamma_1,\Gamma_3,c(\Delta)=a:A} \\ &\overset{(\text{IH})}{\equiv} |b|_{\Gamma_1,\Gamma_2,\Gamma_3,c(\Delta)=a:A} \\ &\equiv |c(\Delta){=}a{:}A \text{ IN } b|_{\Gamma_1,\Gamma_2,\Gamma_3}. \end{aligned}$$

⊠

Lemma 10.39 *Assume that $\langle\Gamma_1, \Gamma_2\rangle$ is sound, and* $\text{DOM}(a) \subseteq \text{DOM}(\Gamma_1, \Gamma_2)$, $\text{DOM}(b) \subseteq \text{DOM}(\Gamma_1)$ *and* $x \notin \text{DOM}(\Gamma_1)$. *Then*

$$|a|_{\Gamma_1,\Gamma_2}[x{:=}|b|_{\Gamma_1}] \equiv |a[x{:=}b]|_{\Gamma_1,\Gamma_2[x:=b]}.$$

PROOF: Induction on the definition of $|a|_{\Gamma_1,\Gamma_2}$. We treat only a few non-trivial cases:

- $a \equiv c(b_1, \ldots, b_n)$ and $\Gamma_2 \equiv \Gamma_{21}, c(\Delta){=}c'{:}C, \Gamma_{22}$.

$$\begin{aligned} &|a|_{\Gamma_1,\Gamma_2}[x{:=}|b|_{\Gamma_1}] \\ &\equiv |c'|_{\Gamma_1,\Gamma_{21},\Delta}[x_i{:=}|b_i|_{\Gamma_1,\Gamma_2}]_{i=1}^n[x{:=}|b|_{\Gamma_1}] \\ &\overset{(10.27)}{\equiv} |c'|_{\Gamma_1,\Gamma_{21},\Delta}[x{:=}|b|_{\Gamma_1}][x_i{:=}|b_i|_{\Gamma_1,\Gamma_2}[x{:=}|b|_{\Gamma_1}]]_{i=1}^n \\ &\overset{(\text{IH})}{\equiv} |c'[x{:=}b]|_{\Gamma_1,\Gamma_{21}[x:=b],\Delta[x:=b]}[x_i{:=}|b_i[x{:=}b]|_{\Gamma_1,\Gamma_2[x:=b]}]_{i=1}^n \\ &\equiv |c(b_1,\ldots,b_n)[x{:=}b]|_{\Gamma_1,\Gamma_2[x:=b]}; \end{aligned}$$

- $a \equiv c(\Delta){=}c'{:}C$ IN d.

$$\begin{array}{rcl} |a|_{\Gamma_1,\Gamma_2}[x{:=}|b|_{\Gamma_1}] & \equiv & |d|_{\Gamma_1,\Gamma_2,c(\Delta)=c':C}[x{:=}|b|_{\Gamma_1}] \\ & \stackrel{\text{(IH)}}{\equiv} & |d[x{:=}b]|_{\Gamma_1,\Gamma_2[x:=b],c(\Delta[x:=b])=c'[x:=b]:C[x:=b]} \\ & \equiv & |(c(\Delta){=}c'{:}C \text{ IN } d)[x{:=}b]|_{\Gamma_1,\Gamma_2[x:=b]}. \end{array}$$

⊠

Lemma 10.40 *If* $\Gamma \vdash d \rightarrow_\delta d'$, Γ *is sound, and* DOM $(d) \subseteq$ DOM (Γ), *then* $|d|_\Gamma \equiv |d'|_\Gamma$ *and* DOM $(d') \subseteq$ DOM (Γ).

PROOF: We prove the following two statements simultaneously by induction on the definition of $|d|_\Gamma$:

- If $\Gamma \vdash d \rightarrow_\delta d'$ then $|d|_\Gamma \equiv |d'|_\Gamma$;
- If $\Gamma \rightarrow_\delta \Gamma'$ then $|d|_\Gamma \equiv |d|_{\Gamma'}$.

We prove the two non-trivial cases:

- $\Gamma \equiv \langle \Gamma_1, c(\Delta){=}a{:}A, \Gamma_2\rangle$, $d \equiv c(b_1, \ldots, b_n)$, and $d' \equiv a[x_i{:=}b_i]_{i=1}^n$.

$$\begin{array}{rcl} |d|_\Gamma & \equiv & |a|_{\Gamma_1,\Delta}[x_i{:=}|b_i|_\Gamma]_{i=1}^n \\ & \stackrel{(10.33)}{\equiv} & |a|_{\Gamma_1}[x_i{:=}|b_i|_\Gamma]_{i=1}^n \\ & \stackrel{(10.38)}{\equiv} & |a|_\Gamma[x_i{:=}|b_i|_\Gamma]_{i=1}^n \\ & \stackrel{(10.39)}{\equiv} & |a[x_i{:=}b_i]_{i=1}^n|_\Gamma \\ & \equiv & |d'|_\Gamma. \end{array}$$

  The several cases that have to be distinguished for $\Gamma \rightarrow_\delta \Gamma'$ are all easy to prove;

- $d \equiv c(\Delta){=}a{:}A$ IN b. Use induction on $\Gamma \vdash d \rightarrow_\delta d'$. We treat two cases:
  - $d' \equiv b$ and $c \notin$ CONS (b).
    Then $|d|_\Gamma \equiv |b|_{\Gamma,c(\Delta)=a:A} \stackrel{(10.38)}{\equiv} |b|_\Gamma \equiv |d'|_\Gamma$;
  - $d' \equiv c(\Delta){=}a{:}A$ IN b' and $\Gamma, c(\Delta){=}a{:}A \vdash b \rightarrow_\delta b'$.
    Then $|d|_\Gamma \equiv |b|_{\Gamma,c(\Delta)=a:A} \stackrel{\text{(IH)}}{\equiv} |b'|_{\Gamma,c(\Delta)=a:A} \equiv |d'|_\Gamma$.

  If $\Gamma \rightarrow_\delta \Gamma'$ then $|d|_\Gamma \equiv |b|_{\Gamma,c(\Delta)=a:A} \stackrel{\text{(IH)}}{\equiv} |b|_{\Gamma',c(\Delta)=a:A} \equiv |d|_{\Gamma'}$.

⊠

Lemma 10.41 *If* Γ *is sound,* $\text{DOM}(d) \subseteq \text{DOM}(\Gamma)$ *and* $d \to_\beta d'$*, then* $|d|_\Gamma \twoheadrightarrow_\beta |d'|_\Gamma$. ⊠

The proof is similar to the proof of Lemma 10.40.

Theorem 10.42 (Confluence for $\to_{\beta\delta}$) *If* $\Gamma \vdash a \twoheadrightarrow_{\beta\delta} b_1$, $\Gamma \vdash a \twoheadrightarrow_{\beta\delta} b_2$ *and* Γ *is sound, then there exists a term* d *such that* $\Gamma \vdash b_1 \twoheadrightarrow_{\beta\delta} d$ *and* $\Gamma \vdash b_2 \twoheadrightarrow_{\beta\delta} d$.

PROOF: The proof is illustrated by the following diagram.

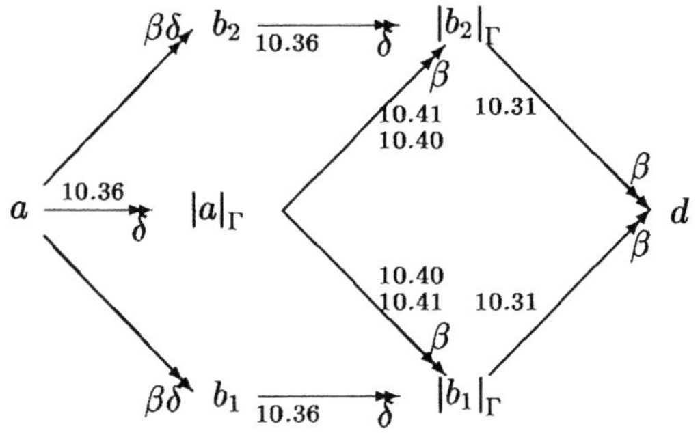

⊠

10b3 Strong normalisation for $\to_\delta$

In [42], van Daalen presents a proof (originally due to de Bruijn) of strong normalisation for a definition system that is at the basis of AUTOMATH. De Vrijer uses a similar technique to prove the finite developments theorem [143]. A similar technique to the one of de Vrijer is also used in [137] to prove strong normalisation for δ-reduction in $\vec{\text{D}}$-PTSs. We extend these techniques to prove strong normalisation for δ-reduction in $\vec{\text{C}}\vec{\text{D}}$-PTSs.

First we define the multiplicity $\mathtt{M}_z(\Gamma, a)$ of a variable z in a term a, depending on a context Γ.

Definition 10.43 For $z \in \mathcal{V}$, $\Gamma \in \mathcal{C}_P$ and $a \in \mathcal{T}_P$ we define a natural number $\mathtt{M}_z(\Gamma, a)$ by induction on the total number of symbols in Γ and a.

$$\begin{array}{rcl} \mathtt{M}_z(\Gamma, z) & = & 1; \\ \mathtt{M}_z(\Gamma, x) & = & 0 \text{ if } x \not\equiv z; \\ \mathtt{M}_z(\Gamma, s) & = & 0 \text{ if } s \in \boldsymbol{S}; \end{array}$$

$$\mathtt{M}_z(\Gamma, c(b_1, \ldots, b_n)) = \begin{cases} \mathtt{M}_z(\langle \Gamma_1, \Delta \rangle, a) + \\ \sum_{i=1}^{n} \mathtt{M}_z(\Gamma, b_i) \cdot \max(1, \mathtt{M}_{x_i}(\langle \Gamma_1, \Delta \rangle, a)) \\ \text{if } \Gamma \equiv \langle \Gamma_1, c(\Delta){=}a{:}A, \Gamma_2 \rangle; \\ \\ \sum_{i=1}^{n} \mathtt{M}_z(\Gamma, b_i) \text{ otherwise}; \end{cases}$$

$$\begin{array}{rcl} \mathtt{M}_z(\Gamma, c(\Delta){=}a{:}A \text{ in } b) & = & \mathtt{M}_z(\langle \Gamma, \Delta \rangle, a) + \mathtt{M}_z(\langle \Gamma, \Delta \rangle, A) + \\ & & \sum_{i=1}^{n} \mathtt{M}_z(\langle \Gamma, \Delta_i \rangle, B_i) + \\ & & \mathtt{M}_z(\langle \Gamma, c(\Delta){=}a{:}A \rangle, b); \end{array}$$

$$\begin{array}{rcl} \mathtt{M}_z(\Gamma, ab) & = & \mathtt{M}_z(\Gamma, a) + \mathtt{M}_z(\Gamma, b); \\ \mathtt{M}_z(\Gamma, \Pi x{:}A.a) & = & \mathtt{M}_z(\langle \Gamma, x{:}A \rangle, a) + \mathtt{M}_z(\Gamma, A); \\ \mathtt{M}_z(\Gamma, \lambda x{:}A.a) & = & \mathtt{M}_z(\langle \Gamma, x{:}A \rangle, a) + \mathtt{M}_z(\Gamma, A). \end{array}$$

Following the line of [143] one can prove the following lemma (using induction on the definitions of $\mathtt{M}_{\_}(-,-)$):

Lemma 10.44

1. *If Γ is sound,* $\text{DOM}(a) \subseteq \text{DOM}(\Gamma)$ *and* $x \notin \text{FV}(a) \cup \text{FV}(\Gamma)$, *then*

$$\mathtt{M}_x(\Gamma, a) = 0;$$

2. *If* $\langle \Gamma_1, \Gamma_3 \rangle$ *is sound and* $\text{DOM}(a) \subseteq \text{DOM}(\langle \Gamma_1, \Gamma_3 \rangle)$, *then*

$$\mathtt{M}_x(\langle \Gamma_1, \Gamma_3 \rangle, a) = \mathtt{M}_x(\langle \Gamma_1, \Gamma_2, \Gamma_3 \rangle, a);$$

3. *If* $\langle \Gamma_1, \Gamma_2 \rangle$ *is sound,* $\text{DOM}(b) \subseteq \text{DOM}(\langle \Gamma_1, \Gamma_2 \rangle)$, $\text{DOM}(a) \subseteq \text{DOM}(\Gamma_1)$, $x \not\equiv z$ *and* $x \notin \text{FV}(\Gamma_1)$, *then*

$$\mathtt{M}_z(\langle \Gamma_1, \Gamma_2[x{:=}a] \rangle, b[x{:=}a]) = \mathtt{M}_z(\langle \Gamma_1, \Gamma_2 \rangle, b) + \mathtt{M}_z(\Gamma_1, a) \cdot \mathtt{M}_x(\langle \Gamma_1, \Gamma_2 \rangle, b).$$

⊠

The following lemma requires a somewhat more complicated proof than in [143], as contexts are involved in our situation.

Lemma 10.45 *Let* Γ *be sound,* $\text{DOM}(a) \subseteq \text{DOM}(\Gamma)$. *If* $\Gamma \vdash a \to_\delta b$, *then* $\mathtt{M}_x(\Gamma, a) \geq \mathtt{M}_x(\Gamma, b)$.

PROOF: We simultaneously prove, using induction on the total number of symbols in Γ and a, the following two statements:

1. If $\Gamma \vdash a \to_\delta b$, then $\mathtt{M}_x(\Gamma, a) \geq \mathtt{M}_x(\Gamma, b)$;
2. If $\Gamma \to_\delta \Gamma'$, then $\mathtt{M}_x(\Gamma, a) \geq \mathtt{M}_x(\Gamma', a)$.

The proof is straightforward, using the lemma above. ⊠

Next we define, for $\Gamma \in \mathcal{C}_P$ and $a \in \mathcal{T}_P$, a natural number $\mathrm{L}_\Gamma(a)$ that decreases with each δ reduction step. It is similar to the mappings defined in [143] (used to prove the finite developments theorem), in [42] and in [137] (used to prove strong normalisation of δ-reduction). This function $\mathrm{L}_{-}(-)$ computes an upper bound for the length of the longest δ-reduction path from a term to its δ-normal form.

Definition 10.46 For $\Gamma \in \mathcal{C}_P$ and $a \in \mathcal{T}_P$ we define $\mathrm{L}_\Gamma(a)$ by induction on the total number of symbols in Γ and a:

$$\begin{array}{rcl}
\mathrm{L}_\Gamma(x) & = & 0 \text{ if } x \in \mathcal{V}; \\
\mathrm{L}_\Gamma(s) & = & 0 \text{ if } s \in \boldsymbol{S};
\end{array}$$

$$\mathrm{L}_\Gamma(c(b_1, \ldots, b_n)) = \begin{cases} \mathrm{L}_{\langle \Gamma_1, \Delta \rangle}(a) + \\ \sum_{i=1}^n \mathrm{L}_\Gamma(b_i) \cdot \max(1, \mathtt{M}_{x_i}(\langle \Gamma_1, \Delta \rangle, a)) + 1 \\ \text{if } \Gamma \equiv \langle \Gamma_1, c(\Delta){=}a{:}A, \Gamma_2 \rangle; \\ \\ \sum_{i=1}^n \mathrm{L}_\Gamma(b_i) \ \text{otherwise}; \end{cases}$$

$$\begin{array}{rcl}
\mathrm{L}_\Gamma(c(\Delta){=}a{:}A \text{ in } b) & = & \mathrm{L}_{\langle \Gamma, \Delta \rangle}(a) + \mathrm{L}_{\langle \Gamma, \Delta \rangle}(A) + \\
& & \sum_{i=1}^n \mathrm{L}_{\langle \Gamma, \Delta_i \rangle}(B_i) + \\
& & \mathrm{L}_{\langle \Gamma, c(\Delta)=a:A \rangle}(b) + 1;
\end{array}$$

$$\begin{array}{rcl}
\mathrm{L}_\Gamma(ab) & = & \mathrm{L}_\Gamma(a) + \mathrm{L}_\Gamma(b); \\
\mathrm{L}_\Gamma(\Pi x{:}A.a) & = & \mathrm{L}_{\langle \Gamma, x:A \rangle}(a) + \mathrm{L}_\Gamma(A); \\
\mathrm{L}_\Gamma(\lambda x{:}A.a) & = & \mathrm{L}_{\langle \Gamma, x:A \rangle}(a) + \mathrm{L}_\Gamma(A).
\end{array}$$

Similar properties as in Lemma 10.44 and Lemma 10.45 hold for $\mathrm{L}_{-}(-)$:

Lemma 10.47

1. If $\langle \Gamma_1, \Gamma_3 \rangle$ is sound, $\text{DOM}(a) \subseteq \text{DOM}(\langle \Gamma_1, \Gamma_3 \rangle)$*, then*

$$\mathrm{L}_{\langle \Gamma_1, \Gamma_3 \rangle}(a) = \mathrm{L}_{\langle \Gamma_1, \Gamma_2, \Gamma_3 \rangle}(a);$$

2. If $\langle \Gamma_1, \Gamma_2 \rangle$ is sound, $\text{DOM}(b) \subseteq \text{DOM}(\langle \Gamma_1, \Gamma_2 \rangle)$, $\text{DOM}(a) \subseteq \text{DOM}(\Gamma_1)$, *and $x \notin$* $\text{FV}(\Gamma_1)$*, then*

$$\mathrm{L}_{\langle \Gamma_1, \Gamma_2[x:=a] \rangle}(b[x:=a]) = \mathrm{L}_{\langle \Gamma_1, \Gamma_2 \rangle}(b) + \mathrm{L}_{\Gamma_1}(a) \cdot \mathtt{M}_x(\langle \Gamma_1, \Gamma_2 \rangle, b). \quad ⊠$$

The lemma above is used to prove the crucial property of $\mathrm{L}_{-}(-)$:

Lemma 10.48 *If Γ is sound, $\mathrm{DOM}(a) \subseteq \mathrm{DOM}(\Gamma)$ and $\Gamma \vdash a \to_{\delta} b$, then* $\mathrm{L}_{\Gamma}(a) > \mathrm{L}_{\Gamma}(b)$.

PROOF: Similar to the proof of Lemma 10.45. ⊠

Theorem 10.49 (Strong Normalisation for δ) *The reduction δ (when restricted to sound contexts Γ and terms a with $\mathrm{DOM}(a) \subseteq \mathrm{DOM}(\Gamma)$) is strongly normalising, i.e. there are no infinite δ-reduction paths.*

PROOF: This follows from lemma 10.48. ⊠

Even weak normalisation fails without the restriction to sound contexts Γ and to terms a with $\mathrm{DOM}(a) \subseteq \mathrm{DOM}(\Gamma)$. For example, take

$$\Gamma \equiv \langle \mathtt{c}() = \mathtt{d}() : \mathtt{A}, \mathtt{d}() = \mathtt{c}() : \mathtt{A} \rangle.$$

The term $\mathtt{c}()$ does not have a Γ-normal form.

10c Properties of legal terms

The properties in this section are proved for all terms that are legal in a Pure Type System with parameters, i.e. for terms a for which there exist A and Γ such that $\Gamma \vdash^{\vec{C}\vec{D}} a : A$ or $\Gamma \vdash^{\vec{C}\vec{D}} A : a$. The main property we prove is that strong normalisation is preserved by certain extensions.

Many of the standard properties of PTSs in [3] hold for $\vec{C}\vec{D}$-PTSs as well. In the same way as in [3] we can prove the Substitution Lemma, Correctness of Types, Subject Reduction (for $\beta\delta$-reduction) and Uniqueness of Types (for singly sorted $\vec{C}\vec{D}$-PTSs):

Theorem 10.50 *Let $\mathfrak{S}$ be a specification. The type system $\lambda^{\vec{C}\vec{D}}(\mathfrak{S})$ has the following properties:*

- *Substitution Lemma;*
- *Correctness of Types;*
- *Subject Reduction (for $\to_{\beta\delta}$).*

Moreover, if $\mathfrak{S}$ is singly sorted then $\lambda^{\vec{C}\vec{D}}(\mathfrak{S})$ has Uniqueness of Types. ⊠

The Generation Lemma is extended with two extra cases:

Lemma 10.51 (Generation Lemma, extension)

1. *If* $\Gamma \vdash^{\vec{C}\vec{D}} c(b_1, \ldots, b_n) : D$ *then there exist* s, Δ *and* A *such that* $\Gamma \vdash D =_{\beta\delta} A[x_i{:=}b_i]_{i=1}^n$, *and* $\Gamma \vdash^{\vec{C}\vec{D}} b_i : B_i[x_j{:=}b_j]_{j=1}^{i-1}$. *Besides we have one of these two possibilities:*

   (a) *Either* $\Gamma = \langle \Gamma_1, c(\Delta){:}A, \Gamma_2 \rangle$ *and* $\Gamma_1, \Delta \vdash^{\vec{C}\vec{D}} A : s$;

   (b) *Or* $\Gamma = \langle \Gamma_1, c(\Delta){=}a{:}A, \Gamma_2 \rangle$ *and* $\Gamma_1, \Delta \vdash^{\vec{C}\vec{D}} a : A$;

2. *If* $\Gamma \vdash^{\vec{C}\vec{D}} c(\Delta){=}a{:}A \text{ IN } b : D$ *then we have two possibilities:*

   (a) *Either* $\Gamma, c(\Delta){=}a{:}A \vdash^{\vec{C}\vec{D}} b : B$, $\Gamma \vdash^{\vec{C}\vec{D}} (c(\Delta){=}a{:}A \text{ IN } B) : s$ *and* $\Gamma \vdash D =_{\beta\delta} c(\Delta){=}a{:}A \text{ IN } B$;

   (b) *Or* $\Gamma, c(\Delta){=}a{:}A \vdash^{\vec{C}\vec{D}} b : s$ *and* $\Gamma \vdash D =_{\beta\delta} s$.

⊠

In case 1(b) we do not necessarily have $\Gamma_1, \Delta \vdash^{\vec{C}\vec{D}} A : s$. For instance, in the $\vec{C}\vec{D}$-PTSs of the Barendregt cube one can abbreviate terms a of type $\Box$, whilst $\Box$ is not typable in these systems.

Also Correctness of Contexts has some extra cases compared to usual PTSs. Recall that Γ is *legal* if there are b, B such that $\Gamma \vdash^{\vec{C}\vec{D}} b : B$.

Lemma 10.52 (Correctness of Contexts)

1. *If* $\Gamma, x{:}A, \Gamma'$ *is legal then there exists a sort* s *such that* $\Gamma \vdash^{\vec{C}\vec{D}} A : s$;

2. *If* $\Gamma, c(\Delta){:}A, \Gamma'$ *is legal then* $\Gamma, \Delta \vdash^{\vec{C}\vec{D}} A : s$;

3. *If* $\Gamma, c(\Delta){=}a{:}A, \Gamma'$ *is legal then* $\Gamma, \Delta \vdash^{\vec{C}\vec{D}} a : A$. ⊠

Again, in case 3 we do not necessarily have $\Gamma, \Delta \vdash^{\vec{C}\vec{D}} A : s$.

Now we prove that $\lambda^{\vec{C}\vec{D}}(\mathfrak{S})$ is $\beta\delta$-strongly normalising if a slightly larger PTS $\lambda(\mathfrak{S}')$ is β-strongly normalising. The proof follows the same ideas of [137] to prove that a PTS extended with definitions is $\beta\delta$-strongly normalising.

For legal terms $a \in \mathcal{T}_P$ in a context Γ, we define a lambda term $||a||_\Gamma$ without definitions and without parameters. If a is typable in a $\vec{C}\vec{D}$-PTS $\lambda^{\vec{C}\vec{D}}(\mathfrak{S})$, then $||a||_\Gamma$ will be typable in a PTS $\lambda(\mathfrak{S}')$, where $\mathfrak{S}'$ is a so-called *completion* (see Definition 10.60) of the specification $\mathfrak{S}$. Moreover, we take care that if $a \to_\beta a'$, then $||a||_\Gamma \twoheadrightarrow_\beta^+ ||a'||_\Gamma$ (that is: $||a||_\Gamma \twoheadrightarrow_\beta ||a'||_\Gamma$ and $||a||_\Gamma \not\equiv ||a'||\,\Gamma$). Together with strong normalisation of δ-reduction (Theorem 10.49), this guarantees that $\lambda^{\vec{C}\vec{D}}(\mathfrak{S})$ is $\beta\delta$-strongly normalising whenever $\lambda(\mathfrak{S}')$ is β-strongly normalising.

We suppose that $\mathcal{V} \cup \mathcal{C}$, the set of variables and constants that are used to define $\mathcal{T}_P$, is included in the set of variables that is used to define $\mathcal{T}$, the set of terms used for the PTS $\lambda(\mathfrak{S}')$.

Δ still denotes a list of variables $x_1{:}B_1, \ldots, x_n{:}B_n$ and Δ_i is an abbreviation for $x_1{:}B_1, \ldots, x_{i-1}{:}B_{i-1}$. $\lambda\,\Delta.a$ denotes $\lambda_{i=1}^n\, x_i{:}B_i.a$ and $\prod \Delta.A$ denotes $\prod_{i=1}^n x_i{:}B_i.A$.

Definition 10.53 For $a \in \mathcal{T}_P$ and $\Gamma \in \mathcal{C}_P$ we define $\|a\|_\Gamma$ as follows:

$$\begin{array}{rcl}
\|s\|_\Gamma & \equiv & s \text{ if } s \in \boldsymbol{S}; \\
\|x\|_\Gamma & \equiv & x \text{ if } x \in \mathcal{V}; \\
\|c(b_1,\ldots,b_n)\|_\Gamma & \equiv & \begin{cases} \|\lambda\,\Delta.a)\|_{\Gamma_1}\, \|b_1\|_\Gamma \cdots \|b_n\|_\Gamma \\ \text{if } \Gamma = \langle \Gamma_1, c(\Delta){=}a{:}A, \Gamma_2\rangle; \\ \\ c\,\|b_1\|_\Gamma, \ldots, \|b_n\|_\Gamma \text{ otherwise;} \end{cases} \\
\|ab\|_\Gamma & \equiv & \|a\|_\Gamma\, \|b\|_\Gamma\,; \\
\|\lambda x{:}A.b\|_\Gamma & \equiv & \lambda x{:}\, \|A\|_\Gamma \,.\, \|b\|_{\Gamma,x:A}\,; \\
\|\Pi x{:}A.b\|_\Gamma & \equiv & \Pi x{:}\, \|A\|_\Gamma \,.\, \|b\|_{\Gamma,x:A}\,; \\
\|c(\Delta){=}a{:}A \text{ IN } b\|_\Gamma & \equiv & \left(\lambda c{:}(\|\prod \Delta.A\|_\Gamma).\, \|b\|_{\Gamma,c(\Delta)=a:A}\right) \|\lambda\,\Delta.a\|_\Gamma\,.
\end{array}$$

The mapping $\|\_\|_\_$ is slightly different from the mapping $|\_|_\_$. This is because we want $\|\_\|_\_$ to maintain β-reductions. In a term $c(\Delta){=}a{:}A$ IN b, there may be β-redexes in Δ, a or A.
These redexes may be lost in $|c(\Delta){=}a{:}A \text{ IN } b|_\Gamma \equiv |b|_{\Gamma,c(\Delta)=a:A}$.

Due to the extra λ-abstraction in the definition of $\|c(\Delta){=}a{:}A \text{ IN } b\|_\Gamma$, the possible β-redexes in Δ, a and A are maintained.

The mapping $\|\_\|_\_$ is extended to contexts.

Definition 10.54 For a context $\Gamma \in \mathcal{C}_P$ we define $\|\Gamma\|$ as follows:

$$\begin{array}{rcl}
\|\varnothing\| & \equiv & \varnothing; \\
\|\Gamma, x{:}A\| & \equiv & \|\Gamma\|\,, x{:}\, \|A\|_\Gamma\,; \\
\|\Gamma, c(\Delta){:}A\| & \equiv & \|\Gamma\|\,, c{:}\,(\|\prod \Delta.A\|_\Gamma)\,; \\
\|\Gamma, c(\Delta){=}a{:}A\| & \equiv & \|\Gamma\|\,, c{:}\,(\|\prod \Delta.A\|_\Gamma)\,.
\end{array}$$

We have similar properties for $\|\_\|_\_$ as for $|\_|_\_$:

Lemma 10.55 *If* $\langle \Gamma_1, \Gamma_3\rangle$ *is sound and* $\text{DOM}\,(a) \subseteq \text{DOM}\,(\langle \Gamma_1, \Gamma_3\rangle)$, *then* $\|a\|_{\Gamma_1,\Gamma_2,\Gamma_3} \equiv \|a\|_{\Gamma_1,\Gamma_3}$. ⊠

The proof is similar to the proof of Lemma 10.38.

Lemma 10.56 *Assume* $\langle\Gamma_1,\Gamma_2\rangle$ *is sound, and* $\text{DOM}(a) \subseteq \text{DOM}(\langle\Gamma_1,\Gamma_2\rangle)$, $\text{DOM}(b) \subseteq \text{DOM}(\Gamma_1)$, *and* $x \notin \text{DOM}(\Gamma_1)$. *Then*

$$\|a\|_{\Gamma_1,\Gamma_2}[x{:=}\|b\|_{\Gamma_1}] \equiv \|a[x{:=}b]\|_{\Gamma_1,\Gamma_2[x:=b]}\,. \qquad \boxtimes$$

The proof is similar to the proof of Lemma 10.39.

We now show that $\|\_\|_\_$ translates a δ-reduction into zero or more β-reductions, and that it translates a β-reduction into one or more β-reductions.

Lemma 10.57 *Let* Γ *be sound, and assume* $\text{DOM}(a) \subseteq \text{DOM}(\Gamma)$. *If* $a \to_\delta b$ *then* $\|a\|_\Gamma \twoheadrightarrow_\beta \|b\|_\Gamma$.

PROOF: Using induction on the definition of $\Gamma \vdash a \to_\delta b$, we simultaneously prove:

1. If $a \to_\delta b$ then $\|a\|_\Gamma \twoheadrightarrow_\beta \|b\|_\Gamma$;
2. If $\Gamma \to_\delta \Gamma'$ then $\|a\|_\Gamma \twoheadrightarrow_\beta \|a\|_{\Gamma'}$.

We only treat two non-trivial cases.

- $\Gamma_1, c(\Delta){=}a{:}A, \Gamma_2 \vdash c(b_1,\ldots,b_n) \to_\delta a[x_i{:=}b_i]_{i=1}^n$. Observe:

$$\begin{array}{rcl}
\|c(b_1,\ldots,b_n)\|_\Gamma & \equiv & \|\lambda\,\Delta.a\|_{\Gamma_1}\,\|b_1\|_\Gamma\cdots\|b_n\|_\Gamma \\
& \equiv & \left(\overset{n}{\underset{i=1}{\lambda}}\, x_i{:}\,\|B_i\|_{\Gamma_1,\Delta_i}\,.\,\|a\|_{\Gamma,\Delta}\right)\|b_1\|_\Gamma\cdots\|b_n\|_\Gamma \\
& \twoheadrightarrow_\beta^n & \|a\|_{\Gamma_1,\Delta}[x_i{:=}\|b_i\|_\Gamma]_{i=1}^n \\
& \overset{(10.55)}{\equiv} & \|a\|_{\Gamma_1}[x_i{:=}\|b_i\|_\Gamma]_{i=1}^n \\
& \overset{(10.55)}{\equiv} & \|a\|_\Gamma[x_i{:=}\|b_i\|_\Gamma]_{i=1}^n \\
& \overset{(10.56)}{\equiv} & \|a[x_i{:=}b_i]_{i=1}^n\|_\Gamma\,;
\end{array}$$

- $\Gamma \vdash c(\Delta){=}a{:}A \text{ IN } b \to_\delta b$ because $c \notin \text{CONS}(b)$. Then $c \notin \text{FV}(\|b\|_\Gamma)$. Hence

$$\begin{array}{rcl}
\|c(\Delta){=}a{:}A \text{ IN } b\|_\Gamma & \equiv & \left(\lambda c : \|\prod\Delta.A\|_\Gamma\,.\,\|b\|_{\Gamma,c(\Delta)=a:A}\right)\|\lambda\,\Delta.a\|_\Gamma \\
& \to_\beta & \|b\|_{\Gamma,c(\Delta)=a:A}[c{:=}\|\lambda\,\Delta.a\|_\Gamma] \\
& \equiv & \|b\|_\Gamma\,.
\end{array}$$

$\boxtimes$

Lemma 10.58 *Let* Γ *be sound, and assume* $\text{DOM}(a) \subseteq \text{DOM}(\Gamma)$. *If* $a \to_\beta b$ *then* $\|a\|_\Gamma \twoheadrightarrow_\beta^+ \|b\|_\Gamma$.

PROOF: The following two statements are proved simultaneously by induction on the structure of a.

1. If $a \rightarrow_\beta b$ then $\|a\|_\Gamma \twoheadrightarrow^+_\beta \|b\|_\Gamma$;
2. If $\Gamma \rightarrow_\beta \Gamma'$ then $\|a\|_\Gamma \twoheadrightarrow_\beta \|a\|_{\Gamma'}$.

(IH 1) refers to the induction hypothesis on 1, (IH 2) to the induction hypothesis on 2. We do not treat all cases, and only prove the first statement.

- $c(b_1,\ldots,b_n) \rightarrow_\beta c(b_1,\ldots,b'_j,\ldots,b_n)$, where $b_j \rightarrow_\beta b'_j$, and $\Gamma \equiv \langle \Gamma_1, c(\Delta){=}a{:}A, \Gamma_2\rangle$. We have:

$$\begin{array}{rcl} \|c(b_1,\ldots,b_n)\|_\Gamma & \equiv & \|\lambda\,\Delta.a\|_{\Gamma_1}\,\|b_1\|_\Gamma\cdots\|b_n\|_\Gamma \\ & \overset{\text{(IH 1)}}{\twoheadrightarrow^+_\beta} & \|\lambda\,\Delta.a\|_{\Gamma_1}\,\|b_1\|_\Gamma\cdots\left\|b'_j\right\|_\Gamma\cdots\|b_n\|_\Gamma \\ & \equiv & \left\|c(b_1,\ldots,b'_j,\ldots,b_n)\right\|_\Gamma ; \end{array}$$

- $(\lambda x{:}p.q)r \rightarrow_\beta q[x{:=}r]$. Observe:

$$\begin{array}{rcl} \|(\lambda x{:}p.q)r\|_\Gamma & \equiv & \left(\lambda x{:}\,\|p\|_\Gamma\,.\,\|q\|_{\Gamma,x:p}\right)\|r\|_\Gamma \\ & \rightarrow_\beta & \|q\|_{\Gamma,x:p}\,[x{:=}\,\|r\|_\Gamma] \\ & \overset{(10.56)}{\equiv} & \|q[x{:=}r]\|_\Gamma ; \end{array}$$

- $c(\Delta){=}a{:}A$ IN $b \rightarrow_\beta c(\Delta'){=}a{:}A$ IN b, where
  - $\Delta' \equiv x_1{:}B_1,\ldots,x_j{:}B'_j,\ldots,x_n : B_n$;
  - $B_j \rightarrow_\beta B'_j$.

  Write $C_i \equiv B_i$ if $i \neq j$ and $C_j \equiv B'_j$.
  Furthermore, let $\Delta'_i \equiv x_1{:}C_1,\ldots,x_{i-1}{:}C_{i-1}$. Observe that

$$\begin{array}{rl} \multicolumn{2}{l}{\|c(\Delta){=}a{:}A \text{ IN } b\|_\Gamma} \\ \equiv & \left(\lambda c : (\|\prod\Delta.A\|_\Gamma)\,.\,\|b\|_{\Gamma,c(\Delta)=a:A}\right) \\ & (\|\lambda\,\Delta.a\|_\Gamma) \\ \overset{\text{(IH 1)}}{\twoheadrightarrow^+_\beta} & \left(\lambda c : (\|prod\Delta'.A\|_\Gamma)\,.\,\|b\|_{\Gamma,c(\Delta)=a:A}\right) \\ & (\|\lambda\,\Delta'.a\|_\Gamma) \\ \overset{\text{(IH 2)}}{\twoheadrightarrow_\beta} & \left(\lambda c : (\|\prod\Delta'.A\|_\Gamma)\,.\,\|b\|_{\Gamma,c(\Delta')=a:A}\right) \\ & (\|\lambda\,\Delta'.a\|_\Gamma) \\ \equiv & \|c(\Delta'){=}a{:}A \text{ IN } b\|_\Gamma\,. \end{array}$$

⊠

The proof for the cases $c(b_1, \ldots, b_n)$ and $c(\Delta){=}a{:}A$ IN b shows that this lemma does not hold if we use $|\_|\_$ instead of $||\_||\_$. The proof for the case $c(\Delta){=}a{:}A$ IN b shows the need to prove that $||a||_\Gamma \twoheadrightarrow_\beta ||a||_{\Gamma'}$ if $\Gamma \to_\beta \Gamma'$.

Definition 10.59 The specification $\mathfrak{S} = (\boldsymbol{S}, \boldsymbol{A}, \boldsymbol{R})$ is called *quasi full* if for all $s_1, s_2 \in \boldsymbol{S}$ there exists $s_3 \in \boldsymbol{S}$ such that $(s_1, s_2, s_3) \in \boldsymbol{R}$.

Definition 10.60 A specification $\mathfrak{S}' = (\boldsymbol{S}', \boldsymbol{A}', \boldsymbol{R}')$ is a *completion* of $\mathfrak{S} = (\boldsymbol{S}, \boldsymbol{A}, \boldsymbol{R})$ if

1. $\boldsymbol{S} \subseteq \boldsymbol{S}'$, $\boldsymbol{A} \subseteq \boldsymbol{A}'$, and $\boldsymbol{R} \subseteq \boldsymbol{R}'$;
2. S' is quasi full;
3. for all $s \in \boldsymbol{S}$ there is a sort $s' \in \boldsymbol{S}'$ such that $(s, s') \in \boldsymbol{A}'$.

Theorem 10.61 *Let $\mathfrak{S} = (\boldsymbol{S}, \boldsymbol{A}, \boldsymbol{R})$ and $\mathfrak{S}' = (\boldsymbol{S}', \boldsymbol{A}', \boldsymbol{R}')$ be such that $\mathfrak{S}'$ is a completion of $\mathfrak{S}$. If $\Gamma \vdash_{\mathfrak{S}}^{\vec{C}\vec{D}} a : A$ then*

$$||\Gamma|| \vdash_{\mathfrak{S}'} ||a||_\Gamma : ||A||_\Gamma .$$

PROOF: Induction on the derivation of $\Gamma \vdash_{\mathfrak{S}}^{\vec{C}\vec{D}} a : A$. The rules of normal PTSs do not cause any problem, and the proof for the rules for parametric constants are simplifications of the proofs for the rules for parametric definitions. We therefore only focus on the rules for parametric definitions.

- δ-application:

$$\frac{\begin{array}{ll}\Gamma_1, c(\Delta){=}a{:}A, \Gamma_2 \vdash^{\vec{C}\vec{D}} b_i : B_i[x_j{:=}b_j]_{j=1}^{i-1} & (i = 1, \ldots, n) \\ \Gamma_1, c(\Delta){=}a{:}A, \Gamma_2 \vdash^{\vec{C}\vec{D}} a : A & (\text{if } n = 0)\end{array}}{\Gamma_1, c(\Delta){=}a{:}A, \Gamma_2 \vdash^{\vec{C}\vec{D}} c(b_1, \ldots, b_n) : A[x_j{:=}b_j]_{j=1}^n}.$$

  Write $\Gamma \equiv \Gamma_1, c(\Delta){=}a{:}A, \Gamma_2$.

  - If $n = 0$ then we know by induction that

  $$||\Gamma|| \vdash_{\mathfrak{S}'} ||a||_\Gamma : ||A||_\Gamma$$

  and we are done because $||c(b_1, \ldots, b_n)||_\Gamma \equiv ||a||_{\Gamma_1} \overset{(10.55)}{\equiv} ||a||_\Gamma$.

  - If $n > 0$. As we have a derivation of $\Gamma_1, c(\Delta){=}a{:}A, \Gamma_2 \vdash^{\vec{C}\vec{D}} b_1{:}B_1$, we can use Correctness of Contexts to find a (shorter) derivation of $\Gamma_1, \Delta \vdash^{\vec{C}\vec{D}} a{:}A$. By the induction hypothesis, we have

  $$||\Gamma_1, \Delta|| \vdash_{\mathfrak{S}'} ||a||_{\Gamma_1, \Delta} : ||A||_{\Gamma_1, \Delta'} . \qquad (10.1)$$

Moreover, we can use the induction hypothesis to find

$$\|\Gamma_1\|, c{:}\,\|\textstyle\prod \Delta.A\|_{\Gamma_1}, \|\Gamma_2\| \vdash_{\mathfrak{S}'} \|b_i\|_\Gamma : \left\|B_i[x_j{:=}b_j]_{j=1}^{i-1}\right\|_\Gamma . \quad (10.2)$$

We can use Correctness of Types for the PTS $\lambda\mathfrak{S}'$ to find $s \in S'$ with

$$\|\Gamma_1\| \vdash_{\mathfrak{S}'} \|\textstyle\prod \Delta.A\|_{\Gamma_1} : s. \quad (10.3)$$

Using rule (λ), (10.1) and (10.3) result in

$$\|\Gamma_1\| \vdash_{\mathfrak{S}'} \|\lambda\,\Delta.a\|_{\Gamma_1} : \left\|\textstyle\prod \Delta.A\right\|_{\Gamma_1} .$$

By definition of $\|\prod \Delta.A\|_{\Gamma_1}$, this means

$$\|\Gamma\| \vdash_{\mathfrak{S}'} \|\lambda\,\Delta.a\|_{\Gamma_1} : \textstyle\prod_{i=1}^n x_i{:}\,\|B_i\|_{\Gamma_1,\Delta} . \|A\|_{\Gamma_1,\Delta} . \quad (10.4)$$

By Lemma 10.56, $\left\|B_i[x_j{:=}b_j]_{j=1}^{i-1}\right\|_\Gamma \equiv \|B_i\|_{\Gamma,\Delta_i} [x_j{:=}\,\|b_j\|_\Gamma]_{j=1}^{i-1}$.
Using (10.2) and the application rule, we can derive from (10.4) that:

$$\|\Gamma\| \vdash_{\mathfrak{S}'} \left(\|\lambda\,\Delta.a\|_{\Gamma_1} \|b_1\|_\Gamma \cdots \|b_n\|_\Gamma\right) : \left(\|A\|_{\Gamma_1,\Delta} [x_j{:=}\,\|b_j\|_\Gamma]_{j=1}^n\right) .$$

We are done because

$$\|c(b_1,\ldots,b_n)\|_\Gamma \equiv \|\lambda\,\Delta.a\|_{\Gamma_1} \|b_1\|_\Gamma \cdots \|b_n\|_\Gamma$$

and

$$\begin{aligned} \|A\|_{\Gamma_1,\Delta} [x_j{:=}\,\|b_j\|_\Gamma]_{j=1}^n &\overset{(10.55)}{\equiv} \|A\|_{\Gamma,\Delta} [x_j{:=}\,\|b_j\|_\Gamma]_{j=1}^n \\ &\overset{(10.56)}{\equiv} \left\|A[x_j{:=}b_j]_{j=1}^n\right\|_\Gamma ; \end{aligned}$$

- δ-weakening:

$$\frac{\Gamma \vdash^{\vec{C}\vec{D}} b : B \qquad \Gamma, \Delta \vdash^{\vec{C}\vec{D}} a : A}{\Gamma, c(\Delta){=}a{:}A \vdash^{\vec{C}\vec{D}} b : B} .$$

By induction, $\|\Gamma,\Delta\| \vdash_{\mathfrak{S}'} \|a\|_{\Gamma,\Delta} : \|A\|_{\Gamma,\Delta}$, so

$$\|\Gamma\|, x_1{:}\,\|B_1\|_{\Gamma,\Delta_1}, \ldots, x_n{:}\,\|B_n\|_{\Gamma,\Delta_n} \vdash_{\mathfrak{S}'} \|a\|_{\Gamma,\Delta} : \|A\|_{\Gamma,\Delta} . \quad (10.5)$$

By Correctness of Contexts for $\lambda\mathfrak{S}'$, there are $s_1,\ldots,s_n \in S'$ such that

$$\|\Gamma\|, x_1{:}\,\|B_1\|_{\Gamma,\Delta_1}, \ldots, x_{i-1}{:}\,\|B_{i-1}\|_{\Gamma,\Delta_{i-1}} \vdash_{\mathfrak{S}'} \|B_i\|_{\Gamma,\Delta_i} : s_i . \quad (10.6)$$

By Correctness of Types for $\lambda_{\mathfrak{S}}^{\vec{C}\vec{D}}$, there are two possibilities:

– There is $s \in \boldsymbol{S}$ such that $A \equiv s$. As $\mathfrak{S}'$ is a completion of $\mathfrak{S}$, there is $s' \in \boldsymbol{S}'$ such that $\|\Gamma, \Delta\| \vdash_{\mathfrak{S}'} s : s'$.
– There is $s' \in \boldsymbol{S}$ such that $\Gamma, \Delta \vdash^{\vec{C}\vec{D}} A : s'$. Then by induction, $\|\Gamma, \Delta\| \vdash \|A\|_{\Gamma,\Delta} : s'$.

In any case: we can determine $s'_0 \in \boldsymbol{S}'$ such that

$$\|\Gamma\|, x_1{:}\,\|B_1\|_{\Gamma,\Delta_1}, \ldots, x_n{:}\,\|B_n\|_{\Gamma,\Delta_n} \vdash_{\mathfrak{S}'} \|A\|_{\Gamma,\Delta} : s'_0. \tag{10.7}$$

As $\mathfrak{S}'$ is quasi-full, we can subsequently determine $s'_1, \ldots, s'_n$ such that $(s_i, s'_{i-1}, s'_i) \in \boldsymbol{R}'$ for $i = 1, \ldots, n$. This allows us to apply Π-formation n times, with as premises (10.6) and (10.8), and as conclusion:

$$\|\Gamma\| \vdash_{\mathfrak{S}'} \textstyle\prod_{i=1}^{n} x_i{:}\,\|B_i\|_{\Gamma,\Delta_i}.\,\|A\|_{\Gamma,\Delta} : s'_n.$$

Notice that $\prod_{i=1}^{n} x_i{:}\,\|B_i\|_{\Gamma,\Delta_i}.\,\|A\|_{\Gamma,\Delta} \equiv \|\prod \Delta.A\|_\Gamma$. As the induction hypothesis gives us also $\|\Gamma\| \vdash_{\mathfrak{S}'} \|b\|_\Gamma : \|B\|_\Gamma$, we can use the weakening rule of $\lambda\mathfrak{S}'$ to obtain

$$\|\Gamma\|, c{:}\,\|\textstyle\prod \Delta.A\|_\Gamma \vdash_{\mathfrak{S}'} \|b\|_\Gamma : \|B\|_\Gamma\,.$$

We are done because $\|b\|_\Gamma \equiv \|b\|_{\Gamma,c(\Delta)=a:A}$ and $\|B\|_\Gamma \equiv \|B\|_{\Gamma,c(\Delta)=a:A}$ (Lemma 10.55);

• δ-formation:

$$\frac{\Gamma_1, c(\Delta){=}a{:}A \vdash^{\vec{C}\vec{D}} B : s}{\Gamma_1 \vdash^{\vec{C}\vec{D}} c(\Delta){=}a{:}A \text{ IN } B : s}.$$

Write $\Gamma \equiv \Gamma_1, c(\Delta){=}a{:}A$.
By the induction hypothesis, we have $\|\Gamma\| \vdash_{\mathfrak{S}'} \|B\|_\Gamma : s$, so

$$\|\Gamma_1\|, c{:}\,\|\textstyle\prod \Delta.A\|_{\Gamma_1} \vdash_{\mathfrak{S}'} \|B\|_\Gamma : s. \tag{10.8}$$

By Correctness of Contexts on (10.8) there is $s_1 \in \boldsymbol{S}'$ such that

$$\|\Gamma_1\| \vdash_{\mathfrak{S}'} \|\textstyle\prod \Delta.A\|_{\Gamma_1} : s_1. \tag{10.9}$$

Moreover: As $\mathfrak{S}'$ is a completion of $\mathfrak{S}$, there is $s_2 \in \boldsymbol{S}'$ such that $(s{:}s_2) \in \boldsymbol{A}'$. By the Start Lemma,

$$\|\Gamma_1\|, c{:}\,\|\textstyle\prod \Delta.A\|_{\Gamma_1} \vdash_{\mathfrak{S}'} s : s_2. \tag{10.10}$$

As $\mathfrak{S}'$ is quasi-full, there is $s_3 \in \boldsymbol{S}'$ such that $(s_1, s_2, s_3) \in \boldsymbol{R}'$. Hence we can apply Π-formation:

$$\|\Gamma_1\| \vdash_{\mathfrak{S}'} \Pi c{:}\,\|\textstyle\prod \Delta.A\|_{\Gamma_1}.s : s_3. \tag{10.11}$$

We can now apply λ-formation on (10.8) and (10.11):

$$\|\Gamma_1\| \vdash_{\mathfrak{S}'} \left(\lambda c{:}\, \|\prod \Delta.A\|_{\Gamma_1} . \|B\|_{\Gamma}\right) : \left(\Pi c{:}\, \|\prod \Delta.A\|_{\Gamma_1} .s\right). \qquad (10.12)$$

As we have a derivation of $\Gamma_1, c(\Delta){=}a{:}A \vdash^{\vec{C}\vec{D}} B : s$, we can apply Correctness of Contexts to find a (shorter) derivation of $\Gamma_1, \Delta \vdash^{\vec{C}\vec{D}} a{:}A$, so by induction:

$$\|\Gamma_1, \Delta\| \vdash_{\mathfrak{S}'} \|a\|_{\Gamma_1,\Delta} : \|A\|_{\Gamma_1,\Delta}.$$

Using (10.9), we can repeatedly apply λ-abstraction and obtain

$$\|\Gamma_1\| \vdash_{\mathfrak{S}'} \|\lambda\, \Delta.a\|_{\Gamma_1} : \|\prod \Delta.A\|_{\Gamma_1}. \qquad (10.13)$$

Using (10.12) and the application rule, we find:

$$\|\Gamma_1\| \vdash_{\mathfrak{S}'} \left(\lambda c{:}\, \|\prod \Delta.A\|_{\Gamma_1} . \|B\|_{\Gamma}\right) \|\lambda\, \Delta.a\|_{\Gamma_1} : s;$$

- δ-introduction:

$$\frac{\Gamma_1, c(\Delta){=}a{:}A \vdash^{\vec{C}\vec{D}} b{:}B \qquad \Gamma_1 \vdash^{\vec{C}\vec{D}} c(\Delta){=}a{:}A \text{ IN } B : s}{\Gamma_1 \vdash^{\vec{C}\vec{D}} (c(\Delta){=}a{:}A \text{ IN } b) : (c(\Delta){=}a{:}A \text{ IN } B)}.$$

Similarly to the previous case, we can find derivations of (10.13) and

$$\|\Gamma_1\| \vdash_{\mathfrak{S}'} \left(\lambda c{:}\, \|\prod \Delta.A\|_{\Gamma_1} . \|b\|_{\Gamma}\right) : \left(\Pi c{:}\, \|\prod \Delta.A\|_{\Gamma_1} . \|B\|_{\Gamma}\right). \qquad (10.14)$$

Using (10.13), (10.14) and the application rule, we find

$$\|\Gamma_1\| \vdash_{\mathfrak{S}'} \left(\lambda c{:}\, \|\prod \Delta.A\|_{\Gamma_1} . \|b\|_{\Gamma}\right) \|\lambda\, \Delta.a\|_{\Gamma_1} : \|B\|_{\Gamma} \left[c{:=}\, \|\lambda\, \Delta.a\|_{\Gamma_1}\right].$$

By the induction hypothesis,

$$\|\Gamma_1\| \vdash_{\mathfrak{S}'} \left(\lambda c{:}\, \|\prod \Delta.A\|_{\Gamma_1} . \|B\|_{\Gamma}\right) \|\lambda\, \Delta.a\|_{\Gamma_1} : s,$$

so we can apply the conversion rule to find

$$\|\Gamma_1\| \vdash_{\mathfrak{S}'} \|c(\Delta){=}a{:}A \text{ in } b\|_{\Gamma_1} : \|c(\Delta){=}a{:}A \text{ in } B\|_{\Gamma_1};$$

- δ-conversion:

$$\frac{\Gamma \vdash^{\vec{C}\vec{D}} b : B \qquad \Gamma \vdash^{\vec{C}\vec{D}} B' : s \qquad \Gamma \vdash B =_{\delta} B'}{\Gamma \vdash^{\vec{C}\vec{D}} b : B'}.$$

By induction, $\|\Gamma\| \vdash_{\mathfrak{S}'} \|b\|_{\Gamma} : \|B\|_{\Gamma}$ and $\|\Gamma\| \vdash_{\mathfrak{S}'} \|B'\|_{\Gamma} : s$. By Lemma 10.57, $\|B\|_{\Gamma} =_{\beta} \|B'\|_{\Gamma}$. By Conversion, $\|\Gamma\| \vdash_{\mathfrak{S}'} \|b\|_{\Gamma} : \|B'\|_{\Gamma}$.

⊠

Theorem 10.62 *Let $\mathfrak{S} = (\boldsymbol{S}, \boldsymbol{A}, \boldsymbol{R})$ and $\mathfrak{S}' = (\boldsymbol{S}', \boldsymbol{A}', \boldsymbol{R}')$ be such that $\mathfrak{S}'$ is a completion of $\mathfrak{S}$. If the PTS $\lambda(\mathfrak{S}')$ is β-strongly normalising, then the $\vec{C}\vec{D}$-PTS $\lambda^{\vec{C}\vec{D}}(\mathfrak{S})$ is $\beta\delta$-strongly normalising.*

PROOF: Suppose that $\lambda(\mathfrak{S}')$ is β-strongly normalising, and suppose towards a contradiction that $\lambda^{\vec{C}\vec{D}}(\mathfrak{S})$ is not $\beta\delta$-strongly normalising, i.e. there is an infinite $\beta\delta$-reduction sequence $a_1 \to_{\beta\delta} a_2 \to_{\beta\delta} \ldots$, starting at $a \equiv a_1$ and $\Gamma \vdash^{\vec{C}\vec{D}} a : A$.

Observe that the number of β-reductions in this sequence is infinite. Otherwise there would be $n \in \mathbb{N}$ such that $\Gamma \vdash a_n \to_\delta a_{n+1} \to_\delta a_{n+2} \ldots$, which contradicts the fact that δ-reduction is strongly normalising (Theorem 10.49).

We conclude that the reduction sequence is of the form

$$\Gamma \vdash a \twoheadrightarrow_\delta a_{n_1} \to_\beta a_{n_2} \twoheadrightarrow_\delta a_{n_3} \to_\beta a_{n_4} \twoheadrightarrow_\delta \cdots$$

By lemmas 10.57 and 10.58 there is an infinite β-reduction sequence starting at $\|a\|_\Gamma$:

$$\|a\|_\Gamma \twoheadrightarrow_\beta \|a_{n_1}\|_\Gamma \twoheadrightarrow^+_\beta \|a_{n_2}\|_\Gamma \twoheadrightarrow_\beta \|a_{n_3}\|_\Gamma \twoheadrightarrow^+_\beta \|a_{n_4}\|_\Gamma \twoheadrightarrow_\beta \cdots$$

and by Theorem 10.61, $\|\Gamma\| \vdash_{\mathfrak{S}'} \|a\|_\Gamma : \|A\|_\Gamma$, which contradicts the assumption that $\lambda(\mathfrak{S}')$ is β-strongly normalising. ⊠

10d Restrictive use of parameters

In the extension of PTSs to $\vec{C}\vec{D}$-PTSs presented in Sections 10a-10c, we did not put any serious restrictions on the use of parameters:

1. If $\mathfrak{S} = (\boldsymbol{S}, \boldsymbol{A}, \boldsymbol{R})$ is a specification, then the introduction of a parametric constant c in $\lambda^{\vec{C}\vec{D}}(\mathfrak{S})$ only requires that its intended type A is of type s, for some sort $s \in \boldsymbol{S}$. Similarly, for the introduction of a parametric definition we only require that its definiens a is of a certain type A. By correctness of types, either $A \equiv s$, or A has type s, for some $s \in \boldsymbol{S}$;

2. Similarly, if $\Gamma \equiv \Gamma_1, c(\Delta){=}a{:}A, \Gamma_2$, or $\Gamma \equiv \Gamma_1, c(\Delta){:}A, \Gamma_2$, the only restrictions we put on Δ are that Δ must contain only variable declarations, and that Γ_1, Δ must be legal. There are no additional restrictions on the types B_i of the declarations $x_i{:}B_i$ in Δ.

Something similar is the case with Π-formation rules in a (parameter-free) PTS in which there is no restriction on the use of Π-formation rules: $(s_1, s_2, s_3) \in \boldsymbol{R}$ for any $s_1, s_2, s_3 \in \boldsymbol{S}$. In the specific situation that $\boldsymbol{S} = \{(*, \Box)\}$ and $\boldsymbol{A} = \{*{:}\Box\}$, this would give the system λC, which is on top of the Barendregt cube. The other systems of the Barendregt cube cannot be constructed if we do not put restrictions on the rules that are allowed. It is the variation in the set of Π-formation rules that makes it possible to distinguish the various type systems in the cube (and the various logical systems that are behind it, via the PAT-isomorphism).

In this section we study $\vec{C}\vec{D}$-PTSs in which we put restrictions on the types of parametric constants and definitions, and their parameters. These restrictions can be described in a set $\boldsymbol{P}$ of parametric rules, just as the restrictions on Π-formation rules is described in $\boldsymbol{R}$. The effect of the rules in $\boldsymbol{P}$ is as follows.

- Assume we have a constant declaration $c(\Delta) : A$ that is part of a legal context Γ. By Correctness of Contexts, A has type s for some $s \in \boldsymbol{S}$. Similarly, for each declaration $x_i{:}B_i$ in Δ there is a sort s_i such that B_i has type s_i. The use of parameters is restricted by demanding that $(s_i, s) \in \boldsymbol{P}$ for $i = 1, \ldots, n$;

- In principle, the same holds for a definition declaration $c(\Delta){=}a{:}A$. However, there is a small difference on this point. It is not necessary that A has type s for some sort $s \in \boldsymbol{S}$: it can be the case that $A \equiv s$ and that $s : s'$ does not hold for any $s' \in \boldsymbol{S}$. This is a feature that occurs in the DPTSs of [137]. To keep our system compatible with the DPTSs, we want to maintain this feature.

  To cover this case, we do not only introduce rules of the form (s_i, s), but also rules of the form (s_i, TOP). If the use of parameters is restricted by a set $\boldsymbol{P}$, then either $(s_i, s) \in \boldsymbol{P}$ for $i = 1, \ldots, n$, or A is a topsort, and $(s_i, \text{TOP}) \in \boldsymbol{P}$ for $i = 1, \ldots, n$.

In the specific case of the Barendregt cube, the combination of $\boldsymbol{R}$ and $\boldsymbol{P}$ leads to a refinement of the cube, thus making it possible to classify more type systems within one and the same framework.

The similarity of restricting the use of parameters by a set $\boldsymbol{P}$ with restricting the use of Π-formation by a set $\boldsymbol{R}$ gives us a theoretical motivation for the work in this section. But there are also some practical motivations, as several type systems can be described using restriction of parameters.

Example 10.63 Consider the Pascal function `double` that was presented in the Introduction to this chapter.

- Remark that `double` only takes object variables as parameters. In Pascal, it is not possible to have functions with type variables as parameters;

- Moreover, `double` returns an object. It is not possible in Pascal to construct functions that return a type as result.

So the use of parameters is restricted to the object level.

Other examples (ML, LF, AUTOMATH) are discussed in Section 10e.

10d1 $\vec{\mathrm{C}}\vec{\mathrm{D}}$-PTSs with restricted parameters

We now give a formal definition of Pure Type Systems with restricted parameters and restricted parametric definitions.

Definition 10.64 (Parametric Specification) A *parametric specification* is a quadruple $(\boldsymbol{S}, \boldsymbol{A}, \boldsymbol{R}, \boldsymbol{P})$ such that $(\boldsymbol{S}, \boldsymbol{A}, \boldsymbol{R})$ is a specification (cf. Definition 4.18), and $\boldsymbol{P} \subseteq \boldsymbol{S} \times (\boldsymbol{S} \cup \{\text{TOP}\})$. The parametric specification is called *singly sorted* if the specification $(\boldsymbol{S}, \boldsymbol{A}, \boldsymbol{R})$ is singly sorted.

The set $\boldsymbol{P}$ enables us to present a restricted version of the $\vec{\mathrm{C}}$-weakening rule of Definition 9.5. We call this rule *restricted* $\vec{\mathrm{C}}$-weakening ($\hat{\mathrm{C}}$-weak):

$$\frac{\Gamma \vdash^{\hat{C}} b : B \qquad \Gamma, \Delta_i \vdash^{\hat{C}} B_i : s_i \qquad \Gamma, \Delta \vdash^{\hat{C}} A : s}{\Gamma, c(\Delta) : A \vdash^{\hat{C}} b : B} \qquad (s_i, s) \in \boldsymbol{P}$$

The condition $(s_i, s) \in \boldsymbol{P}$ must holds for all $i \in \{1, \ldots, n\}$. However, it is not necessary that all the s_i are equal: in one application of rule ($\hat{\mathrm{C}}$-weak) it is possible to rely on more than one element of $\boldsymbol{P}$.

Definition 10.65 The *typing relation* $\vdash^{\hat{C}}$ is the smallest relation on $\mathcal{C}_P \times \mathcal{T}_P \times \mathcal{T}_P$ closed under the rules in Definition 4.21, ($\vec{\mathrm{C}}$-app) of Definition 9.5 and ($\hat{\mathrm{C}}$-weak).

Similarly, we present a restricted version of the δ-weakening rule of Definition 10.21. We call this rule *restricted* $\vec{\mathrm{D}}$-weakening ($\hat{\mathrm{D}}$-weak):

$$\frac{\Gamma \vdash^{\hat{D}} b : B \qquad \Gamma, \Delta_i \vdash^{\hat{D}} B_i : s_i \qquad \Gamma, \Delta \vdash^{\hat{D}} a : A : s}{\Gamma, c(\Delta){=}a{:}A \vdash^{\hat{D}} b : B} \qquad (s_i, s) \in \boldsymbol{P}$$

Again, $(s_i, s) \in \boldsymbol{P}$ must hold for all $i \in \{1, \ldots, n\}$, and again it is not necessary that all the s_i are equal.

For the case that A is a topsort, we present a special version of this rule. By $A : \text{TOP}$ we denote that $A \equiv s$ and that there is no $s' \in \boldsymbol{S}$ such that $(s : s') \in \boldsymbol{A}$.

$$\frac{\Gamma \vdash^{\hat{D}} b : B \qquad \Gamma, \Delta_i \vdash^{\hat{D}} B_i : s_i \qquad \Gamma, \Delta \vdash^{\hat{D}} a : A : \text{TOP}}{\Gamma, c(\Delta){=}a{:}A \vdash^{\hat{D}} b : B} \qquad (s_i, \text{TOP}) \in \boldsymbol{P}$$

For all $i \in \{1, \ldots, n\}$, $(s_i, \text{TOP}) \in \boldsymbol{P}$ must hold, but the s_i may, again, be different.

Definition 10.66 The *typing relation* $\vdash^{\hat{D}}$ is the smallest relation on $\mathcal{C}_P \times \mathcal{T}_P \times \mathcal{T}_P$ closed under the rules in Definition 4.21, both versions of ($\hat{\mathrm{D}}$-weak) and ($\vec{\mathrm{D}}$-app), ($\vec{\mathrm{D}}$-form), ($\vec{\mathrm{D}}$-intro), and ($\vec{\mathrm{D}}$-conv) of Definition 10.21.

Definition 10.67 The *typing relation* $\vdash^{\hat{C}\hat{D}}$ is obtained from the relation $\vdash^{\vec{C}\vec{D}}$ by replacing rule ($\vec{\text{C}}$-weak) by rule ($\hat{\text{C}}$-weak) and rule ($\vec{\text{D}}$-weak) by rules ($\hat{\text{D}}$-weak).

Definition 10.68 (Pure Type Systems with Restricted Parameters and Restricted Parametric Definitions) Let $\mathfrak{S}$ be a parametric specification. The Pure Type System with restricted parameters and restricted parametric definitions ($\hat{\text{C}}\hat{\text{D}}$-PTS) and parametric specification $\mathfrak{S}$ is denoted $\lambda^{\hat{C}\hat{D}}(\mathfrak{S})$. The system consists of the set of terms $\mathcal{T}_P$, the set of contexts $\mathcal{C}_P$, β-reduction, δ-reduction, and the typing relation $\vdash^{\hat{C}\hat{D}}$.

We do not extensively discuss the various meta-properties of $\hat{\text{C}}\hat{\text{D}}$-PTSs. This is because a $\hat{\text{C}}\hat{\text{D}}$-PTS with parametric specification $(\boldsymbol{S}, \boldsymbol{A}, \boldsymbol{R}, \boldsymbol{P})$ is a subsystem of the $\vec{\text{C}}\vec{\text{D}}$-PTS with specification $(\boldsymbol{S}, \boldsymbol{A}, \boldsymbol{R})$. We only give a stronger formulation of the extension of the Generation Lemma 10.51:

Lemma 10.69 (Generation Lemma, second extension)
If $\Gamma \vdash^{\hat{C}\hat{D}} c(b_1, \ldots, b_n) : D$ *then there exist* s, Δ *and* A *such that* $\Gamma \vdash D =_{\beta\delta} A[x_i{:=}b_i]_{i=1}^n$, *and* $\Gamma \vdash b_i : B_i[x_j{:=}b_j]_{j=1}^{i-1}$. *Besides we have one of these three possibilities:*

1. *Either we have that* $\Gamma = \langle \Gamma_1, c(\Delta){:}A, \Gamma_2\rangle$ *and* $\Gamma_1, \Delta \vdash^{\hat{C}\hat{D}} A : s$, *and for each* i *there is* s_i *with* $(s_i, s) \in \boldsymbol{P}$ *and* $\Gamma, \Delta_i \vdash^{\hat{C}\hat{D}} B_i : s_i$;

2. *Or we have that* $\Gamma = \langle \Gamma_1, c(\Delta){=}a{:}A and \Gamma_2\rangle$, $\Gamma_1, \Delta \vdash^{\hat{C}\hat{D}} a : A : s$, *and for each* i *there is* s_i *with* $(s_i, s) \in \boldsymbol{P}$ *and* $\Gamma, \Delta_i \vdash^{\hat{C}\hat{D}} B_i : s_i$;

3. *Or we have that* $\Gamma = \langle \Gamma_1, c(\Delta){=}a{:}A and \Gamma_2\rangle$, $\Gamma_1, \Delta \vdash^{\hat{C}\hat{D}} a : A : \text{TOP}$, *and for each* i *there is* s_i *with* $(s_i, \text{TOP}) \in \boldsymbol{P}$ *and* $\Gamma, \Delta_i \vdash^{\hat{C}\hat{D}} B_i : s_i$. ⊠

An important observation is the following one.

Remark 10.70 Our systems with restricted parameters cover the PTSs with definitions (D-PTSs) that were introduced in [137]. Let $\mathfrak{S} = (\boldsymbol{S}, \boldsymbol{A}, \boldsymbol{R})$ a specification, and observe the parametric specification $\mathfrak{S}' = (\boldsymbol{S}, \boldsymbol{A}, \boldsymbol{R}, \varnothing)$. The fact that the set of parametric rules is empty does not exclude the existence of definitions: it is still possible to apply the rules $\hat{\text{D}}$-weak for $n = 0$. In that case, we obtain only definitions without parameters, and the rules of the parametric system reduces precisely to the rules of a D-PTS with specification $\mathfrak{S}$.[2]

[2] The parametric system with specification $\mathfrak{S}'$ has a $\hat{\text{C}}$-weakening rule while the systems of [137] does not. But the $\hat{\text{C}}$-weakening rule can only be used for $n = 0$, and in that case $\hat{\text{C}}$-weakening can be imitated by the normal weakening rule of PTSs: a parametric constant with zero parameters is in fact a parameter-free constant, and for such a constant one can use a variable as well.

For the comparison of $\hat{C}\hat{D}$-PTSs with other PTSs, we introduce some terminology.

We argued earlier on in this chapter, that a parameter mechanism can be seen as a system for abstraction and application that is weaker than the λ-calculus mechanism. We will make this precise by proving (in Theorem 10.79) that a D-PTS with specification $(\boldsymbol{S}, \boldsymbol{A}, \boldsymbol{R})$ is as powerful as any $\hat{C}\hat{D}$-PTS with parametric specification $(\boldsymbol{S}, \boldsymbol{A}, \boldsymbol{R}, \boldsymbol{P})$ for which $(s_1, s_2) \in \boldsymbol{P}$ implies $(s_1, s_2, s_2) \in \boldsymbol{R}$. We call such a $\hat{C}\hat{D}$-PTS *parametrically conservative*:

Definition 10.71 Let $\mathfrak{S} = (\boldsymbol{S}, \boldsymbol{A}, \boldsymbol{R}, \boldsymbol{P})$ be a parametric specification. We say that $\mathfrak{S}$ is *parametrically conservative* if for all $s_1, s_2 \in \boldsymbol{S}$, $(s_1, s_2) \in \boldsymbol{P}$ implies $(s_1, s_2, s_2) \in \boldsymbol{R}$.

Each $\hat{C}\hat{D}$-PTS can be extended to a parametrically conservative one by taking its *parametric closure*:

Definition 10.72 Let $\mathfrak{S} = (\boldsymbol{S}, \boldsymbol{A}, \boldsymbol{R}, \boldsymbol{P})$ be a parametric specification. We define $\mathrm{CL}(\mathfrak{S})$, the *parametric closure* of $\mathfrak{S}$, by $(\boldsymbol{S}, \boldsymbol{A}, \boldsymbol{R}', \boldsymbol{P})$, where
$\boldsymbol{R}' = \boldsymbol{R} \cup \{(s_1, s_2, s_2) \mid (s_1, s_2) \in \boldsymbol{P}\}$.

The Lemma below follows immediately from the definitions above.

Lemma 10.73 *Let $\mathfrak{S}$ be a parametric specification.*

1. $\mathrm{CL}(\mathfrak{S})$ *is parametrically conservative;*
2. $\mathrm{CL}(\mathrm{CL}(\mathfrak{S})) = \mathrm{CL}(\mathfrak{S})$. ⊠

10d2 Imitating parameters by λ-abstractions

Let $\mathfrak{S} = (\boldsymbol{S}, \boldsymbol{A}, \boldsymbol{R}, \boldsymbol{P})$ be a parametric specification. If $\mathfrak{S}$ is parametrically conservative, then each parametric rule (s_1, s_2) of $\mathfrak{S}$ has an equivalent Π-formation rule (s_1, s_2, s_2). In this section we show that this Π-formation rule can indeed take over the role of the parametric rule (s_1, s_2). This means that $\mathfrak{S}$ has the same "power" (see Theorem 10.79) as $(\boldsymbol{S}, \boldsymbol{A}, \boldsymbol{R}, \varnothing)$. With Remark 10.70 in mind, this even means that $\mathfrak{S}$ has the same power as the D-PTS with specification $(\boldsymbol{S}, \boldsymbol{A}, \boldsymbol{R})$.

In order to compare $\mathfrak{S} = (\boldsymbol{S}, \boldsymbol{A}, \boldsymbol{R}, \boldsymbol{P})$ with $\mathfrak{S}' = (\boldsymbol{S}, \boldsymbol{A}, \boldsymbol{R}, \varnothing)$, we need to remove the parameters from the syntax of $\lambda^{\hat{C}\hat{D}}(\mathfrak{S})$. This is easy:

- The parametric application in a term $c(b_1, \ldots, b_n)$ is replaced by function application $cb_1 \cdots b_n$;
- A local parametric definition is translated by a parameter-free local definition, and the parameters are replaced by λ-abstractions;

- A global parametric definition is translated by a parameter-free global definition, and the parameters are replaced by λ-abstractions.

This leads to the following definitions:

Definition 10.74 We define the parameter-free translation $\{t\}$ of a term $t \in \mathcal{T}_P$ as follows:

$$\begin{array}{rcl} \{x\} & \equiv & x; \\ \{s\} & \equiv & s; \\ \{c(b_1,\ldots,b_n)\} & \equiv & c\,\{b_1\}\cdots\{b_n\}\,; \\ \{ab\} & \equiv & \{a\}\,\{b\}\,; \\ \{\lambda x{:}A.b\} & \equiv & \lambda x{:}\,\{A\}\,.\,\{b\}\,; \\ \{\Pi x{:}A.B\} & \equiv & \Pi x{:}\,\{A\}\,.\,\{B\}\,; \\ \{c(\Delta){=}a{:}A \text{ IN } b\} & \equiv & c(){=}\,\{\lambda\,\Delta.a\}\,{:}\,\{\prod\Delta.A\}\ \text{IN}\ \{b\}\,. \end{array}$$

Definition 10.75 We extend the definition of $\{\_\}$ to contexts:

$$\begin{array}{rcl} \{\langle\rangle\} & \equiv & \langle\rangle; \\ \{\Gamma,x{:}A\} & \equiv & \{\Gamma\}\,,x{:}\,\{A\}\,; \\ \{\Gamma,c(\Delta){:}A\} & \equiv & \{\Gamma\}\,,c(){:}\,\{\prod\Delta.A\}\,; \\ \{\Gamma,c(\Delta){=}a{:}A\} & \equiv & \{\Gamma\}\,,c(){=}\,\{\lambda\,\Delta.a\}\,{:}\,\{\prod\Delta.A\}\,. \end{array}$$

To demonstrate the behaviour of $\{\_\}$ under $\beta\delta$-reduction, we need a lemma that shows how to manipulate with substitutions and $\{\_\}$. The proof is straightforward, using induction on the structure of a.

Lemma 10.76 *For $a,b \in \mathcal{T}_P$: $\{a[x{:=}b]\} \equiv \{a\}\,[x{:=}\,\{b\}]$.* ⊠

The mapping $\{\_\}$ maintains β-reduction. A δ-reduction is translated into a δ-reduction followed by zero or more β-reductions. These β-reductions take over the n substitutions that are needed in a δ-reduction

$$c(b_1,\ldots,b_n) \rightarrow_\delta a[x_i{:=}b_i]_{i=1}^n.$$

Lemma 10.77

1. *If $a \rightarrow_\beta a'$ then $\{a\} \twoheadrightarrow^+_\beta \{a'\}$;*

2. *If $\Gamma \vdash a \rightarrow_\delta a'$ then there is a'' such that $\{\Gamma\} \vdash \{a\} \twoheadrightarrow^+_\delta a'' \twoheadrightarrow_\beta \{a'\}$;*

3. *If $\Gamma \vdash a \twoheadrightarrow_{\beta\delta} a'$ then $\{\Gamma\} \vdash \{a\} \twoheadrightarrow_{\beta\delta} \{a'\}$.*

PROOF: (1) follows easily by induction on the structure of a, and Lemma 10.76. (3) follows from (1) and (2). We only show (2), using induction on the definition of $\Gamma \vdash a \to_\delta a'$, and treating only the most important case.

Assume $\Gamma \equiv \Gamma_1, c(\Delta){=}a{:}A, \Gamma_2$ and $\Gamma \vdash c(b_1, \ldots, b_n) \to_\delta a[x_i{:=}b_i]_{i=1}^n$. Observe that $\{\Gamma\} \equiv \{\Gamma_1\}, c(){=}\{\lambda\,\Delta.a\} : \{\prod \Delta.A\}, \{\Gamma_2\}$, so

$$\begin{array}{rcl} \{\Gamma\} \vdash c()\,\{b_1\} \cdots \{b_n\} & \to_\delta & \{\lambda\,\Delta.a\}\,\{b_1\} \cdots \{b_n\} \\ & \twoheadrightarrow_\beta & \{a\}\,[x_i{:=}\,\{b_i\}]_{i=1}^n \\ & \stackrel{(10.76)}{\equiv} & \{a[x_i{:=}b_i]_{i=1}^n\}\,. \end{array}$$

⊠

Remark 10.78 In 10.77.1, we cannot replace $\twoheadrightarrow_\beta^+$ by $\to_\beta$. This has to do with the definition of $\{c(\Delta){=}a{:}A \text{ IN } b\}$. One β-reduction in Δ gives rise to (at least) two β-reductions in $c(){=}\{\lambda\,\Delta.a\} : \{\prod \Delta.A\}$ IN $\{b\}$.

Similarly, we cannot replace the $\twoheadrightarrow_\delta^+$ in 10.77.2 by $\to_\delta$.

Now we show that $\{\_\}$ embeds the $\hat{C}\hat{D}$-PTS with parametric specification $\mathfrak{S} = (\boldsymbol{S}, \boldsymbol{A}, \boldsymbol{R}, \boldsymbol{P})$ in the $\hat{C}\hat{D}$-PTS with parametric specification $\mathfrak{S}' = (\boldsymbol{S}, \boldsymbol{A}, \boldsymbol{R}, \varnothing)$, provided that $\mathfrak{S}$ is parametrically conservative.

Theorem 10.79 *Let $\mathfrak{S} = (\boldsymbol{S}, \boldsymbol{A}, \boldsymbol{R}, \boldsymbol{P})$ be a parametric specification. Assume $\mathfrak{S}$ is parametrically conservative. Let $\mathfrak{S}' = (\boldsymbol{S}, \boldsymbol{A}, \boldsymbol{R}, \varnothing)$. Then*

$$\Gamma \vdash_{\mathfrak{S}}^{\hat{C}\hat{D}} a : A \Longrightarrow \{\Gamma\} \vdash_{\mathfrak{S}'}^{\hat{C}\hat{D}} \{a\} : \{A\}\,.$$

PROOF: Induction on the derivation of $\Gamma \vdash_{\mathfrak{S}}^{\hat{C}\hat{D}} a : A$. With the help of Lemma 10.76 and Lemma 10.77.3, all cases are straightforward except for the ($\hat{C}$-weak) and ($\hat{D}$-weak) rules. We only treat the ($\hat{D}$-weak) rule; the proof for ($\hat{C}$-weak) is similar. So: assume the last step of the derivation was

$$\frac{\Gamma \vdash_{\mathfrak{S}}^{\hat{C}\hat{D}} b : B \qquad \Gamma, \Delta_i \vdash_{\mathfrak{S}}^{\hat{C}\hat{D}} B_i : s_i \qquad \Gamma, \Delta \vdash_{\mathfrak{S}}^{\hat{C}\hat{D}} a : A : s}{\Gamma, c(\Delta){=}a{:}A \vdash_{\mathfrak{S}}^{\hat{C}\hat{D}} b : B}\,(s_i, s) \in \boldsymbol{P}.$$

By the induction hypothesis, we have:

$$\{\Gamma\} \vdash_{\mathfrak{S}'}^{\hat{C}\hat{D}} \{b\} : \{B\}; \tag{10.15}$$

$$\{\Gamma, \Delta_i\} \vdash_{\mathfrak{S}'}^{\hat{C}\hat{D}} \{B_i\} : s_i; \tag{10.16}$$

$$\{\Gamma, \Delta\} \vdash_{\mathfrak{S}'}^{\hat{C}\hat{D}} \{a\} : \{A\}; \tag{10.17}$$

$$\{\Gamma, \Delta\} \vdash_{\mathfrak{S}'}^{\hat{C}\hat{D}} \{A\} : s. \tag{10.18}$$

$\mathfrak{S}$ is parametrically conservative, so $(s_i, s, s) \in \boldsymbol{R}$ for $i = 1, \ldots, n$. Therefore, we can repeatedly use the Π-formation rule, starting with (10.18) and (10.16), obtaining

$$\{\Gamma\} \vdash_{\mathfrak{S}'}^{\hat{C}\hat{D}} \prod_{i=1}^{n} x_i{:}\,\{B_i\}\,.\,\{A\} : s. \tag{10.19}$$

Notice: $\prod_{i=1}^{n} x_i{:}\,\{B_i\}\,.\,\{A\} \equiv \{\prod \Delta.A\}$. Repeatedly using λ-formation, using (10.17) and (10.19), results in

$$\{\Gamma\} \vdash_{\mathfrak{S}'}^{\hat{C}\hat{D}} \lambda_{i=1}^{n} x_i{:}\,\{B_i\}\,.\,\{a\} : \{\prod \Delta.A\}\,. \tag{10.20}$$

Similarly, $\lambda_{i=1}^{n} x_i{:}\,\{B_i\}\,.\,\{a\} \equiv \{\lambda\,\Delta.a\}$. Using ($\hat{\mathrm{D}}$-weak) (for the specification $\mathfrak{S}'$) on (10.15), (10.16), (10.19) and (10.20) results in

$$\{\Gamma\}\,, c() {=} \{\lambda\,\Delta.a\} : \{\prod \Delta.A\} \vdash_{\mathfrak{S}'}^{\hat{C}\hat{D}} \{b\} : \{B\}\,.$$

⊠

Remark 10.80 The results in Section 10d2 were presented for $\hat{\mathrm{C}}\hat{\mathrm{D}}$-PTSs. The same result, however, can be obtained for $\hat{\mathrm{C}}$-PTSs, that is: for PTSs with restricted parameters, but without definitions. We can also give an alternative formulation of Remark 10.70, stating that a $\hat{\mathrm{C}}$-PTS with specification $(\boldsymbol{S}, \boldsymbol{A}, \boldsymbol{R}, \varnothing)$ is in fact nothing more than a C-PTS with specification $(\boldsymbol{S}, \boldsymbol{A}, \boldsymbol{R})$.

10d3 Refined Barendregt cubes

Theorem 10.79 has important consequences. The mapping $\{\_\}$ is fairly simple. It only translates some parametric abstractions and applications into λ-calculus style abstractions and applications. Hence a $\hat{\mathrm{C}}\hat{\mathrm{D}}$-PTS with parametric specification $\mathfrak{S} = (\boldsymbol{S}, \boldsymbol{A}, \boldsymbol{R}, \varnothing)$ can be extended with any set of parametric rules without extending its logical power, as long as the parametric specification obtained remains parametrically conservative.

In this section, we will apply the insight obtained in Section 10d2 to a concrete situation: the Barendregt cube. The Barendregt cube (Figure 4.2 on page 119) is a three-dimensional presentation of eight well-known PTSs.

This cube can be constructed not only for PTSs, but also for C-PTSs, D-PTSs, $\vec{\mathrm{C}}$-PTSs, $\vec{\mathrm{D}}$-PTSs, and their combinations (see Figure 10.1 on page 263).

With Theorem 10.79, we can place certain $\hat{\mathrm{C}}\hat{\mathrm{D}}$-PTSs in the cube of D-PTSs (and, with Remark 10.80 in mind, certain $\hat{\mathrm{C}}$-PTSs can also be placed in the cube of C-PTSs). Let us, for example, have a look at the following parametric specifications:

$$(\boldsymbol{S}, \boldsymbol{A}, \{(*,*,*), (*,\Box,\Box)\}, \varnothing);$$
$$(\boldsymbol{S}, \boldsymbol{A}, \{(*,*,*), (*,\Box,\Box)\}, \{(*,*)\});$$
$$(\boldsymbol{S}, \boldsymbol{A}, \{(*,*,*), (*,\Box,\Box)\}, \{(*,\Box)\});$$
$$(\boldsymbol{S}, \boldsymbol{A}, \{(*,*,*), (*,\Box,\Box)\}, \{(*,*), (*,\Box)\}).$$

where $\boldsymbol{S} = \{*, \Box\}$ and $\boldsymbol{A} = \{(*, \Box)\}$. By Theorem 10.79, the $\hat{C}\hat{D}$-PTSs with the above specifications are all equal in power, and according to Remark 10.70, they are all equal in power to the D-PTS with the specification of λP.

Now look at the parametric specification

$$\mathfrak{S} = (\boldsymbol{S}, \boldsymbol{A}, \{(*, *, *)\}, \{(*, *), (*, \Box)\}).$$

The $\hat{C}$-PTS $\lambda^{\hat{C}}(\mathfrak{S})$ is clearly stronger than the PTS $\lambda_{\rightarrow}$, as in $\lambda^{\hat{C}}(\mathfrak{S})$ it is possible (in a restricted way) to talk about predicates. For instance, we can have the following context:

$$\begin{array}{rcl} \alpha & : & *, \\ \mathtt{eq}(\mathtt{x}{:}\alpha, \mathtt{y}{:}\alpha) & : & *, \\ \mathtt{refl}(\mathtt{x}{:}\alpha) & : & \mathtt{eq}(\mathtt{x}, \mathtt{x}), \\ \mathtt{symm}(\mathtt{x}{:}\alpha, \mathtt{y}{:}\alpha, \mathtt{p}{:}\mathtt{eq}(\mathtt{x}, \mathtt{y})) & : & \mathtt{eq}(\mathtt{y}, \mathtt{x}), \\ \mathtt{trans}(\mathtt{x}{:}\alpha, \mathtt{y}{:}\alpha, \mathtt{z}{:}\alpha, \mathtt{p}{:}\mathtt{eq}(\mathtt{x}, \mathtt{y}), \mathtt{q}{:}\mathtt{eq}(\mathtt{y}, \mathtt{z})) & : & \mathtt{eq}(\mathtt{x}, \mathtt{z}) \end{array}$$

This context introduces an equality predicate eq on objects of type α, and axioms `refl`, `symm`, `trans` for the reflexivity, symmetry and transitivity of eq. It is not possible to introduce such a predicate eq in the PTS $\lambda_{\rightarrow}$ without any parameter mechanism. On the other hand, $\lambda^{\hat{C}}(\mathfrak{S})$ is weaker than the PTS λP: in λP we can construct the type $\Pi x{:}\alpha.\Pi y{:}\alpha.*$, which allows us to introduce variables eq of type $\Pi x{:}\alpha.\Pi y{:}\alpha.*$. This makes it possible to speak about *any* binary predicate, instead of one fixed predicate eq. It also gives us the possibility to speak about the term eq without the need to apply two terms of type α to it.

Altogether, this puts the $\hat{C}$-PTS $\lambda^{\hat{C}}(\mathfrak{S})$ clearly in between the PTSs $\lambda_{\rightarrow}$ and λP. Similarly, the $\hat{C}\hat{D}$-PTS $\lambda^{\hat{C}\hat{D}}(\mathfrak{S})$ is in between the D-PTSs $\lambda_{\rightarrow}$ and λP. We can illustrate this in the Barendregt cube by putting the specification $\boldsymbol{R}$ in the middle of the edge that connects the systems $\lambda_{\rightarrow}$ and λP.

This idea can be generalised to obtain a refinement of the Barendregt cube. We start with the system $\lambda_{\rightarrow}$. Adding an extra Π-formation rule (s_1, s_2, s_2) to $\lambda_{\rightarrow}$ corresponds to moving in one dimension (to the right, upward, or backward) in the cube. We add the possibility of moving in one dimension in the cube, but stopping half-way the cube, and we let this movement correspond to extending the system with the parameter rule (s_1, s_2). This "going only half-way" is in line with Theorem 10.79, which says that Π-formation rule (s_1, s_2, s_2) can mimic the parameter rule (s_1, s_2). In other words, the system obtained by "going all the way" is at least as strong as the system obtained by "going only half-way".

10e Systems in the refined Barendregt cube

In this section, we show that the refined Barendregt cube enables us to compare some well-known type systems with systems from the Barendregt cube. In particular, we show that ML, LF, λ68, and λQE can be seen as systems in the refined Barendregt cube. This is depicted in Figure 10.2 on page 302, and motivated in the four subsections below.

10e1 ML

In ML (see for instance [108]) one can define the polymorphic identity as follows (we use the notation of this chapter. In ML, the types and the parameters are left implicit):

$$\mathtt{Id}(\alpha{:}*) = (\lambda \mathtt{x}{:}\alpha.\mathtt{x}) : (\alpha \to \alpha).$$

But it is not possible to make an explicit λ-abstraction over $\alpha{:}*$: the expression

$$\mathtt{Id} = (\lambda\alpha{:}*.\lambda \mathtt{x}{:}\alpha.\mathtt{x}) : (\Pi\alpha{:}*.\alpha \to \alpha)$$

cannot be constructed in ML, as the type $\Pi\alpha{:}*.\alpha \to \alpha$ does not belong to the language of ML. Therefore, we can state that ML does not have a Π-formation rule $(\square, *, *)$, but that it does have the parametric rule $(\square, *)$.

Similarly, one can introduce the type of lists together with some elementary operations in ML as follows:

$$\begin{aligned} \mathtt{List}(\alpha{:}*) &: *; \\ \mathtt{nil}(\alpha{:}*) &: \mathtt{List}(\alpha); \\ \mathtt{cons}(\alpha{:}*) &: \alpha \to \mathtt{List}(\alpha) \to \mathtt{List}(\alpha), \end{aligned}$$

but the expression $\Pi\alpha{:}*.*$ does not belong to ML, so introducing `List` by

$$\mathtt{List} : \Pi\alpha{:}*.*$$

is not possible in ML. We conclude that ML does not have a Π-formation rule $(\square, \square, \square)$, but only the parametric rule $(\square, \square)$. Together with the fact that ML has a Π-formation rule $(*, *, *)$, this places ML in the middle of the left side of the refined Barendregt cube, exactly in between $\lambda_\to$ and $\lambda\omega$.

10e2 LF

Geuvers [59] initially describes the system LF (see [64]) as the PTS λP. However, the use of the Π-formation rule $(*, \square, \square)$ is quite restrictive in most applications of LF. Geuvers splits the λ-formation rule in two rules:

$$(\lambda_0) \quad \frac{\Gamma, x{:}A \vdash M : B \qquad \Gamma \vdash \Pi x{:}A.B : *}{\Gamma \vdash \lambda_0 x{:}A.M : \Pi x{:}A.B};$$

$$(\lambda_P) \quad \frac{\Gamma, x{:}A \vdash M : B \qquad \Gamma \vdash \Pi x{:}A.B : \square}{\Gamma \vdash \lambda_P x{:}A.M : \Pi x{:}A.B}.$$

System LF without rule (λ_P) is called LF$^-$. β-reduction is split into β_0-reduction and β_P-reduction:

$$(\lambda_0 x{:}A.M)N \to_{\beta_0} M[x{:=}N];$$
$$(\lambda_P x{:}A.M)N \to_{\beta_P} M[x{:=}N].$$

Geuvers then shows that

- If $M : *$ or $M : A : *$ in LF, then the β_P-normal form of M contains no λ_P;
- If $\Gamma \vdash_{\mathrm{LF}} M : A$, and Γ, M, A do not contain a λ_P, then $\Gamma \vdash_{\mathrm{LF}^-} M : A$;
- If $\Gamma \vdash M : A(: *)$, all in β_P-normal form, then $\Gamma \vdash_{\mathrm{LF}^-} M : A(: *)$.

This means that the only real need for a type $\Pi x{:}A.B : \square$ is to be able to declare a variable in it. The only point at which this is really done is where the bool-style implementation of PAT is made (see Section 4a4): the construction of the type of the operator Prf (in an unparameterised form) has to be made as follows:

$$\frac{\texttt{prop}{:}* \vdash \texttt{prop}{:}\, * \qquad \texttt{prop}{:}*, \alpha{:}\texttt{prop} \vdash *{:}\square}{\texttt{prop}{:}* \vdash (\Pi\alpha{:}\texttt{prop}.*) : \square}.$$

In the practical use of LF, this is the only point where the Π-formation rule $(*, \square, \square)$ is used. No λ_P-abstractions are used, either, and the term Prf is only used when it is applied to a term $p{:}\texttt{prop}$. This means that the practical use of LF would not be restricted if we introduced Prf in a parametric form, and replaced the Π-formation rule $(*, \square, \square)$ by a parameter rule $(*, \square)$. This puts (the practical applications of) LF in between the systems $\lambda_\to$ and λP in the refined Barendregt cube.

10e3 λ68 and AUT-68

Looking back at the system AUT-68 of Section 7a and its λ-calculus variant λ68 that was constructed and discussed in Sections 7b-7c, we remark that AUT-68 has a parameter mechanism and a mechanism for global parametric definitions:

- A line $(\Gamma; k; \mathrm{PN}; \texttt{type})$ in a book is nothing more that the declaration of a parametric constant $k(\Gamma){:}*$, and a line $(\Gamma; k; \Sigma_1; \texttt{type})$ is the declaration of a global parametric definition $k(\Gamma){=}\Sigma_1{:}*$. There are no demands on the context Γ, and this means that for a declaration $x{:}A \in \Gamma$ we can have either $A \equiv \texttt{type}$ (in PTS-terminology: $A \equiv *$, so $A : \square$) or $A{:}\texttt{type}$ (in PTS-terminology: $A : *$). We conclude that AUT-68 has the parameter rules $(*, \square)$ and $(\square, \square)$;
- Similarly, lines like $(\Gamma; k; \mathrm{PN}; \Sigma_2)$ and $(\Gamma; k; \Sigma_1; \Sigma_2)$, where $\Sigma_2{:}\texttt{type}$, represent parametric constants and global parametric definitions that are constructed using the parameter rules $(*, *)$ and $(\square, *)$.

Moreover, AUT-68 has a λ-calculus mechanism with as only Π-formation rule $(*, *, *)$.

This suggests that AUT-68 can be represented by a $\hat{C}\hat{D}$-PTS with specification

$$\mathfrak{S}_{68} = (\boldsymbol{S}, \boldsymbol{A}, \{(*, *, *)\}, \boldsymbol{S} \times \boldsymbol{S})$$

where $\boldsymbol{S} = \{*, \square\}$ and $\boldsymbol{A} = \{(*, \square)\}$. This system can be found in the exact middle of the refined Barendregt cube.

As for the structure of abstraction and application, this gives a good description of AUT-68. The position of AUT-68 in the refined Barendregt cube gives a far better idea of the force of AUT-68 than, for instance, the description of AUT-68 in [3], where it cannot be clearly positioned in the Barendregt cube. Another advantage is that $\lambda^{\hat{C}\hat{D}}(\mathfrak{S}_{68})$ has parameters. Thus, it is closer to the original system AUT-68 than the system $\lambda 68$ that was described in Chapter 7, and in [3] (though in Theorem 7.63 and Remark 7.64, we minutely described the way in which the parameter mechanism appears in $\lambda 68$).

On the other hand, we should not say that AUT-68 is exactly the system $\lambda^{\hat{C}\hat{D}}(\mathfrak{S}_{68})$. There are several differences:

- DPTSs have global and local definitions. AUTOMATH has only global definitions;
- In DPTSs, the type B of a definition $x{=}T{:}B$ does not have to be typable itself. In AUTOMATH, B has to be typable;
- The D-reduction of DPTSs is not substitutive; δ-reduction of AUTOMATH is substitutive;

These differences can also be found between AUT-68 and the DPTSs of Chapter 8 (see Section 8b1).

10e4 λQE and AUT-QE

In λQE we have a Π-formation rule $(*, \square, \square)$ additionally to the rules of $\lambda 68$. This means that the applicational and abstractional behaviour can be described by the $\hat{C}\hat{D}$-PTS with Π-formation rules $(*, *, *)$ and $(*, \square, \square)$, and parametric rules (s_1, s_2) for $s_1, s_2 \in \boldsymbol{S}$. This system is located in the middle of the right side of the refined Barendregt cube, exactly in between λC and λP. Again, this is not the exact representation of AUT-QE; there are differences that are similar to those described in Section 8b1. Moreover, AUT-QE has a rule of type inclusion (see the Conclusion of Chapter 7), which is not taken into account in $\hat{C}\hat{D}$-PTSs.

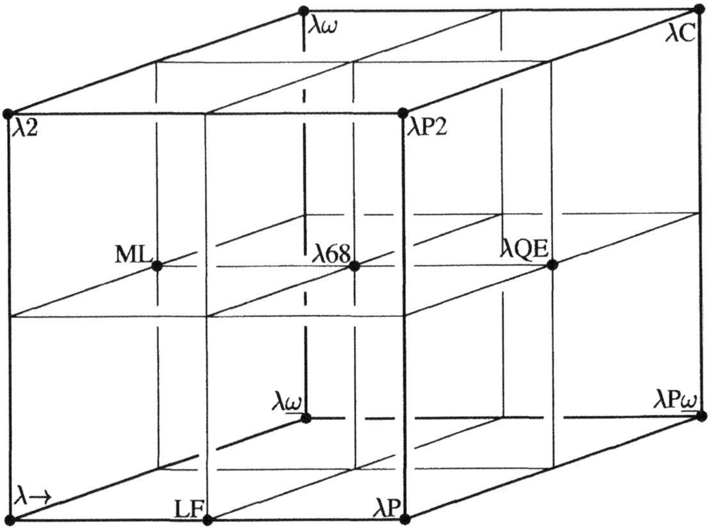

Figure 10.2: LF, ML, λ68, and λQE in the refined Barendregt cube

10e5 PAL

The AUTOMATH languages are all based on two concepts: typed λ-calculus and a parameter/definition mechanism. Both concepts can be isolated: it is possible to study λ-calculus without a parameter/definition mechanism (for instance via the format of Pure Type Systems), but one can also isolate the parameter/definition mechanism from AUTOMATH. One then obtains a language that is called PAL, the "Primitive AUTOMATH Language". It cannot be described within the refined Barendregt cube (as all the systems in that cube have at least some basic λ-calculus in it), but it can be described as a $\hat{C}\hat{D}$-PTS with the following parametric specification:

$$\begin{array}{rcl} \boldsymbol{S} & = & \{*, \Box\} \\ \boldsymbol{A} & = & \{(*, \Box)\} \\ \boldsymbol{R} & = & \varnothing \\ \boldsymbol{P} & = & \{(*,*), (*,\Box), (\Box,*), (\Box,\Box)\} \end{array}$$

This parametric specification corresponds to the parametric specifications that were given for the AUTOMATH systems above, from which the Π-formation rules are removed.

10f First-order predicate logic

A standard way to code first-order predicate logic in PAT-style (Curry-Howard variant) uses a type system that looks familiar to λP. It is due to Berardi, and presented in Definition 5.4.5 of [3].

To keep objects and object types separated from proofs and propositions, the sorts $*$ and $\Box$ of λP are replaced by $*_s, *_p, *_f, \Box_s$ and $\Box_p$. Here, $*_s$ and $\Box_s$ handle the objects and object types, whilst $*_p, \Box_p$ are used for propositions and their proofs. The sort $*_f$ is used to store the types of the function symbols of the first-order language. For the construction of logical implication and universal quantification, the Π-formation rules $(*_p, *_p, *_p)$ and $(*_s, *_p, *_p)$ are used. The Π-formation rule $(*_s, *_s, *_f)$ allows the formation of a function space between object types, and the Π-formation rule $(*_s, *_f, *_f)$ makes it possible to form functions of several arguments between object types. There is no sort $\Box_f$, as free variables for function spaces are not allowed. The construction of relation symbols requires Π-formation rule $(*_s, \Box_p, \Box_p)$.

Thus, we find a PTS (or a D-PTS) with the following specification:

$$\begin{aligned} \boldsymbol{S} &= \{*_s, *_p, *_f, \Box_s, \Box_p\}; \\ \boldsymbol{A} &= \{(*_s, \Box_s), (*_p, \Box_p)\}; \\ \boldsymbol{R} &= \{(*_s, *_s, *_f), (*_s, *_f, *_f), (*_s, *_p, *_p), (*_p, *_p, *_p), (*_s, \Box_p, \Box_p)\}. \end{aligned}$$

Due to the Π-formation rule $(*_s, \Box_p, \Box_p)$ in the PTS-representation of first-order logic, there are types that are not in β-normal form:

Example 10.81 For a term $A : *_s$ we can form the type $\Pi x{:}A.*_p$. If b is a term of type $*_p$ in which a variable $x{:}A$ may occur free, we can form $\lambda x{:}A.b$ of type $\Pi x{:}A.*_p$. Applying this term to a term a of type A results in $(\lambda x{:}A.b)a$ of type $*_p$. This term is a type (because it has type $*_p$) and is not in β-normal form.

If a PTS has types that are not in β-normal form, it is possible that there are applications of the conversion rule

$$\frac{\Gamma \vdash A : B \qquad \Gamma \vdash B' : s \qquad B =_\beta B'}{\Gamma \vdash A : B'}$$

in a deduction in such a PTS. The conversion rule has as a disadvantage that its implementation in computer systems makes the system slow. This is because it may be very time-consuming (or memory-consuming) to establish whether two λ-terms are β-equal or not. Hence, it would be useful to have a type system in which all types are in β-normal form.

In the above formulation of first-order predicate logic, $(*_s, \Box_p, \Box_p)$ is the only rule which allows to form types that are not in β-normal form. We show this as follows. Assume, $\Gamma \vdash P : s$, P is not in β-normal form, and all the subterms P' of P that are a type are in β-normal form. Then P cannot be a sort or a variable. As P has type s, P cannot be of the form $\lambda x{:}P_1.P_2$, either. If $P \equiv \Pi x{:}P_1.P_2$ then either P_1 or P_2 are not in β-normal form. As P_1 and P_2 are both types, this does not occur. So P must be an application term $P_1 P_2$. By the Generation Lemma for PTSs, there is a type A and a sort s such that $\Gamma \vdash P_1 : (\Pi x{:}A.s)$. By Correctness of Types, there is a sort s' such that $\Gamma \vdash (\Pi x{:}A.s) : s'$. By the Generation Lemma, there is $(s_1, s_2, s') \in \boldsymbol{R}$ such that $\Gamma, x{:}A \vdash s : s_2$ and $\Gamma \vdash A : s_1$. This means that (s, s_2) is an axiom, and therefore $s_2 \in \{\Box_s, \Box_p\}$. Hence, $(s_1, s_2, s_3) = (*_s, \Box_p, \Box_p)$.

We conclude that implementations of first-order predicate logic in type theory would be more efficient if it were possible to avoid rule $(*_s, \Box_p, \Box_p)$. With the use of parameters, it is easy to avoid that rule. This is because rule $(*_s, \Box_p, \Box_p)$ is only necessary to type the relation symbols of the first-order language. And as relation symbols in a first-order language are always introduced with parameters, it is no restriction to introduce them in the type system in a parametrised way. This can be done with the parameter-rule $(*_s, \Box_p)$: if we want to introduce a n-ary relation symbol R with arguments of type $U_1, \ldots, U_n$ (where the U_is are of type $*_s$), we apply $\vec{\mathrm{C}}$-weakening (let $\Delta \equiv x_1{:}U_1, \ldots, x_n{:}U_n$ and $\Delta_i \equiv x_1{:}U_1, \ldots, x_{i-1}{:}U_{i-1}$):

$$\frac{\Gamma \vdash b{:}B \qquad \Gamma, \Delta_i \vdash U_i{:}\, *_s \qquad \Gamma, \Delta \vdash *_p{:}\Box_p}{\Gamma, R(\Delta) : *_p \vdash b : B}.$$

This involves the use of the parameter-rule $(*_s, \Box_p)$.

Hence, replacing rule $(*_s, \Box_p, \Box_p)$ by parameter-rule $(*_s, \Box_p)$ enables one to remove the conversion rule in the type-theoretic representation of first-order predicate logic, making it more efficient (see the forthcoming Theorem 10.84). It is reasonable to replace more rules by parameter-rules in the case of first-order predicate logic, as we presently explain.

Function symbols in a first-order language are also of a parametric nature. The sort $*_f$, the Π-formation rules $(*_s, *_s, *_f)$ and $(*_s, *_f, *_f)$ are only used to construct the types of these function symbols. We can introduce these function symbols in a more realistic way by using parametric rule $(*_s, *_s)$ instead of the Π-formation rules $(*_s, *_s, *_f)$ and $(*_s, *_f, *_f)$:

$$\frac{\Gamma \vdash b{:}B \qquad \Gamma, \Delta_i \vdash U_i{:}\, *_s \qquad \Gamma, \Delta \vdash U{:}*_s}{\Gamma, f(\Delta) : U \vdash b : B}.$$

We have now obtained a $\hat{\mathrm{C}}$-PTS with $\mathfrak{S}' = (\boldsymbol{S}', \boldsymbol{A}', \boldsymbol{R}', \boldsymbol{P}')$ as its parametric specification where:

$$\begin{array}{rcl} \boldsymbol{S}' & = & \{*_s, *_p, \Box_s, \Box_p\}; \\ \boldsymbol{A}' & = & \{(*_s, \Box_s), (*_p, \Box_p)\}; \\ \boldsymbol{R}' & = & \{(*_s, *_p, *_p), (*_p, *_p, *_p)\}; \\ \boldsymbol{P}' & = & \{(*_s, *_s), (*_s, \Box_p)\}. \end{array}$$

We now prove that types in this Ĉ-PTS are always in β-normal form. For the proof we need as a lemma that any object term (that is: a term P such that there is Q with $P : Q : *_s$) is in β-normal form.

Lemma 10.82 *If* $\Gamma \vdash^{\hat{C}}_{\mathfrak{S}'} P : Q : *_s$ *then* P *is in* β*-normal form.*

PROOF: Induction on the structure of P.

- The cases $P \in \mathcal{V}$ and $P \in \boldsymbol{S}'$ are trivial;
- If $P \equiv c(b_1, \ldots, b_n)$ then we use the second extension of the Generation Lemma, 10.69, and determine $B_1, \ldots, B_n, B$ and $s_1, \ldots, s_n, s$ such that $\Gamma \vdash^{\hat{C}}_{\mathfrak{S}'} b_i{:}B_i[x_j{:=}b_j]_{j=1}^{i-1}$ and $\Gamma, x_1{:}B_1, \ldots x_{i-1}{:}B_{i-1} \vdash^{\hat{C}}_{\mathfrak{S}'} B_i{:}s_i$, and $(s_i, s) \in \boldsymbol{P}'$. Due to the definition of $\boldsymbol{P}'$, we get $s_i \equiv *_s$ for all i. By the Substitution Lemma, we have $\Gamma \vdash B_i[x_j{:=}b_j]_{j=1}^{i-1} : *_s$, and therefore $\Gamma \vdash b_i{:}B_i[x_j{:=}b_j]_{j=1}^{i-1} : *_s$. By the induction hypothesis, the b_i are in β-normal form. Therefore, $c(b_1, \ldots, b_n)$ is in β-normal form;
- If $P \equiv P_1 P_2$ then by the the Generation Lemma, there are R_1, R_2 such that $\Gamma \vdash^{\hat{C}}_{\mathfrak{S}'} P_1 : \Pi x{:}R_1.R_2$, and $Q =_\beta R_2[x{:=}P_2]$. By Correctness of Types there is $s \in \boldsymbol{S}'$ such that $\Gamma \vdash^{\hat{C}}_{\mathfrak{S}'} (\Pi x{:}R_1.R_2) : s$. By the Generation Lemma and the definition of $\boldsymbol{R}'$, $\Gamma, x{:}R_1 \vdash^{\hat{C}}_{\mathfrak{S}'} R_2{:}*_p$. By the Substitution Lemma, $\Gamma \vdash^{\hat{C}}_{\mathfrak{S}'} R_2[x{:=}P_2]{:}*_p$. Let Q' be a common β-reduct of Q and $R_2[x{:=}P_2]$. By Subject Reduction, $\Gamma \vdash^{\hat{C}}_{\mathfrak{S}'} Q' : *_s$ and $\Gamma \vdash^{\hat{C}}_{\mathfrak{S}'} Q' : *_p$, which contradicts Unicity of Types. We conclude that the case $P \equiv P_1 P_2$ does not occur;
- If $P \equiv \lambda x{:}P_1.P_2$ then there are R_1, R_2 such that $Q =_\beta \Pi x{:}R_1.R_2$. Let Q' be a common β-reduct of Q and $\Pi x{:}R_1.R_2$. There are R_1', R_2' such that $Q' \equiv \Pi x{:}R_1'.R_2'$. By Subject Reduction, $\Gamma \vdash^{\hat{C}}_{\mathfrak{S}'} \Pi x{:}R_1'.R_2' : *_s$. By the Generation Lemma, there are s_1, s_2 such that $(s_1, s_2, *_s) \in \boldsymbol{R}'$. This is not the case. So the case $P \equiv \lambda x{:}P_1.P_2$ does not occur;
- If $P \equiv \Pi x{:}P_1.P_2$ then there is s such that $Q \equiv s$. By the Generation Lemma, this would mean that $s : *_s$ is an axiom, which is not the case. So the case $P \equiv \Pi x{:}P_1.P_2$ does not occur.

⊠

Remark 10.83 The proof of this lemma not only shows that a P for which we have $P : Q : *_s$ is always in normal form. It also shows that P can only be a variable or an expression of the form $c(b_1, \ldots, b_n)$ such that there are $B_1, \ldots, B_n$ with $b_i : B_i : *_s$. This corresponds exactly to the definition of terms in first-order logic. We conclude that our specification $\mathfrak{S}'$ results in an exact description of the terms of first-order logic.

Theorem 10.84 *Assume* $\Gamma \vdash^{\hat{C}}_{\mathfrak{S}'} P : s$. *Then* P *is in* β*-normal form.*

PROOF: Induction on the structure of P.

- The cases $P \in \mathcal{V}$ and $P \in \boldsymbol{S}'$ are trivial;
- $P \equiv c(b_1, \ldots, b_n)$. By the second extension of the Generation Lemma 10.69, there are sorts $s_1, \ldots, s_n$ and terms $B_1, \ldots, B_n$ such that $(s_i, s) \in \boldsymbol{P}'$, $\Gamma \vdash^{\hat{C}}_{\mathfrak{S}'} b_i : B_i[x_j{:=}b_j]_{j=1}^{i-1}$ and $\Gamma, x_1{:}B_1, \ldots, x_{i-1}{:}B_{i-1} \vdash^{\hat{C}}_{\mathfrak{S}'} B_i{:}s_i$. By the definition of $\boldsymbol{P}'$, $s_i \equiv *_s$ for all i. By the Substitution Lemma, $\Gamma \vdash^{\hat{C}}_{\mathfrak{S}'} b_i : B_i[x_j{:=}b_j]_{j=1}^{i-1} : *_s$. By Lemma 10.82, the b_i are in β-normal form. Therefore $c(b_1, \ldots, b_n)$ is in β-normal form;
- $P \equiv P_1 P_2$. By the Generation Lemma, there are R_1, R_2 such that $\Gamma \vdash^{\hat{C}}_{\mathfrak{S}'} P_1 : \Pi x{:}R_1.R_2$ and $s =_\beta R_2[x{:=}P_2]$. By Correctness of Types, the Generation Lemma and the definition of $\boldsymbol{R}'$, $\Gamma, x{:}R_1 \vdash^{\hat{C}}_{\mathfrak{S}'} R_2 : *_p$. By the Substitution Lemma, $\Gamma \vdash^{\hat{C}}_{\mathfrak{S}'} R_2[x{:=}P_2] : *_p$. By Subject Reduction, $\Gamma \vdash^{\hat{C}}_{\mathfrak{S}'} s : *_p$. This means that $(s, *_p)$ is an axiom, which is not the case. We conclude that the case $P \equiv P_1 P_2$ does not occur;
- $P \equiv \lambda x{:}P_1.P_2$. By the Generation Lemma, $s =_\beta \Pi x{:}R_1.R_2$ for some R_1, R_2. This is impossible. We conclude that the case $P \equiv \lambda x{:}P_1.P_2$ does not occur;
- $P \equiv \Pi x{:}P_1.P_2$. By the Generation Lemma, there are s_1, s_2 such that $\Gamma \vdash^{\hat{C}}_{\mathfrak{S}'} P_1 : s_1$ and $\Gamma, x{:}P_1 \vdash^{\hat{C}}_{\mathfrak{S}'} P_2 : s_2$. By the induction hypothesis, P_1 and P_2 are in β-normal form. So P is in β-normal form.

⊠

We conclude that replacing the Π-formation rules

$$(*_s, *_s, *_f) \qquad (*_s, *_f, *_f) \qquad (*_s, \Box_p, \Box_p)$$

by parametric rules

$$(*_s, *_s) \qquad (*_s, \Box_p)$$

makes the implementations of first-order languages in type theory

- easier to implement (as the conversion rule becomes superfluous);

- more realistic (it gives, for example, an exact description of the terms in first-order logic, something that cannot be done in the parameter-free PTS proposed by Berardi).

Conclusions: Yet another extension of PTSs?

Since PTSs have been introduced, many extensions have been proposed (see [4] for a non-exhaustive list). The reader may wonder why yet another extension of PTSs is proposed in this chapter, and whether it is more interesting than those other extensions or not. In this section we give an answer to these questions.

Practical motivation

- Parameters and parametric definitions occur in many implementations of type systems, and more general, in programming languages. The Pascal-function `double` that was introduced at the beginning of this chapter can be described in our formalism by the context declaration

  `double(z:Int)=z+z:Int;`

- The AUTOMATH systems, which form the basis for most modern proof checkers that are based on type theory, can be described in our system. The description given in Chapter 7 is precise, but it is not a description that looks natural. The separate abstractors ¶ and § do their job as well as possible in a type system without parameters, but a description of AUTOMATH that includes parameters does more justice to that system. Moreover, it places AUTOMATH in a more general framework, so that it can easily be compared with other type systems (see Figure 10.2 on page 302);
- Modern type systems, like LF and ML, have already been described as one of the systems of the Barendregt cube (Figure 10.2 on page 302). But in Section 10e we showed that a more detailed description can be given in the refined Barendregt cube of Figure 10.2;
- As argued in Section 10f, parameters are useful when describing first-order logic in type theory. Compared to the traditional PTS-representation (systems related to λP of the Barendregt cube) of first-order logic, parametric representations are
  - easier to implement (because the conversion rule is not needed);
  - closer to the original first-order language and therefore closer to the intuition;

- As argued in the beginning of this chapter, parameters make it possible to distinguish the attitude of users and developers of a system. Often, the user only needs a (partially) parametrised version of the system, whilst the developer wants to have the possibilities of full λ-abstractions.

But "parameters give a better description of the type theory that is used in practice" is not the only argument in favour of the system of this chapter. There is more than that.

The heart of type theory

As explained before, the explicit and formal use of types (and thus an early form of what is presently called "type theory") was originally intended to prevent the paradoxes that occurred in logic and mathematics at the end of the 19th and the beginning of the 20th century. But it was not the only method developed for this purpose. Another tool was the fine-tuning of Cantor's Set Theory [23, 24] by Zermelo [147]. The approach of type theory however, is completely different from the set-theoretical approach. Type theory focuses on the notion of *function* in logic and mathematics, and throughout the history of type theory, functions have remained one of the main objects of study for type theorists.

In the abstract theory of functions, there are only two constructions:

- *Functionalisation* which is the process in which a function is constructed out of an expression. For example: the construction of the function $x \mapsto x + 3$ from the expression $2 + 3$. This function is denoted in the λ-calculus by $\lambda x.(x + 3)$.
- *Instantiation* which is the process in which a function value is calculated when a suitable argument is assigned to a function. For example: the construction of the term $2 + 3$ by applying the function $x \mapsto x + 3$ to the argument 2.

Clearly these two processes are each others inverses.

It appears that the functionalisation process can be split up into two parts:

- *Abstraction from a subexpression.* We replace a subexpression by a variable. For instance, in the expression $2 + 3$ we can replace the subexpression 2 by a variable x, thus obtaining $x + 3$.
- *Function construction.* The expression $x+3$ stands for: the addition of *some* number x to the number 3. It does not denote one specific natural number, but if we replace x by a natural number, then the resulting expression represents a natural number. This gives rise to an algorithm: we feed a natural

number to the algorithm, and the algorithm adds 3 to that number. This algorithm is called a *function*. In the λ-calculus, it is denoted by $\lambda x.(x+3)$.

Similarly, instantiation can be split up into two parts:

- *Application construction.* Juxtaposing a function to an argument denotes an *intended* function application. For example, applying the function $\lambda x.(x+3)$ to the argument 2 leads to the intended function application $(\lambda x.(x+3))2$.

- *Concretisation to a subexpression.* Calculating the result of this intended application. In our example: 2 is substituted for x, for every occurrence of x in the expression $x+3$. The result is: $2+3$.

Both functionalisation and instantiation are present in many important type theories and programming languages, and in many important logical systems. But different theories/systems focus on different parts of these processes. In the functionalisation process, for example, Frege [49] focuses mainly on abstraction from a subexpression, whilst the λ-calculus pays more attention to function construction. After our exploration of type theory throughout the present work, we still think that functionalisation and instantiation are the heart of type theory, for more than one reason:

- Functionalisation and instantiation stood at the cradle of type theory. The story of type theory began with Frege's abstraction principles (instantiation was not explicitly defined, but definitely present in an implicit form), and the logical paradoxes that arose if one does not use these principles carefully. Type theory made a careful use of Frege's principles possible;

- An important application of modern type theory is logic. This is due to the PAT-principle, which on its turn is based on the interpretation of $\rightarrow$ and $\forall$ as function types. And function types exist because of functionalisation and instantiation.

The parameter mechanism shows us a new, different form of functionalisation and instantiation and therefore makes the theory of functions richer and more interesting:

- It gives us a better idea of the possibilities of the traditional forms of functionalisation and instantiation;

- It places these traditional forms in a broader perspective by showing that these forms are not the only possible forms of functionalisation and instantiation.

In this light, the parameter mechanism is not only an *extension* of Pure Type Systems (as depicted in Figure 10.1 on page 263), but also (and particularly) a *refinement* of this framework, resulting in refinements of parts of it, like the Barendregt cube (Figure 9.1 on page 251).

Future work

There are several things concerning parametric type systems that deserve to be studied in the future:

- The meta-theoretical properties may have easier proofs than the ones presented in this chapter. In particular, the proof of strong normalisation for a parametric type system is based on strong normalisation for a PTS that may have more Π-formation rules. It would be interesting to know whether (and to what extent) these rather strong demands can be weakened;

- In the systems proposed in this chapter, it is not possible to have a parametric constant (or definition) that takes a parametric function as a parameter. For example: We want to formulate the property $\mathtt{Ref}(B)$ for binary relations B over type T, indicating that this relation is reflexive. In our current system, B cannot be a parametric function $b(x{:}T, y{:}T)$, because $b(x{:}T, y{:}T)$ is not a term. We must make the full λ-abstraction $\lambda x{:}T.\lambda y{:}T.b(x, y)$ (which is a term) if we want to give it as an argument to $\mathtt{Ref}$.

  It may be useful to design a system in which the parametric function $b(x{:}T, y{:}T)$ could be substituted for B without the need of making the λ-abstractions.

- There may be a relation between the parameter mechanism of this chapter and AUTOMATH, and the use of parameters in the representation of higher order propositional functions in the ramified theory of types of Russell and Whitehead.

Appendix A

Type systems in this book

Aa Pure Type Systems

Definition A.1 (Terms of PTSs, 4.16) Let $\mathbb{V}$ be a set of variables and $\mathbb{C}$ a set of constants. Assume that $\mathbb{V}$ and $\mathbb{C}$ are countably infinite and are disjoint. The set $\mathcal{T}(\mathbb{V}, \mathbb{C})$ (shorthand: $\mathcal{T}$) is given by:

$$\mathcal{T} ::= \mathbb{V} \mid \mathbb{C} \mid \mathcal{T}\mathcal{T} \mid \lambda\mathbb{V}{:}\mathcal{T}.\mathcal{T} \mid \Pi\mathbb{V}{:}\mathcal{T}.\mathcal{T}.$$

Definition A.2 (Reduction, 4.17) The relation $\rightarrow_\beta$ is described by the contraction rule

$$(\lambda x{:}A_1.A_2)B \rightarrow_\beta A_2[x{:=}B]$$

and the usual compatibility rules.

$=_\beta$ is the smallest reflexive, symmetric and transitive relation that includes $\rightarrow_\beta$.

Definition A.3 (PTSs Specifications, 4.18) A *specification* for PTSs is a triple $(\boldsymbol{S}, \boldsymbol{A}, \boldsymbol{R})$, such that $\boldsymbol{S} \subseteq \mathbb{C}$, $\boldsymbol{A} \subseteq \boldsymbol{S} \times \boldsymbol{S}$ and $\boldsymbol{R} \subseteq \boldsymbol{S} \times \boldsymbol{S} \times \boldsymbol{S}$. The specification is called *singly sorted* if $\boldsymbol{A}$ is a (partial) function $\boldsymbol{S} \rightarrow \boldsymbol{S}$, and $\boldsymbol{R}$ is a (partial) function $\boldsymbol{S} \times \boldsymbol{S} \rightarrow \boldsymbol{S}$. $\boldsymbol{S}$ is called the set of *sorts*, $\boldsymbol{A}$ is the set of *axioms*, and $\boldsymbol{R}$ is the set of (Π-formation) *rules* of the specification.

Definition A.4 (Contexts of PTSs, 4.19) A *context* is a finite (possibly empty) list $x_1{:}A_1, \ldots, x_n{:}A_n$ (shorthand: $\vec{x}{:}\vec{A}$) of variable declarations. $\{x_1, \ldots, x_n\}$ is called the *domain* $\text{DOM}\left(\vec{x}{:}\vec{A}\right)$ of the context. The *empty context* is denoted $\langle\rangle$. If Γ, Δ are contexts then we write $\Gamma \subseteq \Delta$ if all declarations in Γ are also in Δ.

Definition A.5 (Pure Type Systems, 4.21) Let $\mathfrak{S} = (\boldsymbol{S}, \boldsymbol{A}, \boldsymbol{R})$ be a specification. The Pure Type System $\lambda\mathfrak{S}$ describes in which ways judgements $\Gamma \vdash_{\mathfrak{S}} A : B$ (or

$\Gamma \vdash A : B$, if it is clear which $\mathfrak{S}$ is used) can be derived. $\Gamma \vdash A : B$ states that A has type B in context Γ.

(axiom) $$\langle\rangle \vdash s_1 : s_2 \qquad (s_1, s_2) \in \boldsymbol{A}$$

(start) $$\frac{\Gamma \vdash A : s}{\Gamma, x{:}A \vdash x : A} \qquad x \notin \text{DOM}(\Gamma)$$

(weak) $$\frac{\Gamma \vdash A : B \qquad \Gamma \vdash C : s}{\Gamma, x{:}C \vdash A : B} \qquad x \notin \text{DOM}(\Gamma)$$

(Π) $$\frac{\Gamma \vdash A : s_1 \qquad \Gamma, x{:}A \vdash B : s_2}{\Gamma \vdash (\Pi x{:}A.B) : s_3} \qquad (s_1, s_2, s_3) \in \boldsymbol{R}$$

(λ) $$\frac{\Gamma, x{:}A \vdash b : B \qquad \Gamma \vdash (\Pi x{:}A.B) : s}{\Gamma \vdash (\lambda x{:}A.b) : (\Pi x{:}A.B)}$$

(appl) $$\frac{\Gamma \vdash F : (\Pi x{:}A.B) \qquad \Gamma \vdash a : A}{\Gamma \vdash Fa : B[x{:=}a]}$$

(conv) $$\frac{\Gamma \vdash A : B \qquad \Gamma \vdash B' : s \qquad B =_\beta B'}{\Gamma \vdash A : B'}$$

Ab The Barendregt cube

Definition A.6 (Terms of the Barendregt cube) Let $\mathbb{V}$ be a set of variables and $\mathbb{C}$ a set of constants. Assume that $\mathbb{V}$ and $\mathbb{C}$ are countably infinite and are disjoint. The set $\mathcal{T}(\mathbb{V}, \mathbb{C})$ (shorthand: $\mathcal{T}$) is given by:

$$\mathcal{T} ::= \mathbb{V} \mid \mathbb{C} \mid \mathcal{T}\mathcal{T} \mid \lambda\mathbb{V}{:}\mathcal{T}.\mathcal{T} \mid \Pi\mathbb{V}{:}\mathcal{T}.\mathcal{T}.$$

Definition A.7 (Reduction in the Barendregt cube)) The relation $\to_\beta$ is described by the contraction rule

$$(\lambda x{:}A_1.A_2)B \to_\beta A_2[x{:=}B]$$

and the usual compatibility rules.

$=_\beta$ is the smallest reflexive, symmetric and transitive relation that includes $\to_\beta$.

Definition A.8 (Specifications in the cube, 4.23) Let (s_1, s_2) denote (s_1, s_2, s_2). A *cube specification* is a triple $(\boldsymbol{S}, \boldsymbol{A}, \boldsymbol{R})$, such that

- $\boldsymbol{S} = \{*, \square\} \subseteq \mathbb{C}$,
- $\boldsymbol{A} = \{* : \square\} \subseteq \boldsymbol{S} \times \boldsymbol{S}$ and

- $\boldsymbol{R} \subseteq \{(*,*),(*,\square),(\square,*),(\square,\square)\} \subseteq \boldsymbol{S}\times\boldsymbol{S}\times\boldsymbol{S}$ such that $(*,*)\in\boldsymbol{R}$.

Note that, as all the systems of the cube have the same sets of sorts $\boldsymbol{S}$ and axioms $\boldsymbol{A}$, it is enough to represent each system by its set of rules $\boldsymbol{R}$ instead of using all the specification $(\boldsymbol{S},\boldsymbol{A},\boldsymbol{R})$.

Definition A.9 (Contexts of the cube) A *context* is a finite (possibly empty) list $x_1{:}A_1,\ldots,x_n{:}A_n$ (shorthand: $\vec{x}{:}\vec{A}$) of variable declarations. $\{x_1,\ldots,x_n\}$ is called the *domain* $\text{DOM}\left(\vec{x}{:}\vec{A}\right)$ of the context. The *empty context* is denoted $\langle\rangle$. If Γ,Δ are contexts then we write $\Gamma\subseteq\Delta$ if all declarations in Γ are also in Δ.

Definition A.10 (Systems of the Barendregt cube, 4.24)
Let $(\boldsymbol{S},\boldsymbol{A},\boldsymbol{R})$ be a cube specification. The type system $\lambda\boldsymbol{R}$ describes in which ways judgments $\Gamma\vdash_{\boldsymbol{R}} A:B$ (or $\Gamma\vdash A:B$, if it is clear which $\boldsymbol{R}$ is used) can be derived. $\Gamma\vdash A:B$ states that A has type B in context Γ. The typing rules are inductively defined as follows:

(axiom) $$\langle\rangle\vdash *:\square$$

(start) $$\frac{\Gamma\vdash A:s}{\Gamma,x{:}A\vdash x:A}\qquad x\notin\text{DOM}(\Gamma)$$

(weak) $$\frac{\Gamma\vdash A:B\qquad \Gamma\vdash C:s}{\Gamma,x{:}C\vdash A:B}\qquad x\notin\text{DOM}(\Gamma)$$

(Π) $$\frac{\Gamma\vdash A:s_1\qquad \Gamma,x{:}A\vdash B:s_2}{\Gamma\vdash(\Pi x{:}A.B):s_2}\qquad (s_1,s_2)\in\boldsymbol{R}$$

(λ) $$\frac{\Gamma,x{:}A\vdash b:B\qquad \Gamma\vdash(\Pi x{:}A.B):s}{\Gamma\vdash(\lambda x{:}A.b):(\Pi x{:}A.B)}$$

(appl) $$\frac{\Gamma\vdash F:(\Pi x{:}A.B)\qquad \Gamma\vdash a:A}{\Gamma\vdash Fa:B[x{:=}a]}$$

(conv) $$\frac{\Gamma\vdash A:B\qquad \Gamma\vdash B':s\qquad B=_\beta B'}{\Gamma\vdash A:B'}$$

The dependencies between these systems can be depicted in the Barendregt cube given in the following figure:

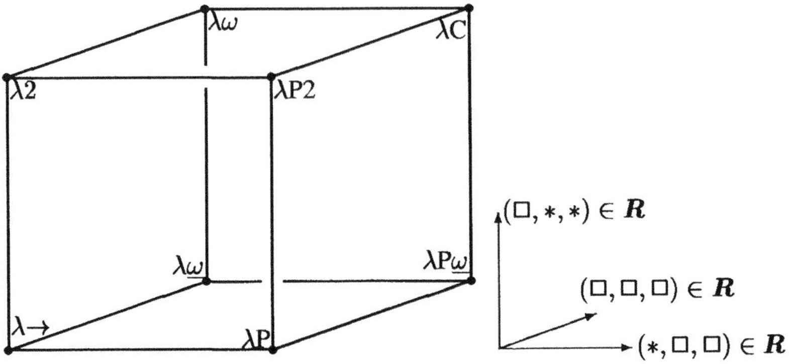

Ac The Ramified Theory of Types

Ac1 RTT

We assume

- A set $\mathcal{A}$ of *individual symbols* (the basic signs);
- A set $\mathcal{V}$ of *variables* (the signs that indicate replaceable objects);
- A set $\mathcal{R}$ of *relation symbols* together with a map $\mathfrak{a} : \mathcal{R} \to \mathbb{N}^+$ indicating the *arity* of each relation-symbol (these are used to form the basic propositions).

We assume that $\mathcal{A}$ and $\mathcal{V}$ are countably infinite and that for each $n \in \mathbb{N}^+$, the set $\{R \in \mathcal{R} \mid \mathfrak{a}(R) = n\}$ is also countably infinite.
We also assume that $\{\mathtt{x}, \mathtt{y}, \mathtt{z}, \mathtt{x}_1, \ldots\} \subseteq \mathcal{V}$ and that $\mathcal{V}$ is ordered as follows:

$$\mathtt{x} < \mathtt{x}_1 < \ldots \mathtt{y} < \mathtt{y}_1 < \ldots \mathtt{z} < \mathtt{z}_1 \ldots$$

We let $a_1, a_2, \ldots$ range over $\mathcal{A}$, $x, y, z, x_1, \ldots$ range over $\mathcal{V}$ and $R, S, \ldots$ range over $\mathcal{R}$.

Definition A.11 (Atomic propositions, 2.2) A list of symbols of the form $R(a_1, \ldots, a_{\mathfrak{a}(R)})$ is called an *atomic proposition*.

Definition A.12 (Propositional functions, 2.3) We define a collection $\mathcal{P}$ of *propositional functions* (pfs), and for each element f of $\mathcal{P}$ we simultaneously define the collection $\mathrm{FV}(f)$ of *free variables* of f:

1. If $i_1, \ldots, i_{\mathfrak{a}(R)} \in \mathcal{A} \cup \mathcal{V}$ then $R(i_1, \ldots, i_{\mathfrak{a}(R)}) \in \mathcal{P}$.
   $\mathrm{FV}\big(R(i_1, \ldots, i_{\mathfrak{a}(R)})\big) \stackrel{\text{def}}{=} \{i_1, \ldots, i_{\mathfrak{a}(R)}\} \cap \mathcal{V}$;
2. If $f, g \in \mathcal{P}$ then $f \vee g \in \mathcal{P}$ and $\neg f \in \mathcal{P}$.
   $\mathrm{FV}(f \vee g) \stackrel{\text{def}}{=} \mathrm{FV}(f) \cup \mathrm{FV}(g)$; $\mathrm{FV}(\neg f) \stackrel{\text{def}}{=} \mathrm{FV}(f)$;

3. If $f \in \mathcal{P}$ and $x \in \text{FV}(f)$ then $\forall x[f] \in \mathcal{P}$.
   $\text{FV}(\forall x[f]) \stackrel{\text{def}}{=} \text{FV}(f) \setminus \{x\}$;

4. If $n \in \mathbb{N}$ and $k_1, \ldots, k_n \in \mathcal{A} \cup \mathcal{V} \cup \mathcal{P}$, then $z(k_1, \ldots, k_n) \in \mathcal{P}$.
   $\text{FV}(z(k_1, \ldots, k_n)) \stackrel{\text{def}}{=} \{z, k_1, \ldots, k_n\} \cap \mathcal{V}$.
   If $n = 0$ then we write $z()$ in order to distinguish the pf $z()$ from the variable z;

5. All pfs can be constructed by using the construction-rules 1, 2, 3 and 4 above.

Definition A.13 (Propositions, 2.4) A propositional function f is a *proposition* if $\text{FV}(f) = \varnothing$.

Definition A.14 (Ramified types, 2.36)

1. 0^0 is a ramified type;

2. If $t_1^{a_1}, \ldots, t_n^{a_n}$ are ramified types, and $a \in \mathbb{N}$, $a > \max(a_1, \ldots, a_n)$, then $(t_1^{a_1}, \ldots, t_n^{a_n})^a$ is a ramified type (if $n = 0$ then take $a \geq 0$);

3. All ramified types can be constructed using the rules 1 and 2.

If t^a is a ramified type, then a is called the *order* of t^a.

Definition A.15 (Predicative types, 2.40)

1. 0^0 is a predicative type;

2. If ${t_1}^{a_1}, \ldots, {t_n}^{a_n}$ are predicative types, and $a = 1 + \max(a_1, \ldots, a_n)$ (take $a = 0$ if $n = 0$), then $(t_1^{a_1}, \ldots, t_n^{a_n})^a$ is a predicative type;

3. All predicative types can be constructed using the rules 1 and 2 above.

Definition A.16 (Contexts, 2.42) Let $x_1, \ldots, x_n \in \mathcal{V}$ be distinct variables, and assume $t_1^{a_1}, \ldots, t_n^{a_n}$ are ramified types. Then $\{x_1{:}t_1^{a_1}, \ldots, x_n{:}t_n^{a_n}\}$ is a *context*. The set $\{x_1, \ldots, x_n\}$ is called the *domain* of the context and is denoted by $\text{dom}(\{x_1{:}t_1^{a_1}, \ldots, x_n{:}t_n^{a_n}\})$.

Definition A.17 (Ramified Theory of Types: RTT, 2.44)
The *judgement* $\Gamma \vdash f : t^a$ is inductively defined as follows:

1. **(start)** For all a:

$$\vdash a : 0^0.$$

   For all atomic pfs f:

$$\vdash f : ()^0;$$

2. **(connectives)** Assume $\Gamma \vdash f{:}(t_1^{a_1}, \ldots, t_n^{a_n})^a$, $\Delta \vdash g{:}(u_1^{b_1}, \ldots, u_m^{b_m})^b$, and $x < y$ for all $x \in \mathrm{dom}(\Gamma)$ and $y \in \mathrm{dom}(\Delta)$. Then
$$\Gamma \cup \Delta \vdash f \vee g : \left(t_1^{a_1}, \ldots, t_n^{a_n}, u_1^{b_1}, \ldots, u_m^{b_m}\right)^{\max(a,b)};$$
$$\Gamma \vdash \neg f : (t_1^{a_1}, \ldots, t_n^{a_n})^a;$$

3. **(abstraction from parameters)** If $\Gamma \vdash f : (t_1^{a_1}, \ldots, t_m^{a_m})^a$, $t_{m+1}^{a_{m+1}}$ is a *predicative* type, $g \in \mathcal{A} \cup \mathcal{P}$ is a parameter of f, $\Gamma \vdash g : t_{m+1}^{a_{m+1}}$, and $x < y$ for all $x \in \mathrm{dom}(\Gamma)$, then
$$\Gamma' \vdash h : (t_1^{a_1}, \ldots, t_{m+1}^{a_{m+1}})^{\max(a, a_{m+1}+1)}.$$
Here, h is a pf obtained by replacing all parameters g' of f which are α_Γ-equal to g by y. Moreover, Γ' is the subset of the context $\Gamma \cup \{y : t_{m+1}^{a_{m+1}}\}$ such that $\mathrm{dom}(\Gamma')$ contains exactly all the variables that occur in h;

4. **(abstraction from pfs)** If $\Gamma \vdash f : (t_1^{a_1}, \ldots, t_m^{a_m})^a$, $(t_1^{a_1}, \ldots, t_m^{a_m})^a$ is a *predicative* type, $x < z$ for all $x \in \mathrm{dom}(\Gamma)$, and $y_1 < \cdots < y_n$ are the free variables of f, then
$$\Gamma' \vdash z(y_1, \ldots, y_n) : (t_1^{a_1}, \ldots, t_m^{a_m}, (t_1^{a_1}, \ldots, t_m^{a_m})^a)^{a+1},$$
where Γ' is the subset of $\Gamma \cup \{z{:}(t_1^{a_1}, \ldots, t_m^{a_m})^a\}$ such that $\mathrm{dom}(\Gamma') = \{y_1, \ldots, y_n, z\}$;

5. **(weakening)** If Γ, Δ are contexts, $\Gamma \subseteq \Delta$, and $\Gamma \vdash f : t^a$, then also $\Delta \vdash f : t^a$;

6. **(substitution)** If y is the ith free variable in f (according to the order on variables), and $\Gamma \cup \{y : t_i^{a_i}\} \vdash f : (t_1^{a_1}, \ldots, t_n^{a_n})^a$, and $\Gamma \vdash k : t_i^{a_i}$ then
$$\Gamma' \vdash f[y{:=}k] : (t_1^{a_1}, \ldots, t_{i-1}^{a_{i-1}}, t_{i+1}^{a_{i+1}}, \ldots, t_n^{a_n})^b.$$
Here, $b = 1 + \max(a_1, \ldots, a_{i-1}, a_{i+1}, \ldots, a_n, c)$, and
$$c = \max\{j \mid \forall x{:}t^j \text{ occurs in } f[y{:=}k]\}$$
(if $n = 1$ and $\{j \mid \forall x{:}t^j \text{ occurs in } f[y{:=}k]\} = \varnothing$ then take $b = 0$) and once more, Γ' is the subset of $\Gamma \cup \{y : t_i^{a_i}\}$ such that $\mathrm{dom}(\Gamma')$ contains exactly all the variables that occur in $f[y{:=}k]$;

7. **(permutation)** If y is the ith free variable in f (according to the order on variables), and $\Gamma \cup \{y{:}t_i^{a_i}\} \vdash f : (t_1^{a_1}, \ldots, t_n^{a_n})^a$, and $x < y'$ for all $x \in \mathrm{dom}(\Gamma)$, then
$$\Gamma' \vdash f[y{:=}y'] : (t_1^{a_1}, \ldots, t_{i-1}^{a_{i-1}}, t_{i+1}^{a_{i+1}}, \ldots, t_n^{a_n}, t_i^{a_i})^a.$$

Γ' is the subset of $\Gamma \cup \{y{:}t_i^{a_i}, y'{:}t_i^{a_i}\}$ such that domΓ' contains exactly all the variables that occur in $f[y{:=}y']$;

8. **(quantification)** If y is the ith free variable in f (according to the order on variables), and $\Gamma \cup \{y{:}t_i^{a_i}\} \vdash f : (t_1^{a_1}, \ldots, t_n^{a_n})^a$, then

$$\Gamma \vdash \forall y{:}t_i^{a_i}[f] : (t_1^{a_1}, \ldots, t_{i-1}^{a_{i-1}}, t_{i+1}^{a_{i+1}}, \ldots, t_n^{a_n})^a.$$

Ac2 λRTT

Definition A.18 (Terms of λRTT, 5.1) Let $\mathcal{A}$, $\mathcal{V}$ and $\mathcal{R}$ be as in Section Ac1 where $\mathcal{A}$, $\mathcal{V}$ and $\mathcal{R}$ are mutually disjoint. Let $\bot$, ι, $*_s$, $*_n$ and $\Box_n$ for $n \in \mathbb{N}^+$ be new constants (not already in $\mathcal{A}$, $\mathcal{V}$ or $\mathcal{R}$) which are all different.

Define the set $\mathcal{T}$ of terms of λRTT by:

$$\begin{aligned}\mathcal{T} \quad ::= \quad & *_s \mid *_{\mathbb{N}^+} \mid \Box_{\mathbb{N}^+} \mid \mathcal{A} \mid \mathcal{V} \mid \mathcal{R} \mid \iota \mid \bot \mid \\ & \mathcal{T}\mathcal{T} \mid \lambda\mathcal{V}{:}\mathcal{T}.\mathcal{T} \mid \Pi\mathcal{V}{:}\mathcal{T}.\mathcal{T}.\end{aligned}$$

Define $\boldsymbol{S} = \{*_s, \Box_1, \Box_2, \ldots, *_1, *_2, \ldots\}$, $\mathbb{C} = \boldsymbol{S} \cup \{\bot\} \cup \{\iota\} \cup \mathcal{A} \cup \mathcal{R}$ and $\mathbb{V} = \mathcal{V}$. Then obviously, $\mathcal{T} = \mathcal{T}(\mathbb{V}, \mathbb{C})$.

Definition A.19 (Specification of λRTT) We take the specification of λRTT to be the triple $(\boldsymbol{S}, \boldsymbol{A}, \boldsymbol{R})$ where

- $\boldsymbol{S} = \{*_s, \Box_1, \Box_2, \ldots, *_1, *_2, \ldots\} \subseteq \mathbb{C}$ over which s, s_1, s_2 range.

- $\boldsymbol{A} = \{*_n : \Box_n, \bot : *_1, \iota : *_s, a : \iota$ for $a \in \mathcal{A}$,
  $R : \underbrace{\iota \to \cdots \to \iota}_{\mathfrak{a}(R) \text{ times } \iota} \to *_1$ for $R \in \mathcal{R}\}$.

- $\boldsymbol{R} = \{(*_s, \Box_n, \Box_n) \mid n \geq 1\} \cup$
  $\{(\Box_m, \Box_n, \Box_n) \mid 1 \leq m < n\} \cup$
  $\{(*_m, *_n, *_{\max(m,n)}) \mid m, n \geq 1\} \cup$
  $\{(*_s, *_n, *_n) \mid n \geq 1\} \cup$
  $\{(\Box_m, *_n, *_n) \mid 1 \leq m < n\}$.

Definition A.20 (Derivation Rules for λRTT, 5.7) Take the specification of λRTT given in Definition A.19. The derivation rules for λRTT are as follows:

(Axioms)

$$\vdash *_n : \Box_n \qquad (n \geq 1)$$

$$\vdash \iota : *_s$$

$$\vdash \bot : *_1$$

$$\vdash a : \iota \qquad (a \in \mathcal{A})$$

$$\vdash R : \underbrace{\iota \to \cdots \to \iota}_{\mathfrak{a}(R) \text{ times } \iota} \to *_1 \qquad (R \in \mathcal{R})$$

(Start)

$$\frac{\Gamma \vdash A : s}{\Gamma, x{:}A \vdash x{:}A}$$

(Weak)

$$\frac{\Gamma \vdash M : N \qquad \Gamma \vdash A : s}{\Gamma, x{:}A \vdash M : N}$$

(Π-form)

$$\frac{\Gamma \vdash A : s_1 \qquad \Gamma, x{:}A \vdash B : s_2}{\Gamma \vdash (\Pi x{:}A.B) : s_3} \qquad (s_1, s_2, s_3) \in \boldsymbol{R}$$

(Π-in)

$$\frac{\Gamma, x{:}A \vdash b{:}B \qquad \Gamma \vdash (\Pi x{:}A.B) : s}{\Gamma \vdash (\lambda x{:}A.b) : (\Pi x{:}A.B)}$$

(Π-el)

$$\frac{\Gamma \vdash M : \Pi x{:}A.B \qquad \Gamma \vdash N : A}{\Gamma \vdash MN : B[x{:=}N]}$$

(Conv)

$$\frac{\Gamma \vdash A : B \qquad \Gamma \vdash B' : s \qquad B =_\beta B'}{\Gamma \vdash A : B'}$$

(Incl)

$$\frac{\Gamma \vdash A : *_n}{\Gamma \vdash A : *_{n+1}}$$

Ad The Simple Theory of Types

Ad1 STT

See Section 3b1.

We adopt for STT, the same sets $\mathcal{A}$, $\mathcal{V}$, and $\mathcal{R}$ of RTT. Moreover, we adopt the definitions of atomic propositions, propositional functions and propositions of RTT (see Definitions A.11, A.12 and A.13).

Definition A.21 (Ramsey's simple types, 2.30)

1. 0 is a simple type;

2. If $t_1, \ldots, t_n$ are simple types, then also $(t_1, \ldots, t_n)$ is a simple type. $n = 0$ is allowed: then we obtain the simple type ();

3. All simple types can be constructed using the rules 1 and 2.

Definition A.22 (Contexts) Let $x_1, \ldots, x_n \in \mathcal{V}$ be distinct variables, and assume $t_1, \ldots, t_n$ are ramified types. Then $\{x_1{:}t_1, \ldots, x_n{:}t_n\}$ is a *context*.
The set $\{x_1, \ldots, x_n\}$ is called the *domain* of the context and is denoted by $\mathrm{dom}(\{x_1{:}t_1, \ldots, x_n{:}t_n\})$.

Definition A.23 (Simple Type Theory: STT**)** The *judgement* $\Gamma \vdash f : t$ is inductively defined as follows:

1. **(start)** For all a:

$$\vdash a : 0.$$

For all atomic pfs f:

$$\vdash f : ();$$

2. **(connectives)** Assume $\Gamma \vdash f{:}(t_1, \ldots, t_n)$, $\Delta \vdash g{:}(u_1, \ldots, u_m)$, and $x < y$ for all $x \in \mathrm{dom}(\Gamma)$ and $y \in \mathrm{dom}(\Delta)$. Then

$$\Gamma \cup \Delta \vdash f \vee g : (t_1, \ldots, t_n, u_1, \ldots, u_m);$$

$$\Gamma \vdash \neg f : (t_1, \ldots, t_n);$$

3. **(abstraction from parameters)** If $\Gamma \vdash f : (t_1, \ldots, t_m)$, $g \in \mathcal{A} \cup \mathcal{P}$ is a parameter of f, $\Gamma \vdash g : t_{m+1}$, and $x < y$ for all $x \in \mathrm{dom}(\Gamma)$, then

$$\Gamma' \vdash h : (t_1, \ldots, t_{m+1}).$$

Here, h is a pf obtained by replacing all parameters g' of f which are α_Γ-equal to g by y. Moreover, Γ' is the subset of the context $\Gamma \cup \{y : t_{m+1}\}$ such that $\mathrm{dom}(\Gamma')$ contains exactly all the variables that occur in h;

4. **(abstraction from pfs)** If $\Gamma \vdash f : (t_1, \ldots, t_m)$, $x < z$ for all $x \in \mathrm{dom}(\Gamma)$, and $y_1 < \cdots < y_n$ are the free variables of f, then

$$\Gamma' \vdash z(y_1, \ldots, y_n) : (t_1, \ldots, t_m, (t_1, \ldots, t_m)),$$

where Γ' is the subset of $\Gamma \cup \{z{:}(t_1, \ldots, t_m)\}$ such that $\mathrm{dom}(\Gamma') = \{y_1, \ldots, y_n, z\}$;

5. **(weakening)** If Γ, Δ are contexts, $\Gamma \subseteq \Delta$, and $\Gamma \vdash f : t$, then also $\Delta \vdash f : t$;

6. **(substitution)** If y is the ith free variable in f (according to the order on variables), and $\Gamma \cup \{y : t_i\} \vdash f : (t_1, \ldots, t_n)$, and $\Gamma \vdash k : t_i$ then

$$\Gamma' \vdash f[y{:=}k] : (t_1, \ldots, t_{i-1}, t_{i+1}, \ldots, t_n).$$

Here, $b = 1 + \max(a_1, \ldots, a_{i-1}, a_{i+1}, \ldots, a_n, c)$, and

$$c = \max\{j \mid \forall x{:}t \text{ occurs in } f[y{:=}k]\}$$

(if $n = 1$ and $\{j \mid \forall x{:}t \text{ occurs in } f[y{:=}k]\} = \varnothing$ then take $b = 0$) and once more, Γ' is the subset of $\Gamma \cup \{y : t_i\}$ such that $\mathrm{dom}(\Gamma')$ contains exactly all the variables that occur in $f[y{:=}k]$;

7. **(permutation)** If y is the ith free variable in f (according to the order on variables), and $\Gamma \cup \{y{:}t_i\} \vdash f : (t_1, \ldots, t_n)$, and $x < y'$ for all $x \in \mathrm{dom}(\Gamma)$, then
$$\Gamma' \vdash f[y{:=}y'] : (t_1, \ldots, t_{i-1}, t_{i+1}, \ldots, t_n, t_i).$$
Γ' is the subset of $\Gamma \cup \{y{:}t_i, y'{:}t_i\}$ such that $\mathrm{dom}\Gamma'$ contains exactly all the variables that occur in $f[y{:=}y']$;

8. **(quantification)** If y is the ith free variable in f (according to the order on variables), and $\Gamma \cup \{y{:}t_i\} \vdash f : (t_1, \ldots, t_n)$, then
$$\Gamma \vdash \forall y{:}t_i[f] : (t_1, \ldots, t_{i-1}, t_{i+1}, \ldots, t_n).$$

Ad2 λSTT

See Section 5b

Definition A.24 (Terms of λSTT) Let $\mathcal{A}$, $\mathcal{V}$ and $\mathcal{R}$ be as in Section Ac1 where $\mathcal{A}$, $\mathcal{V}$ and $\mathcal{R}$ are mutually disjoint. Let $\bot$, ι, $*_s$, $*$ and $\Box$ be new constants (not already in $\mathcal{A}$, $\mathcal{V}$ or $\mathcal{R}$) which are all different.

Define the set $\mathcal{T}$ of terms of λSTT by:

$$\begin{array}{rcl} \mathcal{T} & ::= & *_s \mid * \mid \Box \mid \mathcal{A} \mid \mathcal{V} \mid \mathcal{R} \mid \iota \mid \bot \mid \\ & & \mathcal{T}\mathcal{T} \mid \lambda\mathcal{V}{:}\mathcal{T}.\mathcal{T} \mid \Pi\mathcal{V}{:}\mathcal{T}.\mathcal{T}. \end{array}$$

Define $\boldsymbol{S} = \{*_s, \Box, *\}$. $\mathbb{C} = \boldsymbol{S} \cup \{\bot\} \cup \{\iota\} \cup \mathcal{A} \cup \mathcal{R}$ and $\mathbb{V} = \mathcal{V}$. Then obviously, $\mathcal{T} = \mathcal{T}(\mathbb{V}, \mathbb{C})$.

Definition A.25 (Specification of λSTT) We take the specification of λSTT to be the triple $(\boldsymbol{S}, \boldsymbol{A}, \boldsymbol{R})$ where

- $\boldsymbol{S} = \{*_s, \Box, *\} \subseteq \mathbb{C}$ over which s, s_1, s_2 range.
- $\boldsymbol{A} = \{* : \Box, \bot : *, \iota : *_s, a : \iota \text{ for } a \in \mathcal{A}, R : \underbrace{\iota \to \cdots \to \iota}_{a(R) \text{ times } \iota} \to * \text{ for } R \in \mathcal{R}\}$.
- $\boldsymbol{R} = \{(*_s, \Box, \Box), (\Box, \Box, \Box), (*, *, *), (*_s, *, *), (\Box, *, *)\}$.

Definition A.26 (Derivation Rules for λSTT) Take the specification of λSTT given in Definition A.25. The derivation rules for λRTT are as follows:

(Axioms)

$$\vdash * : \square$$
$$\vdash \iota : *_s$$
$$\vdash \bot : *$$
$$\vdash a : \iota \qquad (a \in \mathcal{A})$$
$$\vdash R : \underbrace{\iota \to \cdots \to \iota}_{\mathfrak{a}(R) \text{ times } \iota} \to * \qquad (R \in \mathcal{R})$$

(Start) $$\frac{\Gamma \vdash A : s}{\Gamma, x{:}A \vdash x{:}A}$$

(Weak) $$\frac{\Gamma \vdash M : N \qquad \Gamma \vdash A : s}{\Gamma, x{:}A \vdash M : N}$$

(Π-form) $$\frac{\Gamma \vdash A : s_1 \qquad \Gamma, x{:}A \vdash B : s_2}{\Gamma \vdash (\Pi x{:}A.B) : s_3} \qquad (s_1, s_2, s_3) \in \boldsymbol{R}$$

(Π-in) $$\frac{\Gamma, x{:}A \vdash b{:}B \qquad \Gamma \vdash (\Pi x{:}A.B) : s}{\Gamma \vdash (\lambda x{:}A.b) : (\Pi x{:}A.B)}$$

(Π-el) $$\frac{\Gamma \vdash M : \Pi x{:}A.B \qquad \Gamma \vdash N : A}{\Gamma \vdash MN : B[x{:=}N]}$$

(Conv) $$\frac{\Gamma \vdash A : B \qquad \Gamma \vdash B' : s \qquad B =_\beta B'}{\Gamma \vdash A : B'}$$

Ae Church's simply typed λ-calculus $\lambda_{\to\text{Church}}$

Definition A.27 (Types of $\lambda_{\to\text{Church}}$, 3.6) The types of $\lambda_{\to\text{Church}}$ are defined as follows:

- ι and o are types;
- If α and β are types, then so is $\alpha \to \beta$.

We denote the set of simple types by $\mathcal{T}$.

Definition A.28 (Terms of $\lambda_{\to\text{Church}}$, 3.7) The terms of $\lambda_{\to\text{Church}}$ are the following:

- $\neg$, $\wedge$, $\forall_\alpha$ for each type α, and $\imath_\alpha$ for each type α, are terms;
- A variable is a term;
- If A, B are terms, then so is AB;

- If A is a term, and x a variable, then $\lambda x{:}\alpha.A$ is a term.

Definition A.29 (Contexts of $\lambda_{\to\mathbf{Church}}$, 3.8) A *context* in $\lambda_{\to\text{Church}}$ is a set $\{x_1{:}\alpha_1, \ldots x_n{:}\alpha_n\}$ where the x_i are distinct variables and the α_i are types.

Definition A.30 (Typing rules of $\lambda_{\to\mathbf{Church}}$, 3.9) The judgement $\Gamma \vdash A : \alpha$ holds if it can be derived using the following rules:

- $\Gamma \vdash \neg : o \to o$;
  $\Gamma \vdash \wedge : o \to o \to o$;
  $\Gamma \vdash \forall_\alpha : (\alpha \to o) \to o$;
  $\Gamma \vdash \iota_\alpha : (\alpha \to o) \to \alpha$;
- $\Gamma \vdash x : \alpha$ if $x{:}\alpha \in \Gamma$;
- If $\Gamma, x{:}\alpha \vdash A : \beta$ then $\Gamma \vdash (\lambda x{:}\alpha.A) : \alpha \to \beta$;
- If $\Gamma \vdash A : \alpha \to \beta$ and $\Gamma \vdash B : \alpha$ then $\Gamma \vdash (AB) : \beta$.

Af A fragment of Nuprl in PTS-style

Definition A.31 (Terms, 6.1) Let $\mathbb{V}$ be a countably infinite set of variables, $\mathbb{Z}$ be the set of integers over which $m, n, \ldots$ ranges, and let $\perp$ represent the undefined or a contradiction. We take the set of sorts $\boldsymbol{S} = \{*_1, *_2, \ldots\}$ and assume that $\mathbb{V}$, $\boldsymbol{S}, \mathbb{Z}, \{\mathbb{Z}\}$ and $\{\perp\}$ are mutually disjoint. We take $\mathbb{C} = \boldsymbol{S} \cup \{\perp\} \cup \mathbb{Z} \cup \{\mathbb{Z}\}$. Note that the sets $\mathbb{C}$ and $\mathbb{V}$ are disjoint and are countably infinite. The set of terms $\mathcal{T}$ is defined by the following abstract syntax:

$$\mathcal{T} ::= \mathbb{V} \mid \mathbb{C} \mid \mathcal{T}\mathcal{T} \mid \lambda\mathbb{V}{:}\mathcal{T}.\mathcal{T} \mid \Pi\mathbb{V}{:}\mathcal{T}.\mathcal{T} \mid \mathcal{T} \times \mathcal{T} \mid \langle \mathcal{T}, \mathcal{T} \rangle \mid \pi_1(\mathcal{T}) \mid \pi_2(\mathcal{T})$$

Definition A.32 (Reduction in Nuprl, A.32) We take the usual relation $\to_\beta$ of PTSs (see Definition 4.17). In addition, we take the relation $\to_{<>}$ described by the contraction rules

$$\pi_1(\langle A, B\rangle) \to_{<>} A \text{ and } \pi_2(\langle A, B\rangle) \to_{<>} B$$

and the usual compatibility rules.

We define $\to_{\beta<>}$, $\twoheadrightarrow_{<>}$ and $\twoheadrightarrow_{\beta<>}$ in the obvious way and take $=_{\beta<>}$ to be the smallest reflexive, symmetric and transitive relation that includes $\to_{\beta<>}$.

Definition A.33 (Specification of Nuprl, 6.3) The *specification* of Nuprl is a triple $(\boldsymbol{S}, \boldsymbol{A}, \boldsymbol{R})$, such that

- $\boldsymbol{S} = \{*_1, *_2, \ldots\}$,

- $\boldsymbol{A} = \{\bot : *_1, *_n : *_{n+1}$ for $n \in \mathbb{N}^+, \mathbb{Z} : *_1, n : \mathbb{Z}$ for $n \in \mathbb{Z}\}$ and

- $\boldsymbol{R} = \{(*_n, *_n, *_n)\}$.

Definition A.34 (Derivable statements in Nuprl,6.4) A statement $\Gamma \vdash A : B$ is *derivable* if it can be deduced by repeated application of the rules below:

(Axioms) $\quad \vdash \bot : *_1 \qquad \vdash *_n : *_{n+1} \quad (n \in \mathbb{N})$

$\qquad\qquad\quad \vdash \mathbb{Z} : *_1 \qquad \vdash n : \mathbb{Z} \quad (n \in \mathbb{Z})$

(Start) $$\frac{\Gamma \vdash A : *_n}{\Gamma, x{:}A \vdash x{:}A} \qquad (x \text{ is } \Gamma\text{-fresh})$$

(Weak) $$\frac{\Gamma \vdash M : N \qquad \Gamma \vdash A : *_n}{\Gamma, x{:}A \vdash M : N} \qquad (x \text{ is } \Gamma\text{-fresh})$$

(Π-form) $$\frac{\Gamma \vdash A : *_n \qquad \Gamma, x{:}A \vdash B : *_n}{\Gamma \vdash (\Pi x{:}A.B) : *_n}$$

(λ) $$\frac{\Gamma, x{:}A \vdash b : B \qquad \Gamma \vdash (\Pi x{:}A.B) : *_n}{\Gamma \vdash (\lambda x{:}A.b) : (\Pi x{:}A.B)}$$

((App) $$\frac{\Gamma \vdash M : (\Pi x{:}A.B) \qquad \Gamma \vdash N : A}{\Gamma \vdash MN : B[x{:=}N]}$$

(×-form) $$\frac{\Gamma \vdash A : *_n \qquad \Gamma \vdash B : *_n}{\Gamma \vdash (A \times B) : *_n}$$

(Pairs) $$\frac{\Gamma \vdash a : A \qquad \Gamma \vdash b : B \qquad \Gamma \vdash (A \times B) : *_m}{\Gamma \vdash \langle a, b\rangle : (A \times B)}$$

(Left) $$\frac{\Gamma \vdash M : (A \times B)}{\Gamma \vdash \pi_1(M) : A}$$

(Right) $$\frac{\Gamma \vdash M : (A \times B)}{\Gamma \vdash \pi_2(M) : B}$$

(Conv) $$\frac{\Gamma \vdash M : A \qquad \Gamma \vdash B : *_n \qquad A =_{\beta<>} B}{\Gamma \vdash M : B}$$

(⊆) $$\frac{\Gamma \vdash A : *_n}{\Gamma \vdash A : *_{n+1}}$$

Ag AUTOMATH

Ag1 AUT-68

For writing lines and books in AUT-68 we need

- The symbol `type`;
- A set $\mathcal{V}$ of variables;
- A set $\mathcal{C}$ of constants;
- The symbols ( ) [ ] : — , .

We assume that $\mathcal{V}$ and $\mathcal{C}$ are infinite, or at least offer us as many different elements as needed. We also assume that $\mathcal{V} \cap \mathcal{C} = \varnothing$ and that $\mathtt{type} \notin \mathcal{V} \cup \mathcal{C}$.

Definition A.35 (Expressions, 7.1) We define the set $\mathcal{E}$ of AUT-*68-expressions* inductively:

(variable) If $x \in \mathcal{V}$ then $x \in \mathcal{E}$;

(parameter) If $a \in \mathcal{C}$, $n \in \mathbb{N}$ ($n = 0$ is allowed) and $\Sigma_1, \ldots, \Sigma_n \in \mathcal{E}$ then $a(\Sigma_1, \ldots, \Sigma_n) \in \mathcal{E}$;

(abstraction) If $x \in \mathcal{V}$, $\Sigma \in \mathcal{E} \cup \{\mathtt{type}\}$ and $\Omega \in \mathcal{E}$ then $[x{:}\Sigma]\Omega \in \mathcal{E}$;

(application) If $\Sigma_1, \Sigma_2 \in \mathcal{E}$ then $\langle\Sigma_2\rangle\Sigma_1 \in \mathcal{E}$.

We define also $\mathcal{E}^+ \stackrel{\text{def}}{=} \mathcal{E} \cup \{\mathtt{type}\}$.

Definition A.36 (Books and lines, 7.7) An AUT-68-*book* is a finite list (possibly empty) of (AUT-68)-lines (to be defined next). If $l_1, \ldots, l_n$ are the lines of book $\mathfrak{B}$, we write $\mathfrak{B} \equiv l_1, \ldots, l_n$.

An AUT-68-*line* is a 4-tuple $(\Gamma; k; \Sigma_1; \Sigma_2)$. Here,

- Γ is a context, i.e. a finite (possibly empty) list $x_1{:}\alpha_1, \ldots, x_n{:}\alpha_n$, where the x_is are different elements of $\mathcal{V}$ and the α_is are elements of $\mathcal{E}^+$;
- k is an element of $\mathcal{V} \cup \mathcal{C}$;
- Σ_1 can be (only):
  - The symbol — (if $k \in \mathcal{V}$);
  - The symbol PN (if $k \in \mathcal{C}$);
  - An element of $\mathcal{E}$ (if $k \in \mathcal{C}$);

- Σ_2 is an element of $\mathcal{E}^+$.

Definition A.37 (Definitional equality $=_{\beta\delta}$ or $=_{\beta \mathrm{d}}$)

- We take the usual relation $\to_\beta$ of PTSs (see Definition 4.17) but written in the notation of AUTOMATH as follows:

$$\langle\Sigma\rangle[x{:}\Omega_2]\Omega_1 \to_\beta \Omega_1[x{:=}\Sigma]$$

- Let $\mathfrak{B}$ be a book, Γ a correct context with respect to $\mathfrak{B}$, and Σ a correct expression with respect to $\mathfrak{B};\Gamma$. We define $\Sigma \to_\delta \Omega$ by the usual compatibility rules, and

  (δ) If $\Sigma = b(\Sigma_1,\ldots,\Sigma_n)$, and $\mathfrak{B}$ contains a line $(x_1{:}\alpha_1,\ldots,x_n{:}\alpha_n;b;\Xi_1;\Xi_2)$ where $\Xi_1 \in \mathcal{E}$, then

$$\Sigma \to_\delta \Xi_1[x_1,\ldots,x_n{:=}\Sigma_1,\ldots,\Sigma_n].$$

- For $r \in \{\beta,\delta\}$, we use notations like $\twoheadrightarrow_r$, $\twoheadrightarrow_r^+$ and $=_r$ as usual.
- $=_{\beta\delta}$ is the smallest equivalence relation that contains both $=_\beta$ and $=_\delta$.
- Σ_1 and Σ_2 are called *definitionally equal* (with respect to a book $\mathfrak{B}$) if $\Sigma_1 =_{\beta\delta} \Sigma_2$.
- Note that $=_{\beta\delta}$ is the same relation as $=_{\beta \mathrm{d}}$ where d is given in Definition 7.13.

Definition A.38 (Correct books and contexts, 7.10) A book $\mathfrak{B}$ and a context Γ are *correct* if $\mathfrak{B};\Gamma \vdash \text{OK}$ can be derived with the following rules

(axiom) $$\varnothing;\varnothing \vdash \text{OK}$$

(context ext.) $$\frac{\mathfrak{B}_1,(\Gamma;x;\text{—};\alpha),\mathfrak{B}_2;\Gamma \vdash \text{OK}}{\mathfrak{B}_1,(\Gamma;x;\text{—};\alpha),\mathfrak{B}_2;\Gamma,x{:}\alpha \vdash \text{OK}}$$

(book ext.: var1) $$\frac{\mathfrak{B};\Gamma \vdash \text{OK}}{\mathfrak{B},(\Gamma;x;\text{—};\texttt{type});\varnothing \vdash \text{OK}}$$

(book ext.: var2) $$\frac{\mathfrak{B};\Gamma \vdash \Sigma_2 : \texttt{type}}{\mathfrak{B},(\Gamma;x;\text{—};\Sigma_2);\varnothing \vdash \text{OK}}$$

(book ext.: pn1) $$\frac{\mathfrak{B};\Gamma \vdash \text{OK}}{\mathfrak{B},(\Gamma;k;\text{PN};\texttt{type});\varnothing \vdash \text{OK}}$$

(book ext.: pn2) $$\frac{\mathfrak{B};\Gamma \vdash \Sigma_2 : \texttt{type}}{\mathfrak{B},(\Gamma;k;\text{PN};\Sigma_2);\varnothing \vdash \text{OK}}$$

(book ext.: def1) $$\frac{\mathfrak{B};\Gamma \vdash \Sigma_1 : \texttt{type}}{\mathfrak{B},(\Gamma;k;\Sigma_1;\texttt{type});\varnothing \vdash \text{OK}}$$

(book ext.: def2)
$$\frac{\mathfrak{B};\Gamma \vdash \Sigma_2 : \texttt{type} \qquad \mathfrak{B};\Gamma \vdash \Sigma_1 : \Sigma_2' \qquad \mathfrak{B};\Gamma \vdash \Sigma_2 =_{\beta \mathrm{d}} \Sigma_2'}{\mathfrak{B},(\Gamma;k;\Sigma_1;\Sigma_2);\varnothing \vdash \text{OK}}$$

For the (book ext.) rules, we assume that the introduced identifiers $x \in \mathcal{V}$ and $k \in \mathcal{C}$ do not occur anywhere in $\mathfrak{B}$ and Γ.

Definition A.39 (Correct statements, 7.11) A statement $\mathfrak{B}; \Gamma \vdash \Sigma : \Omega$ is *correct* if it can be derived with the rules below (the start rule uses the notions of correct context and correct book as given in Definition 7.10).

(start) $$\frac{\mathfrak{B}; \Gamma_1, x{:}\alpha, \Gamma_2 \vdash \text{OK}}{\mathfrak{B}; \Gamma_1, x{:}\alpha, \Gamma_2 \vdash x{:}\alpha}$$

(parameters) $$\frac{\begin{array}{c}\mathfrak{B} \equiv \mathfrak{B}_1, (x_1{:}\alpha_1, \ldots, x_n{:}\alpha_n; b; \Omega_1; \Omega_2), \mathfrak{B}_2 \\ \mathfrak{B}; \Gamma \vdash \Sigma_i{:}\alpha_i[x_1, \ldots, x_{i-1}{:=}\Sigma_1, \ldots, \Sigma_{i-1}]\,(i = 1, \ldots, n)\end{array}}{\mathfrak{B}; \Gamma \vdash b(\Sigma_1, \ldots, \Sigma_n) : \Omega_2[x_1, \ldots, x_n{:=}\Sigma_1, \ldots, \Sigma_n]}$$

(abstr.1) $$\frac{\mathfrak{B}; \Gamma \vdash \Sigma_1{:}\texttt{type} \qquad \mathfrak{B}; \Gamma, x{:}\Sigma_1 \vdash \Omega_1{:}\texttt{type}}{\mathfrak{B}; \Gamma \vdash [x{:}\Sigma_1]\Omega_1 : \texttt{type}}$$

(abstr.2) $$\frac{\mathfrak{B}; \Gamma \vdash \Sigma_1{:}\texttt{type} \qquad \mathfrak{B}; \Gamma, x{:}\Sigma_1 \vdash \Omega_1{:}\texttt{type} \qquad \mathfrak{B}; \Gamma, x{:}\Sigma_1 \vdash \Sigma_2{:}\Omega_1}{\mathfrak{B}; \Gamma \vdash [x{:}\Sigma_1]\Sigma_2 : [x{:}\Sigma_1]\Omega_1}$$

(application) $$\frac{\mathfrak{B}; \Gamma \vdash \Sigma_1 : [x{:}\Omega_1]\Omega_2 \qquad \mathfrak{B}; \Gamma \vdash \Sigma_2 : \Omega_1}{\mathfrak{B}; \Gamma \vdash \langle\Sigma_2\rangle\Sigma_1 : \Omega_2[x{:=}\Sigma_2]}$$

(conversion) $$\frac{\mathfrak{B}; \Gamma \vdash \Sigma : \Omega_1 \qquad \mathfrak{B}; \Gamma \vdash \Omega_2{:}\texttt{type} \qquad \mathfrak{B}; \Gamma \vdash \Omega_1 =_{\beta d} \Omega_2}{\mathfrak{B}; \Gamma \vdash \Sigma : \Omega_2}$$

When using the parameter rule, we assume that $\mathfrak{B}; \Gamma \vdash \text{OK}$, even if $n = 0$.

Ag2 λ68

Definition A.40 (Terms of λ68, 7.22.1) The terms of λ68 form a set $\mathcal{T}$ defined by

$$\mathcal{T} ::= \mathcal{V} \mid \mathcal{C} \mid \boldsymbol{S} \mid \mathcal{T}\mathcal{T} \mid \lambda\mathcal{V}{:}\mathcal{T}.\mathcal{T} \mid \S\mathcal{V}{:}\mathcal{T}.\mathcal{T} \mid \Pi\mathcal{V}{:}\mathcal{T}.\mathcal{T} \mid \P\mathcal{V}{:}\mathcal{T}.\mathcal{T},$$

where $\boldsymbol{S}$ is the set of sorts $\{*, \Box, \Delta\}$.

Let $\mathbb{V} = \mathcal{C} \cup \mathcal{V}$ and assume that $\boldsymbol{S} \cap \mathbb{V} = \varnothing$.

(Recall that $\mathcal{V} \cap \mathcal{C} = \varnothing$ from the start of Section Ag1).

Definition A.41 (Contexts in λ68, 7.22.2) We define the notion of context inductively:

- $\varnothing; \varnothing$ is a context; $\text{DOM}\,(\varnothing; \varnothing) = \varnothing$;

- If $\Delta;\Gamma$ is a context, $x \in \mathcal{V}$, x does not occur in $\Delta;\Gamma$ and $A \in \mathcal{T}$, then $\Delta;\Gamma,x{:}A$ is a context (x is a newly introduced variable);
  $\textsc{dom}(\Delta;\Gamma) = \textsc{dom}(\Delta;\Gamma) \cup \{x\}$;
- If $\Delta;\Gamma$ is a context, $b \in \mathcal{C}$, b does not occur in $\Delta;\Gamma$ and $A \in \mathcal{T}$ then $\Delta,b{:}A;\Gamma$ is a context (in this case b is a *primitive* constant; cf. the primitive notions of AUTOMATH in Section 7a1);
  $\textsc{dom}(\Delta,b{:}A;\Gamma) = \textsc{dom}(\Delta;\Gamma) \cup \{b\}$;
- If $\Delta;\Gamma$ is a context, $b \in \mathcal{C}$, b does not occur in $\Delta;\Gamma$, $A \in \mathcal{T}$, and $T \in \mathcal{T}$, then $\Delta,b{:=}T{:}A;\Gamma$ is a context (in this case b is a *defined* constant; cf. the definitions of AUTOMATH in Section 7a1);
  $\textsc{dom}(\Delta,b{:=}T{:}A;\Gamma) = \textsc{dom}(\Delta;\Gamma) \cup \{b\}$.

Definition A.42 (Reduction in λ68)

- We use the usual notion of β-reduction;
- We define the notion of δ-reduction on terms. Let Δ be the left part of a context. If $(b{:=}(\S_{i=1}^{n}\, x_i{:}A_i.T) : (\P_{i=1}^{n}\, x_i{:}A_i.B)) \in \Delta$, where B is not of the form $\P y{:}B_1.B_2$, then

  $$\Delta \vdash bX_1 \cdots X_n \to_\delta T[x_1,\ldots,x_n{:=}X_1,\ldots,X_n]$$

  for all $X_1,\ldots X_n \in \mathcal{T}$.

  We also have the usual compatibility rules on δ-reduction.
- For $r \in \{\beta,\delta\}$, we use notations like $\twoheadrightarrow_r$, $\twoheadrightarrow_r^+$, $=_r$ as usual.

Definition A.43 (Derivation rules in λ68, 7.22.3) † denotes *where* $s_1 \equiv *, \Box$

(Axiom) $\quad ;\vdash * : \Box$

(Start: v) $$\frac{\Delta;\Gamma \vdash A : s_1}{\Delta;\Gamma,x{:}A \vdash x : A} \quad \dagger$$

(Start: pc) $$\frac{\Delta;\Gamma \vdash B : s_1 \qquad \Delta;\vdash \P\Gamma.B : s_2}{\Delta,b{:}\P\Gamma.B;\vdash b : \P\Gamma.B} \quad \dagger$$

(Start: dc) $$\frac{\Delta;\Gamma \vdash T : B : s_1 \qquad \Delta;\vdash \P\Gamma.B : s_2}{\Delta,b{:=}(\S\Gamma.T){:}(\P\Gamma.B);\vdash b : \P\Gamma.B} \quad \dagger$$

(Weak: v) $$\frac{\Delta;\Gamma \vdash M : N \qquad \Delta;\Gamma \vdash A : s_1}{\Delta;\Gamma,x{:}A \vdash M : N} \quad \dagger$$

(Weak: pc) $$\frac{\Delta;\vdash M : N \qquad \Delta;\Gamma \vdash B : s_1 \qquad \Delta;\vdash \P\Gamma.B : s_2}{\Delta,b{:}\P\Gamma.B;\vdash M : N} \quad \dagger$$

(Weak: dc) $$\frac{\Delta;\vdash M : N \qquad \Delta;\Gamma \vdash T : B : s_1 \qquad \Delta;\vdash \P\Gamma.B : s_2}{\Delta,b{:=}(\S\Gamma.T){:}(\P\Gamma.B);\vdash M : N} \quad \dagger$$

(Π-form) $$\frac{\Delta;\Gamma \vdash A : * \qquad \Delta;\Gamma, x{:}A \vdash B : *}{\Delta;\Gamma \vdash (\Pi x{:}A.B) : *}$$

(¶-form) $$\frac{\Delta;\Gamma \vdash A : s_1 \qquad \Delta;\Gamma, x{:}A \vdash B : s_2}{\Delta;\Gamma \vdash (\P x{:}A.B) : \Delta} \qquad \dagger$$

(λ) $$\frac{\Delta;\Gamma \vdash \Pi x{:}A.B : * \qquad \Delta;\Gamma, x{:}A \vdash F : B}{\Delta;\Gamma \vdash (\lambda x{:}A.F) : (\Pi x{:}A.B)}$$

(App₁) $$\frac{\Delta;\Gamma \vdash M : \Pi x{:}A.B \qquad \Delta;\Gamma \vdash N : A}{\Delta;\Gamma \vdash MN : B[x{:=}N]}$$

(App₂) $$\frac{\Delta;\Gamma \vdash M : \P x{:}A.B \qquad \Delta;\Gamma \vdash N : A}{\Delta;\Gamma \vdash MN : B[x{:=}N]}$$

(Conv) $$\frac{\Delta;\Gamma \vdash M : A \qquad \Delta;\Gamma \vdash B : s \qquad \Delta \vdash A =_{\beta\delta} B}{\Delta;\Gamma \vdash M : B}.$$

The newly introduced variables in the Start-rules and Weakening-rules are assumed to be fresh. Moreover, when introducing a variable b with a "pc"-rule or a "dc"-rule, we assume $b \in \mathcal{C}$, and when introducing x via a "v"-rule, we assume $x \in \mathcal{V}$.

Ah Pure Type Systems with definitions

Ah1 PTSs with definitions in contexts

Definition A.44 (Terms, 4.3) Let $\mathbb{V}$ be a set of variables and $\mathbb{C}$ a set of constants. Assume that $\mathbb{V}$ and $\mathbb{C}$ are countably infinite and are disjoint. The set $\mathcal{T}(\mathbb{V},\mathbb{C})$ (shorthand: $\mathcal{T}$) is given by:

$$\mathcal{T} ::= \mathbb{V} \mid \mathbb{C} \mid \mathcal{T}\mathcal{T} \mid \lambda\mathbb{V}{:}\mathcal{T}.\mathcal{T} \mid \Pi\mathbb{V}{:}\mathcal{T}.\mathcal{T}.$$

Definition A.45 (Reduction, 4.14)

- The relation $\to_\beta$ is described by the contraction rule

  $$(\lambda x{:}A_1.A_2)B \to_\beta A_2[x{:=}B]$$

  and the usual compatibility rules.

  $=_\beta$ is the smallest reflexive, symmetric and transitive relation that includes $\to_\beta$.

- $\Gamma \vdash B \overset{\text{def}}{=} B'$ is defined on $\mathcal{T}$ as the smallest equivalence relation closed under:

  – If $B =_\beta B'$ then $\Gamma \vdash B \overset{\text{def}}{=} B'$

- If $x = D : C \in \Gamma$ and B' arises from B by substituting one particular free occurrence of x in B by D then $\Gamma \vdash B \stackrel{\text{def}}{=} B'$.

Definition A.46 (Specification, 4.18) A *specification* is a triple $(\boldsymbol{S}, \boldsymbol{A}, \boldsymbol{R})$, such that $\boldsymbol{S} \subseteq C$, $\boldsymbol{A} \subseteq \boldsymbol{S} \times \boldsymbol{S}$ and $\boldsymbol{R} \subseteq \boldsymbol{S} \times \boldsymbol{S} \times \boldsymbol{S}$. The specification is called *singly sorted* if $\boldsymbol{A}$ is a (partial) function $\boldsymbol{S} \to \boldsymbol{S}$, and $\boldsymbol{R}$ is a (partial) function $\boldsymbol{S} \times \boldsymbol{S} \to \boldsymbol{S}$. $\boldsymbol{S}$ is called the set of *sorts*, $\boldsymbol{A}$ is the set of *axioms*, and $\boldsymbol{R}$ is the set of (Π-formation) *rules* of the specification.

Definition A.47 (declarations, definitions, contexts, 8.2)

1. A *declaration* d is as in Definition 4.19.
2. A *definition* d is of the form $x = B : A$ and defines x of type A to be B. We define $\mathtt{var}(d)$, $\mathtt{type}(d)$ and $\mathtt{ab}(d)$ to be x, A, and B respectively. We define $\text{FV}(d) \equiv \text{FV}(A) \cup \text{FV}(B)$.
3. We use $d, d', d_1, \ldots$ to range over declarations and definitions.
4. A *context* Γ is a (possibly empty) concatenation of declarations and definitions $d_1, d_2, \cdots, d_n$ such that if $i \neq j$, then $\mathtt{var}(d_i) \not\equiv \mathtt{var}(d_j)$. We define $\text{DOM}(\Gamma) = \{\mathtt{var}(d) \mid d \in \Gamma\}$, $\Gamma\text{-}\mathtt{decl} = \{d \in \Gamma \mid d \text{ is a declaration }\}$ and $\Gamma\text{-}\mathtt{abb} = \{d \in \Gamma \mid d \text{ is a definition }\}$.
   Note that $\text{DOM}(\Gamma) = \{\mathtt{var}(d) \mid d \in \Gamma\text{-}\mathtt{decl} \cup \Gamma\text{-}\mathtt{abb}\}$.
   We use $\Gamma, \Delta, \Gamma', \Gamma_1, \Gamma_2, \ldots$ to range over contexts.

Definition A.48 The new typing relation $\vdash$ is obtained by adding three new rules to the typing rules of Definition 4.21: (start-def), (weak-def), and (def) below, and by replacing the (conv) rule by (new-conv) as follows:

(axiom)	$\langle\rangle \vdash s_1 : s_2$	$(s_1, s_2) \in \boldsymbol{A}$
(start)	$\dfrac{\Gamma \vdash A : s}{\Gamma, x{:}A \vdash x : A}$	$x \notin \text{DOM}(\Gamma)$
(weak)	$\dfrac{\Gamma \vdash A : B \qquad \Gamma \vdash C : s}{\Gamma, x{:}C \vdash A : B}$	$x \notin \text{DOM}(\Gamma)$
(Π)	$\dfrac{\Gamma \vdash A : s_1 \qquad \Gamma, x{:}A \vdash B : s_2}{\Gamma \vdash (\Pi x{:}A.B) : s_3}$	$(s_1, s_2, s_3) \in \boldsymbol{R}$
(λ)	$\dfrac{\Gamma, x{:}A \vdash b : B \qquad \Gamma \vdash (\Pi x{:}A.B) : s}{\Gamma \vdash (\lambda x{:}A.b) : (\Pi x{:}A.B)}$	
(appl)	$\dfrac{\Gamma \vdash F : (\Pi x{:}A.B) \qquad \Gamma \vdash a : A}{\Gamma \vdash Fa : B[x{:=}a]}$	

(start-def) $\dfrac{\Gamma \vdash A : s \qquad \Gamma \vdash B : A}{\Gamma, x = B{:}A \vdash x : A}$ $\quad x \notin \text{DOM}(\Gamma)$

(weak-def) $\dfrac{\Gamma \vdash A : B \qquad \Gamma \vdash C : s \quad \Gamma \vdash D : C}{\Gamma, x = D{:}C \vdash A : B}$ $\quad x \notin \text{DOM}(\Gamma)$

(def) $\dfrac{\Gamma, x = B{:}A \vdash C : D}{\Gamma \vdash (\pi x : A.C)B : D[x := B]}$ $\quad \text{for } \pi \in \{\lambda, \Pi\}$

(new-conv) $\dfrac{\Gamma \vdash A : B \quad \Gamma \vdash B' : s \quad \Gamma \vdash B \stackrel{\text{def}}{=} B'}{\Gamma \vdash A : B'}$

Ah2 PTSs with definitions in the terms and the contexts

Definition A.49 (Terms) Let $\mathbb{V}$ be a set of variables and $\mathbb{C}$ a set of constants. Assume that $\mathbb{V}$ and $\mathbb{C}$ are countably infinite and are disjoint. The set $\mathcal{T}(\mathbb{V}, \mathbb{C})$ (shorthand: $\mathcal{T}$) is given by:

$$\mathcal{T} ::= \mathbb{V} \mid \mathbb{C} \mid \mathcal{T}\mathcal{T} \mid \lambda\mathbb{V}{:}\mathcal{T}.\mathcal{T} \mid \Pi\mathbb{V}{:}\mathcal{T}.\mathcal{T} \mid \mathbb{V}{=}\mathcal{T}{:}\mathcal{T} \text{ in } \mathcal{T}$$

The expression $x{=}a{:}A$ in B means, let x by equal to a of type A in term B.

Definition A.50 (Reduction, 4.14)

- The relation $\rightarrow_\beta$ is described by the contraction rule

  $$(\lambda x{:}A_1.A_2)B \rightarrow_\beta A_2[x{:=}B]$$

  and the usual compatibility rules.

  $=_\beta$ is the smallest reflexive, symmetric and transitive relation that includes $\rightarrow_\beta$.

- D-reduction is defined by the following rules:

  $$\Gamma_1, x{=}a{:}A, \Gamma_2 \vdash x \rightarrow_{\text{D}} a;$$

  $$\Gamma \vdash (x{=}a{:}A \text{ in } b) \rightarrow_{\text{D}} b \qquad (x \notin \text{FV}(b));$$

  $$\frac{\Gamma, x{=}a{:}A \vdash b \rightarrow_{\text{D}} b'}{\Gamma \vdash (x{=}a{:}A \text{ in } b) \rightarrow_{\text{D}} (x{=}a{:}A \text{ in } b')}$$

  and the usual compatibility rules. $=_{\text{D}}$ is the smallest reflexive, symmetric and transitive relation that includes D-reduction.

Definition A.51 (Specification, 4.18) A *specification* is a triple $(\boldsymbol{S}, \boldsymbol{A}, \boldsymbol{R})$, such that $\boldsymbol{S} \subseteq C$, $\boldsymbol{A} \subseteq \boldsymbol{S} \times \boldsymbol{S}$ and $\boldsymbol{R} \subseteq \boldsymbol{S} \times \boldsymbol{S} \times \boldsymbol{S}$. The specification is called *singly sorted* if $\boldsymbol{A}$ is a (partial) function $\boldsymbol{S} \rightarrow \boldsymbol{S}$, and $\boldsymbol{R}$ is a (partial) function $\boldsymbol{S} \times \boldsymbol{S} \rightarrow \boldsymbol{S}$. $\boldsymbol{S}$ is called the set of *sorts*, $\boldsymbol{A}$ is the set of *axioms*, and $\boldsymbol{R}$ is the set of (Π-formation) *rules* of the specification.

Definition A.52 (declarations, definitions, contexts, 8.2)

1. A *declaration* d is as in Definition 4.19.

2. A *definition* d is of the form $x = B : A$ and defines x of type A to be B. We define $\mathtt{var}(d)$, $\mathtt{type}(d)$ and $\mathtt{ab}(d)$ to be x, A, and B respectively. We define $\mathrm{FV}(d) \equiv \mathrm{FV}(A) \cup \mathrm{FV}(B)$.

3. We use $d, d', d_1, \ldots$ to range over declarations and definitions.

4. A *context* Γ is a (possibly empty) concatenation of declarations and definitions $d_1, d_2, \cdots, d_n$ such that if $i \neq j$, then $\mathtt{var}(d_i) \not\equiv \mathtt{var}(d_j)$. We define $\mathrm{DOM}\,(\Gamma) = \{\mathtt{var}(d) \mid d \in \Gamma\}$, $\Gamma\text{-}\mathtt{decl} = \{d \in \Gamma \mid d \text{ is a declaration}\}$ and $\Gamma\text{-}\mathtt{abb} = \{d \in \Gamma \mid d \text{ is a definition}\}$.
   Note that $\mathrm{DOM}\,(\Gamma) = \{\mathtt{var}(d) \mid d \in \Gamma\text{-}\mathtt{decl} \cup \Gamma\text{-}\mathtt{abb}\}$.
   We use $\Gamma, \Delta, \Gamma', \Gamma_1, \Gamma_2, \ldots$ to range over contexts.

Definition A.53 The new typing relation $\vdash$ is obtained by adding to the typing rules of Definition 4.21 the five rules (D-start), (D-weak), (D-form), (D-intro) and (D-conv) as follows:

(axiom) $\quad \langle\rangle \vdash s_1 : s_2 \qquad (s_1, s_2) \in \boldsymbol{A}$

(start) $$\frac{\Gamma \vdash A : s}{\Gamma, x{:}A \vdash x : A} \qquad x \notin \mathrm{DOM}\,(\Gamma)$$

(weak) $$\frac{\Gamma \vdash A : B \qquad \Gamma \vdash C : s}{\Gamma, x{:}C \vdash A : B} \qquad x \notin \mathrm{DOM}\,(\Gamma)$$

(Π) $$\frac{\Gamma \vdash A : s_1 \qquad \Gamma, x{:}A \vdash B : s_2}{\Gamma \vdash (\Pi x{:}A.B) : s_3} \qquad (s_1, s_2, s_3) \in \boldsymbol{R}$$

(λ) $$\frac{\Gamma, x{:}A \vdash b : B \qquad \Gamma \vdash (\Pi x{:}A.B) : s}{\Gamma \vdash (\lambda x{:}A.b) : (\Pi x{:}A.B)}$$

(appl) $$\frac{\Gamma \vdash F : (\Pi x{:}A.B) \qquad \Gamma \vdash a : A}{\Gamma \vdash Fa : B[x{:=}a]}$$

(D-start) $$\frac{\Gamma \vdash a : A}{\Gamma, x{=}a{:}A \vdash x : A}$$

(D-weak) $$\frac{\Gamma \vdash b : B \qquad \Gamma \vdash a : A}{\Gamma, x{=}a{:}A \vdash b : B}$$

(D-form) $$\frac{\Gamma, x{=}a{:}A \vdash B : s}{\Gamma \vdash (x{=}a{:}A \text{ in } B) : s}$$

(D-intro) $$\frac{\Gamma, x{=}a{:}A \vdash b : B \qquad \Gamma \vdash (x{=}a{:}A \text{ in } B) : s}{\Gamma \vdash (x{=}a{:}A \text{ in } b) : (x{=}a{:}A \text{ in } B)}$$

(D-conv) $$\frac{\Gamma \vdash b : B \qquad \Gamma \vdash B' : s \qquad \Gamma \vdash B =_{\mathrm{D}} B'}{\Gamma \vdash b : B'}$$

Ai Pure Type Systems with parametric constants

Definition A.54 (Terms, 9.1) The set $\mathcal{T}_P$ of *parametric terms* is defined together with the set $\mathcal{L}_V$ of *lists of variables* and the set $\mathcal{L}_T$ of *lists of terms* as follows:

$$\begin{array}{lll} \mathcal{T}_P & ::= & \mathcal{V} \mid \boldsymbol{S} \mid \mathcal{C}(\mathcal{L}_T) \mid \mathcal{T}_P\mathcal{T}_P \mid \lambda\mathcal{V}{:}\mathcal{T}_P.\mathcal{T}_P \mid \Pi\mathcal{V}{:}\mathcal{T}_P.\mathcal{T}_P; \\ \mathcal{L}_T & ::= & \varnothing \mid \langle \mathcal{L}_T, \mathcal{T}_P \rangle. \end{array}$$

where, as usual, $\mathcal{V}$ is a set of variables, $\mathcal{C}$ is a set of constants, and $\boldsymbol{S} = \{*, \Box\}$ is a set of sorts. We assume that $\mathcal{V}$, $\mathcal{C}$ and $\boldsymbol{S}$ are mutually disjoint and that $\mathcal{V}$ and $\mathcal{C}$ are countable (possibly infinite). Formally, lists of terms are of the form $\langle \ldots \langle\langle \varnothing, A_1\rangle, A_2\rangle \ldots A_n\rangle$. We usually write $\langle A_1, \ldots, A_n\rangle$ or even $A_1, \ldots, A_n$. In a parametric term of the form $c(b_1, \ldots, b_n)$, the subterms $b_1, \ldots, b_n$ are called the *parameters* of the term.

Definition A.55 (Reduction, 4.17) The relation $\rightarrow_\beta$ is described by the contraction rule

$$(\lambda x{:}A_1.A_2)B \rightarrow_\beta A_2[x{:=}B]$$

and the usual compatibility rules which are now also extended for parametric terms as follows:

if $b_i \rightarrow_\beta b_i'$ then $c(b_1, \ldots, b_i, \ldots, b_n) \rightarrow_\beta c(b_1, \ldots, b_i', \ldots, b_n)$.

$=_\beta$ is the smallest reflexive, symmetric and transitive relation that includes $\rightarrow_\beta$.

Definition A.56 (Contexts, 9.3) Given the set of parametric terms, we define the set $\mathcal{C}_P$ of *parametric contexts* (which we denote by $\Gamma, \Gamma', \ldots$) and the set $\mathcal{L}_V$ of *lists of variable declarations* as follows:

$$\begin{array}{lll} \mathcal{C}_P & ::= & \varnothing \mid \langle \mathcal{C}_P, \mathcal{V}{:}\mathcal{T}_P \rangle \mid \langle \mathcal{C}_P, \mathcal{C}(\mathcal{L}_V){:}\mathcal{T}_P \rangle \\ \mathcal{L}_V & ::= & \varnothing \mid \langle \mathcal{L}_V, \mathcal{V}{:}\mathcal{T}_P \rangle. \end{array}$$

Definition A.57 (Specification) Let (s_1, s_2) denote (s_1, s_2, s_2). A *parametric cube specification* is a quadruple $(\boldsymbol{S}, \boldsymbol{A}, \boldsymbol{R}, \boldsymbol{P})$, such that:

- $\boldsymbol{S} = \{*, \Box\}$,
- $\boldsymbol{A} = \{* : \Box\} \subseteq \boldsymbol{S} \times \boldsymbol{S}$,

- $\boldsymbol{R} \subseteq \{(*,*),(*,\Box),(\Box,*),(\Box,\Box)\} \subseteq \boldsymbol{S} \times \boldsymbol{S} \times \boldsymbol{S}$ such that $(*,*) \in \boldsymbol{R}$, and
- $\boldsymbol{P} \subseteq \{(*,*),(*,\Box),(\Box,*),(\Box,\Box)\}$, such that $(*,*) \in \boldsymbol{P}$.

We use the same notations as Definition 4.18. Note that the difference between one parametric cube specification and another is in the set of rules $\boldsymbol{R}$ and in the set of parametric rules $\boldsymbol{P}$.

Definition A.58 (The Barendregt cube with parametric constants)
Let $(\boldsymbol{S},\boldsymbol{A},\boldsymbol{R},\boldsymbol{P})$ be a parametric cube specification as in Definition A.57. The judgments that are derivable in $\lambda\boldsymbol{RP}$ are determined by the rules for $\lambda\boldsymbol{R}$ of Definition 4.24 and the following two rules where $\Delta \equiv x_1{:}B_1,\ldots,x_n{:}B_n$ and $\Delta_i \equiv x_1{:}B_1,\ldots,x_{i-1}{:}B_{i-1}$:

$$(\vec{\mathrm{C}}\text{-weak})\quad \frac{\Gamma\vdash b:B \qquad \Gamma,\Delta_i\vdash B_i:s_i \qquad \Gamma,\Delta\vdash A:s}{\Gamma,c(\Delta):A\vdash b:B}\quad (s_i,s)\in\boldsymbol{P}$$

$$(\vec{\mathrm{C}}\text{-app})\quad \frac{\begin{array}{lcll}\Gamma_1,c(\Delta){:}A,\Gamma_2 & \vdash & b_i{:}B_i[x_j{:=}b_j]_{j=1}^{i-1} & (i=1,\ldots,n)\\ \Gamma_1,c(\Delta){:}A,\Gamma_2 & \vdash & A:s & (\text{if } n=0)\end{array}}{\Gamma_1,c(\Delta){:}A,\Gamma_2\vdash c(b_1,\ldots,b_n):A[x_j{:=}b_j]_{j=1}^{n}}$$

where the c that is introduced in the $\vec{\mathrm{C}}$-weakening rule is assumed to be Γ-fresh.

Aj A $\vec{C}\vec{D}$-PTS and its subsystems

Aj1 PTSs with parameters and definitions

Definition A.59 (Terms, 9.1) The set $\mathcal{T}_P$ of *parametric terms* is defined together with the set $\mathcal{L}_V$ of *lists of variables* and the set $\mathcal{L}_T$ of *lists of terms*:

$$\begin{array}{lcl}\mathcal{T}_P & ::= & \mathcal{V}\mid \boldsymbol{S}\mid \mathcal{C}(\mathcal{L}_T)\mid \mathcal{T}_P\mathcal{T}_P\mid \lambda\mathcal{V}{:}\mathcal{T}_P.\mathcal{T}_P\mid\\ & & \Pi\mathcal{V}{:}\mathcal{T}_P.\mathcal{T}_P\mid \mathcal{C}(\mathcal{L}_V){=}\mathcal{T}_P{:}\mathcal{T}_P \text{ IN } \mathcal{T}_P;\\ \mathcal{L}_V & ::= & \varnothing\mid\langle\mathcal{L}_V,\mathcal{V}{:}\mathcal{T}_P\rangle;\\ \mathcal{L}_T & ::= & \varnothing\mid\langle\mathcal{L}_T,\mathcal{T}_P\rangle.\end{array}$$

where $\mathcal{V}$ is a set of variables, $\mathcal{C}$ is a set of constants, and $\boldsymbol{S}$ is a set of sorts. We assume that $\mathcal{V}$, $\mathcal{C}$ and $\boldsymbol{S}$ are mutually disjoint and that $\mathcal{V}$ and $\mathcal{C}$ are countable (possibly infinite).

Definition A.60 (Contexts, 10.6) The set of *contexts* is given by

$$\mathcal{C}_P \;::=\; \varnothing\mid\langle\mathcal{C}_P,\mathcal{V}{:}\mathcal{T}_P\rangle\mid\langle\mathcal{C}_P,\mathcal{C}(\mathcal{L}_V){=}\mathcal{T}_P{:}\mathcal{T}_P\rangle\mid\langle\mathcal{C}_P,\mathcal{C}(\mathcal{L}_V){:}\mathcal{T}_P\rangle.$$

Definition A.61

- β-reduction is defined as before and we use $\rightarrow_\beta$, $\twoheadrightarrow_\beta$, $\twoheadrightarrow_\beta^+$, and $=_\beta$ as usual.
- δ-reduction is defined as the smallest relation $\rightarrow_\delta$ on $\mathcal{C}_P \times \mathcal{T}_P \times \mathcal{T}_P$ closed under the rules ($\delta 1$), ($\delta 2$) and ($\delta 3$) below and under the adapted compatibility rules (see Definition 10.15).

$$\Gamma_1, c(\Delta){=}a{:}A, \Gamma_2 \vdash c(b_1, \ldots, b_n) \rightarrow_\delta a[x_i{:=}b_i]_{i=1}^n \qquad (\delta 1)$$

$$\Gamma \vdash c(\Delta){=}a{:}A \text{ IN } b \rightarrow_\delta b \text{ if } c \notin \text{CONS}(b) \qquad (\delta 2)$$

$$\frac{\Gamma, c(\Delta){=}a{:}A \vdash b \rightarrow_\delta b'}{\Gamma \vdash c(\Delta){=}a{:}A \text{ IN } b \rightarrow_\delta c(\Delta){=}a{:}A \text{ IN } b'} \qquad (\delta 3)$$

Definition A.62 ($\vec{\mathbf{C}}$: parametric constants, 9.5) The *typing relation* $\vdash^{\vec{C}}$ is the smallest relation on $\mathcal{C}_P \times \mathcal{T}_P \times \mathcal{T}_P$ closed under the rules in Definition 4.21 and the following ones (we write $\Delta \equiv x_1{:}B_1, \ldots, x_n{:}B_n$):

($\vec{\mathbf{C}}$-weak)
$$\frac{\Gamma \vdash^{\vec{C}} b : B \qquad \Gamma, \Delta \vdash^{\vec{C}} A : s}{\Gamma, c(\Delta) : A \vdash^{\vec{C}} b : B}$$

($\vec{\mathbf{C}}$-app)
$$\frac{\begin{array}{lll}\Gamma_1, c(\Delta){:}A, \Gamma_2 & \vdash^{\vec{C}} & b_i{:}B_i[x_j{:=}b_j]_{j=1}^{i-1} \quad (i = 1, \ldots, n) \\ \Gamma_1, c(\Delta){:}A, \Gamma_2 & \vdash^{\vec{C}} & A : s \quad (\text{if } n = 0)\end{array}}{\Gamma_1, c(\Delta){:}A, \Gamma_2 \vdash^{\vec{C}} c(b_1, \ldots, b_n) : A[x_j{:=}b_j]_{j=1}^n}$$

where $s \in \mathcal{S}$ and the c that is introduced in the $\vec{\text{C}}$-weakening rule is assumed to be Γ-fresh.

Definition A.63 ($\vec{\mathbf{D}}$: parametric definitions 10.21) The *typing relation* $\vdash^{\vec{D}}$ is the smallest relation on $\mathcal{C}_P \times \mathcal{T}_P \times \mathcal{T}_P$ closed under the rules in Definition 4.21 and the following ones:

($\vec{\mathbf{D}}$-weak)
$$\frac{\Gamma \vdash^{\vec{D}} b : B \qquad \Gamma, \Delta \vdash^{\vec{D}} a : A}{\Gamma, c(\Delta){=}a{:}A \vdash^{\vec{D}} b : B}$$

($\vec{\mathbf{D}}$-app)
$$\frac{\begin{array}{ll}\Gamma_1, c(\Delta){=}a{:}A, \Gamma_2 \vdash^{\vec{D}} b_i : B_i[x_j{:=}b_j]_{j=1}^{i-1} & (i = 1, \ldots, n) \\ \Gamma_1, c(\Delta){=}a{:}A, \Gamma_2 \vdash^{\vec{D}} a : A & (\text{if } n = 0)\end{array}}{\Gamma_1, c(\Delta){=}a{:}A, \Gamma_2 \vdash^{\vec{D}} c(b_1, \ldots, b_n) : A[x_j{:=}b_j]_{j=1}^n}$$

($\vec{\mathbf{D}}$-form)
$$\frac{\Gamma, c(\Delta){=}a{:}A \vdash^{\vec{D}} B : s}{\Gamma \vdash^{\vec{D}} c(\Delta){=}a{:}A \text{ IN } B : s}$$

$$(\vec{\mathbf{D}}\textbf{-intro}) \quad \frac{\Gamma, c(\Delta){=}a{:}A \vdash^{\vec{D}} b : B \qquad \Gamma \vdash^{\vec{D}} c(\Delta){=}a{:}A \text{ IN } B : s}{\Gamma \vdash^{\vec{D}} c(\Delta){=}a{:}A \text{ IN } b : c(\Delta){=}a{:}A \text{ IN } B}$$

$$(\vec{\mathbf{D}}\textbf{-conv}) \quad \frac{\Gamma \vdash^{\vec{D}} b : B \qquad \Gamma \vdash^{\vec{D}} B' : s \qquad \Gamma \vdash B =_{\delta} B'}{\Gamma \vdash^{\vec{D}} b : B'}$$

where $s \in \boldsymbol{S}$, and the c that is introduced in the $\vec{\text{D}}$-weakening rule is assumed to be Γ-fresh.

Definition A.64 (Pure Type Systems with (parametric) constants and (parametric) definitions, 10.22) Let $\mathfrak{S}$ be a specification.

- A *Pure Type System with (parametric) constants* $\vec{\text{C}}$-PTS is denoted as $\lambda^{\vec{C}}(\mathfrak{S})$ and consists of a set of terms $\mathcal{T}_P$, a set of contexts $\mathcal{C}_P$, the β-reduction rule and the typing relation $\vdash^{\vec{C}}$;
- A *Pure Type System with (parametric) definitions* $\vec{\text{D}}$-PTS is denoted as $\lambda^{\vec{D}}(\mathfrak{S})$ and consists of a set of terms $\mathcal{T}_P$, a set of contexts $\mathcal{C}_P$, the β- and δ-reduction rules and the typing relation $\vdash^{\vec{D}}$;
- A *Pure Type System with (parametric) constants and (parametric) definitions* $\vec{\text{C}}\vec{\text{D}}$-PTS is denoted as $\lambda^{\vec{C}\vec{D}}(\mathfrak{S})$ and consists of a set of terms $\mathcal{T}_P$, a set of contexts $\mathcal{C}_P$, the β- and δ-reduction rules and the typing relation $\vdash^{\vec{C}\vec{D}}$, which is the smallest relation on $\mathcal{C}_P \times \mathcal{T}_P \times \mathcal{T}_P$ that is closed under the rules of Definition 4.21 and the rules of $\vdash^{\vec{C}}$ and $\vdash^{\vec{D}}$.

Aj2 PTSs with restricted parameters and definitions

Definition A.65 (Parametric Specification, 10.64) A *parametric specification* is a quadruple $(\boldsymbol{S}, \boldsymbol{A}, \boldsymbol{R}, \boldsymbol{P})$ such that $(\boldsymbol{S}, \boldsymbol{A}, \boldsymbol{R})$ is a specification (cf. Definition 4.18), and $\boldsymbol{P} \subseteq \boldsymbol{S} \times (\boldsymbol{S} \cup \{\text{TOP}\})$. The parametric specification is called *singly sorted* if the specification $(\boldsymbol{S}, \boldsymbol{A}, \boldsymbol{R})$ is singly sorted.

Definition A.66 ($\hat{\text{C}}$: restricted constants, 10.65) Let $(\boldsymbol{S}, \boldsymbol{A}, \boldsymbol{R}, \boldsymbol{P})$ be a parametric specification. The *typing relation* $\vdash^{\hat{C}}$ is obtained from the relation $\vdash^{\vec{C}}$ by replacing rule ($\vec{\text{C}}$-weak) by the following rule ($\hat{\text{C}}$-weak):

$$\frac{\Gamma \vdash^{\hat{C}} b : B \qquad \Gamma, \Delta_i \vdash^{\hat{C}} B_i : s_i \qquad \Gamma, \Delta \vdash^{\hat{C}} A : s}{\Gamma, c(\Delta) : A \vdash^{\hat{C}} b : B} \qquad (s_i, s) \in \boldsymbol{P}.$$

Definition A.67 ($\hat{\text{D}}$: restricted definitions, 10.66) Let $(\boldsymbol{S}, \boldsymbol{A}, \boldsymbol{R}, \boldsymbol{P})$ be a parametric specification. The *typing relation* $\vdash^{\hat{D}}$ is obtained from the relation $\vdash^{\vec{D}}$ by replacing rule ($\vec{\text{D}}$-weak) by the following rules ($\hat{\text{D}}$-weak):

$$\frac{\Gamma \vdash^{\hat{D}} b : B \qquad \Gamma, \Delta_i \vdash^{\hat{D}} B_i : s_i \qquad \Gamma, \Delta \vdash^{\hat{D}} a : A : s}{\Gamma, c(\Delta){=}a{:}A \vdash^{\hat{D}} b : B} \qquad (s_i, s) \in \boldsymbol{P};$$

$$\frac{\Gamma \vdash^{\hat{D}} b : B \qquad \Gamma, \Delta_i \vdash^{\hat{D}} B_i : s_i \qquad \Gamma, \Delta \vdash^{\hat{D}} a : A : \text{TOP}}{\Gamma, c(\Delta){=}a{:}A \vdash^{\hat{D}} b : B} \quad (s_i, \text{TOP}) \in \boldsymbol{P}.$$

Definition A.68 Let $(\boldsymbol{S}, \boldsymbol{A}, \boldsymbol{R}, \boldsymbol{P})$ be a parametric specification. The *typing relation* $\vdash^{\hat{C}\hat{D}}$ is obtained from the relation $\vdash^{\vec{C}\vec{D}}$ by replacing rule ($\vec{\text{C}}$-weak) by rule ($\hat{\text{C}}$-weak) and rule ($\vec{\text{D}}$-weak) by rules ($\hat{\text{D}}$-weak).

Definition A.69 (Pure Type Systems with Restricted Parameters and Restricted Parametric Definitions, 10.68) Let $(\boldsymbol{S}, \boldsymbol{A}, \boldsymbol{R}, \boldsymbol{P})$ be a parametric specification. The Pure Type System with restricted parameters and restricted parametric definitions ($\hat{\text{C}}\hat{\text{D}}$-PTS) and parametric specification $\mathfrak{S}$ is denoted $\lambda^{\hat{C}\hat{D}}(\mathfrak{S})$. The system consists of the set of terms $\mathcal{T}_P$, the set of contexts $\mathcal{C}_P$, β-reduction, δ-reduction, and the typing relation $\vdash^{\hat{C}\hat{D}}$.

Bibliography

[1] S. Abramsky, Dov M. Gabbay, and T.S.E. Maibaum, editors. *Handbook of Logic in Computer Science, Volume 2: Background: Computational Structures*. Oxford University Press, 1992.

[2] H.P. Barendregt. *The Lambda Calculus: its Syntax and Semantics*. Studies in Logic and the Foundations of Mathematics **103**. North-Holland, Amsterdam, revised edition, 1984.

[3] H.P. Barendregt. Lambda calculi with types. In [1], pages 117–309. Oxford University Press, 1992.

[4] G. Barthe. Extensions of pure type systems. In M. Dezani-Ciancaglini and G. Plotkin, editors, *Second International Conference on Typed Lambda Calculi and Applications*, pages 16–31, Edinburgh, 1995. Springer Verlag, Heidelberg.

[5] P. Benacerraf and H. Putnam, editors. *Philosophy of Mathematics*. Cambridge University Press, second edition, 1983.

[6] L.S. van Benthem Jutting. A Translation of Landau's "Grundlagen" in AUTOMATH. Technical report, Eindhoven University of Technology, 1976.

[7] L.S. van Benthem Jutting. *Checking Landau's "Grundlagen" in the Automath system*. PhD thesis, Eindhoven University of Technology, 1977. Published as Mathematical Centre Tracts nr. 83 (Amsterdam, Mathematisch Centrum, 1979).

[8] L.S. van Benthem Jutting. Description of AUT-68. Technical Report 12, Eindhoven University of Technology, 1981. Also in [112], pp. 251–273.

[9] L.S. van Benthem Jutting. Typing in pure type systems. *Information and Computation*, 105:30–41, 1993.

[10] S. Berardi. Towards a mathematical analysis of the Coquand-Huet calculus of constructions and the other systems in Barendregt's cube. Technical report, Dept. of Computer Science, Carnegie-Mellon University and Dipartimento Matematica, Universita di Torino, 1988.

[11] E.W. Beth. *The Foundations of Mathematics*. Studies in Logic and the Foundations of Mathematics. North-Holland, Amsterdam, 1959.

[12] R. Bloo. *Preservation of Termination for Explicit Substitutions*. PhD thesis, Eindhoven University of Technology, 1997.

[13] Roel Bloo, Fairouz Kamareddine, Twan Laan, and Rob Nederpelt. Parameters in Pure Type Systems. In S. Rajsbaum, editor, *Proc. Latin American Symposium on Theor. INformatics*, volume 2286 of *Lecture Notes in Computer Science*, pages 371–385, 2002.

[14] Roel Bloo, Fairouz Kamareddine, and Rob Nederpelt. The Barendregt Cube with Definitions and Generalised Reduction. *Information and Computation*, 126 (2):123–143, 1996.

[15] G. Boolos. The iterative conception of set. *Philosophy*, LXVIII:215–231, 1971.

[16] V.A.J. Borghuis. *Coming to Terms with Modal Logic: On the interpretation of modalities in typed λ-calculus*. PhD thesis, Technische Universiteit Eindhoven, 1994.

[17] L.E.J. Brouwer. *Over de Grondslagen der Wiskunde*. PhD thesis, Universiteit van Amsterdam, 1907. Dutch; English translation in [69].

[18] N.G. de Bruijn. AUTOMATH, a language for mathematics. Technical Report 68-WSK-05, T.H.-Reports, Eindhoven University of Technology, 1968.

[19] N.G. de Bruijn. The mathematical language AUTOMATH, its usage and some of its extensions. In M. Laudet, D. Lacombe, and M. Schuetzenberger, editors, *Symposium on Automatic Demonstration*, pages 29–61, IRIA, Versailles, 1968. Springer Verlag, Berlin, 1970. Lecture Notes in Mathematics **125**; also in [112], pages 73–100.

[20] N.G. de Bruijn. The Mathematical Vernacular, a language for mathematics with typed sets. In P. Dybjer et al., editors, *Proceedings of the Workshop on Programming Languages*. Marstrand, Sweden, 1987. Reprinted in [112] in combination with *Formalizing the Mathematical Vernacular* (formerly unpublished, 1982).

[21] N.G. de Bruijn. Reflections on Automath. Eindhoven University of Technology, 1990. Also in [112], pages 201–228.

[22] C. Burali-Forti. Una questione sui numeri transfiniti. *Rendiconti del Circolo Matematico di Palermo*, 11:154–164, 1897. English translation in [67], pages 104–112.

[23] G. Cantor. Beiträge zur Begründung der transfiniten Mengenlehre (Erster Artikel). *Mathematische Annalen*, 46:481–512, 1895.

[24] G. Cantor. Beiträge zur Begründung der transfiniten Mengenlehre (Zweiter Artikel). *Mathematische Annalen*, 49:207–246, 1897.

[25] F. Cardone and J.R. Hindley. History of lambda calculus and combinatory logic. To appear.

[26] A.-L. Cauchy. *Cours d'Analyse de l'Ecole Royale Polytechnique*. Debure, Paris, 1821. Also as *Œuvres Complètes* (2), volume III, Gauthier-Villars, Paris, 1897.

[27] A. Church. A set of postulates for the foundation of logic (1). *Annals of Mathematics*, 33:346–366, 1932.

[28] A. Church. A set of postulates for the foundation of logic (2). *Annals of Mathematics*, 34:839–864, 1933.

[29] A. Church. A formulation of the simple theory of types. *The Journal of Symbolic Logic*, 5:56–68, 1940.

[30] A. Church. *The Calculi of Lambda Conversion*. Princeton University Press, 1941.

[31] A. Church. Comparison of Russell's resolution of the semantic antinomies with that of Tarski. *The Journal of Symbolic Logic*, 41:747–760, 1976.

[32] N.B. Cocchiarella. Frege's double correlation thesis and Quine's set theories NF and ML. *Philosophical Logic*, 13, 1984.

[33] N.B. Cocchiarella. Philosophical perspectives on formal theories of predication. *Handbook of Philosophical Logic*, 4, 1986.

[34] A.B. Compagnoni. *Higher-Order Subtyping with Intersection Types*. PhD thesis, Katholieke Universiteit Nijmegen, 1995.

[35] R.L. Constable et al. *Implementing Mathematics with the Nuprl Proof Development System*. Prentice-Hall, New Jersey, 1986.

[36] T. Coquand. An analysis of Girard's paradox. In *Proceedings of the Symposium on Logic in Computing Science*, Cambridge, Massachusetts, 1986. IEEE.

[37] T. Coquand and G. Huet. The calculus of constructions. *Information and Computation*, 76:95–120, 1988.

[38] H.B. Curry. Functionality in combinatory logic. *Proceedings of the National Academy of Science of the USA*, 20:584–590, 1934.

[39] H.B. Curry. *Foundations of Mathematical Logic*. McGraw-Hill Series in Higher Mathematics. McGraw-Hill Book Company, Inc., 1963.

[40] H.B. Curry and R. Feys. *Combinatory Logic I*. Studies in Logic and the Foundations of Mathematics. North-Holland, Amsterdam, 1958.

[41] D.T. van Daalen. A description of Automath and some aspects of its language theory. In P. Braffort, editor, *Proceedings of the Symposium APLASM*, volume I, pages 48–77, 1973. Also in [112], pages 101–126.

[42] D.T. van Daalen. *The Language Theory of Automath*. PhD thesis, Eindhoven University of Technology, 1980.

[43] R. Dedekind. *Stetigkeit und irrationale Zahlen*. Vieweg & Sohn, Braunschweig, 1872.

[44] G. Dowek et al. The Coq Proof Assistant Version 5.6, Users Guide. Technical Report 134, INRIA, Le Chesney, 1991.

[45] Euclid. The Elements. 325 B.C. English translation in [66].

[46] S. Feferman. Systems of predicative analysis. *Symbolic Logic*, 29:1–30, 1964.

[47] S. Feferman. Toward useful type-free theories I. *Journal of Symbolic Logic*, 49:75–111, 1984.

[48] A. Fraenkel, Y. Bar-Hillel, and A. Levy. *Foundations of Set Theory*. North-Holland, 1973.

[49] G. Frege. *Begriffsschrift, eine der arithmetischen nachgebildete Formelsprache des reinen Denkens*. Nebert, Halle, 1879. Also in [67], pages 1–82.

[50] G. Frege. *Grundlagen der Arithmetik, eine logisch-mathematische Untersuchung über den Begriff der Zahl*. Breslau, 1884.

[51] G. Frege. *Funktion und Begriff, Vortrag gehalten in der Sitzung vom 9. Januar der Jenaischen Gesellschaft für Medicin und Naturwissenschaft*. Hermann Pohle, Jena, 1891. English translation in [107], pages 137–156.

[52] G. Frege. *Grundgesetze der Arithmetik, begriffsschriftlich abgeleitet*, volume I. Pohle, Jena, 1892. Reprinted 1962 (Olms, Hildesheim).

[53] G. Frege. Über Sinn und Bedeutung. *Zeitschrift für Philosophie und philosophische Kritik*, new series, 100:25–50, 1892. English translation in [107], pages 157–177.

[54] G. Frege. Ueber die Begriffschrift des Herrn Peano und meine eigene. *Berichte über die Verhandlungen der Königlich Sächsischen Gesellschaft der Wissenschaften zu Leipzig, Mathematisch-physikalische Klasse 48*, pages 361–378, 1896. English translation in [107], pages 234–248.

[55] G. Frege. Letter to Russell. English translation in [67], pages 127–128, 1902.

[56] G. Frege. *Grundgesetze der Arithmetik, begriffsschriftlich abgeleitet*, volume II. Pohle, Jena, 1903. Reprinted 1962 (Olms, Hildesheim).

[57] M.H. Sorensen G. Barthe. Domain-free pure type systems. In *Logical Foundations of Computer Science*, volume 1234 of *Lecture Notes in Computer Science*, pages 9–20. Springer, 1997.

[58] C.I. Gerhardt, editor. Die philosophischen Schriften von Gottfried Wilhelm Leibniz. Berlin, 1890.

[59] J.H. Geuvers. *Logics and Type Systems*. PhD thesis, Catholic University of Nijmegen, 1993.

[60] J.-Y. Girard. *Interprétation fonctionelle et élimination des coupures dans l'arithmétique d'ordre supérieur*. PhD thesis, Université Paris VII, 1972.

[61] K. Gödel. Über formal unentscheidbare Sätze der Principia Mathematica und verwandter Systeme I. *Monatshefte für Mathematik und Physik*, 38:173–198, 1931. German; English translation in [67], pages 592–618.

[62] K. Gödel. Russell's mathematical logic. In P.A. Schlipp, editor, *The Philosophy of Bertrand Russell*. Evanston & Chicago, Northwestern University, 1944. Also in [5], pages 447–469.

[63] I. Grattan-Guinness. *The Search for Mathematical Roots, 1870-1930*. Princeton University Press, 2001.

[64] R. Harper, F. Honsell, and G. Plotkin. A framework for defining logics. In *Proceedings Second Symposium on Logic in Computer Science*, pages 194–204, Washington D.C., 1987. IEEE.

[65] R. Harper and R. Pollack. Type checking with universes. *Theoretical Computer Science*, 89:107–136, 1991.

[66] T.L. Heath. *The Thirteen Books of Euclid's Elements*. Dover Publications, Inc., New York, 1956.

[67] J. van Heijenoort, editor. *From Frege to Gödel: A Source Book in Mathematical Logic, 1879–1931*. Harvard University Press, Cambridge, Massachusetts, 1967.

[68] A. Heyting. *Mathematische Grundlagenforschung. Intuitionismus. Beweistheorie*. Ergebnisse der Mathematik und ihrer Grenzgebiete. Springer Verlag, Berlin, 1934.

[69] A. Heyting, editor. *Brouwer: Collected Works*, volume 1. North-Holland, Amsterdam, 1975.

[70] D. Hilbert and W. Ackermann. *Grundzüge der Theoretischen Logik*. Die Grundlehren der Mathematischen Wissenschaften in Einzeldarstellungen, Band XXVII. Springer Verlag, Berlin, first edition, 1928.

[71] J.R. Hindley and J.P. Seldin. *Introduction to Combinators and λ-calculus*, volume 1 of *London Mathematical Society Student Texts*. Cambridge University Press, 1986.

[72] R. Holmes. Systems of combinatory logic related to predicative and "mildly impredicative" fragments of Quine's "new foundations". *Annals of Pure and Applied Logic*, 59:45–53, 1993.

[73] R. Holmes. Subsystems of Quine's "new foundations" with predicativity restrictions. *Notre Dame Journal of Formal Logic*, 40(2):183–196, 1999.

[74] R. Holmes. Polymorphic type checking for Principia Mathematica. In Fairouz Kamareddine, editor, *Thirty Five Years of Automating Mathematics*, volume 28 of *Applied Logic series*. Kluwer Academic Publishers, 2003.

[75] W.A. Howard. The formulas-as-types notion of construction. In [136], pages 479–490, 1980.

[76] P.B. Jackson. *Enhancing the Nuprl Proof Development System and Applying it to Computational Abstract Algebra*. PhD thesis, Cornell University, Ithaca, New York, 1995.

[77] P.B. Jackson. The Nuprl proof development system, Version 4.1 reference manual and user's guide. Cornell University, Department of Computing Science, Ithaca, New York, 1995.

[78] R.B. Jensen. On the consistency of a slight modification of Quine's NF. *Synthese*, 19:250–263, 1969.

[79] Fairouz Kamareddine. Postponement, conservation and preservation of strong normalisation for generalised reduction. *Logic and Computation*, 10 (5):721–738, 2000.

[80] Fairouz Kamareddine. On functions and types: A tutorial. In *Proceedings of SOFSEM 2002*, volume 2540 of *Lecture Notes in Computer Science*. Springer Verlag, 2002.

[81] Fairouz Kamareddine, Roel Bloo, and Rob Nederpelt. On π-conversion in the λ-cube and the combination with abbreviations. *Annals of Pure and Applied Logics*, 97:27–45, 1999.

[82] Fairouz Kamareddine and Twan Laan. A reflection on Russell's ramified types and Kripke's hierarchy of truths. *Journal of the Interest Group in Pure and Applied Logic*, 4(2):195–213, 1996.

[83] Fairouz Kamareddine and Twan Laan. A correspondence between Martin-Löf type theory, the ramified theory of types and pure type systems. *Logic, Language and Information*, 10(3):375–402, 2001.

[84] Fairouz Kamareddine, Twan Laan, and Rob Nederpelt. Refining the Barendregt cube using parameters. *Fifth International Symposium on Functional and Logic Programming, FLOPS 2001*, Lecture Notes in Computer Science:375–389, 2001.

[85] Fairouz Kamareddine, Twan Laan, and Rob Nederpelt. Types in logic and mathematics before 1940. *Bulletin of Symbolic Logic*, 8(2):185–245, 2002.

[86] Fairouz Kamareddine, Twan Laan, and Rob Nederpelt. De Bruijn's Automath and Pure Type Systems. In Fairouz Kamareddine, editor, *Thirty Five Years of Automating Mathematics*, Applied Logic series. Kluwer Academic Publishers, 2003.

[87] Fairouz Kamareddine, Twan Laan, and Rob Nederpelt. Revisiting the notion of function. *Logic and Algebraic Programming*, 54(1-2):65–107, 2003.

[88] Fairouz Kamareddine and Rob Nederpelt. On stepwise explicit substitution. *International Journal of Foundations of Computer Science*, 4(3):197–240, 1993.

[89] Fairouz Kamareddine and Rob Nederpelt. Generalising reduction in the λ-calculus. *Journal of Functional Programming*, 5(4):637–651, 1995.

[90] Fairouz Kamareddine and Rob Nederpelt. Canonical typing and Π-conversion in the Barendregt Cube. *Journal of Functional Programming*, 6(2):245–267, 1996.

[91] Fairouz Kamareddine and Rob Nederpelt. A useful λ-notation. *Theoretical Computer Science*, 155:85–109, 1996.

[92] S.C. Kleene and J.B. Rosser. The inconsistency of certain formal logics. *Annals of Mathematics*, 36:630–636, 1935.

[93] J.W. Klop. Term rewriting systems. In [1], pages 1–116. Oxford University Press, 1992.

[94] G.T. Kneebone. *Mathematical Logic and the Foundations of Mathematics*. D. Van Nostrand Comp., London, New York, Toronto, 1963.

[95] A.N. Kolmogorov. Zur Deutung der Intuitionistischen Logik. *Mathematisches Zeitschrift*, 35:58–65, 1932.

[96] S. Kripke. Outline of a theory of truth. *Journal of Philosophy*, 72:690–716, 1975.

[97] T. Laan. A formalization of the Ramified Type Theory. Technical Report 94-33, TUE Computing Science Reports, Eindhoven University of Technology, 1994.

[98] T. Laan. *The Evolution of Type Theory in Logic and Mathematics*. PhD thesis, Eindhoven University of Technology, 1997.

[99] Twan Laan and Michael Franssen. Parameters for first order logic. *Logic and Computation*, 2001.

[100] E. Landau. *Grundlagen der Analysis*. Leipzig, 1930.

[101] G. Landini. *Russell's hidden substitutional theory*. Oxford University Press, 1998.

[102] D. Leivant. Finitely stratified polymorphism. *Information and Computation*, 93, 1991. Selections from the 1989 *IEEE Symposium on Logic in Computer Science*.

[103] G. Longo and E. Moggi. Constructive natural deduction and its modest interpretation. Technical Report CMU-CS-88-131, Carnegie Mellon University, Pittsburgh, USA, 1988.

[104] P. Martin-Löf. An intuitionistic theory of types: predicative part. In H.E. Rose and J.C. Shepherdson, editors, *Logic Colloquium '73*, pages 73–118, Amsterdam, 1975. North-Holland. Studies in Logic and the Foundations of Mathematics **80**.

[105] P. Martin-Löf. Constructive mathematics and computer programming. In *Sixth International Congress for Logic, Methodology and Philosophy of Science*, pages 153–175, Amsterdam, 1982. North-Holland.

[106] P. Martin-Löf. *Intuitionistic Type Theory*. Studies in Proof Theory. Bibliopolis, Napoli, 1984.

[107] B. McGuinness, editor. *Gottlob Frege: Collected Papers on Mathematics, Logic, and Philosophy*. Basil Blackwell, Oxford, 1984.

[108] R. Milner, M. Tofte, and R. Harper. *Definition of Standard ML*. MIT Press, Cambridge (Massachusetts)/London, 1990.

[109] C. Murthy. *Extracting Constructive Content from Classical Proofs*. PhD thesis, Cornell University, Ithaca, New York, 1990.

[110] R.P. Nederpelt. *Strong Normalization in a Typed Lambda Calculus with Lambda Structured Types*. PhD thesis, Eindhoven University of Technology, 1973. Also in [112], pages 389–468.

[111] R.P. Nederpelt. Presentation of natural deduction. *Recueil des travaux de l'Institut Mathématique, Nouvelle série*, 2(10):115–126, 1977. Symposium: Set Theory. Foundations of Mathematics, Beograd 1977.

[112] R.P. Nederpelt, J.H. Geuvers, and R.C. de Vrijer, editors. *Selected Papers on Automath*. Studies in Logic and the Foundations of Mathematics **133**. North-Holland, Amsterdam, 1994.

[113] M.J. O'Donnell. *Computing in Systems Described by Equations*, volume 58 of *Lecture Notes in Computer Science*. Springer Verlag, 1977.

[114] E. Palmgren. *On fixed point operators, inductive definitions and universe in Martin-Lof type theory*. PhD thesis, Uppsala University, 1991.

[115] M. Parigot. Lambda-mu-calculus: an algorithmic interpretation of classical natural deduction. In A. Voronkov, editor, *Logic Programming and Automated Reasoning: International Conference LPAR '92*, pages 190–201. Springer-Verlag, 1992.

[116] G. Peano. *Arithmetices principia, nova methodo exposita*. Bocca, Turin, 1889. English translation in [67], pages 83–97.

[117] G. Peano. *Formulaire de Mathématique*. Bocca, Turin, 1894–1908. 5 successive versions; the final edition issued as *Formulario Mathematico*.

[118] W. Peremans. Ups and downs of type theory. Technical Report 94-14, TUE Computing Science Notes, Eindhoven University of Technology, 1994.

[119] H. Poincaré. Du rôle de l'intuition et de la logique en mathématiques. *C.R. du IIme Cong. Intern. des Math., Paris 1990*, pages 200–202, 1902.

[120] W. Van Orman Quine. New foundations for mathematical logic. *American Mathematical Monthly*, 44:70–80, 1937. Also in [122], pages 80–101.

[121] W. Van Orman Quine. *Mathematical Logic*. Norton, New York, 1940. Revised edition Cambridge, Harvard University Press, 1951.

[122] W. Van Orman Quine. *From a Logical Point of View: 9 Logico-Philosophical Essays*. Harvard University Press, Cambridge, Massachusetts, second edition, 1961.

[123] W. Van Orman Quine. *Set Theory and its Logic*. Harvard University Press, Cambridge, Massachusetts, 1963.

[124] F.P. Ramsey. The foundations of mathematics. *Proceedings of the London Mathematical Society*, 2nd series, 25:338–384, 1926.

[125] G.R. Renardel de Lavalette. Strictness analysis via abstract interpretation for recursively defined types. *Information and Computation*, 99:154–177, 1991.

[126] J.C. Reynolds. *Towards a theory of type structure*, volume 19 of *Lecture Notes in Computer Science*, pages 408–425. Springer, 1974.

[127] A.C.M. van Rooij. *Analyse voor Beginners*. Epsilon Uitgaven, Utrecht, 1986.

[128] J.B. Rosser. Highlights of the history of the lambda-calculus. *Annals of the History of Computing*, 6(4):337–349, 1984.

[129] B. Russell. Letter to Frege. English translation in [67], pages 124–125, 1902.

[130] B. Russell. *The Principles of Mathematics*. Allen & Unwin, London, 1903.

[131] B. Russell. Mathematical logic as based on the theory of types. *American Journal of Mathematics*, 30:222–262, 1908. Also in [67], pages 150–182.

[132] M. Schönfinkel. Über die Bausteine der mathematischen Logik. *Mathematische Annalen*, 92:305–316, 1924. Also in [67], pages 355–366.

[133] K. Schütte. *Beweistheorie*. Die Grundlehren der Mathematischen Wissenschaften in Einzeldarstellungen, Band 103. Springer Verlag, Berlin, 1960.

[134] K. Schütte. *Proof Theory*. Die Grundlehren der mathematischen Wissenschaften 225. Springer-Verlag, 1977.

[135] J.P. Seldin. Personal communication, 1996.

[136] J.P. Seldin and J.R. Hindley, editors. *To H.B. Curry: Essays on Combinatory Logic, Lambda Calculus and Formalism*. Academic Press, New York, 1980.

[137] P. Severi and E. Poll. Pure type systems with definitions. In A. Nerode and Yu.V. Matiyasevich, editors, *Proceedings of LFCS'94 (LNCS* **813***)*, pages 316–328, New York, 1994. LFCS'94, St. Petersburg, Russia, Springer Verlag.

[138] E.P. Specker. The axiom of choice in Quine's New Foundations for Mathematical Logic. *Proc. Nat. Acad. Sci. USA.*, 39:972–975, 1953.

[139] T. Streicher. *Semantics of Type Theory*. Birkhäuser, 1991.

[140] W.W. Tait. Infinitely long terms of transfinite type. In J.N. Crossley and M.A.E. Dummett, editors, *Formal Systems and Recursive Functions*, Amsterdam, 1965. North-Holland.

[141] A. Tarski. Der Wahrheitsbegriff in den formalisierten Sprachen. *Studia Philosophica*, 1:261–405, 1936. German translation by L. Blauwstein from the Polish original (1933) with a postscript added.

[142] J. Terlouw. Een nadere bewijstheoretische analyse van GSTT's. Technical report, Department of Computer Science, University of Nijmegen, 1989.

[143] R. de Vrijer. A direct proof of the finite developments theorem. *The Journal of Symbolic Logic*, 50(2):339–343, 1985.

[144] H. Weyl. *Das Kontinuum*. Veit, Leipzig, 1918. German; also in: Das Kontinuum und andere Monographien, Chelsea Pub.Comp., New York, 1960.

[145] A.N. Whitehead and B. Russell. *Principia Mathematica*, volume I, II, III. Cambridge University Press, 1910^1, 1927^2. All references are to the first volume, unless otherwise stated.

[146] R.L. Wilder. *The Foundations of Mathematics*. Robert E. Krieger Publishing Company, Inc., New York, second edition, 1965.

[147] E. Zermelo. Untersuchungen über die Grundlagen der Mengenlehre. *Math. Annalen*, 65:261–281, 1908.

[148] J. Zucker. Formalization of classical mathematics in Automath. In *Colloque International de Logique,* Clermont-Ferrand, pages 135–145, Paris, CNRS, 1977. Colloques Internationaux du Centre National de la Recherche Scientifique, 249.

Subject Index

Name Index

List of Figures

APPLIED LOGIC SERIES

1. D. Walton: *Fallacies Arising from Ambiguity.* 1996 ISBN 0-7923-4100-7
2. H. Wansing (ed.): *Proof Theory of Modal Logic.* 1996 ISBN 0-7923-4120-1
3. F. Baader and K.U. Schulz (eds.): *Frontiers of Combining Systems.* First International Workshop, Munich, March 1996. 1996 ISBN 0-7923-4271-2
4. M. Marx and Y. Venema: *Multi-Dimensional Modal Logic.* 1996 ISBN 0-7923-4345-X
5. S. Akama (ed.): *Logic, Language and Computation.* 1997 ISBN 0-7923-4376-X
6. J. Goubault-Larrecq and I. Mackie: *Proof Theory and Automated Deduction.* 1997 ISBN 0-7923-4593-2
7. M. de Rijke (ed.): *Advances in Intensional Logic.* 1997 ISBN 0-7923-4711-0
8. W. Bibel and P.H. Schmitt (eds.): *Automated Deduction - A Basis for Applications.* Volume I. Foundations - Calculi and Methods. 1998 ISBN 0-7923-5129-0
9. W. Bibel and P.H. Schmitt (eds.): *Automated Deduction - A Basis for Applications.* Volume II. Systems and Implementation Techniques. 1998 ISBN 0-7923-5130-4
10. W. Bibel and P.H. Schmitt (eds.): *Automated Deduction - A Basis for Applications.* Volume III. Applications. 1998 ISBN 0-7923-5131-2
    (Set vols. I-III: ISBN 0-7923-5132-0)
11. S.O. Hansson: *A Textbook of Belief Dynamics.* Theory Change and Database Updating. 1999 Hb: ISBN 0-7923-5324-2; Pb: ISBN 0-7923-5327-7
    Solutions to exercises. 1999. Pb: ISBN 0-7923-5328-5
    Set: (Hb): ISBN 0-7923-5326-9; (Pb): ISBN 0-7923-5329-3
12. R. Pareschi and B. Fronhöfer (eds.): *Dynamic Worlds from the Frame Problem to Knowledge Management.* 1999 ISBN 0-7923-5535-0
13. D.M. Gabbay and H. Wansing (eds.): *What is Negation?* 1999 ISBN 0-7923-5569-5
14. M. Wooldridge and A. Rao (eds.): *Foundations of Rational Agency.* 1999 ISBN 0-7923-5601-2
15. D. Dubois, H. Prade and E.P. Klement (eds.): *Fuzzy Sets, Logics and Reasoning about Knowledge.* 1999 ISBN 0-7923-5911-1
16. H. Barringer, M. Fisher, D. Gabbay and G. Gough (eds.): *Advances in Temporal Logic.* 2000 ISBN 0-7923-6149-0
17. D. Basin, M.D. Agostino, D.M. Gabbay, S. Matthews and L. Viganò (eds.): *Labelled Deduction.* 2000 ISBN 0-7923-6237-3
18. P.A. Flach and A.C. Kakas (eds.): *Abduction and Induction.* Essays on their Relation and Integration. 2000 ISBN 0-7923-6250-0
19. S. Hölldobler (ed.): *Intellectics and Computational Logic.* Papers in Honor of Wolfgang Bibel. 2000 ISBN 0-7923-6261-6

APPLIED LOGIC SERIES

20. P. Bonzon, M. Cavalcanti and Rolf Nossum (eds.): *Formal Aspects of Context.* 2000 ISBN 0-7923-6350-7
21. D.M. Gabbay and N. Olivetti: *Goal-Directed Proof Theory.* 2000 ISBN 0-7923-6473-2
22. M.-A. Williams and H. Rott (eds.): *Frontiers in Belief Revision.* 2001 ISBN 0-7923-7021-X
23. E. Morscher and A. Hieke (eds.): *New Essays in Free Logic.* In Honour of Karel Lambert. 2001 ISBN 1-4020-0216-5
24. D. Corfield and J. Williamson (eds.): *Foundations of Bayesianism.* 2001 ISBN 1-4020-0223-8
25. L. Magnani, N.J. Nersessian and C. Pizzi (eds.): *Logical and Computational Aspects of Model-Based Reasoning.* 2002 Hb: ISBN 1-4020-0712-4; Pb: ISBN 1-4020-0791-4
26. D.J. Pym: *The Semantics and Proof Theory of the Logic of Bunched Implications.* 2002 ISBN 1-4020-0745-0
27. P.B. Andrews: *An Introduction to Mathematical Logic and Type Theory: To Truth Through Proof.* Second edition. 2002 ISBN 1-4020-0763-9
28. F.D. Kamareddine: *Thirty Five Years of Automating Mathematics.* 2003 ISBN 1-4020-1656-5
29. F. Kamareddine, T. Laan and R. Nederpelt: *A Modern Perspective on Type Theory.* From its Origins until Today. 2004 ISBN 1-4020-2334-0

KLUWER ACADEMIC PUBLISHERS – DORDRECHT / BOSTON / LONDON

GPSR Compliance

The European Union's (EU) General Product Safety Regulation (GPSR) is a set of rules that requires consumer products to be safe and our obligations to ensure this.

If you have any concerns about our products, you can contact us on ProductSafety@springernature.com

In case Publisher is established outside the EU, the EU authorized representative is:

Springer Nature Customer Service Center GmbH
Europaplatz 3
69115 Heidelberg, Germany

Batch number: 10755284

Printed by Printforce, the Netherlands